Flip Chip, Hybrid Bonding, Fan-In, and Fan-Out Technology

John H. Lau

Flip Chip, Hybrid Bonding, Fan-In, and Fan-Out Technology

Springer

John H. Lau
Unimicron Technology Corporation
Palo Alto, CA, USA

ISBN 978-981-97-2139-9 ISBN 978-981-97-2140-5 (eBook)
https://doi.org/10.1007/978-981-97-2140-5

This Springer imprint is published by the registered company Springer Nature Singapore Pte Ltd.
The registered company address is: 152 Beach Road, #21-01/04 Gateway East, Singapore 189721, Singapore

Paper in this product is recyclable.

Preface

Flip chip is one of the chip-level interconnects and has been used for high-performance and high-density applications such as central processing unit, graphic processing unit, and memory. Hybrid bonding is one of the flip chip technologies, except it is for much smaller pad pitch and higher performance applications. Wafer-level packaging such as fan-in and fan-out have been used for smartphones, tablets, etc.

For the next few years, we will see more implementations of flip chip, hybrid bonding, and wafer-level fan-in and fan-out technologies, whether it is for yield, cost, time-to-market, performance, form factor, or power consumption. Unfortunately, for most of the practicing engineers and managers, as well as scientists and researchers, these flip chip, hybrid bonding, and wafer-level fan-in and fan-out technologies are not well understood. Thus, there is an urgent need, both in industry and research institutes, to create a comprehensive book on the current state of knowledge of these flip chip, hybrid bonding, and wafer-level fan-in and fan-out technologies. This book is written so that readers can quickly learn about the basics of problem-solving methods and understand the trade-offs inherent in making system-level decisions.

There are six chapters in this book, namely: (1) Flip Chip Technology (2) Cu-Cu Hybrid Bonding, (3) Fan-In Technology, (4) Fan-Out Technology, (5) Chiplet Communications (Bridges), and (6) Co-packaged Optics.

Chapter 1 presents the flip chip technology. Wafer bumping (the mother of flip chip technology) of the C4 (controlled collapsed chip connection) bump and C2 (chip connection) bump will be briefly mentioned first. It is followed by various substrates, e.g., organic, fan-out RDLs (redistribution-layers), TSV (through-silicon via) interposer, Si-bridge embedded in organic substrate, Si-bridge embedded in fan-out epoxy molding compound with RDLs for flip chip applications. Various flip chip assemblies will be presented, and it is followed by underfill and reliability. Finally, a high throughput flip chip process will be provided.

Chapter 2 presents the Cu-Cu hybrid bonding. The recent advances and trends of the hybrid bonding technology by Samsung, SK Hynix, Micron, Mitsui, Sony, CEA-Leti, TEL, Applied Materials, Intel, TSMC, MKS, Resonac, Adeia, Technische Universitaet Dresden, EVG, imec, and UCLA will be briefly mentioned and discussed.

Chapter 3 presents the wafer-level fan-in (wafer-level chip-scale package) technology. Besides the conventional wafer-level chip-scale packages, the material, process, fabrications, and reliability of a six-sides molded panel-level chip-scale package will be provided and discussed.

Chapter 4 presents the wafer/panel-level fan-out packaging. The following eight topics will be discussed: (1) chip-first with die face-down for heterogeneous integration of 4 chips and 4 capacitors, (2) chip-first with die face-down for heterogeneous integration of mini-LED RGB display, (3) chip-first with die face-up for a large 10 mm x 10 mm chip, (4) panel-level fan-out RDL-first packaging for heterogeneous integration, (5) hybrid substrate with PID by RDL-first panel-level packaging, (6) hybrid substrate with ABF by RDL-first panel-level packaging, (7) hybrid substrate with ultralarge organic interposer for heterogeneous integration, and (8) hybrid substrate with interconnect-layer for heterogeneous integration.

Chapter 5 presents the communications (bridges) between chiplets. The application of flip chip, hybrid bonding, and fan-out technologies on the communications between chiplets will be presented. Intel's EMIB (embedded multi-die interconnect bridge), IBM's DBHi (direct bonded heterogeneous integration), TSMC's CoWoS-L (chip on wafer on substrate-L), SPIL's FO-EB-T (fan-out embedded bridge with TSV), and UCIe (Universal Chiplet Interconnect Express) consortium will be briefly discussed.

Chapter 6 presents the co-packaged optics—heterogeneous integration of photonic IC (PIC) and electronic IC (EIC). The application of flip chip, hybrid bonding, and fan-out technologies on the heterogeneous integration of PIC and EIC will be presented. Emphasis is placed on the 2D/3D heterogeneous integration of ASIC switch, PIC, and EIC w/o bridge and 2D/3D heterogeneous integration of ASIC switch, PIC, and EIC on glass substrate.

For whom is this book intended? Undoubtedly it will be of great interest to three groups of specialists: (1) those who are active or intend to become active in research and development of flip chip, hybrid bonding, and wafer-level fan-in and fan-out technologies, (2) those who have encountered practical flip chip, hybrid bonding, and wafer-level fan-in and fan-out problems and wish to understand and learn more methods for solving such problems; and (3) those who have to choose a reliable, creative, high performance, high density, low power consumption, and cost-effective technique for their products. This book can also be used as a text for college and graduate students who have the potential to become our future leaders, scientists, and engineers in the electronics and optoelectronics industry.

I hope that this book will serve as a valuable reference source for all those faced with the challenging problems created by the ever-increasing interest in flip chip, hybrid bonding, and wafer-level fan-in and fan-out technologies. I also hope that it will aid in stimulating further research and development on key enabling technologies and more sound applications to flip chip, hybrid bonding, and wafer-level fan-in and fan-out products. The organizations that learn how to design and manufacture flip chip, hybrid bonding, and wafer-level fan-in and fan-out in their semiconductor packaging systems have the potential to make major advances in the electronics and optoelectronics industry, and to gain great benefits in cost, performance, functionality, density, power, bandwidth, quality, size, and weight. It is my hope that the information presented in this book may assist in removing roadblocks, avoiding unnecessary false starts, and accelerating design, materials, process, and manufacturing development of flip chip, hybrid bonding, and wafer-level fan-in and fan-out technologies.

Palo Alto, CA, USA John H. Lau

Acknowledgments

Development and preparation of Semiconductor Advanced Packaging was facilitated by the efforts of a number of dedicated people. I would like to thank them all, with special mention of Rammohan Kishnamurthy, Springer Nature Scientific Publishing Services (P) Ltd., for their unswerving support and advocacy. My special thanks go to Ms. Jasmine Dou, Springer Singapore, who made my dream of this book come true by effectively sponsoring the project and solving many problems that arose during the book's preparation. It has been a great pleasure and fruitful experience to work with all of them in transferring my messy manuscripts into a very attractive printed book.

The material in this book clearly has been derived from many sources, including individuals, companies, and organizations, and I have attempted to acknowledge by citations in the appropriate parts of the book the assistance that I have been given. It would be quite impossible for me to express my thanks to everyone concerned for their cooperation in producing this book, but I would like to extend due gratitude. Also, I would like to thank several professional societies and publishers for permitting me to reproduce some of their illustrations and information in this book, including the American Society of Mechanical Engineers (ASME) conference proceedings (e.g., International Intersociety Electronic Packaging Conference) and transactions (e.g., Journal of Electronic Packaging), the Institute of Electrical and Electronic Engineers (IEEE) conference proceedings (e.g., Electronic Components and Technology Conference and Electronics Packaging and Technology Conference) and transactions (e.g., Components, Packaging, and Manufacturing Technologies), and the International Microelectronics and Packaging Society (IMAPS) conference proceedings (e.g., International Symposium on Microelectronics), and transactions (e.g., International Journal of Microcircuits and Electronic Packaging).

I would like to thank my former employers, ASM (HK), Industrial Technology Research Institute (ITRI), the Hong Kong University of Science and Technology (HKUST), the Institute of Microelectronics (IME), Agilent, and HP, for providing me with excellent working environments that have nurtured me as a human being, fulfilled my need for job satisfaction, and enhanced my professional reputation. Also,

I would like to thank Dr. Don Rice (HP), Dr. Steve Erasmus (Agilent), Prof. Dim-Lee Kwong (IME), Prof. Ricky Lee (HKUST), Dr. Ian Yi-Jen Chan (ITRI), and Mr. Lee Wai Kwong (ASM) for their kindness and friendship while I was at their organizations. Furthermore, I would like to thank Mr. Tzyy-Jang Tseng (Chairman of Unimicron Technology Corporation) for his trust, respect, and support of my work at Unimicron. Finally, I would like to thank the following colleagues for their stimulating discussions and significant contributions to this book: Ming Li, DeWen Tian, YiuMing Cheung, Lei Yang, Chris Chang, S. W. Ricky Lee, Chien Ouyang, Qinglong Zhang, Kai Ming Yeung, Nelson Fan, Yam Mo Wong, Michael Zahn, Max Koh, Curry Lin, Hsing-Ning Liu, Kai-Ming Yang, Tim Xia, Cheng-Ta Ko, Bruce Puru Lin, Yu-Ling Chuang, R. Chen, M. Ma, Tzyy-Jang Tseng, K. Leung, J. J. Chen, Patrick Po-Chun Huang, Hsing-Ning Liu, Stephen H. Pan, Chia-Yu Peng, Tim Xia, Eagle Lin, Leo Chang, Hsing Ning Liu, Curry Lin, David Cheng, Winnie Lu, Yan-Jun Fan, Margie Li Qingqian, Tony Chen, Iris Xu, Qing Xiang Yong, Zhong Cheng, Eric Kuah, Zhang Li, Kim Hwee Tan, Eric Ng, Penny Lo, Wu Kai, Ji Hao, Koh Sau Wee, Jiang Ran, Cao Xi, Rozalia Beica, Sze Pei Lim, N. C. Lee, Yu-Hua Chen, Mian Tao, Jeffery Lo, Yang Lei, Jiang Ran, Henry Yang, Y. H. Chen, Winsons Bao, Tzu Nien Lee, Jason Wong, Mike Ma, K. Leung, Yu-Ling Chuang, R. Chen, Channing Cheng-Lin Yang, Andy Yan-Jia Peng, Gary Chang-Fu Chen, Vincent Teng, Jones Yu-Cheng Huang, Y. H. Chen, and Show May Chiu.

Definitely, I would like to thank my eminent colleagues (the enumeration of whom would not be practical here) at Unimicron, ASM, ITRI, HKUST, IME, Agilent, EPS, HP, and throughout the electronics industry for their useful help, strong support, and stimulating discussions. Working and socializing with them has been a privilege and an adventure. I learned a lot about life and flip chip, hybrid bonding, and fan-in and fan-out wafer-level packaging from them.

Lastly, I would like to thank my daughter Judy and my wife Teresa for their love, consideration, and patience by allowing me to work peacefully on this book. Their simple belief that I am making a contribution to the electronics industry was a strong motivation for me. Thinking that Judy and her supportive husband (Bill) have been doing very well in the electronics industry and their two adorable kids (Allison and James) have been raising in a happy and loving environment, and Teresa and I are in good health, I want to thank God for His generous blessings.

Palo Alto, CA, USA John H. Lau

Contents

About the Author

John H. Lau Ph.D., P.E. was the CTO from July 2019 to July 2021 and has been a Senior Special Project Assistant since August 2021 of Unimicron in Taiwan. Prior to that, he was a Senior Technical Advisor at ASM Pacific Technology in Hong Kong for five years; a specialist of the Industrial Technology Research Institute in Taiwan for five years; a Visiting Professor at Hong Kong University of Science and Technology for one year; the Director of the Microsystems, Modules, and Components Laboratory at the Institute of Microelectronics in Singapore for two years; and a Senior Scientist/MTS at Hewlett-Packard Laboratory/Agilent in California for more than 25 years.

His professional competences are design, analysis, materials, process, manufacturing, qualification, reliability, testing, and thermal management of electronic, optoelectronic, LED, CIS, and MEMS components and systems, with emphases on solder mechanics and manufacturing, RoHS-compliant products, SMT, flip chip, hybrid bonding, fan-in and fan-out wafer/panel-level packaging, SiP, chiplet design and heterogeneous integration packaging, chiplet communications (bridges), co-packaged optics, and TSV and other enabling technologies for 3D IC integration.

With more than 40 years of R&D and manufacturing experience, he has authored or coauthored more than 535 peer-reviewed articles (out of which 380 are the principal investigator), invented more than 50 issued or pending US patents (out of which 30 are the principal inventor), and given more than 335 lectures/workshops/keynotes worldwide. He has authored or coauthored 23 textbooks (all are the first author) on semiconductor advanced packaging, fan-in and fan-out wafer-level packaging, chiplet design and 3D IC heterogeneous integration packaging, TSV for 3D integration, advanced MEMS packaging, reliability of 2D and 3D IC interconnects, flip chip, WLP, MCM, area-array packages, WLCSP, high-density PCB, SMT, DCA, TAB, lead-free materials, soldering, manufacturing, and solder joint reliability.

He earned a Ph.D. degree in theoretical and applied mechanics from the University of Illinois at Urbana–Champaign, an M.A.Sc. degree in structural engineering from the University of British Columbia, a second M.S. degree in engineering mechanics from the University of Wisconsin at Madison, and a third M.S. degree in management science from Fairleigh Dickinson University in New Jersey. He also has a B.E. degree in civil engineering from National Taiwan University.

He has received many awards from the American Society of Mechanical Engineers (ASME), the Institute of Electrical and Electronics Engineers (IEEE), the Society of Manufacturing Engineers (SME) and other societies, including for the best IEEE/ECTC proceedings paper (1989), outstanding IEEE/EPTC paper (2009), best ASME transactions paper (Journal of Electronic Packaging, 2000), best IEEE transactions paper (CPMT, 2010 and CPMT, 2021), the ASME/EEP Outstanding Technical Achievement Award (1998), IEEE/CPMT Manufacturing Award (1994), IEEE/CPMT Outstanding Contribution Award (2000), IEEE/CPMT Outstanding Sustained Technical Contribution Award (2010), SME Total Excellence in Electronics Manufacturing Award (2001), Pan Wen Yuan Distinguished Research Award (2011), IEEE Meritorious Achievement in Continuing Education Award (2000), IEEE Components, Packaging, and Manufacturing Technology Field Award (2013), and ASME Worcester Reed Warner Medal (2015). He is an elected ASME fellow, IEEE fellow, and IMAPS fellow, and has been actively participating in industry/academy/society meetings/conferences to contribute, learn, and share.

Chapter 1
Flip Chip Technology

1.1 Introduction

Figure 1.1 shows the key elements of a flip chip assembly [1–4]. It consists of a device wafer, wafer bumping, dicing, flip chip assembly directly on a PCB (printed circuit board) or on a package substrate, and then underfilled. Today, the applications of flip chip technology have been extended to [5–10] chip-to-chip, face-to-face, and face-to-back. Figure 1.2 shows Amkor's Double-POSSUMTM package [10]. It can be seen that the package is actually defined by two levels of nesting die. The three daughter dies are flip chip attached to the larger mother die which is then attached to the largest grandma die. The grandma die is then flip chip attached to the package substrate. The bumps between the daughter dies and the mother die are microbumps (Cu-pillar with solder cap) or C2 (chip connection) chumps. C4 (controlled collapse chip connection) bumps are used between the mother die and grandma die, and between the grandma die and package substrate.

The flip chip technology was Introduced by IBM In the early 1960s for their solid logic technology, which became the logical foundation of the IBM System/360 computer line [11, 12]. Figure 1.3a shows the first IBM flip chip with three terminal transistors, which are Ni/Au plated Cu balls embedded in a Sn–Pb solder bump on the three I/O pads of transistor. A Cr–Cu–Au adhesion/seed layer is deposited between the Al–Si contact pads on the Si chip and the solder bump. Figure 1.3b shows the first IBM flip chip assembly (three chips) on a ceramic substrate. As the I/Os increase, the Cu ball is replaced by solder bump. The so-called C4 technology [12] utilizes high-lead solder bumps deposited on wettable metal terminals on the chip and a matching footprint of solder wettable terminals on the substrate. The solder-bumped flip chip is aligned to the substrate, and all solder joints are made simultaneously by reflowing the solder.

In this book, a flip chip is defined [1–8] as a chip attached to the pads of a substrate or another chip with various interconnect materials (e.g., Sn–Pb, Cu, Au, Ag, Ni, In, and isotropic or anisotropic conductive adhesives) and methods (e.g., mass reflow,

J. H. Lau, *Flip Chip, Hybrid Bonding, Fan-In, and Fan-Out Technology*,
https://doi.org/10.1007/978-981-97-2140-5_1

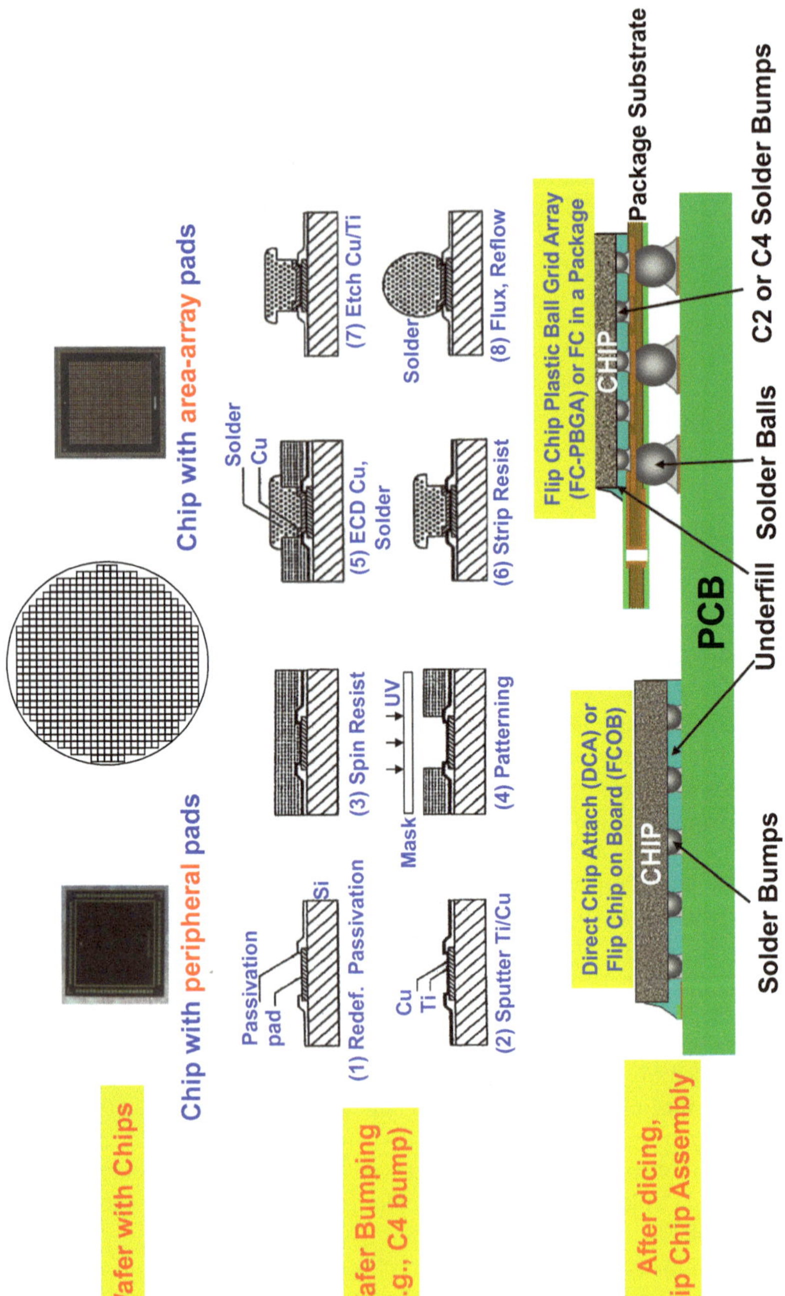

Fig. 1.1 Key elements in flip chip technology

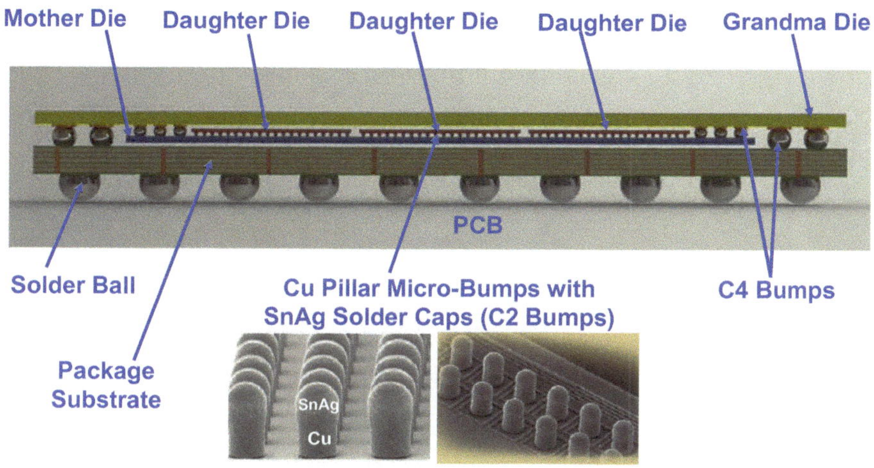

Fig. 1.2 Amkor's double-POSSUMTM package

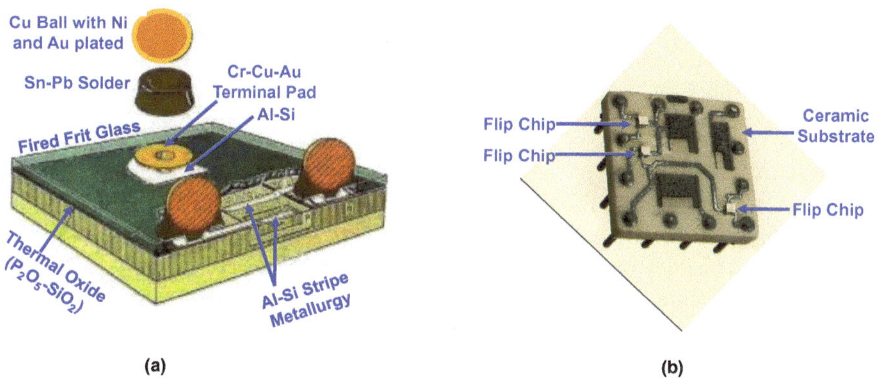

Fig. 1.3 **a** First IBM flip chip technology. **b** IBM flip chip with three chips on a ceramic substrate

thermocompression bonding (TCB), hybrid bonding), as long as the chip surface (active area or I/O-side) is facing the substrate or another chip as shown in Fig. 1.4. Also, in this book, the C4 bumps refer to all kinds of solder.

Flip chip technologies have been used extensively for the processors of mainframe computers, servers, personal computers, notebooks, smartphones, tablets, games, etc., the application specific integrated circuits (ASICs) of networking, telecommunications, etc., and the memories of data storage devices, etc. Most of the flip chip assemblies are mass reflowed. Recently, because of the requirements of higher functionalities of the chips and shrinking the chips' area, the number of pinouts of the processors, ASICs, and memories increases and their pitch (or the spacing between the pin-out pads) decreases. Also, because of the trends of smaller form factors for mobile (e.g., smartphones and tablets) and portable (e.g., notebooks) products, the

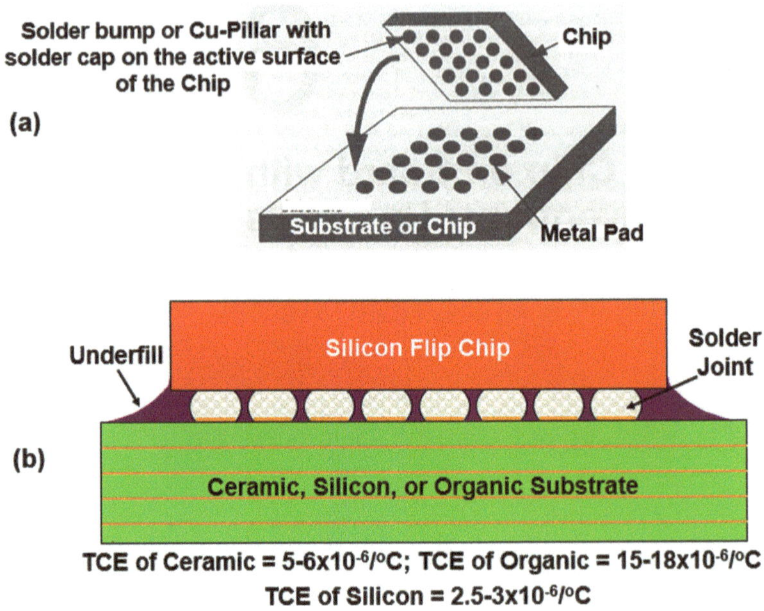

Fig. 1.4 **a** Flip chip technology definition. **b** Flip chip on various substrates with underfill

thickness of the chips and package substrates must be as thin as possible. Higher pin counts, tighter pitches, thinner chips, and thinner package substrates lead to the necessity of the TCB or even the hybrid bonding methods for flip-chip assemblies. In this chapter, besides mass reflow, various TCB techniques are mentioned. A high throughput LPC (liquid phase contact) TCB process is mentioned. Hybrid bonding will be discussed in Chap. 2 of this book.

Recent advances in high-density and low-cost package substrates have promoted more flip chip applications. In this study, the build-up substrate, through-silicon via (TSV)-interposer, TSV-less interposer, build-up substrate with thin-film layers, coreless substrate, bump-on-lead (BOL), embedded-trace-substrate (ETS), and hybrid substrate will be discussed. Also, fan-out redistribution-layer (RDL) substrate (organic interposer) will be mentioned. Furthermore, silicon-bridge embedded in organic substrate and in fan-out RDL substrate are briefly discussed.

In order to enhance the solder joint reliability of flip chip assemblies, underfill is a must, especially for organic package substrate. In this study, the pre-assembly underfill such as the no-flow underfill (NUF), nonconductive paste (NCP), and nonconductive film (NCF) will be discussed. Also, the post-assembly underfill such as the capillary underfill (CUF) and molded underfill (MUF) will be examined. A high-throughput stencil printing of underfill will be discussed. Since wafer bumping is the mother of flip chip technology, it will be briefly mentioned first.

1.2 Wafer Bumping–Stencil Printing

Wafer bumping by the stencil printing method is not only potentially low in cost and high in throughput but is also welcomed by system manufacturers, because stencil printing of solder paste is one of their most important steps in surface mount technology [13]. The only limitation is the very fine pitch of the pads on the chip. Today, stencil printing solder paste on the wafer in volume production is limited to 120 μm pitch [14].

1.2.1 Sample Description and Experiment Setup

The wafer used in this study is a 200 mm-wafer which is shown in Fig. 1.5. The chip size and its critical dimensions are also shown in Fig. 1.5. It can be seen that there are 48 pads for each chip and the pad pitch is 0.75 mm. The pad diameter is 0.33 mm. Figure 1.5c shows a photo of the pad with a copper stud and Fig. 1.5d shows the dimension of the copper stud which is about 60 μm tall. It should be pointed out that the focus of this study is to study the printing characteristics of solder paste on wafers by Taguchi design of experiments. The strength and reliability of solder bumps are beyond the scope of this paper.

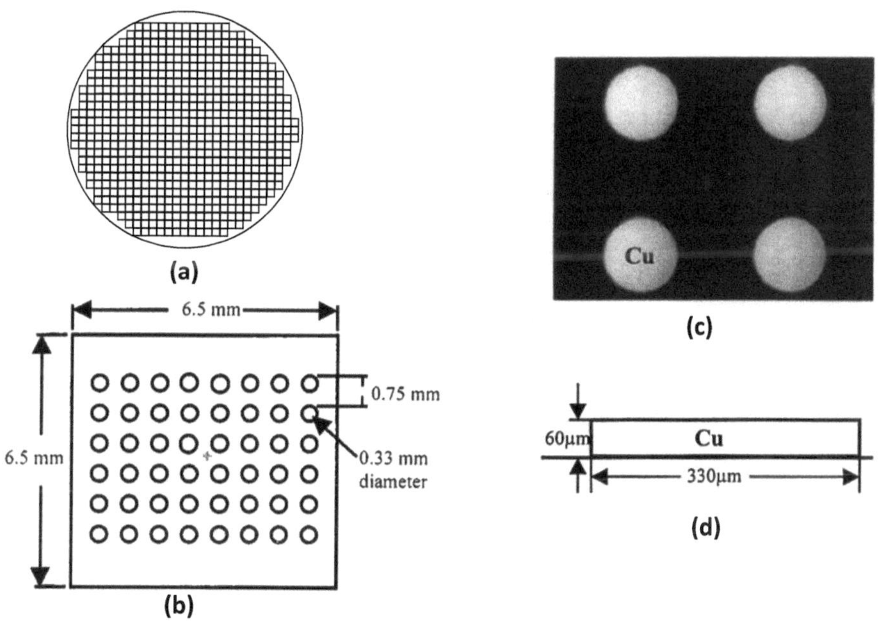

Fig. 1.5 **a** Die layout of a 200 mm-wafer. **b** Chip dimension and pad dimension and pitch. **c** Copper pads. **b** Geometry of copper pad

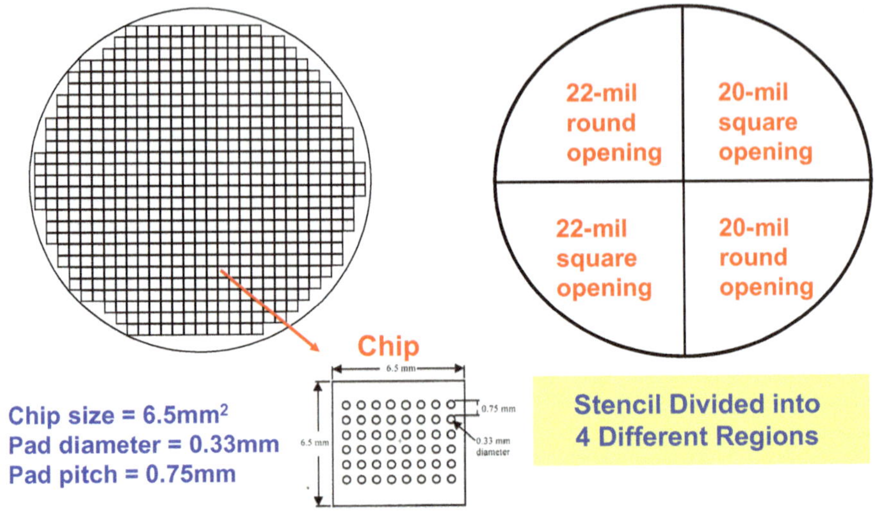

Fig. 1.6 Stencil design with four different openings which divide the 200-mm wafer into four quadrants

The stencil is a 0.2 mm-thick stainless-steel sheet drilled with a phase-quadrupled Nd: YAG laser system made by LPKF Laser & Electronics AG. The stencil is then electro-polished in a 30% potassium hydroxide solution. The stencil contains four different quadrants with varying aperture shapes and sizes as shown in Fig. 1.6. It can be seen that the first quadrant has a 20-mil (0.5 mm) square opening, the second quadrant is with 20-mil round opening, the third quadrant is with 22-mil (0.55 mm) square opening, and the fourth quadrant is with 22-mil round opening.

The MPM SPM-AV auto-vision semiautomatic printer is used for this study. The paste materials are Alpha Metals' type 3 and type 4 (63 wt%Sn-37 wt%Pb) solder alloys. The distributions of powder sizes of Type 3 are: more than 80 wt% of the nominal diameters are between 20 μm and 45 μm, less than 1 wt% of the nominal diameters are larger than 45 μm, no more than 20 wt% of the nominal diameters are less than 20 μm, and none of the powder size is larger than 50 μm. The distributions of powder sizes of Type 4 are: more than 90 wt% of the nominal diameters are between 20 μm and 38 μm, less than 1 wt% of the nominal diameters are larger than 38 μm, no more than 10 wt% of the nominal diameters are less than 20 μm, and none of the powder size is larger than 40 μm. The targeted solder-bump height on the wafer is 13 mils (0.33 mm).

Table 1.1 Control factors of L8 taguchi design experiment*

	Control factors	Level 1	Level 2
1	Paste type	Tape 3	Tape 4
2	Squeeze force	9 lb	12 lb
3	Snap-off height	0 mil	5 mil
4	Aperture shape	Square	Round
5	Aspect ratio	2.50	2.75

* Stencil thickness = 8 mil

1.2.2 Taguchi Design of Experiments

There are eight major categories (operator, environment, printing parameter, printer, stencil, wafer, squeegee, and solder paste) that influence the quality of solder paste printing on wafer. To perform a Taguchi design of experiment with all the parameters is impossible. However, selecting some important parameters such as paste type, squeeze force, snap-off height, aperture shape, and aspect ratio, the characteristics of stencil printing for wafer-level packaging could be analyzed, which is the objective of this study after reviewing references [15–21].

1.2.3 Control Factors in Stencil Printing

Table 1.1 shows the control factors. These are: paste types (type 3 and type 4), squeeze forces (9 and 12 lbs), snap-off heights (0 and 5 mils), aperture shapes (square and circle), and aspect ratios (2.5 and 2.75). The aspect ratio is defined as the stencil opening divided by the stencil thickness. The two-level Taguchi L8 orthogonal array with these control factors is shown in Table 1.2.

1.2.4 Paste Printing and Reflow

The stencil parameters, paste, and process conditions for each test run are shown in Table 1.3. The typical printed solder paste results in each quadrant are shown in Fig. 1.7. The solder pastes on the wafers are reflowed in a DIMA SMT (surface mount technology) Systems NL B. V. reflow oven. The SMT compatible reflow temperature profile is shown in Fig. 1.8. Figure 1.9 shows the solder bumps with different stencil openings, while Fig. 1.10 shows their cross sections. It can be seen that the largest cross sections are from the stencil with 22-mil square openings and the smallest cross sections are from the stencil with 20-mil round openings. The cross sections are about the same for both the 22-mil round openings and the 20-mil square openings.

Table 1.2 Taguchi L8 orthogonal array

Test run	Control factors						
	Paste type	Squeeze force	Snap-off	Shape	Aspect ratio	E1*	E2†
1	1	1	1	1	1	1	1
2	1	1	1	2	2	2	2
3	1	2	2	1	1	2	2
4	1	2	2	2	2	1	1
5	2	1	2	1	2	1	2
6	2	1	2	2	1	2	1
7	2	2	1	1	2	2	1
8	2	2	1	2	1	1	2

* E1 = error 1
† E2 = error 2

Table 1.3 Stencil parameter, paste, and process condition for each run*

Test run	Paste	Force (lb)	Snap-off height (mil)	Shape	Aspect ratio	Dimension (mil)
1	Type 3	9	0	Square	2.5	20
2	Type 3	9	0	Circle	2.75	22
3	Type 3	12	5	Square	2.5	20
4	Type 3	12	5	Circle	2.75	22
5	Type 4	9	5	Square	2.75	22
6	Type 4	9	5	Circle	2.5	20
7	Type 4	12	0	Square	2.75	22
8	Type 4	12	0	Circle	2.5	20

* Stencil thickness = 8 mils. Aspect Ratio = stencil hole size/thickness

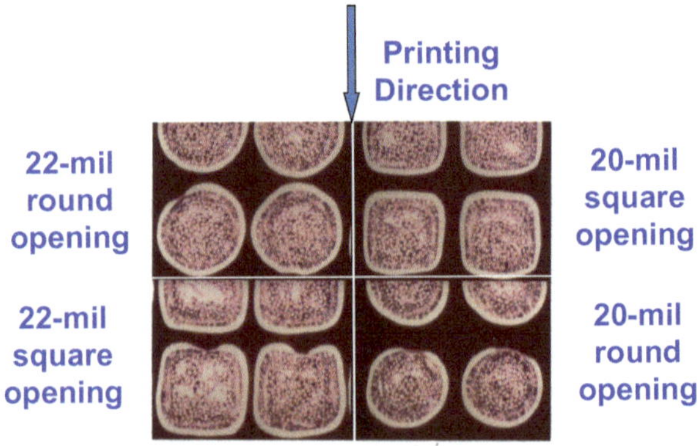

Fig. 1.7 Solder paste geometry before reflowing for four different Stencil opening designs

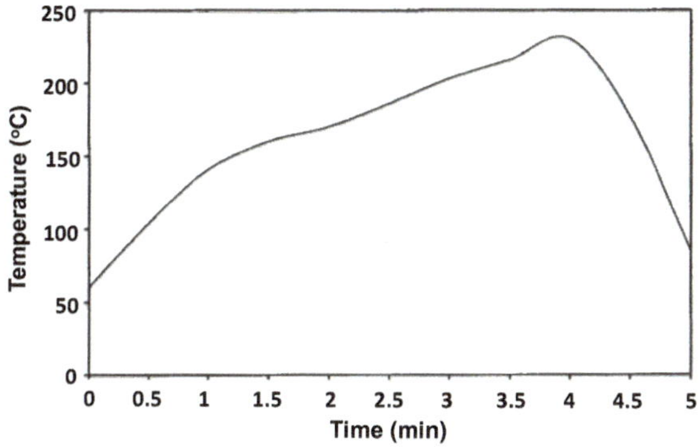

Fig. 1.8 Reflow temperature profile

Fig. 1.9 Solder bump geometry after reflowing

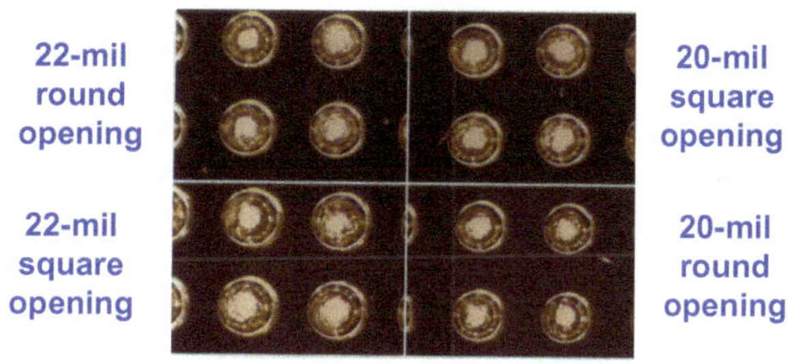

1.2.5 Solder Volume

The solder bump heights and diameters are measured by focusing the bump shadow from its cross section under a high-power microscope. The bump shadow is formed by the rear white light through a transparent molding material. Table 1.4 shows the measured raw data of the solder-bump volume from the two-level L8 experiments. The formula for calculating the volume of the solder bump (i.e., truncated sphere) is: $\pi \times \{h^3/6 + h \times \phi^2/8\}$ and substrate the copper stud volume, where h is the height of the truncated sphere and ϕ is the diameter of the sphere.

The target value (signal to noise ratio: S/N) of the solder bump volume is set to be the Larger-the-Better, which is defined as: $-10\log[MSD]$, in which $MSD = 1/n\sigma(1/y)^2$ is the mean square deviation, y is the measured response, σ is the sample standard deviation, and n is the number of measurements. The larger-the-better means that MSD is minimized. Minimizing the MSD is the same as maximizing the S/N [21].

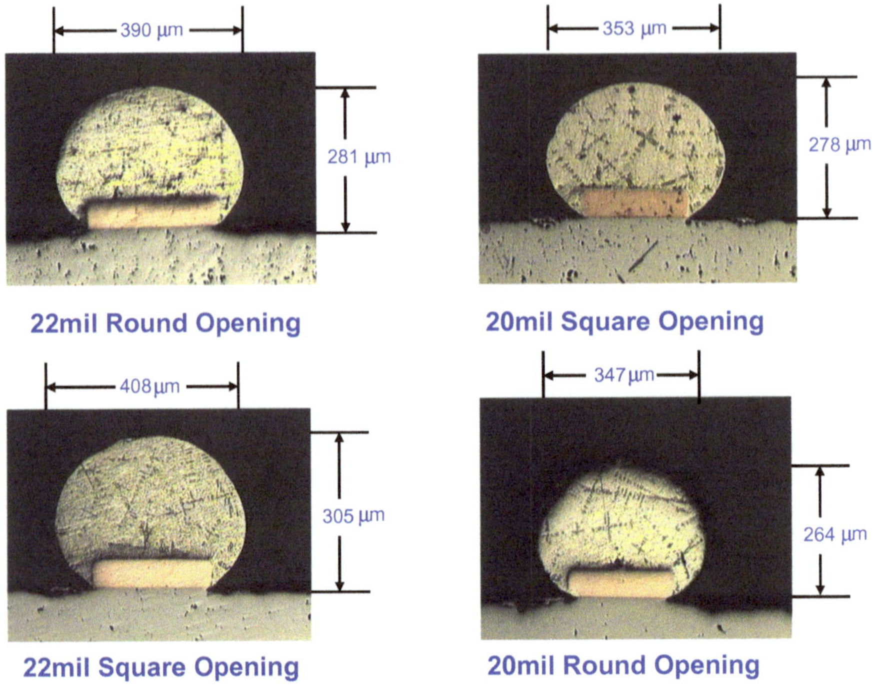

Fig. 1.10 Cross sections from four different stencil openings

Table 1.4 Solder bump volume* raw data measurements for L8 experiment

Test run	Vol. 1	Vol. 2	Vol. 3	Vol. 4	Vol. 5	Vol. 6	Vol. 7	Vol. 8	Vol. 9	Vol. 10	Avg. Vol
1	3.16	3.55	3.12	3.42	3.45	3.42	3.30	3.37	3.22	3.23	3.32
2	3.33	3.17	3.05	3.25	3.25	3.39	3.39	3.44	3.38	3.38	3.30
3	3.35	3.48	3.15	3.41	3.50	3.42	3.57	3.38	3.30	3.42	3.40
4	3.03	2.96	2.99	3.06	2.90	3.03	3.21	3.25	2.99	3.03	3.04
5	4.06	3.97	3.86	4.00	4.00	3.99	3.73	4.14	3.74	3.97	3.94
6	2.55	2.56	2.48	2.33	2.20	2.25	2.33	2.34	2.57	2.58	2.42
7	3.78	3.93	3.82	3.70	3.74	3.77	3.88	3.84	3.76	3.82	3.80
8	2.47	2.39	2.31	2.26	2.55	2.40	2.26	2.34	2.46	2.46	2.39

* Volume unit = $\times 10^7\ \mu m^3$

1.2.6 Data Analysis of the L8 Experiments

The solder-bump volume data from the L8 orthogonal array experiments is analyzed by the analysis of variance (ANOVA™) software. This software is a computational technique that enables someone to quantitatively estimate the relative contribution each control factor makes to the overall measured response and express it as a percentage. This software performs at least two analyses: analysis of mean (raw data) and analysis of variance [signal to noise ratio (S/N ratio)].

Table 1.5 shows the result of analysis of mean (ANOM), where *S* is the sum of squares of source; *V* is the variance of source; *S'* is pure sum of squares of source; *rho%* is the percent contribution of source; and *F* is equal to MS/S_e^2 (MS is the mean square of *V* and S_e^2 is the error variance).

The ranking for *F*-ratio has some physical meanings: 1) $F < 1$: the experimental error outweighs the source (control factor) effect; the source is insignificant and indistinguishable from the experimental error, 2) $F \sim 2$: the source factor has only a moderate effect compared to experimental error, and 3) $F > 4$: the source factor is strong compared to experimental error and is clearly significant. In the present study, for all the source factors, the *F* values in Table 1.5 are larger than the experimental error.

Figure 1.11 shows the effect of control factor (source) for the target value of the experiment. It can be seen that the aperture shape (circle or square) and aspect ratio (2.5 or 2.75, since the stencil thickness is 8 mils, which means the opening sizes are 20 mils or 22 mils) of the stencil have the biggest effects than the other parameters. The 22-mil square-opening stencil produces the largest solder-bump volume (D1 and E2 in Figs. 1.11 and 1.12, and Table 1.5).

In addition to ANOM, the analysis of variance (ANOVA) is performed. The result of ANOVA is shown in Table 1.6. The values indicate that all source factors are significantly larger than the experimental error. Figure 1.12 shows the S/N of parameters for the target value of the experiment. Again, it demonstrates that the dominant factors for solder-bump volume are the opening shape and opening size of the stencil.

Table 1.5 Taguchi L8 analysis of mean (raw data)*

Source (Factors)	*S*	*V*	*F*	S′	rho (%)
A (paste type)	0.003	0.003	1219.512	0.033	1.39
B (squeeze force)	0.016	0.016	591.279	0.016	0.67
C (snap-off height)	2.71E-05	2.71E-05			
D (aperture shape)	1.372	1.372	50,702.14	1.372	57.84
E (aspect ratio)	0.823	0.823	30,413.9	0.823	34.7
e1	0				
e2	0.098	0.001	36.955	0.096	4.05
(e)	2.71E-05	2.71E-05		0.002	0.08

* S = sum of squares of source; **V** = variance of source; *F* = S/(e); S′ = pure sum of squares of source; rho (%) = percent contribution of source; e1 = interaction error; e2 = repetition error; (e) = pooled error

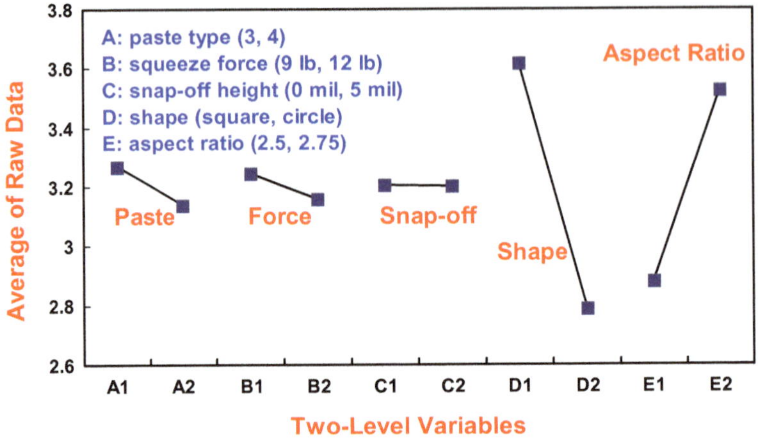

Fig. 1.11 Average of raw data for two-level L8 experiment

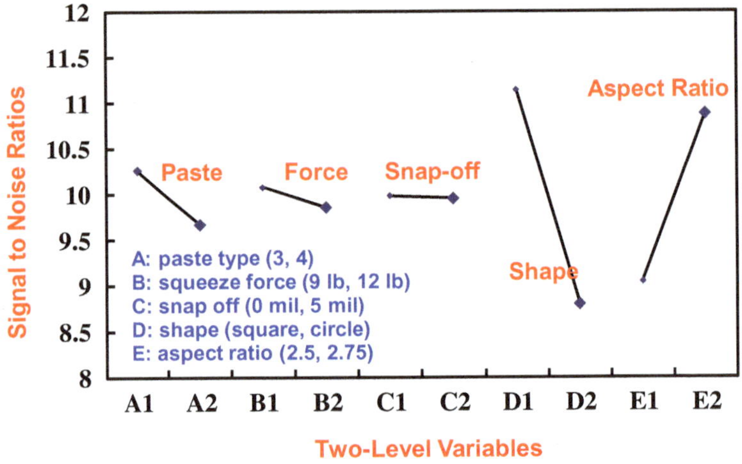

Fig. 1.12 Signal-to-noise ratios for two-level L8 experiment

The role of the other parameters for solder-bump volume is not important under the dominant factors.

1.2.7 *Data Analysis of the L4 Experiments*

Next, a Taguchi experiment is designed to investigate the effects of printing process (squeeze force and snap-off height) and material (solder paste type) on solder-bump

Table 1.6 Taguchi L8 ANOVA (S/N ratio)*

Source (Factors)	S	*V*	*F*	S′	rho (%)
A (paste type)	0.691	0.691	345.5	0.689	3.7
B (squeeze force)	0.101	0.101	50.5	0.099	0.53
C (snap-off height)	0.002	0.002			
D (aperture shape)	10.895	10.895	5447.5	10.893	58.51
E (aspect ratio)	6.714	6.714	3357	6.712	36.05
e1	0				
e2	0				
(e)	2.00E-03	2.00E-03		0.014	0.08

* For the definition of the variances, see the footnote in Table 1.5

volume for the 22-mil square-opening stencil. The target value of solder-bump volume is, again, the Larger-the-Better. Table 1.7 shows the control factors and Table 1.8 shows the L4 orthogonal array. Table 1.9 shows the material and process conditions for the L4 design experiment. Table 1.10 shows the solder-bump volume raw data for the L4 orthogonal array.

The result of ANOM is shown in Table 1.11. The values indicate that the snap-off is the only factor significantly larger than the experimental error. This is also demonstrated in the effect of control factor (parameter) for the target value trend as shown in Fig. 1.13. The snap-off height (5 mil difference) has the biggest effect among the squeeze pressure (3 lb difference) and the solder paste type (type 3 versus type 4).

Table 1.7 Control factors of L4 Taguchi design experiment

Control factors		Level 1	Level 2
1	Paste type	Type 3	Type 4
2	Squeeze force	9 lb	12 lb
3	Snap-off height	0 mil	5 mil

Table 1.8 Taguchi L4 orthogonal array

Test run	Control factors		
	Paste type	Squeeze force	Snap-off height
1	1	1	1
2	1	2	2
3	2	1	2
4	2	2	1

Table 1.9 Paste and process conditions for taguchi l4 design experiment (based on 22-mil square aperture)

Test run	Paste	Force	Snap-off height
1	Type 3	9 lb	0 mil
2	Type 3	12 lb	5 mil
3	Type 4	9 lb	5 mil
4	Type 4	12 lb	0 mil

Table 1.10 Solder bump volume raw data measurements for l4 experiment

Test run	Vol. 1	Vol. 2	Vol. 3	Vol. 4	Vol. 5	Vol. 6	Vol. 7	Vol. 8	Vol. 9	Vol. 10	Avg. Vol
1	3.89	3.70	3.68	3.89	3.78	3.86	3.63	3.71	3.91	3.89	3.79
2	3.81	3.85	3.91	3.82	3.93	3.84	4.10	3.61	4.13	3.58	3.86
3	4.06	3.97	3.86	4.00	4.00	3.99	3.73	4.14	3.74	3.97	3.94
4	3.78	3.93	3.82	3.70	3.74	3.77	3.88	3.84	3.76	3.82	3.80

Volume: $\times 10^7\ \mu m^3$

Table 1.11 TAGUCHI L4 analysis of mean (raw data)

Source (factors)	*S*	*V*	*F*	S′	rho (%)
A (Paste type)	0.002	0.002	2	1.00E-03	2.27
B (Squeeze force)	0.001	1.00E-03			
C (Snap-off height)	0.011	0.011	11	0.01	22.73
e1	0				
e2	0.029	1.00E-03			
(e)	0.03	1.00E-03		0.032	72.73

S, V, F, S′, rho(%), e1, e2, and (e) have been defined in Table 1.5

The result of ANOVA is shown in Table 1.12. Again, the values indicate that the snap-off height is the only factor significantly larger than the experimental error. Figure 1.14 plots the signal to noise ratio of parameters for the target value. Again, this figure shows the effect of 3 lb squeeze force difference and of paste type difference on the solder-bump volume.

1.2.8 Summary

- By choosing the paste type, squeeze pressure, snap-off height, aperture shape, and aspect ratio as the critical parameters, a two-level L8 orthogonal array experiment

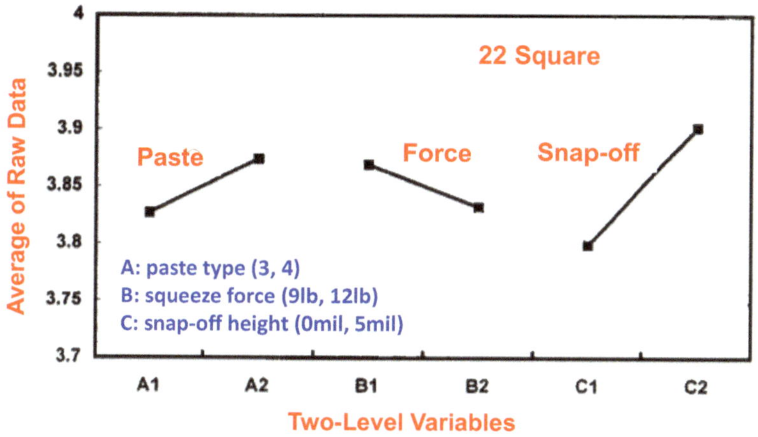

Fig. 1.13 Average of raw data for two-level L4 experiment

Table 1.12 Taguchi L4 analysis of variance (signal-to-noise ratio)

Source (Factors)	*S*	*V*	*F*	S′	rho (%)
A (Paste type)	0.011	0.011	1.833	0.005	7.04
B (Squeeze force)	0.006	0.006			
C (Snap-off height)	0.054	0.054	9	0.048	67.61
e1	0				
e2	0				
(e)	0.006	0.006		0.018	25.35

S, V, F, S′, rho(%), e1, e2, and (e) have been defined in Table 1.5

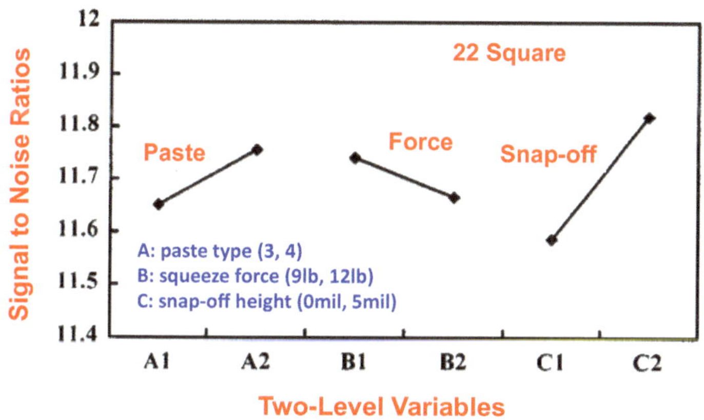

Fig. 1.14 Signal to noise ratios for two-level L4 experiment

has been performed on solder bumping of a 200-mm wafer by stencil printing method.

- Also, by performing the ANOM and ANOVA, the most influential factors have been determined. Furthermore, a two-level L4 orthogonal array experiment has been performed to optimize the other material and process factors. Some important results are summarized as follows.
- For wafer-level packaging, because of the larger-pitch area-arrayed pads on the chip, wafer bumping by stencil printing method is not only low cost but also the logical choice, especially for larger solder bumps without underfill encapsulant.
- From the L8 experiments, it is found that the stencil designs (size and shape) are the dominant factors for solder-bump volume. The 22-mil square-opening stencil produces the largest solder-bump volume without bridging (0.75 mm pitch).
- From the L4 experiments, based on the 22-mil square-opening stencil, it is found that the effect of the snap-off height on the solder-bump volume is larger than the squeeze pressure and paste type.
- Even though alignment is not considered as one of the critical factors in this study, it plays a very important role. Based on the present study, it is found out that good alignment prevents solder-bump bridging. If misalignment occurred, solder paste would be contacted with neighbor copper pad that bridging occurred.

1.3 Wafer Bumping–Electroplating

1.3.1 C4 Bump

Usually, the pad size is equal to 100 μm, and the target bump height is equal to 100 μm. After redefining the passivation opening (usually it is not required), either Ti or TiW (0.1–0.2 μm) are sputtered over the entire surface of the wafer first, followed by 0.3–0.8 μm of Cu. Ti–Cu and TiW–Cu are called under bump metallurgy (UBM). In order to obtain 100 μm bump height, a 40 μm layer of resist is then overlaid on the Ti–Cu or TiW–Cu and a solder bump mask is used to define (ultraviolet exposure) the bump pattern as shown in steps #1–4 in Fig. 1.15. The opening in the resist is 7–10 μm wider than the pad opening in the passivation layer. A 5 μm layer of Cu is then plated over the UBM, followed by electroplating the solder. This is done by applying a static or pulsed current through the plating bath with the wafer as the cathode. In order to plate enough solder to achieve the target (100 μm), the solder is plated over the resist coating by about 15 μm to form a mushroom shape. The resist is then stripped off and the Ti–Cu or TiW–Cu is removed with a hydrogen peroxide or plasma etching. The wafer is then reflowed with flux, which creates smooth truncated spherical solder C4 bumps, Fig. 1.15, due to surface tension as shown in steps #5–8 of Fig. 1.15.

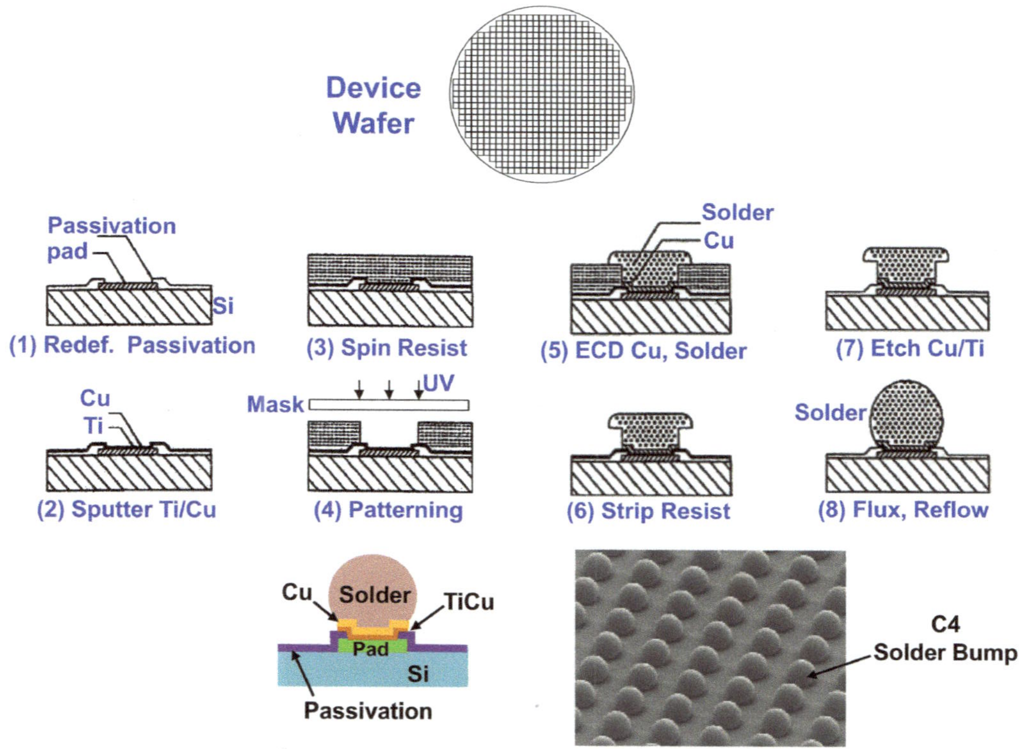

Fig. 1.15 Wafer bumping by ECD or electroplating method for C4 bumps

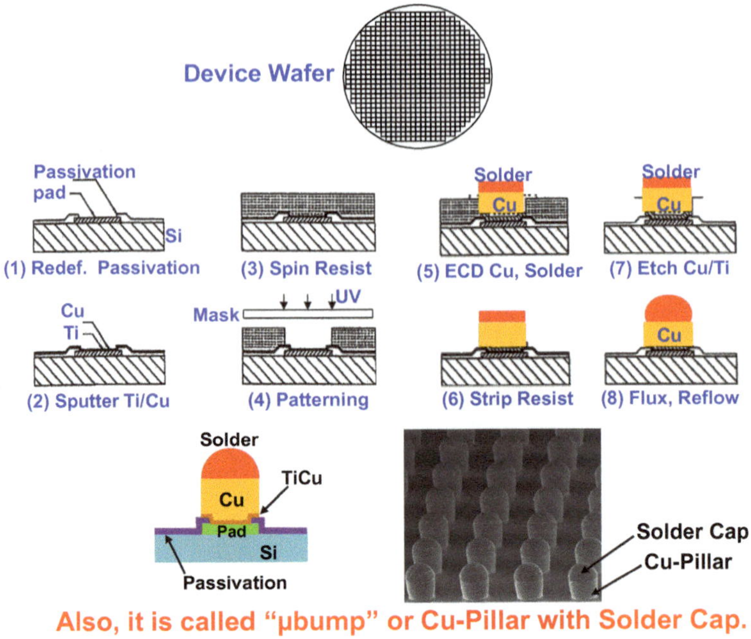

Fig. 1.16 Wafer bumping by ECD or electroplating method for C2 bumps

1.3.2 C2 Bump

Because of higher pin-count and tighter pitch (smaller spacing between pads), there is a possibility of shorting the adjacent solder C4 bumps. Wire interconnects [22] and Cu-pillar with solder cap [23–25] (e.g., Fig. 1.16) can be a solution. The fabrication process is basically the same as that of the C4 bumps except electroplating the Cu instead of solder as shown in step #5 of Fig. 1.16. It is followed by electroplating the solder cap and then reflowing the solder with flux. Because the solder volume is very small compared with the C4 bump, the surface tension is not enough to perform the self-alignment of the Cu pillar with the solder cap bump and therefore, it is sometimes called a C2 (chip connection) bump or microbump.

1.3.3 C4 Bump versus C2 Bump

Besides being able to handle finer pitch, C2 bumps also provide better thermal and electrical performance than C4 bumps. This is because the thermal conductivity (W/m K) and electrical resistivity (μΩ m) of Cu (400 and 0.0172) are superior to those (55–60 and 0.12–0.14) of solder as shown in Fig. 1.17.

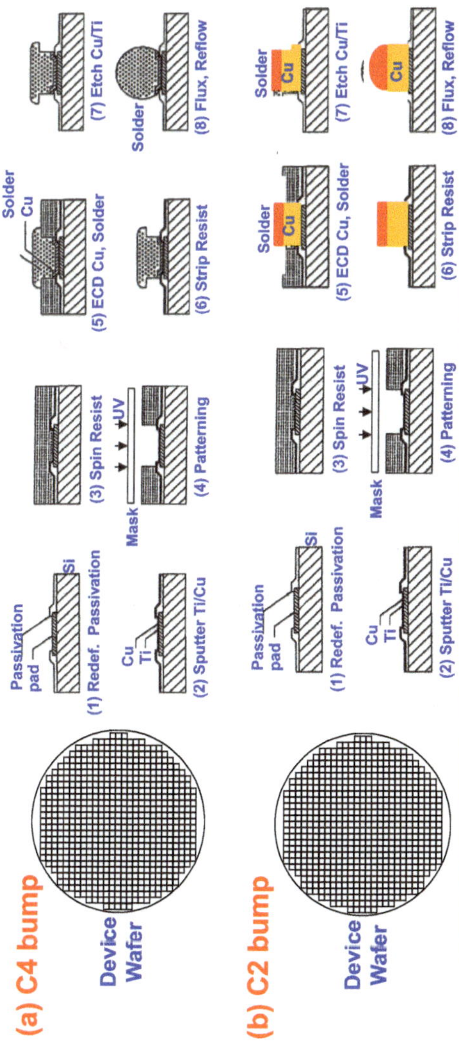

Structure	Thermal conductivity (W/m·K)	Electrical resistivity (μΩ·m)	Bump pitch	Self-alignment
Cu	400	0.0172	-	-
C4 Bump (Solder)	55 - 60	0.12 – 0.14	≥ 50μm	Very good
C2 Bump (Cu Pillar + Solder Cap)	300 (Effective)	0.025 (Effective)	< 50μm	Veery poor

Fig. 1.17 Comparison between C4 and C2 bumps

1.4 Flip Chip Package Substrate

In the past few years, tremendous efforts have been devoted to enhance/advance the capabilities of the conventional low-cost build-up organic package substrates by increasing the number of build-up layers, fabricating thin-film layers on top of the build-up layer, shrinking the dimensions of the metal line width and spacing, reducing the pad size and pitch, eliminating the core, making the BOL, laminating the ETS, and building the hybrid substrate. For silicon substrates, there are the passive TSV-interposer and active TSV-interposer. Also, fan-out RDL substrate (organic interposer) will be more popular. Furthermore, silicon-bridge embedded in organic substrate and in fan-out RDL substrate are getting lots of tractions.

1.5 Flip Chip Package Substrates–Organic

1.5.1 Printed Circuit Boards (PCBs)

A typical PCB is shown in Fig. 1.18a. Most of the PCBs are made from the FR (flame retardant)-4, which is a composite material composed of woven fiberglass cloth with an epoxy resin binder [glass-reinforced epoxy resin laminate as shown in Fig. 1.18b] that is flame resistant. A PCB consists of conductor, trace, track, route, etc. as shown in Fig. 1.19. There are microstrip lines which lay on the surface of the PCB and strip lines which lay in the interior (internal layers) of a PCB. Figure 1.20 shows various kinds of PCB structures such as the single-side, double-sides, multilayers, etc. Nowadays, it is not uncommon to have a PCB with more than 100 layers. However, usually 30 to 60 layers are the maximum.

1.5.2 Build-Up Package Substrates

More than 30 years ago, IBM in Japan at Yasu invented the SLC technology, Fig. 1.21 [26–28], which formed the basis of today's very popular low-cost organic package substrates with build-up layers vertically connected through microvias [29] to support flip chips. There are two parts of the SLC technology: one is the core substrate, and the other is the SLC for the signal wiring. The core substrate is made by the ordinary glass epoxy panel. However, the SLC layers are sequentially built up with the dielectric layers made of photo sensitive epoxy and the conductor plane of copper plating (semi-additive technique). In general, a package substrate with twelve layers (e.g., two core-layers and ten build-up layers (5-2-5)) and 10 μm-line width and spacing is more than adequate to support most of the chips. However, a package substrate with 26 layers (e.g., 12-2-12) is frequently used.

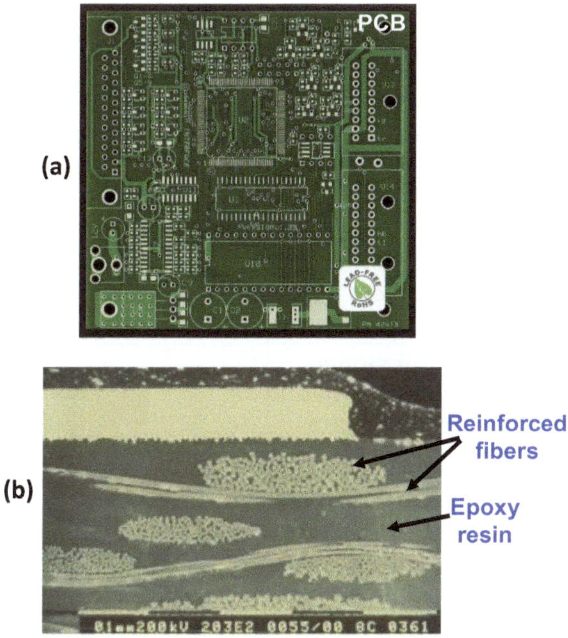

Fig. 1.18 **a** A typical PCB. **b** FR-4 reinforced fibers epoxy resin

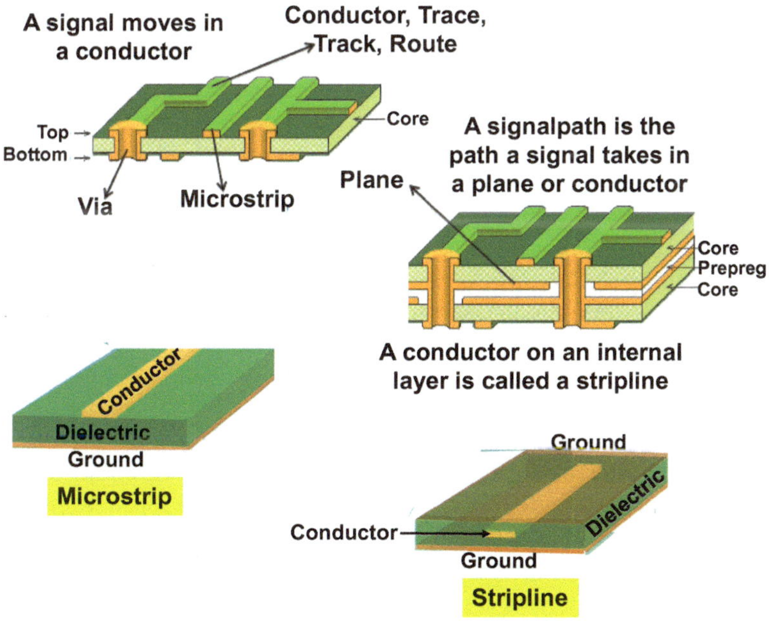

Fig. 1.19 Key elements on a PCB

Pattern	Layer	Schematic
Single-sided	One signal layer	
Double-sided	Two signal layers	
Multilayer	Two signal layers and two power layers	
Multilayer	Four signal layers and two power layers	
Multilayer	Two signal layer and two power layer	

Fig. 1.20 Various kinds of PCBs

1.5.3 Coreless Package Substrates

Coreless substrate was first proposed by Fujitsu [30] in 2006. Figure 1.22 shows the comparison between the conventional organic package substrate with build-up layers and the organic coreless package substrate. It can be seen that the biggest difference is that there is not a core in the coreless package substrate and all the layers of the coreless package substrate are the build-up layers [30–52].

The advantages of the coreless package substrate are [30–52]: (a) because of eliminating the core, the cost of the coreless substrate is lower; (b) by eliminating the core, higher wiring ability can be achieved; (c) better electrical performance because of good high-speed transmission characteristic; and (d) definitely smaller form factor. On the other hand, the disadvantages are [30–52]: (a) because of eliminating the core, the warpage of the coreless substrate is larger; (b) easier to have laminate chipping; (c) poor solder joint yield because of less substrate rigidity; and (d) new manufacturing infrastructure is necessary.

In 2010, Sony manufactured the first coreless package substrate for the cell processor of their PlayStation 3 [42]. Even though coreless substrates have many advantages, they are not popular because of the warpage control issue. One of the key factors affecting the warpage is the coefficient of thermal expansion mismatch of substrate materials. Thus, proper control of this factor will help reduce the warpage issue of coreless substrates. Another factor affecting the warpage is the package

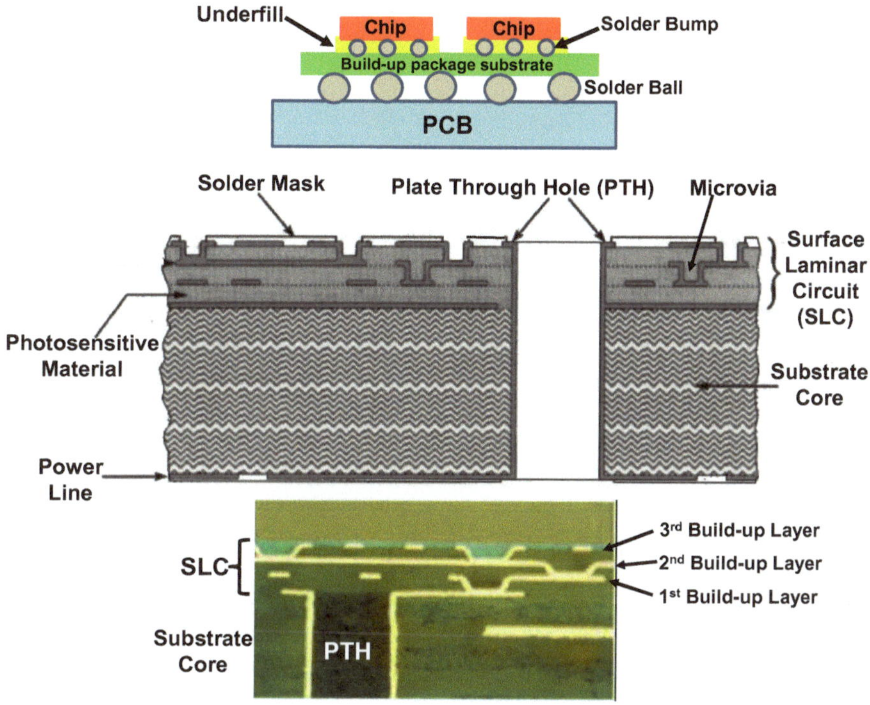

Fig. 1.21 IBM's SLC for flip chip organic build-up package substrate

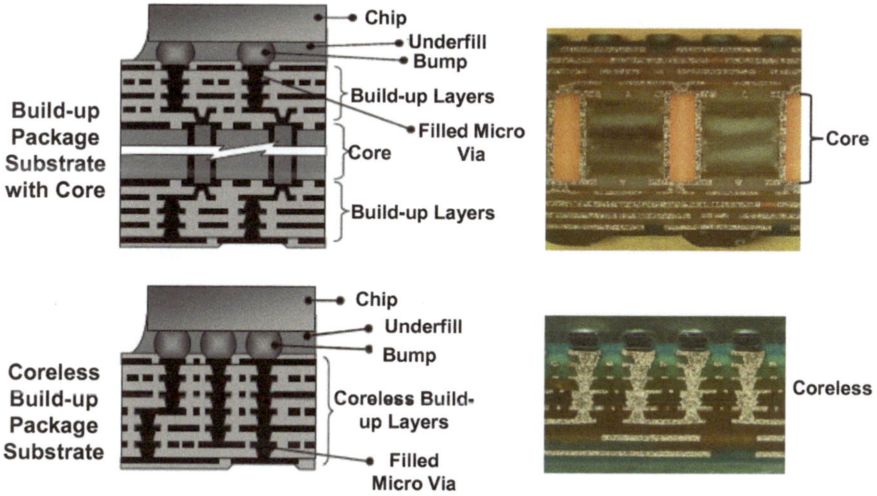

Fig. 1.22 (Top) Flip chip on conventional build-up package substrate. (Bottom) Flip chip on coreless substrate

assembly. Thus, a proper package assembly warpage correction control (with vacuum and pressure) will help improve the warpage problem of coreless substrate.

1.5.4 BOC and BOL Substrates

BOL (bond on lead) was first proposed by STATSChipPac [53–57] and was used by Qualcomm [58] and others [58–61]. A conventional bump-on-capture pad (BOC) or simply bump-on pad (BOP) flip chip organic substrate layout is shown in Fig. 1.23a. It can be seen that the flip chip pads are on a 210 μm area array pitch in a solder mask (SR) defined configuration with one signal escape between bump pads resulting in an effective escape pitch of 105 μm.

The BOL methodology is shown in Fig. 1.23b; here, the landing pad on the substrate is merely the trace (lead) itself, or a slightly widened version of the trace which results in freeing up of enough routing space to allow routing an additional trace between bumps thereby resulting in an effective escape pitch of 70 μm without changing the design rules (trace width and space) of the substrate. The improved BOL structure is shown in Fig. 1.23c. It can be seen that the bump pads are without any solder resist confinement, i.e., open SR [58]. The test vehicles, Cu-column on

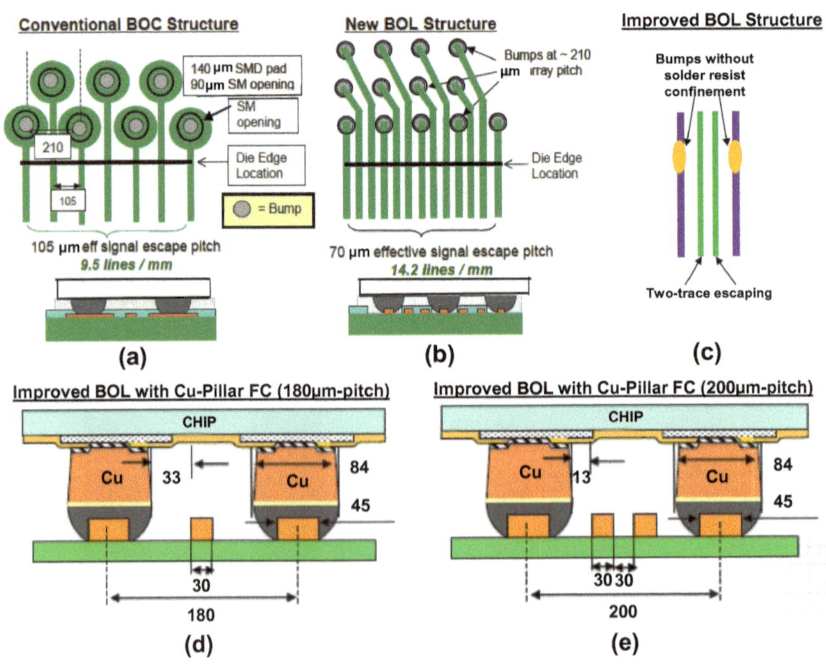

Fig. 1.23 Bump-on-lead (BOL). **a** Conventional BOP. **b** New BOL. **c** Improved BOL, **d** Improved BOL with Cu-pillar FC (180 μm-pitch). **e** Improved BOL with Cu-pillar FC (200 μm-pitch)

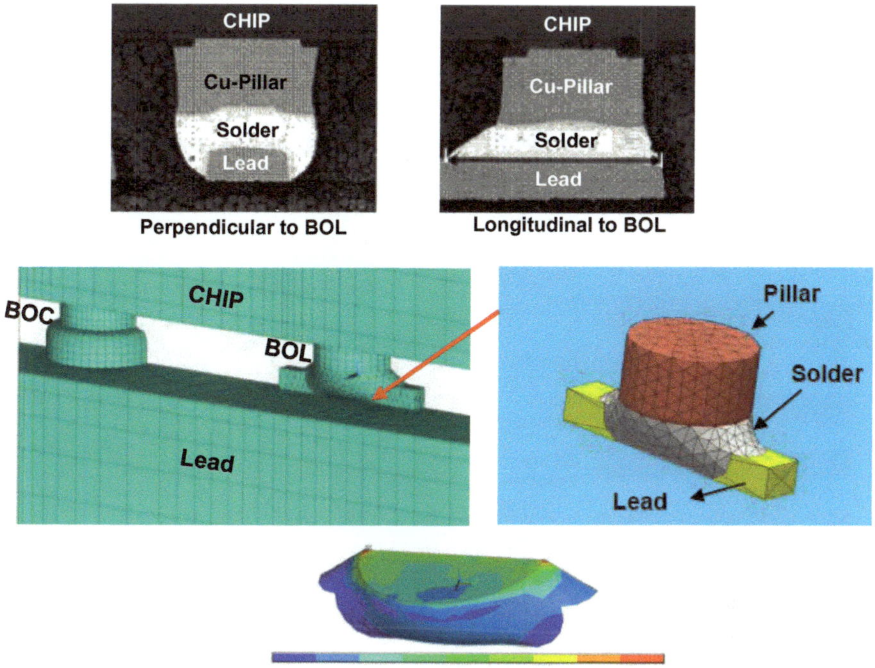

Fig. 1.24 Images of the perpendicular-to-BOL and longitudinal-to-BOL. Finite element models and creep strain contours in the BOL solder joint/

BOL, used in Ref. [58] are shown in Fig. 1.23d and e. It can be seen that one trace between the 180 μm bump pitch and up to two traces with the 200 μm bump pitch can be comfortably routed.

Typical cross sections of the perpendicular-to-BOL and longitudinal-to-BOL are shown in the upper portion of Fig. 1.24. A 3D slide finite element model showing the BOL, BOC (or BOP), and solder joint is shown in the middle of Fig. 1.24. The creep strain contours of the BOL solder joint are shown in the lower portion of Fig. 1.24 [61] and are too small to create solder joint reliability problem under most conditions.

1.5.5 Embedded Trace Substrates (ETS)

ETS is one of the coreless substrates with fine line width/spacing embedding the top metal trace pattern into prepreg layer [62–66]. The process flow of ETS is shown in Fig. 1.25a. It starts from a carrier board with a removable Cu foil. It is followed by using a typical electrolytic copper plating method to form the first layer of copper pattern. Then, laminate a prepreg on the copper pattern. It is followed by laser via drilling, electroless copper coating, dry film laminating, exposing and developing,

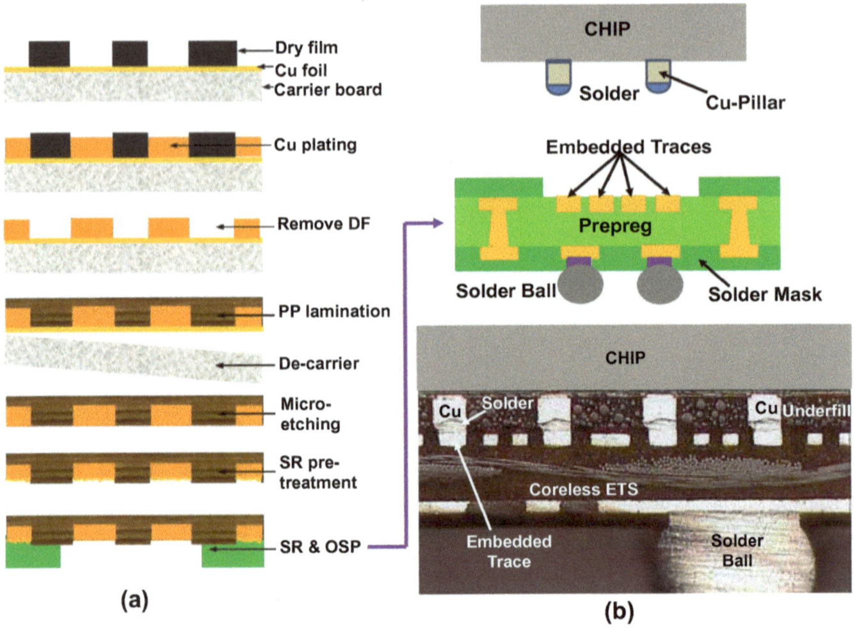

Fig. 1.25 **a** Process flow for fabricating the ETS. **b** Flip chip with C2 bumps on ETS assembly

second layer copper pattern plating, stripping, and micro etching. Once all the copper pattern layers have been completed, the carrier board will be removed. Since the Cu foil is connected to the first copper pattern, micro etching is necessary before SR coating. After the SR opening process, it is completed by metal finishes treatment, e.g., organic solderability preservatives (OSPs). Figure 1.25b shows a cross section of a Cu-pillar flip chip on ETS assembly by SPIL [96]. Most line width/spacing of ETS in use today is 13 μm/13 μm. However, 12 μm/12 μm line width/spacing is in production by Simmtech [66].

1.5.6 Thin-Film Layers on Build-Up Package Substrates

In 2013, Shinko proposed to make thin-film layers on top of the build-up layer of a package substrate. Figure 1.26a shows Shinko's i-THOP (integrated thin film high density organic package) substrate [48, 67] for high performance applications. It is a 4 + (2-2-3) test vehicle, which means there is a two-layer metal core, three build-up metal layers at the bottom (PCB) side, two build-up metal layers on the top (chip) side, and the first number "4" represents that there are four thin-film Cu wiring layers (RDLs) on the surface of the top build-up layer. The thickness, line width, and spacing of the thin-film Cu RDLs can be as small as 2 μm. The thin-film Cu RDLs are vertically connected through a 10 μm via, as shown in Fig. 1.26a. The surface Cu pad-pitch is 40 μm and the Cu pad diameter is 25 μm with a height of 10–12 μm. The i-THOP substrate passed the warpage and reliability tests and there was no via delamination observed [48]. In 2014, Shinko demonstrated that [67] ultrafine pitch flip chips can be successfully assembled on the i-THOP substrate. Figure 1.26b schematically shows the two chips' lateral communications by the 2 μm line width/spacing RDLs of the two thin-film layers, which are built on top of the 1-2-2 build-up organic substrate, i.e., 2 + (1-2-2). Figure 1.26b shows the 40 μm pitch microbumps (Cu-pillar + Ni + SnAg) of the test chips and the 40 μm pitch flip chip bonding pads (25 μm-diameter).

1.6 Flip Chip Package Substrates–Fan-Out EMC/RDL

1.6.1 Fan-Out Chip-First RDL-Substrate

Figure 1.27 schematically shows the cross sections of a typical FOWLP (fan-out wafer-level package). The manufacturing process flow is very simple [68, 69]. First, the device wafer is tested for known-good dies (KGDs) and then singulated into individual dies. This is followed by picking up the KGDs and placing them face-down on a temporary carrier that can be round (wafer) or rectangular (panel) with a double-sided thermal release tape. Then, the reconfigured carrier is over-molded using the compression molding method with epoxy mold compound (EMC) before removing the carrier and the double-sided tape and turning the whole molding (with KGDs) around. Next comes building the RDLs for signals, power, and grounds from the pads. Finally, solder balls (with pitch ranges from 1.27, 1, 0.8, 0.65, 0.5, and 0.4 mm) are mounted and the whole molding (with KGDs, RDLs, and solder balls) is diced into individual packages. Figure 1.27b shows the image of cross-sectional RDL while Fig. 1.27c shows the image of longitudinal RDL. The conventional build-up package substrate has been replaced by the fan-out RDL-substrate.

1.6.2 Fan-Out Chip-Last Hybrid Substrate

Figure 1.28 shows a hybrid substrate supporting multiple chips. The hybrid substrate consists of a build-up package substrate, a fine metal line width and spacing RDL-substrate (or organic interposer), and the solder joints with underfill. The organic substrate is fabricated by a fan-out chip-last process. This hybrid substrate will be elaborated in Chap. 4.

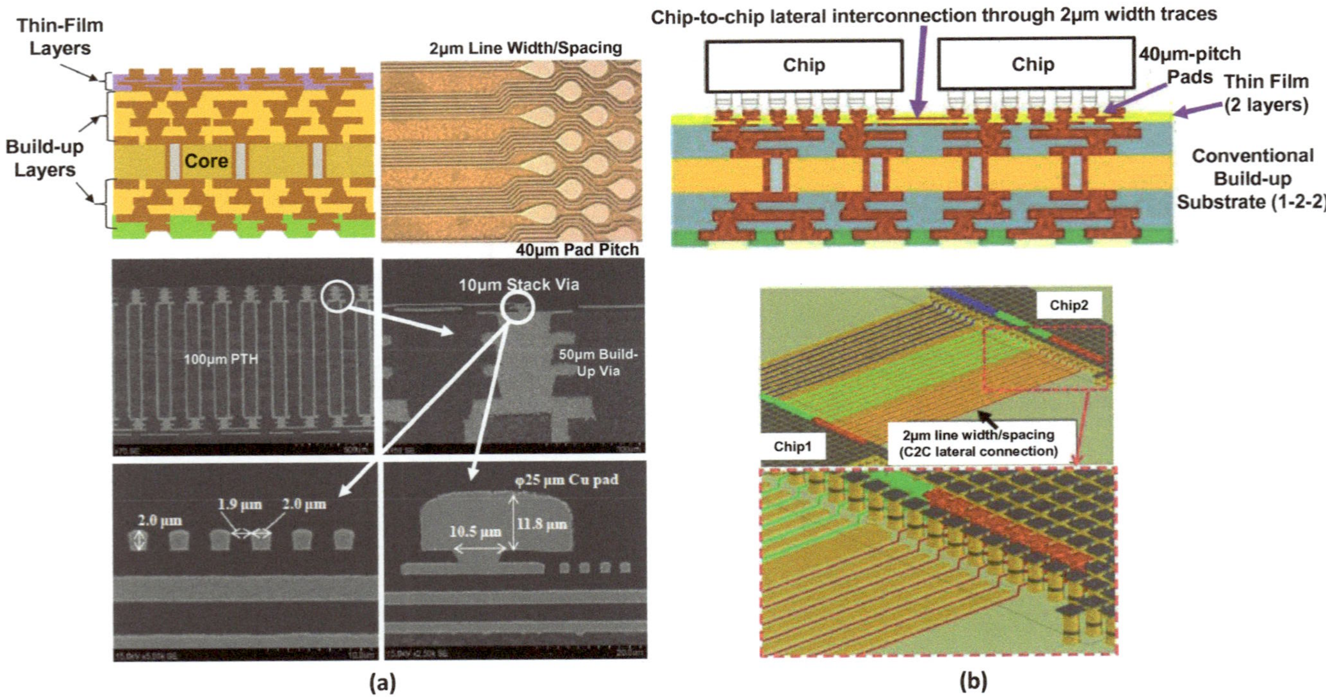

Fig. 1.26 **a** Shinko's i-THOP - a flip chip package substrate with thin film layers on top of the build-up layer. **b** Shinko's i-THOP test vehicle - two thin film layers are built on top of the 1-2-2 package substrate

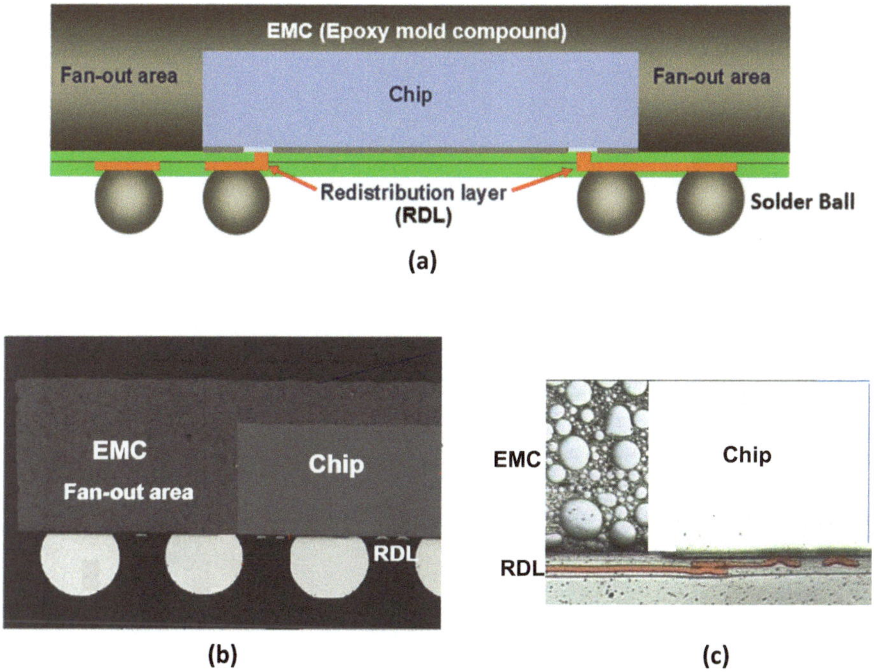

Fig. 1.27 **a** Fan-out wafer-level package. **b** RDL (cross-sectional view), and **c** RDL (longitudinal view)

1.7 Flip Chip Package Substrates—TSV-Interposer

In the past few years, because of the very high-density, high I/Os, and ultrafine pitch requirements such as the field programmable gate array (FPGA), even a fourteen build-up layers (6-2-6) package substrate is not enough to support the chips and a TSV (through-silicon via) interposer is needed [70–82]. There are at least two different kinds of TSV interposer, the passive TSV-interposer and the active TSV-interposer.

1.7.1 Passive TSV-Interposers

The passive TSV-interposer consists of the silicon substrate, RDLs (re-distribution-layers), and TSVs. For example, Fig. 1.29 shows the Xilinx/TSMC's FPBG chip on wafer on substrate (CoWoS) [79–81]. It can be seen that the TSV (10 μm-diameter) interposer (100 μm-deep) has four top RDLs: three Cu damascene layers and one aluminum layer. The 10,000+ of lateral interconnections between FPGA chips are connected mainly by the 0.4 μm-pitch (minimum) RDLs of the interposer. The minimum thickness of the RDLs and passivation is 1 μm. Each FPGA has more

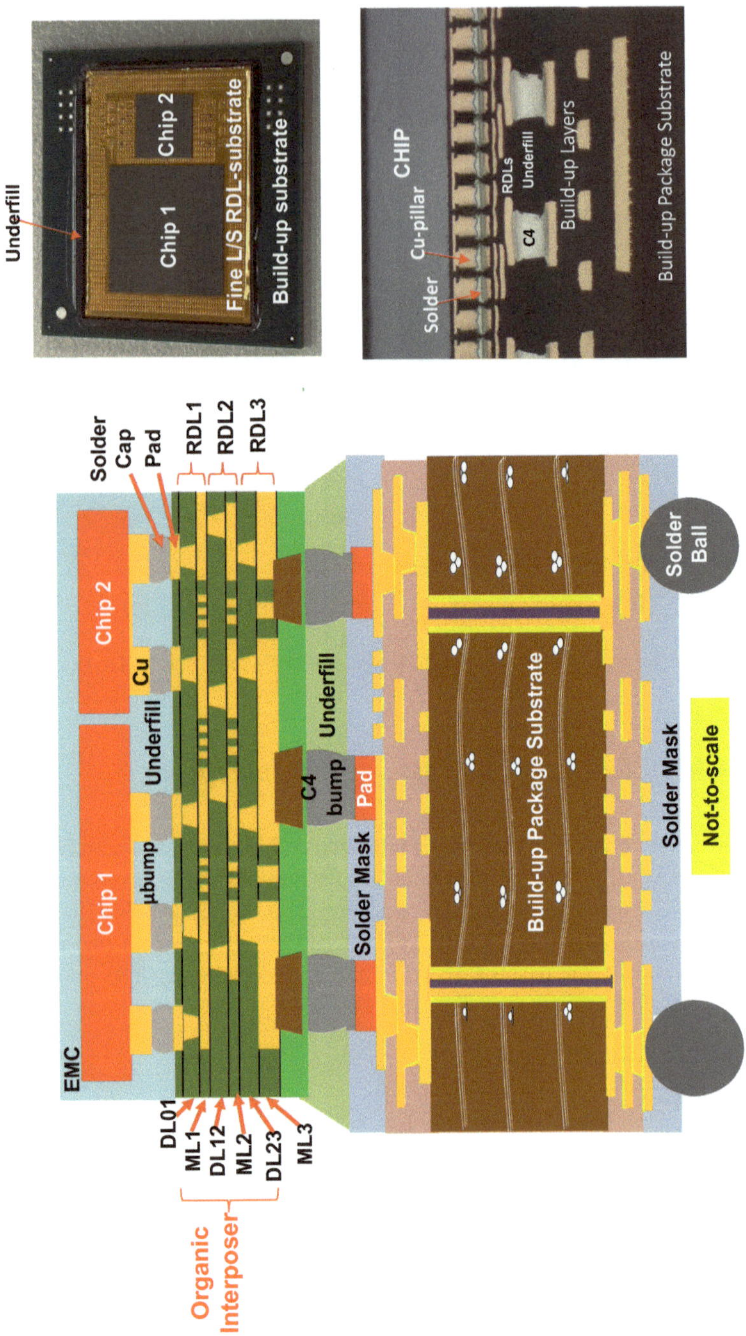

Fig. 1.28 Hybrid substrate (2.3D IC integration)

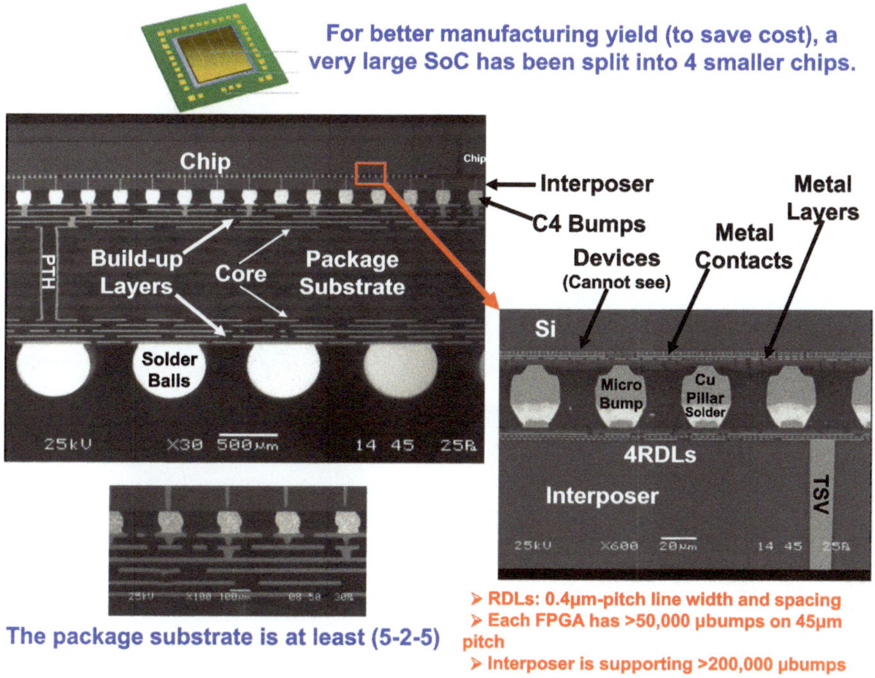

Fig. 1.29 2.5D IC integration with passive TSV-interposer

than 50,000 microbumps (200,000+ microbumps on the interposer) at 45 μm pitch as shown in Fig. 1.29.

1.7.2 Active TSV-Interposers

The active TSV-interposer consists of not only the silicon substrate, RDLs, and TSVs, but also the CMOS (complementary metal–oxide–semiconductor) devices. During IEEE/ECTC2019, Leti/STMicroelectronics demonstrated the feasibility of chiplets on an active TSV-interposer. There are six chiplets on the TSV-interposer with the 64 nm CMOS devices called INTACT (active interposer) [83], Fig. 1.30.

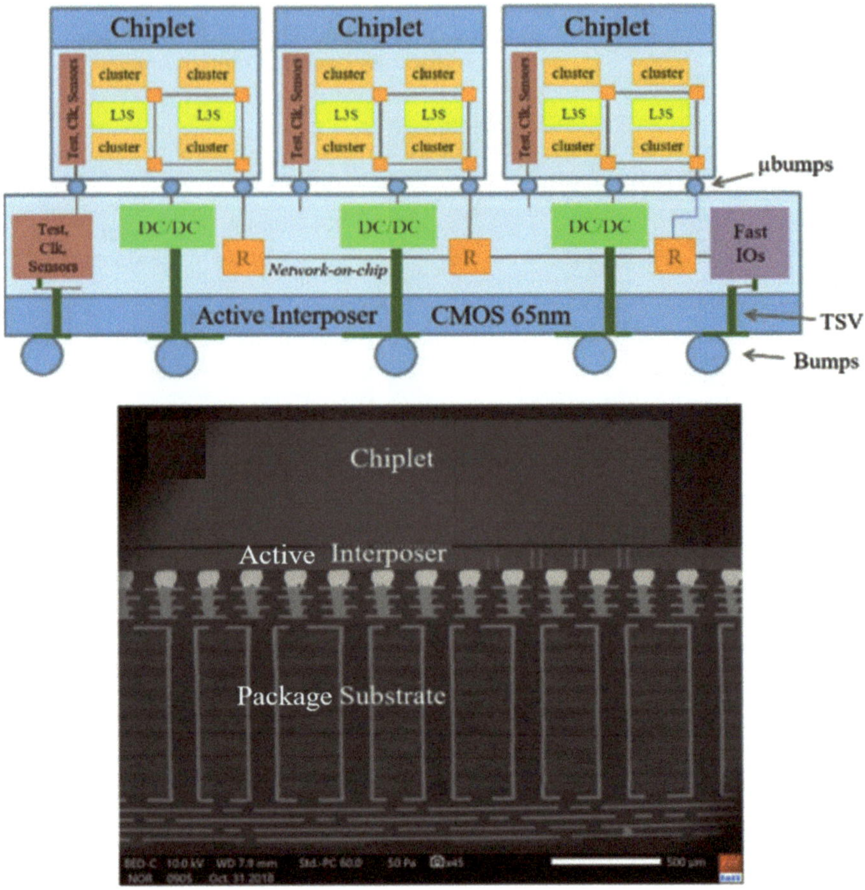

Fig. 1.30 3D IC integration with active TSV-interposer

1.8 Flip Chip Package Substrates–Si-Bridge Embedded in Organic Substrate

1.8.1 Intel's EMIB

In September 2012, Intel proposed an embedded multi-die interconnect bridge (EMIB) [84] to replace the TSV-interposer. The lateral communication between the chips will be taken care of by the silicon embedded bridge and the power/ground and some signals will go through the organic package substrate to the PCB as shown in Fig. 1.31. There are two major tasks in fabricating the organic package substrate with EMIB. One is to make the EMIB, and the other is to make the substrate with EMIB. For making the EMIB, first build the RDLs (including the contact pads) on a Si-wafer by either polymer + ECD (electrochemical deposition) or Cu damascene

+ CMP (chemical–mechanical polishing) methods. Then, thin down the wafer to ~60 μm. Finally, attach the non-RDL side of the Si-wafer to a die-attach film and then singulate the Si-wafer. We have the EMIB.

For making the substrate with EMIB, first place the singulated EMIB with die-attached film on top of the Cu foil in the cavity of the substrate, Fig. 1.31. It is followed by laminating a resin film on the whole organic package substrate. Then, drilling (on epoxy resin) and Cu plating to fill the holes (vias) are done to make connections to the contact pads of the EMIB. Continue Cu plating to make lateral connections of the substrate. Then, it is followed by laminating another resin film on the whole substrate and drilling (on resin) and Cu plating to fill the holes and make contact pads. (Smaller pads on finer pitch are for microbumps while larger pads on gross pitch are for ordinary bumps.) The organic package substrate with EMIB is ready for bonding of the chips as shown in Fig. 1.31. For more information, please read Chap. 5 of this book.

1.8.2 IBM's DBHi

During IEEE/ECTC2021, IBM presented a paper on "Direct Bonded Heterogeneous Integration (DBHi) Si Bridge" [85]. The major differences between Intel's EMIB and IBM's DBHi are as follows. (A) For Intel's EMIB, there are two different (C4 and C2) bumps on the bridge), while for IBM's DBHi, there are C4 bumps on the chiplets and C2 bumps on the bridge (Fig. 1.32). (B) For Intel's EMIB, the bridge is embedded in the cavity of a build-up substrate with a die-attach material and then laminated with another build-up layer on top. Therefore, Intel's EMIB fabrication is very complicated. For IBM's DBHi, the substrate is just a regular build-up substrate with a cavity on top. For more information, please read Chap. 5 of this book.

1.9 Flip Chip Package Substrates–Si-Bridge Embedded in Fan-Out EMC with RDLs

Intel's and IBM's bridges are either embedded in or are on an organic package substrate. There is another class of bridge, which is embedded in the fan-out EMC with RDLs. During IEEE/ECTC (June 2021), there were at least four papers published regarding the application of fan-out packaging technology to embed the rigid bridge in the EMC for the chiplets to perform lateral communications. All four of these papers discuss very similar technologies. In You et al. [86], ASE embedded the bridge in the EMC using the fan-out packaging method and called it stacked Si bridge fan-out chip-on-substrate (sFOCoS). In Lee et al. [87], SPIL called its similar technology fan-out embedded bridge (FO-EB). In Lee et al. [88], Amkor referred

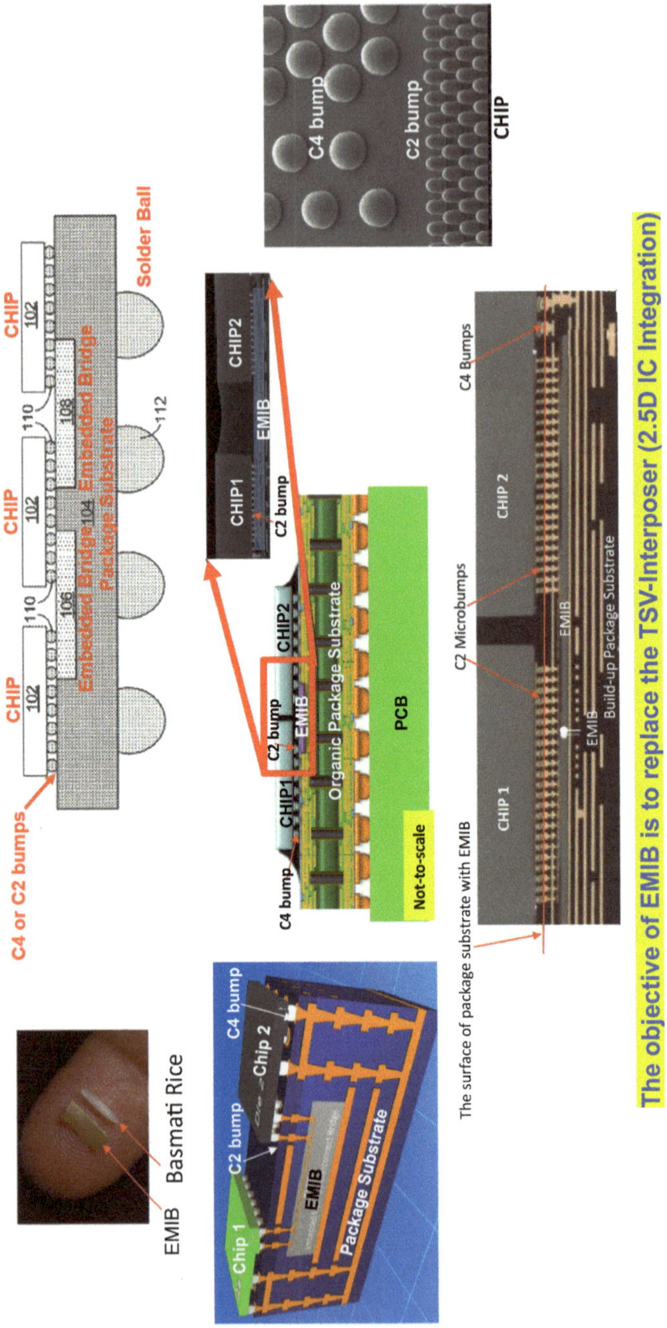

Fig. 1.31 Intel's embedded multi-die interconnect bridge (EMIB)

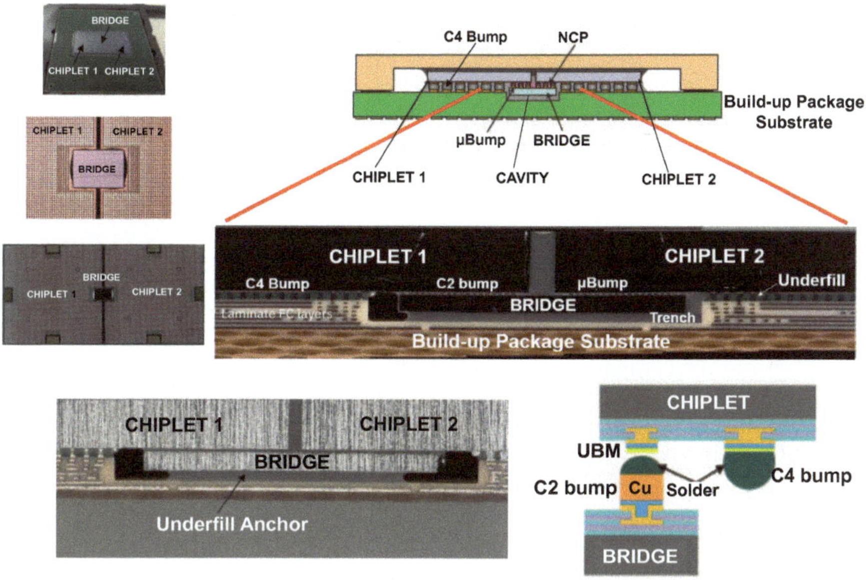

Fig. 1.32 IBM's direct bonded heterogeneous integration (DBHi) Si Bridge

to its comparable technology as S-Connect fan-out interposer. In Chong et al. [89], IME presented its bridge and called it embedded fine interconnect (EFI).

1.9.1 Unimicron's Si-Bridge

Figure 1.33 shows the US patent 11,410,933 filed by Unimicron on May 7, 2021. It shows the process to make a silicon bridge from a silicon wafer. It also shows the silicon bridge which is embedded in the EMC or ABF (Ajinomoto build-up film) with RDLs on top to connect the chiplets and RDLs on the bottom to connect to the package substrate or PCB. The connection between the top RDLs and the bottom RDLs is by TMVs (through mold vias). For more information, please read Chap. 5 of this book.

1.9.2 TSMC's CoWoS-L

During IEEE/ECTC (June 1, 2023), TSMC published two papers [91, 92] on replacing the TSV-interposer of their CoWoS (2.5D IC integration) to the

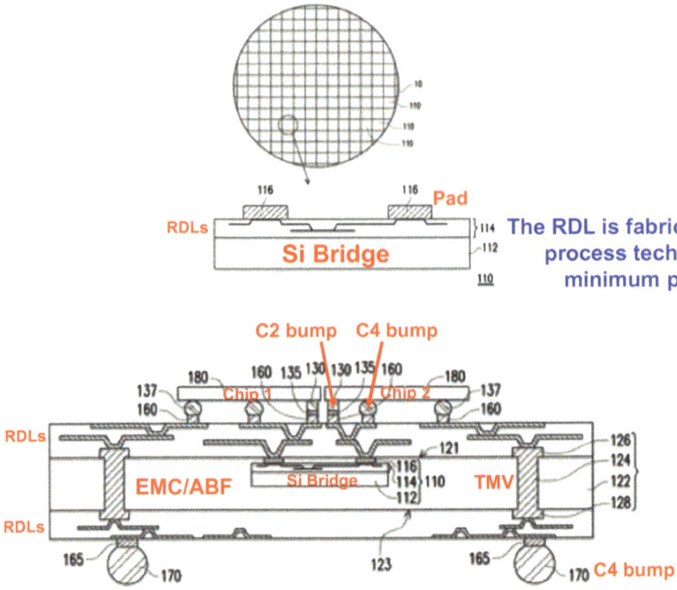

Fig. 1.33 Unimicron's Si bridge embedded in fan-out EMC with RDLs

reconstituted-interposer which consists of silicon bridge, LSI (local silicon interconnect), embedded in EMC with fan-out RDLs as shown in Fig. 1.34. They called it CoWoS-L. For more information, please read Chap. 5 of this book.

1.9.3 SPIL's FO-EB-T

During IEEE/ECTC (June 1, 2023), SPIL published two papers [92, 93] on replacing the TSV-interposer by silicon bridge with RDL and TSVs embedded in the EMC with fan-out RDLs as shown in Fig. 1.35. They called it FO-EB-T. For more information, please read Chap. 5 of this book.

1.10 Flip Chip Assembly

Basically, there are two groups of flip chip assemblies: one is with an intermediate layer between the bonding pads/traces, and the other is not, i.e., nothing! Flip-chip assembly with intermediate layers such as solder for mass reflow and Cu-pillar with solder cap by TCB are called indirect bonding. Cu-to-Cu diffusion bonding, which does not have anything between the bonding pads/traces on the chip/wafer, is therefore, called direct bonding.

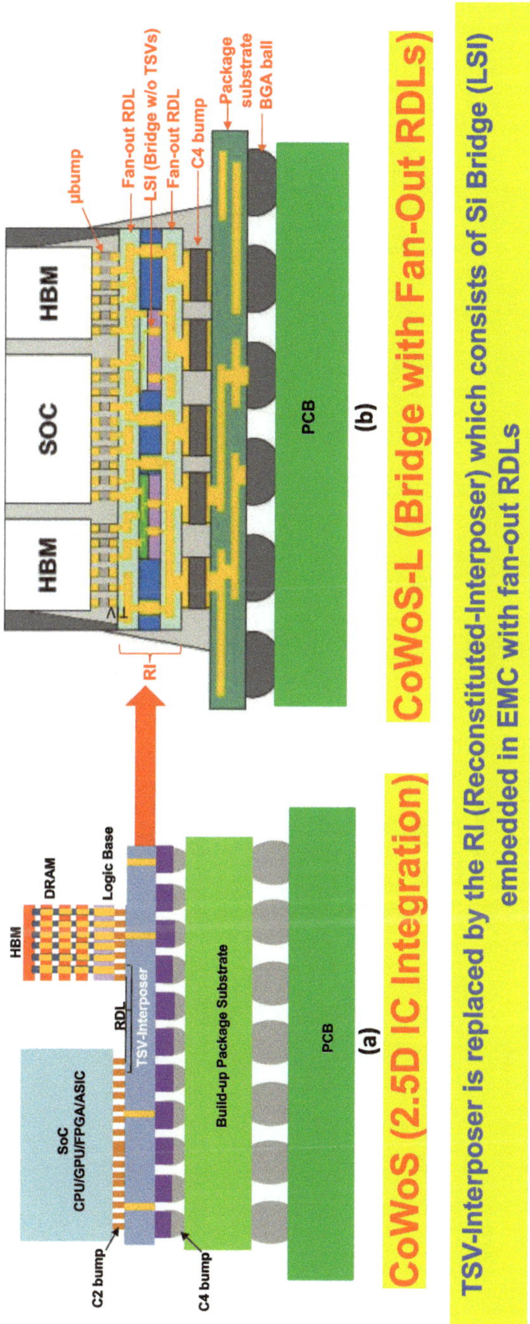

Fig. 1.34 TSMC's CoWoS-L

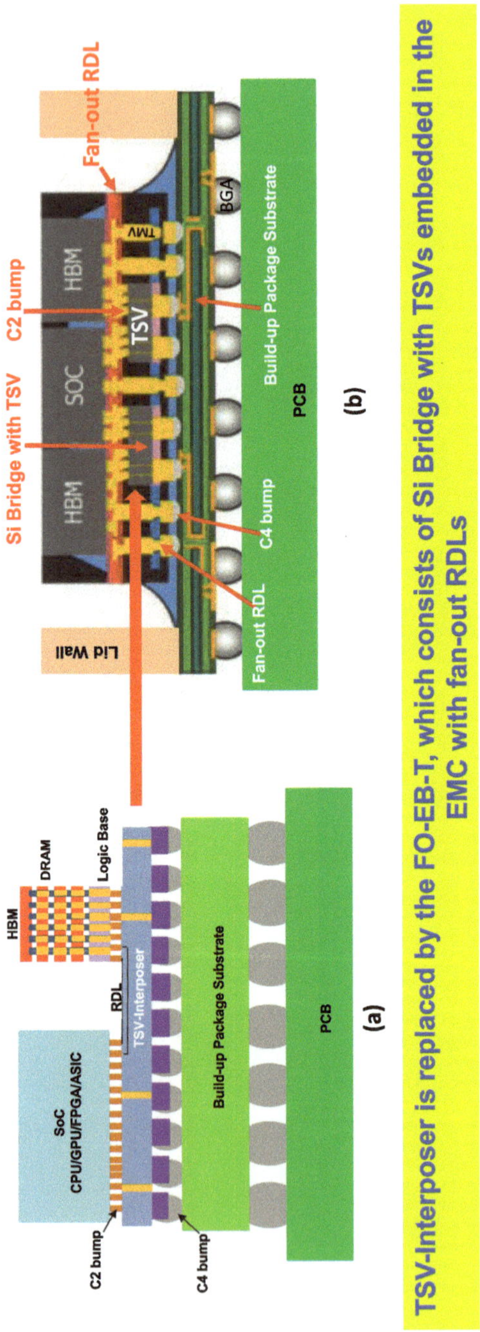

Fig. 1.35 SPIL's FO-EB-T

1.10.1 Cu-to-Cu TCB Direct Bonding

Cu-to-Cu diffusion bonding can go down to ultrafine pitch and pad size (the spacing between pads is 5 μm or less). In order to reduce the tendency to form native oxides that strongly affect the bonding quality and reliability, Cu-to-Cu TCB and usually operates at high temperature (400 °C) and pressure and long process time (60–120 min) [98–100], which are not good for throughput and the device reliability.

1.10.2 Cu-to-Cu Room Temperature Direct (Hybrid) Bonding

On the other hand, Cu-to-Cu diffusion bonding at room temperature [94–96] (also called hybrid bonding) leads to the highest throughput and the least amount of device reliability concerns, as well as very low costs. However, the drawbacks of room temperature bonding are the stringent requirements on: (a) pad/trace/wafer planarization, (b) surface treatment to ensure smooth hydrophilic surfaces for high quality bonding, and (c) the class of clean room (very high required). Cu-to-Cu room temperature hybrid bonding will be discussed in Chap. 2 of this book.

1.10.3 C4 Solder Mass Reflow with CUF

Solder mass reflow has been used for flip-chip assembly for almost 60 years. Most of the solder C4 bumps are mass reflowed on either silicon, ceramic, or organic substrates with capillary underfill (CUF). The assembly process is very simple, Fig. 1.36a: (i) use a lookup and look-down camera to identify the location of the bumps on the chip and the pads on the substrate; (ii) apply flux on either the C4 bumps, or the substrate, or both; and (iii) pick and place the C4 bumped chips on the substrate, then mass reflow with temperatures. Because of the surface tension of the C4 solder bumps during reflow, the process is very robust (self-alignment). Figures 1.37 and 1.38, respectively show the direct chip on board and the chip on package substrate. In general, the spacing between the bumps on the solder mass reflow of C4 bumped chips can be as small as 60 μm.

1.10.4 C2 Solder Mass Reflow with CUF

In the past few years, solder mass reflows of C2 (Cu-pillar with solder cap) bumped chips on either silicon, ceramic, or organic package substrates has been tried for high pin-count and fine-pitch flip-chip assemblies. The assembly process, Fig. 1.36a, is

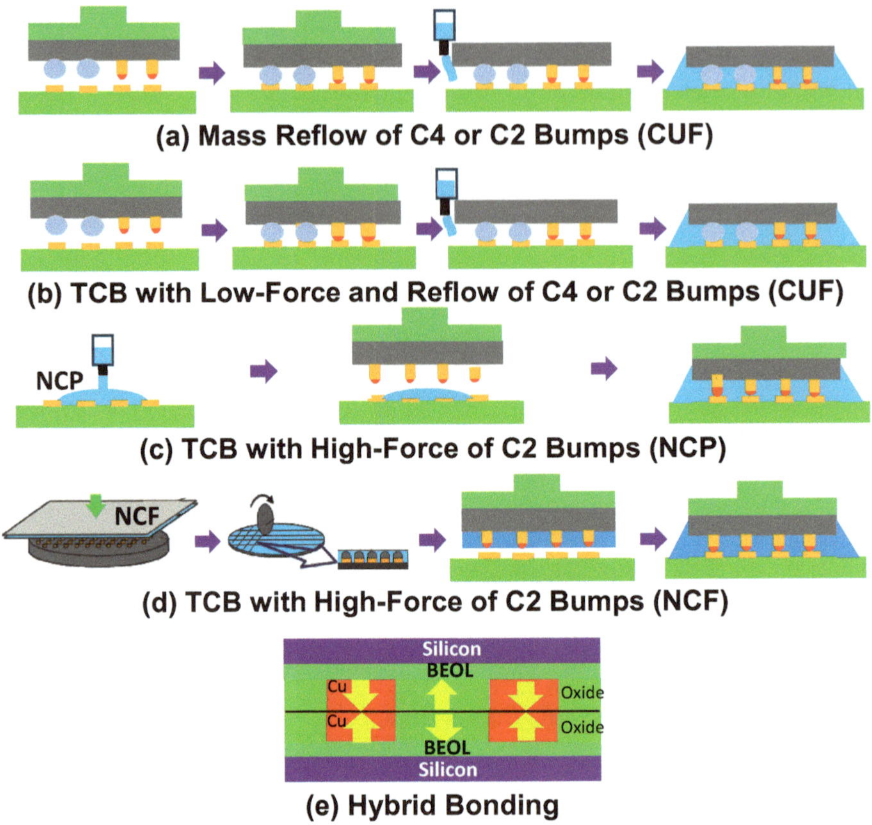

Fig. 1.36 Flip Chip assembly processes. **a** Mass reflows of C4 or C2 bumps (CUF). **b** TCB with low-force and reflow of C4 or C2 bumps (CUF). **c** TCB with high-force of C2 bumps (NCP). **d** TCB with high-force of C2 bumps (NCF). **e** Hybrid bonding

exactly the same as that of the C4 bumps, but the self-alignment characteristic is nowhere near the same, and thus, it is seldom being used.

1.10.5 TCB with Low-Bonding Force with CUF

In the past few years, TCB of chips with an intermediate layer such as C4 and C2 (Cu-pillar with solder cap) bumps on silicon, ceramic, or organic package substrates with CUF, has been attracting attention for high-density and ultrafine pitch flip chip assemblies. Basically, there are two methods, one is with a low-bonding force and the other is with a high-bonding force.

For the one with low bonding force, the assembly process is simple, Fig. 1.36b: (i) first, use the look-up and look-down camera to locate the position of the C2 bumps on

Fig. 1.37 Mass reflow of direct chip-on-board with underfill

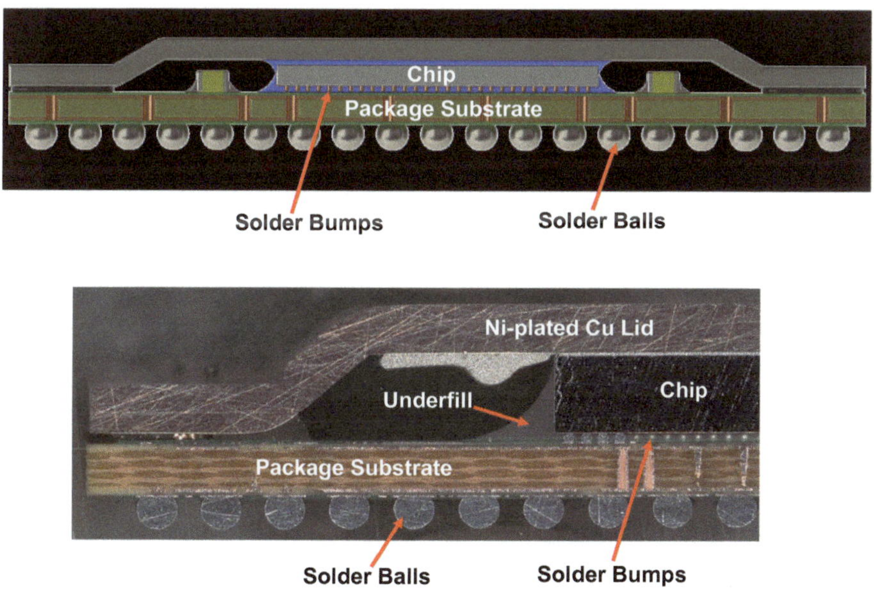

Fig. 1.38 Mass reflow of flip chip on package substrate with underfill

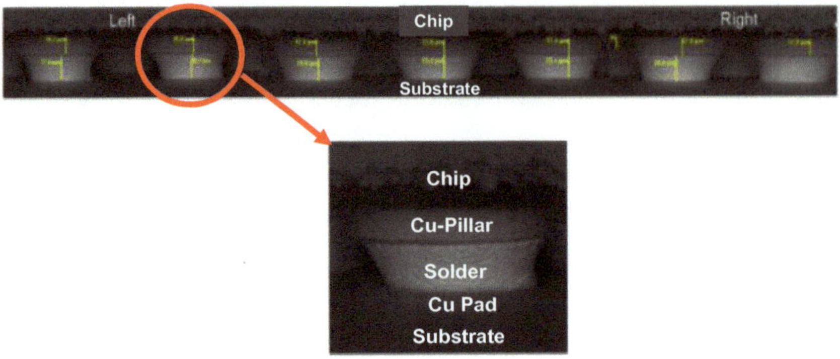

Fig. 1.39 Cross section of a C2 flip chip assembled on an organic package substrate by a TCB with low-force (CUF)

the chip and their corresponding pads on the substrate; (ii) apply flux on the solder cap or on the substrate or both; and (iii) pick-and-place the chip on the substrate and then apply temperature to melt the solder and a low force to hold the chip at a certain distance from the substrate. The above procedure is done one chip at a time and therefore, the throughput is low in comparison with the C2 solder mass reflow process. Figure 1.39 shows a typical cross section of a flip chip assembly with TCB with low force on C2 bumps [97].

1.10.6 C2 TCB with High-Bonding Force (NCP and NCF)

For TCB with a high-bonding force on the C2 bumped chips, the assembly process must be combined with the NCP or NCF underfill, which will be discussed in Sect. 1.11.2.

1.11 Underfill/Reliability

The reliability of flip chip solder joints is enhanced by the application of underfill [98–112], especially on organic substrate. Most underfills consist of low-expansion fillers such as fused silica (SiO_2) and a liquid prepolymer such as thermosetting resin (adhesive) that can be cured to a solid composite.

In 1987, Hitachi showed that with underfill, the thermal fatigue life of the flip chip solder joints on ceramic substrate increased [113]. In 1992, IBM at Yasu proposed the use of the low-cost organic substrate. They showed that with underfill, the large thermal expansion mismatch between the silicon chip (2.5×10^{-6}/°C) and the organic substrate ($15–18 \times 10^{-6}$/°C) is reduced substantially and the solder joints are reliable for most applications. This opened up the doors for today's very popular solder bumped flip chip on low-cost organic substrate packages used, e.g., in the processors of personal computers, notebooks, smartphones, tablets, etc. Basically, there are two different procedures to apply the underfill, namely pre-assembly underfill and post-assembly underfill.

1.11.1 *Post-Assembly Underfill*

For post-assembly underfill, the application of underfill is after the flip chip assembly, i.e., the flip chip is already on the substrate and the solder joints are already mass reflowed (either with C2 or C4 bumps) or low-force TCB with C2 bumps.

For post-assembly underfill, there are basically two methods, namely CUF [114–117] and MUF [118–122]. CUF is the first method that went into volume production [114–117]. For CUF, the underfill is dispensed by a needle or jet w/o vacuum assisted on one (or two) sides of the flip chip on substrate assembly. Because of capillary action, this underfill completely fills the space between the chips, solder joints, and substrates. The chip and the substrate are then firmly bonded by curing the underfill. CUF is performed one chip assembly at a time, thus, throughput is an issue.

Molded underfill was first proposed by Cookson Electronics [118] in 2000 and later by, e.g., Dexter [119], Intel [116], Amkor [120], STATSChipPAC [121], and LETI/STMicroelectronics [122]. For MUF, the modified EMC is transferred molding the chip and filling the gap between the chip, solder joints, and the substrate of the flip chip assembly (Fig. 1.40). The encapsulant of the chip and the underfill are formed at the same time, which will increase the throughputs. However, the challenges of MUF are: (a) the flow of MUF between the chip and the substrate is usually assisted by vacuum, (b) the size of the silica filler of the EMC must be very small for flowability, (c) the cost of EMC for MUF is much higher than that for package molding, (d) package warpage is an issue due to the thermal expansion mismatch between the EMC, chip, and substrate, (e) the molding temperature is limited by the melting point of the solder joints, and (f) the standoff-height and pitch of the solder joints cannot be too small.

In order to increase the throughput of CUF and avoid the drawbacks of MUF, a method of post-assembly underfill has been proposed by Lau et al. [123], where a stencil is designed for printing the underfill material for flip chips on organic-panel and Si-wafer assemblies, which will be presented in Sect. 1.13 of this chapter.

1.11.2 *Pre-assembly Underfill*

For pre-assembly underfill, the application of underfill is either on the substrate or wafer and is before the flip-chip assembly. Solder reflow of the C4 bumps with

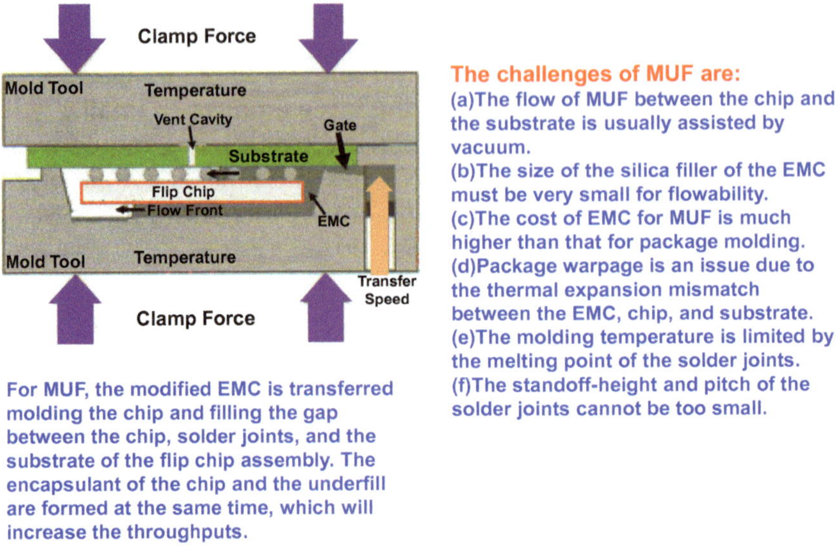

Fig. 1.40 Molded underfill (MUF)

underfill on substrates was first proposed by GIT [124] and is called NUF. High-bonding force TCB of the C2 bumps with nonconductive paste (TC-NCP) underfill on the substrate, Fig. 1.36c was first studied by Amkor [125] and has been used to assemble Qualcomm's SNAPDRAGON application processor for Samsung's Galaxy smartphone as shown in Fig. 1.41. The NUF and NCP underfills can be spun on, dispensed by a needle, or vacuum assisted.

By learning from the chip-on-glass technology, high-bonding force TCB of C2 bumps with nonconductive film (NDF) underfill on wafers have been studied by, e.g., Sanyo [126], Hitachi [127, 128], Tohoku [129, 130], DOW [131], Hynix [132], KAIST/Samsung [133, 134], Amkor/Qualcomm [135], and Toray [136–138] for 2.5D/3D IC integration. Figure 1.42 shows the lamination of NCF on the Cu-pillar with a solder cap bumped wafer.

High-bonding force TCB of the C2 chips with NCF (after singulation from the laminated wafer), Fig. 1.36d, has been in production for 3D IC integration by Samsung on its TSV-based double data rate type 4 dynamic random access memory (DRAM), and by Hynix on the high bandwidth memory (HBM) of AMD's graphic processor unit (GPU) code-named Fiji. The TSV-interposer is supporting the GPU and four-set of HBM, which is a stack of four DRAMs on a base logic), as shown in Fig. 1.42a. This 3D memory cube is stacked by high-force TCB of the C2 chips with NCF one chip at a time and each chip takes ~10 s for the underfill film to gel, the solder to melt, the underfill film to cure, and the solder to solidify. Throughput is a problem!

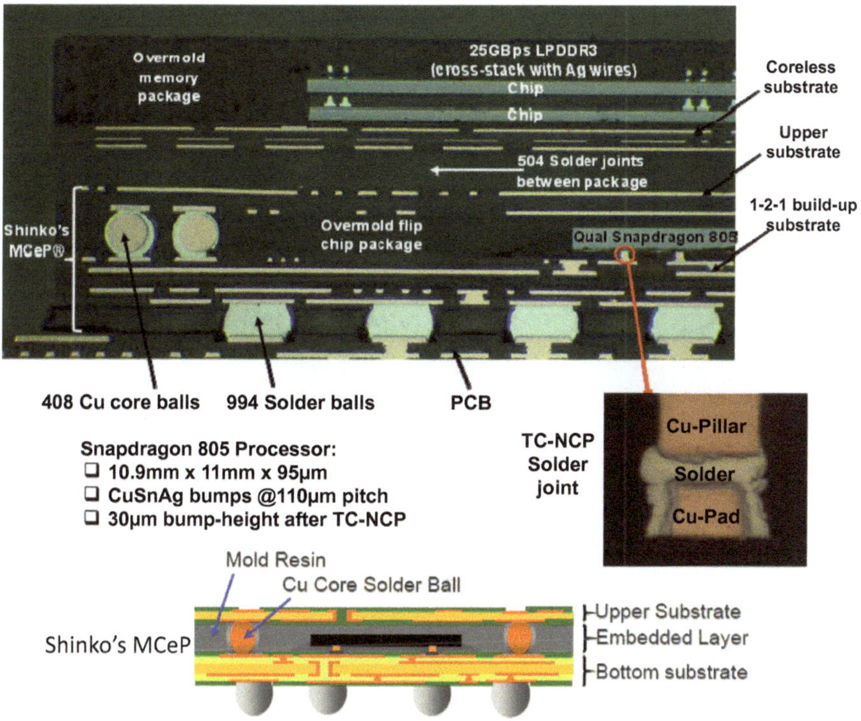

Fig. 1.41 PoP in Samsung's smartphone. The C2 bumped flip chip is TCB with high-force on a package substrate (TC-NCP)

1.12 Reliability–Characterization of Underfill

One of the major reasons why solder bumped flip chip on low-cost organic substrates works is because of the underfill epoxy encapsulant [1–23]. It reduces the effect of the global thermal expansion mismatch between the silicon chip and the organic substrate, i.e., to reduce the stresses and strains in the flip chip solder bumps (since the chip and the substrate are tightly adhered by the underfill), and to redistribute the stresses and strains over the entire chip area that would otherwise be increasingly concentrated near the corner solder bumps of the chip. Other advantages of underfill encapsulant are to protect the chip from moisture, ionic contaminants, radiation and hostile operating environments such as thermal, mechanical pull, shear, twist, and shock/vibration.

In this section, nine different underfill encapsulants with different sizes and contents of filler, and epoxy are studied. Their curing conditions such as time and temperature are measured by a differential scanning calorimeter (DSC) unit. Their material properties such as the TCE (thermal coefficient of expansion), Tg (glass transition temperature), dynamic storage modulus, tangent delta, and moisture content

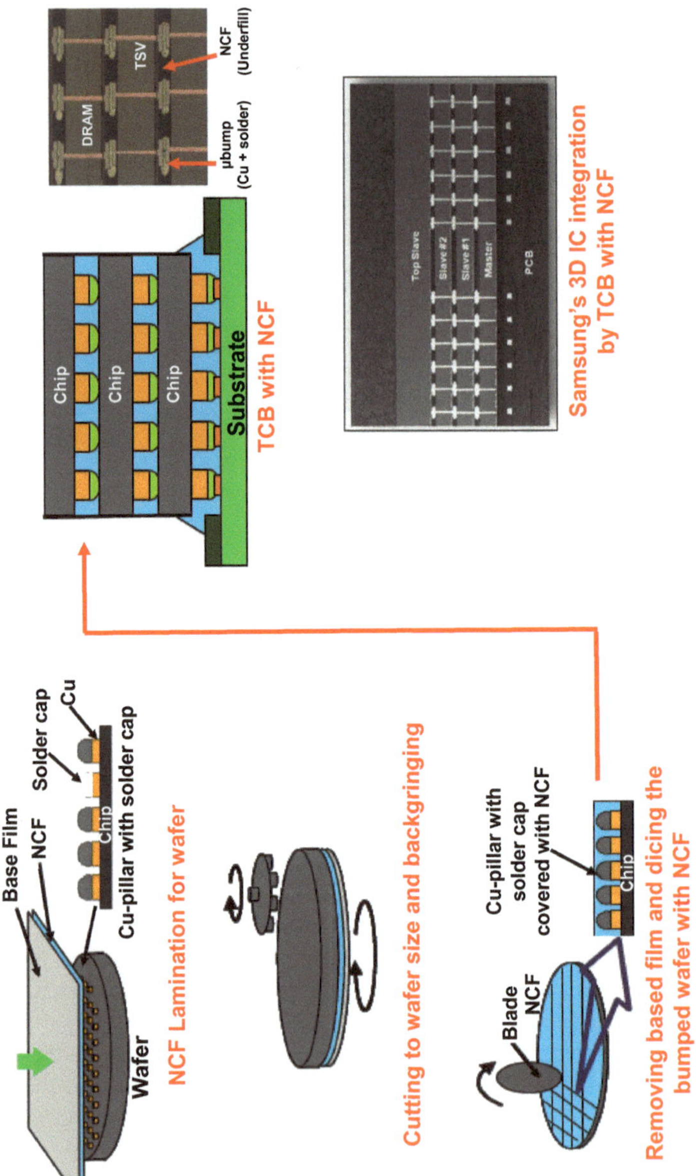

Fig. 1.42 Samsung's HBM. The C2 flip chip is TCB with high-force on another chip (TC-NCF)

are carried out using the thermal mechanical analysis (TMA), dynamic mechanical analysis (DMA), and thermal gravimetric analysis (TGA). Their flow rate and mechanical (shear) strength in a solder bumped flip chip on board are measured. Their effects on the electrical performance (voltage) of a functional flip chip device are determined experimentally. For each test configuration the sample size is three.

1.12.1 Underfill Materials and Applications

There are nine different encapsulant materials (from three different vendors) under consideration, namely, Underfill A, B, C, D1, D2, D3 (three different lots), E, F, G, H, and I, Table 1.13. All of the underfill encapsulants are premixed at the supplier sites, packed in plastic syringes (5 to 10 cc), and then frozen packed at −40 °C to prevent curing. Shipping of these underfill materials requires special handling to maintain the low temperature continuously. Upon receiving the package, one needs to unpack the package, take out the syringes quickly, and then store them into an uninterrupted temperature (−40 °C) control freezer. In this case, most of the underfill encapsulants would have approximately one year storage life.

The filler content and size of Underfill A, B, C, D1, D2, D3, E, F, G, H, and I are shown in Table 1.13. For all these materials their filler is silica. The resin of Underfill A, B, C, D1, D2, D3, and H is bis-phenol type epoxy except Underfill H which is with a mixture of bis-phenol and trade secret resin, and of Underfill E, F, G, and I is a mixture of cycloaliphatic epoxy and trade secret resin.

Before applying the underfill encapsulants, they are removed from the freezer and thawed at room temperature. This takes about an hour to thaw a 5 to 10cc syringes. Once thawed, most of the underfill materials have a pot life of 14 to 16 hours.

Dispensing the underfill encapsulants through a syringe is accomplished with a version system (to locate the edges of chip) and a bumping system (to control the amount of underfill materials). The temperature of the heating plate is around 80 °C. After dispensing the underfill materials in either one side or two adjacent sides of the chip, the chip and substrate should remain on the heating plate for about 30 to 90 seconds (depending on the chip size and gap between the chip and substrate) to enhance the flow rate of the underfill. After the underfill flowed out from the other sides of the chip, the chip on board will be put into a curing oven for 10 to 60 minutes at 130 to 170 °C depending on the curing conditions of the underfill materials.

The most desired features of underfill materials are: (a) low viscosity (fast flow) which can increase throughput, (b) low curing temperature/fast curing time which can reduce cost and be less harmful to other components, (c) low TCE which can reduce thermal expansion mismatch between the chip, solder bumps and substrate, (d) high modulus which leads to good mechanical properties, (e) high glass transition temperature (Tg) which can endure higher temperature environments, (f) low moisture absorption which can extend shelf life, and (g) good adhesion which can improve product life time.

Table 1.13 Average curing conditions of underfill A, B, C, D1, D2, D3, E, F, G, H, and I

Underfill	Filler content and size	Type of epoxy	Curing temp.(°C)	Curing condition temp.(°C) /time (min)			
A	40% < 20 mm	Bis	138.3	150/15	160 / 8	165 / 6	170 / 4
B	60% < 20 mm	Bis	139.7	150/15	160 / 7	165 / 5	170 / 4
C	60% < 20 mm	Bis	135.1	150 / 30	160/13	165 / 9	170 / 6
D1	60% < 20 mm	Bis	140.5	150 / 33	160/ 13	165 / 9	170 / 6
D2	60% < 20 mm	Bis	148.5	150 / 35	160/ 15	165 / 10	170 / 7
D3	60% < 10 mm	Bis	140.9	150/19	160 / 9	165 / 7	170 / 5
E	67% < 5 mm	Cyc*	158.3	150/190	160 / 90	165 / 64	170 / 45
F	67% < 20 mm	Cyc*	163.4	150 / 272	160/128	165 / 90	170 / 63
G	67% < 5 mm	Cyc*	157.2	150 / 528	160/236	165/160	170/109
H	68% < 12 mm	Bis*	141.2	150 / 295	160 / 68	165 / 34	170/ 17
I	67% < 11 mm	Cyc*	145.0	150/140	160 / 39	165 / 21	170/12

Remark: Bis means Bisphenol; Bis* is mixture of Bisphenol and trade secret resin; Cyc* is mixture of cycloaliphatic and trade secret resin

1.12.2 Curing Conditions

For determining the curing conditions of underfill materials, they are put into an aluminum pan (which will form a disc sample with dimensions: 6.4 ± 0.2 mm in diameter and 1.6 ± 0.1 mm in height), weighed, and then put in a DSC equipment. The objective of DSC is to measure the amount of energy (heat) absorbed or released by a sample as it is heated, cooled or held at a constant (isothermal) temperature. The instrument design consists of two independent furnaces, one for the sample and one for the reference, Fig. 1.43. When an exothermic or endothermic change occurs in the sample material, energy is applied or removed to one or both furnaces to compensate for the energy change occurring in the sample. Since the system is always directly measuring energy flow to or from the sample, DSC can directly measure melting temperature, Tg, temperature onset of crystallization, and temperature onset of curing. The kinetic software enables us to analyze a DSC peak to obtain specific kinetic parameters that characterize a reaction process.

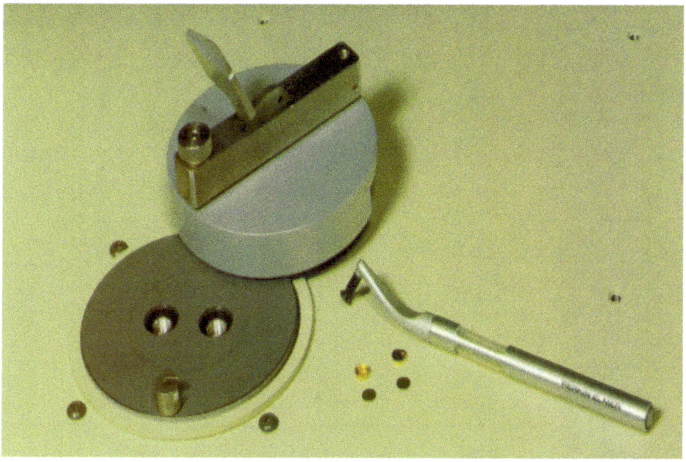

Fig. 1.43 Specimen set-up for DSC (Differential Scanning Calorimeter)

Any material reaction can be represented by the following equation:

$$A \xrightarrow{k} B + \Delta H$$

where A is the material before reaction, B is the material after reaction, ΔH is the heat absorbed or released, and k is the Arrhenius rate constant.

The Arrhenius equation is given by:

$$k = Z \exp(-Ea/RT)$$

where Z is the pre-exponential constant, Ea is the activation energy of the reaction, R is the universal gas constant (8.314j/°C/mole), and T is the absolute temperature in degree Kelvin. The rate of reaction (dx/dt) can be directly measured by DSC and is expressed as:

$$dx/dt = k(1 - x)^n$$

where dx/dt is the rate of reaction, x is the fraction reacted, t is the time, k is the Arrhenius rate constant, and n is the order of reaction. Combining the above equations and assuming a nth order reaction kinetics and constant program rate, activation energy, and pre-exponential constant, then we have

$$dx/dt = Z \exp(-Ea/RT)\,(1 - x)^n$$

The fraction reacted x is directly related to the fractional area of the DSC reaction peak. The kinetic parameters Z, Ea, and n are determined by using an advanced

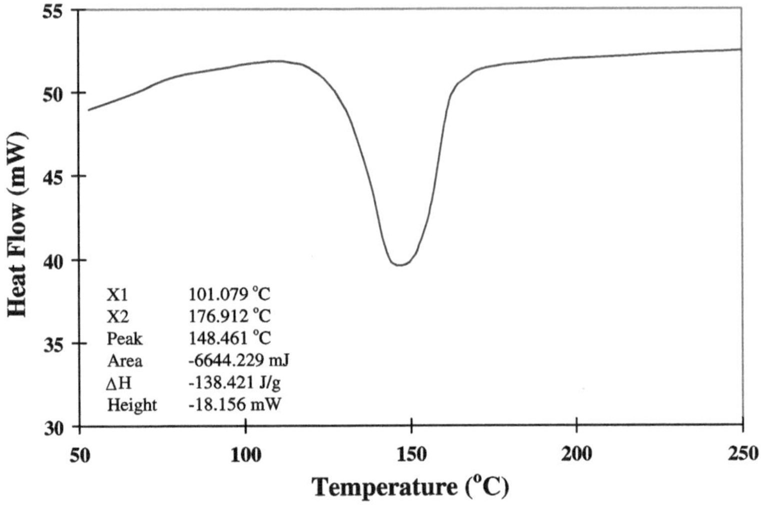

Fig. 1.44 DSC thermal scan curve for Underfill D2

multi-linear regression method (MLR). Thermal scan is carried out at a 5 °C/min heating rate ranging from 40 to 250 °C.

Figure 1.44 shows a typical DSC thermal scan curve and Fig. 1.45 shows a typical degree of conversion (reaction) verses time curves of Underfill D2. It can be seen that the average curing temperature of Underfill D2 is 148.5 °C and that Underfill D2 cannot be 100% cured if the applied temperatures are less than 148.5 °C. Table 1.13 summarizes the average curing conditions of Underfill A, B, C, D1, D2, D3, E, F, G, H, and I. It can be seen that the curing conditions vary with different underfill materials, and they cured faster at higher temperatures. Also, Underfill A, B, C, D1, D2, and D3 cured faster than the others.

1.12.3 Underfill Material Properties

(A) TCE

The TCE of Underfill A, B, C, D1, D2, D3, E, F, G, H, and I (with sample dimensions: 6.4 ± 0.2 mm in diameter and 1.6 ± 0.1 mm in height) is determined by the TMA in an expansion quartz system (50 to 200 °C) at a 5 °C/min heating rate, Fig. 1.46a. The objective of TMA is to measure the change in dimension of a sample (such as expansion or contraction) as the sample is heated, cooled or held at a constant (isothermal) temperature. The instrument design consists of the platinum wound furnace system which can be operated from −170 °C to 1000 °C range. The scan rate is from 0.1 °C to 100 °C/min. The specimen is mounted between a quartz platform and probe, then a static load (Fs), Fig. 1.46b, is applied to specimen while the dimension

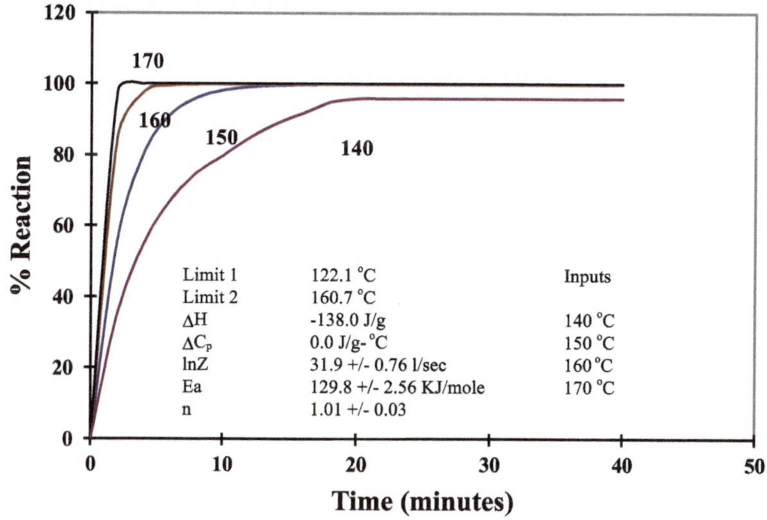

Fig. 1.45 Typical degree of conversion verses time curves of Underfill D2

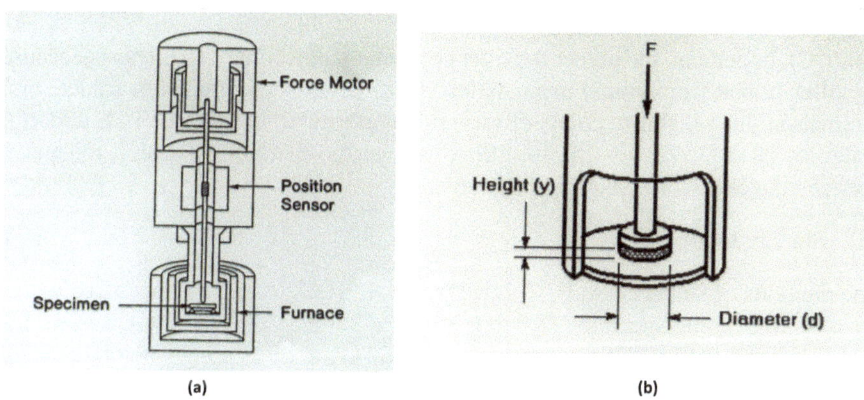

Fig. 1.46 **a** Schematic diagram of a TMA (thermal mechanical analyzer) and DMA (dynamic mechanical analysis). TMA force motor applies static load while DMA force motor applies dynamic load on specimen. **b** Outlook of quartz tube, probe and specimen of TMA for TCE measurement

of the specimen is monitored by the linear variable differential transducer (LVDT) throughout the analysis. The TCE of the material is obtained by the slope of the dimension change verse temperature curve, and Tg of the material is obtained by the onset of the two different slopes of the curve.

Figure 1.47 shows a typical expansion curve of Underfill D2, and Table 1.14 summarizes the average results of Underfill A, B, C, D1, D2, D3, E, F, G, H, and I. It can be seen that the TCE of Underfill A (40% filler content) is the largest (46.5 $\times$ 10^{-6}/°C) and the TCE of Underfill E (67% filler content) is the smallest (21 $\times$

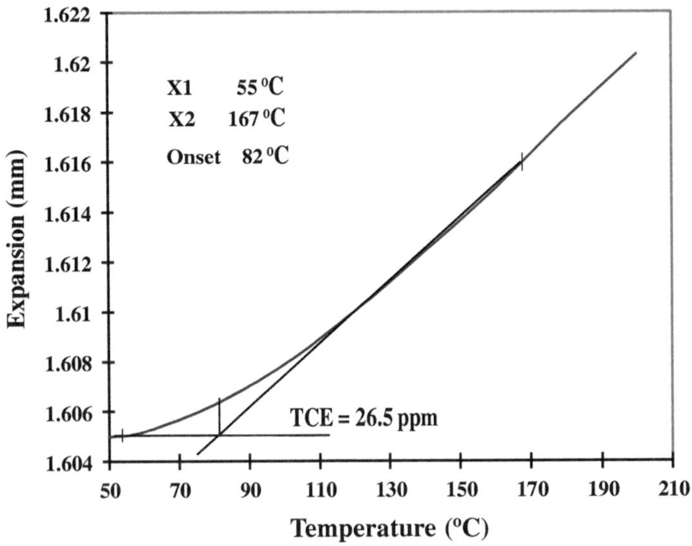

Fig. 1.47 Typical expansion curve of Underfill D2 for determining TCE

10^{-6}/°C). In general, the higher the filler content, the lower the TCE. This is because the filler (made from silica) expands less than the epoxy resin. From solder joint thermal-fatigue reliability points of view it is preferred to have lower TCE underfill materials [1–3, 7, 12, 13, 15, 19, 20] to reduce the thermal expansion mismatch between the chip, solder bumps, and substrate.

(B) Storage Modulus

The modulus of Underfill A, B, C, D1, D2, D3, E, F, G, H, and I is measured with a three-point bending specimen (3.0 ± 0.3 mm × 2.9 ± 0.3 mm × 19 ± 3 mm) in a DMA unit (55 to 200 °C) at a heating rate of 5 °C/min, Fig. 1.48. The objective of DMA is to measure mechanical properties, such as modulus as a function of time, temperature, frequency, stress or combinations of these parameters. The instrument design consists of a force motor which can be programmed to apply constant stress, dynamic stress, or combinations of both. The core rod applies stress to the sample and is held in place using an electromagnetic suspension. The ceramic furnace with platinum furnace element is capable of heating and cooling at very high rate and also can be heated up to 1000 °C. For electronics packaging materials, the flexural properties such as the storage modulus, loss modulus, and tangent delta (tanδ) can be obtained with DMA.

In dynamic stress analysis, flexural storage modulus, Es is a measure of the energy stored per cycle of deformation and can be expressed as:

$$E_s = \frac{F_d x^3 \cos\delta}{4y^3 z\Delta}$$

Table 1.14 Average material properties of Underfill A, B, C, D1, D2, D3, E, F, G, H, and I and their effects on moisture content

Underfill	Filler content and size	TCE	T_g	Storage modulus (GPa)			Moisture content (%)	
		(ppm/°C)	(°C)	25 °C	55 °C	110 °C	Dry	20 h of steam aging
A	40% < 20 mm	46.5	142	3.88	3.8	3.0	0.09	0.33
B	60% < 20 mm	36.2	142	3.3	3.3	2.73	0.05	0.51
C	60% < 20 mm	26.1	158	4.4	3.6	3.42	0.04	0.28
D1	60% < 20 mm	23.4	150	6.45	6.38	5.62	0.07	0.38
D2	60% < 20 mm	26.6	140	6.41	6.24	5.06	0.05	0.28
D3	60% < 10 mm	24.2	150	6.73	6.6	5.84	0.09	0.22
E	67% < 5 mm	21.0	137	4.8	3.3	2.86	0.04	0.52
F	67% < 20 mm	38.6	120	5.4	5.3	2.0	0.08	0.58
G	67% < 5 mm	39.3	140	7.2	6.9	4.2	0.05	0.42
H	68% < 12 mm	28.6	151	7.8	7.67	6.67	0.07	0.28
I	67% < 11 mm	30.0	186	5.09	5.09	4.79	0.09	0.28

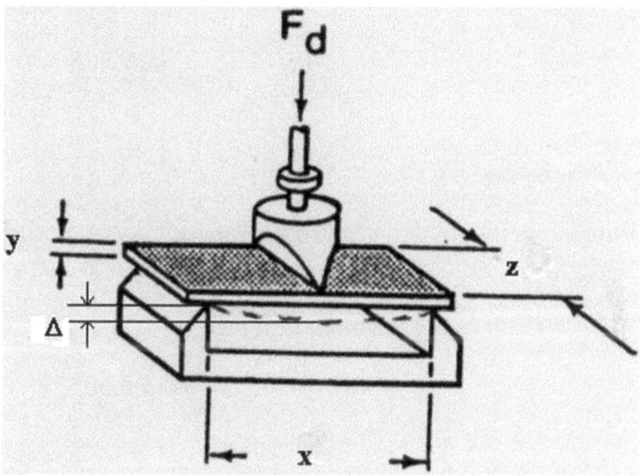

Fig. 1.48 Three-point bending for dynamic modulus and tangent delta

where Fd is the dynamic load, δ is the phase angle, and Δ is the maximum dynamic deflection of specimen (Fig. 1.48). Figure 1.49 shows a typical dynamic flexural storage modulus curve of Underfill D2 as a function of temperatures. The average results of the storage modulus of Underfill A, B, C, D1, D2, D3, E, F, G, H, and I are reported in Table 1.14. It can be seen that the storage modulus of all the underfill materials is temperature dependent, the higher the temperatures the lower the modules. Also, in general, the higher the filler content (act like reinforcement in filler reinforced composites) the larger the storage modulus.

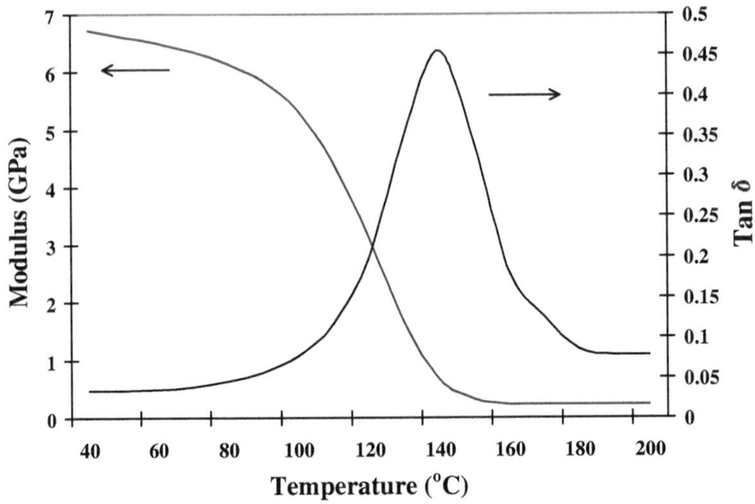

Fig. 1.49 Typical storage modulus and tangent delta curves of Underfill D2

Flexural loss modulus, E_l is a measure of the energy lost per cycle of deformation and can be expressed as:

$$E_l = \frac{F_d x^3 \sin\delta}{4y^3 z\Delta}$$

(C) Tanδ and Tg

Tangent delta (tanδ) which is a measure of material related damping property of Underfill A, B, C, D1, D2, D3, E, F, G, H, and I can be obtained by dividing E_l by E_s or

$$\frac{E_l}{E_s} = \tan\delta$$

The temperature at the peak of a tangent delta (tanδ) curve is often reported in literature as glass transition temperature (Tg). Figure 1.49 shows a typical tangent delta curve of Underfill D2, and the average values of Tg for all the underfills are shown in Table 1.14. In general, the Tg is higher for underfills with bis-phenol epoxy than that with cycloaliphatic epoxy, except Underfill I which is with a high Tg resin.

(D) Moisture Content

Two sets of tests are carried out to determine the moisture content, one is for dry specimen and the other is for steam aging specimen. The steam aging specimen is prepared under steam evaporation for 20 hours in a closed hot water bath. All the specimen dimensions are: 6.4 ± 0.2 mm in diameter and 1.6 ± 0.1 mm in height.

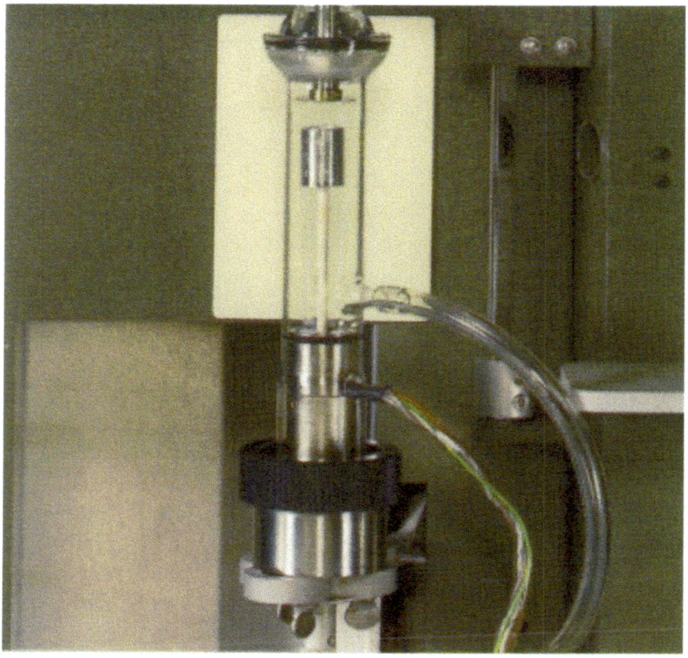

Fig. 1.50 High sensitivity weight measurements with TGA (thermal gravimetric analyzer)

Weight changes of Underfill A, B, C, D1, D2, D3, E, F, G, H, and I are measured with the TGA equipment (Fig. 1.50) under 110 °C for four hours.

The objective of TGA is to measure the change in mass of a sample as the sample is heated, cooled or held at a constant (isothermal) temperature. The instrument design consists of micro-balance which allows the sensitive measurement of weight changes as small as a few micrograms, furnace, and sample holder area. The vertical analytical design also serves to isolate the balance mechanism from the furnace, eliminating temperature fluctuations which cause drifts and non-linearity. This system can measure curie point, decomposition temperature, moisture uptake, and component separation.

The change in mass during thermal scan can be expressed as:

$$\frac{W_f - W_i}{W_i} \times 100\%$$

where W_f is the final weight after thermal scan and W_i is the initial weight before thermal scan. Figure 1.51 shows a typical precent weight change ratio (moisture content) of Underfill D2 before and after 20 hours of steam aging. The average moisture content of all the underfills is shown in Table 1.14. It can be seen that the average moisture content of all the underfills after 20 hours of steam aging is at least three times more than before (the dry condition). In general, this is especially true

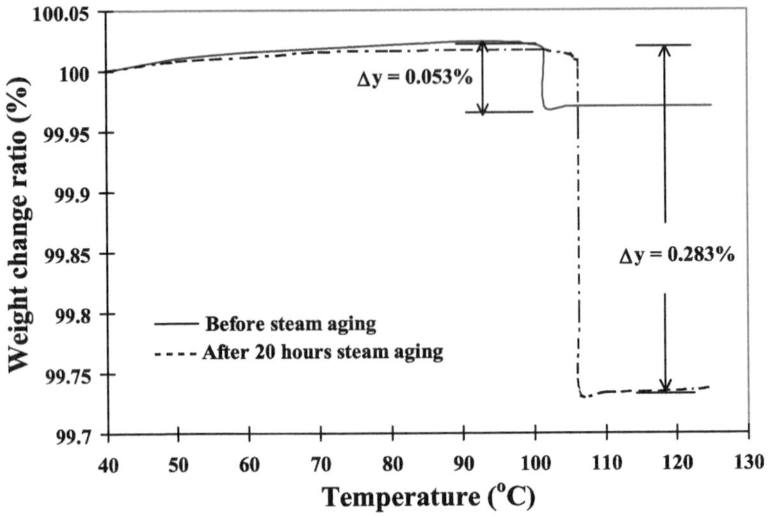

Fig. 1.51 Moisture change ratio of Underfill D2 before and after 20 h of steam aging

for higher filler content underfills, since porosity increases when the amount of filler increases.

1.12.4 *Underfill Flow Rate*

The flow rate of Underfill A, B, C, D1, D2, D3, E, F, G, H, and I are measured underneath a functional solder bumped flip chip on a BT (bismaleimide triazine) substrate. The dimensions of the chip are: 6.3 × 3.6 mm and have 32 bumps (Fig. 1.52). The BT substrate has two metal layers (Fig. 1.53a and b show, respectively, the top and bottom layers) and 32 vias (0.254 mm in diameter). The standoff height of the solder bumps is 0.07 ± 0.02 mm.

The assembled flip chip is placed on a hot plate with 80 °C. Approximately 0.025cc of room temperature underfill is placed around two chip sides in an "L" pattern. The time for the underfill to completely fill the gap is recorded. It should be pointed out that there are not any underfill materials flowing to the bottom side of the substrate. The average flow rate of Underfill A, B, C, D1, D2, D3, E, F, G, H, and I is shown in Table 1.15. It can be seen that, in general, the flow rate is strongly affected by the filler content and size, the more the filler content and the larger the filler size (i.e., the higher viscosity) the smaller the flow rate.

Fig. 1.52 UMC's chip with 32 solder bumps on two opposite sides

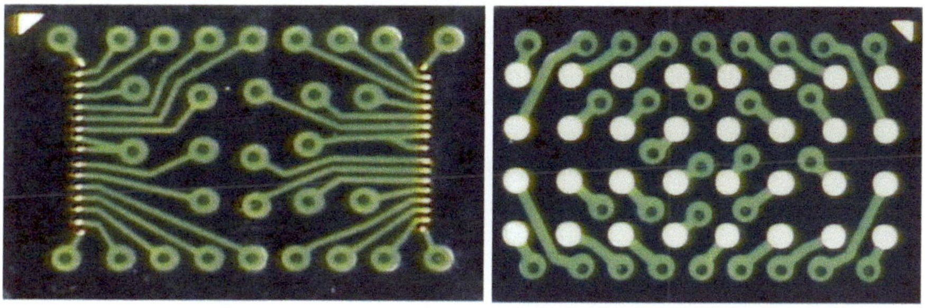

Fig. 1.53 **a** Low-cost substrate with 32 vias and 32 Cu pads on two opposite sides. **b** Low-cost substrate with 32 vias and 32 1 mm pitch land grid array (LGA) Cu pads

Table 1.15 Average flow rate of solder bumped flip chip on board with Underfill A, B, C, D1, D2, D3, E, F, G, H, and I

Underfill	Filler content and size	Flow rate (mm/s)
A	40% < 20 mm	0.167
B	60% < 20 mm	0.133
C	60% < 20 mm	0.003
D1	60% < 20 mm	0.3
D2	60% < 20 mm	0.3
D3	60% < 10 mm	0.3
E	67% < 5 mm	0.118
F	67% < 20 mm	0.105
G	67% < 5 mm	0.105
H	68% < 12 mm	0.63
I	67% < 11 mm	0.42

1.12.5 Mechanical Performance

The Royce Instruments Model 550 is used to perform mechanical shear tests. The shear wedge is placed against one edge of the solder bumped flip chip with underfill on the BT substrate which is clamped on the stage. A push of the wedge is applied to shear the chip/bumps/underfill away. A destructive test. The average test results of Underfill A, B, C, D1, D2, D3, E, F, G, H, and I are shown in Table 1.16, and a typical load-deflection curve of solder bumped flip chip on board with Underfill D2 is shown in Fig. 1.54. It can be seen from Table 1.16 that the average shear force varies with different underfills under consideration. It can also be seen that, in most of the cases, the shear force of the solder bumped flip chip with underfill drops about 25% after 20 hours of steam aging. Thus, the moisture content has significant influence on mechanical performance of underfilled flip chip assemblies.

There are two types of failure mode, one is that the chip is broken into pieces and some of the broken pieces are left on the substrate, Fig. 1.55a. Most of the dry underfills exhibit this type of failure. The other failure mode is that the whole chip is peeled off from the substrate and some underfill is left on the chip, Fig. 1.55b. Most of the underfills after 20 hours of steam aging exhibit this type of failure. The measured shear forces of the first type of failure mode are larger than that of the second type of failure mode. It seems that moisture did degrade interface adhesion and lead to debonding of the interface between the chip and the underfill encapsulant.

Table 1.16 Effects of moisture on the mechanical shear test performance of solder bumped flip chip on board with underfill A, B, C, D1, D2, D3, E, F, G, H, and I

Underfill	Shear force (kgf)		Change
	Dry condition	20 h of steam aging	(%)
A	43.2	33.8	27.8
B	40.5	31.3	29.4
C	43.5	34.2	27.2
D1	55.2	42.3	30.5
D2	65.5	62.2	5.3
D3	62.4	31.2	100
E	41.6	33.2	25.3
F	47.9	32.3	48.3
G	21.7	18.3	18.6
H	43.1	39.9	8.0
I	31.2	25.8	20.9

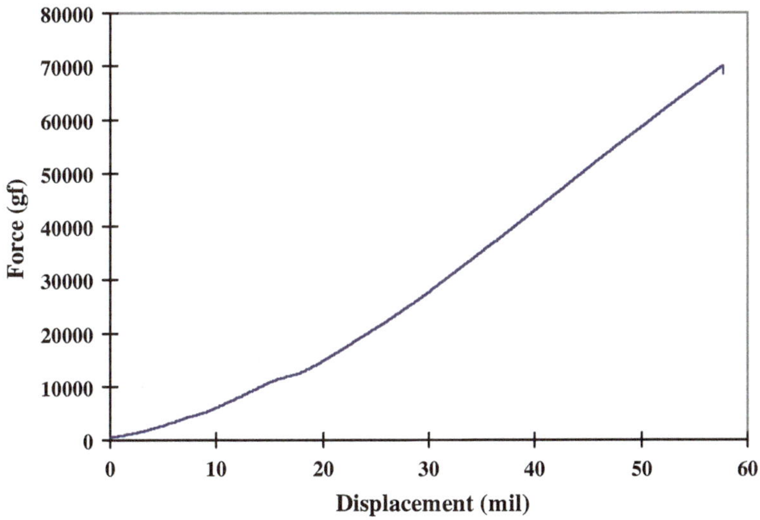

Fig. 1.54 Shear force verses displacement curve of the solder bumped flip chip assembly with Underfill D2

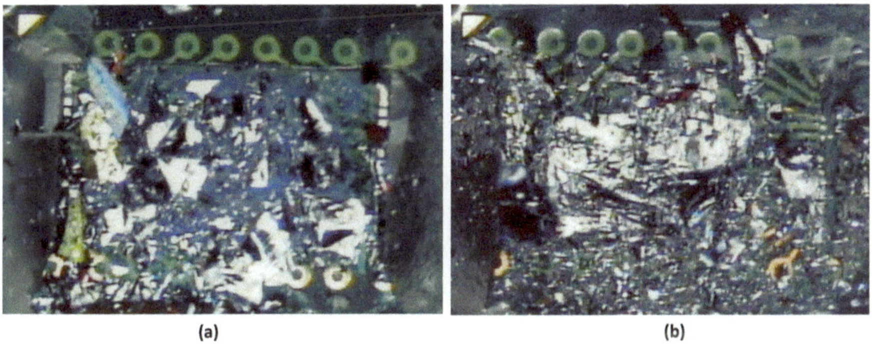

Fig. 1.55 **a** Failure mode of solder bumped flip chip on board with underfill at dry condition. The chip is broken into many small pieces and some remained on the substrate. **b** Failure mode of solder bumped flip chip on board with underfill after 20 h of steam aging. The whole chip is peeled off from the substrate

1.12.6 Electrical Performance

The electrical performance of Underfill A, B, C, D1, D2, D3, E, F, G, H, and I is carried out under a 5 volts DC by using a Kepco power supply. The average voltage readout of the solder-bumped flip chip on board with and without underfills is summarized in Table 1.17. It can be seen that, for all the underfills under consideration, there is almost no difference in voltage readout between the ones with underfill at dry condition and the ones without underfill. Also, for solder-bumped flip chip on board

Table 1.17 Average electrical performance of solder bumped flip chip on board with underfill A, B, C, D1, D2, D3, E, F, G, H, and I

Underfill	Voltage readout (volt)					
	Without underfill		With underfill at dry condition		With underfill after 20 h of steam aging	
	Low	High	Low	High	Low	High
A	3.88	4.05	3.88	4.05	3.87	4.03
B	3.92	4.06	3.94	4.07	3.93	4.05
C	3.91	4.05	3.88	4.05	3.85	4.04
D1	3.90	4.05	3.89	4.05	3.87	4.04
D2	3.91	4.05	3.91	4.06	3.89	4.05
D3	3.91	4.06	3.90	4.06	3.88	4.05
E	3.93	4.08	3.94	4.10	3.92	4.10
F	3.92	4.07	3.93	4.07	3.90	4.08
G	3.94	4.09	3.95	4.10	3.93	4.09
H	3.89	4.05	3.90	4.05	3.88	4.04
I	3.92	4.07	3.92	4.08	3.90	4.06

with underfills, the difference in voltage readout between the ones before and the ones after a 20-hour steam aging is insignificant.

1.12.7 Ranking of Underfill Encapsulants

How to choose an underfill encapsulant? The answer is not easy, since the performance of underfill depends on many factors such as, among others, the curing condition, TCE, storage modulus, Tg, moisture uptake, flow rate, and shear strength (adhesion). However, based on the measurement results of the nine different underfills under consideration, a simple method is proposed to rank (chose) the underfills.

Table 1.18 shows the grading criteria of nine important properties of underfills. It can be seen that these nine properties are: (1) curing time @ 160 °C, (2) TCE, (3) Tg, (4) storage modulus @ 25 °C, (5) storage modulus @ 110 °C, (6) moisture uptake, (7) flow rate, (8) shear force at dry condition, and (9) shear force after 20 hours of steam aging. For example, if the curing time @ 160 °C is 15 minutes then the grading is 5. Also, for example, if the flow rate is 0.1mm/s, then the grading is 6.

The rationales for establishing this table are: (1) the faster the curing time the lower the manufacturing cost and the less harmful to other components, (2) the lower the TCE, the less the mismatch stresses, (3) the higher the Tg the higher temperature the underfill can withstand, (4) and (5) the higher the storage modulus the better the mechanical performance, (6) the lower the moisture uptake the better the shelf life,

Table 1.18 Underfill material properties grading criteria

Properties	Grading (1 = Poor; 3 = Average; 6 = Excellent)*					
	1	2	3	4	5	6
1. Curing time (min) @ 165 °C	>60	40 ~ 59	30 ~ 39	20 ~ 29	10 ~ 19	<10
2. TCE (ppm/ °C)	45 ~ 49	40 ~ 44	35 ~ 39	30 ~ 34	25 ~ 29	20 ~ 24
3. Tg (°C)	120 ~ 129	130 ~ 139	140 ~ 149	150 ~ 159	160 ~ 169	>170
4. Storage modulus (GPa) @ 25 °C, 1 Hz	<3	3.1 ~ 3.9	4 ~ 4.9	5 ~ 5.9	6 ~ 6.9	7 ~ 7.9
5. Storage modulus (GPa) @ 110 °C, 1 Hz	<2	2.1 ~ 2.9	3 ~ 3.9	4 ~ 4.9	5 ~ 5.9	6 ~ 6.9
6. Moisture uptake (%)	0.55 ~ 0.6	0.5 ~ 0.54	0.4 ~ 0.49	0.35 ~ 0.39	0.3 ~ 0.34	<0.3
7. Flow rate (mm/s)	<0.1	0.1 ~ 0.19	0.2 ~ 0.29	0.3 ~ 0.39	0.4 ~ 0.49	>0.5
8. Shear force (kgf) before steam aging	20 ~ 29	30 ~ 39	40 ~ 44	45 ~ 49	50 ~ 59	>60
9. Shear force (kgf) after steam aging	<20	20 ~ 29	30 ~ 34	35 ~ 39	40 ~ 59	>60

* The grading is based on the present test results of Underfill A, B, C, D1, D2, D3, E, F, G, H, and I

(7) the faster the flow rate the higher the through-put, (8) and (9) the higher the shear force the better the adhesion strength.

Table 1.18 shows the ranking of Underfill A, B, C, D1, D2, D3, E, F, G, H, and I based on their measurement results. The way to construct this table is, for example, for Underfill A, the measured curing time @ 165 °C is 6 minutes (Table 1.13), thus in Column 1 it is graded a 6 (<10 minutes in Table 1.18), the measured TCE is 46.5 ppm/°C (Table 1.14), thus in Column 2 it is graded a 1 (45–49 ppm/°C in Table 1.18), the measured Tg is 142 °C (Table 1.14) thus in Column 3 it is graded a 3 (140–149 °C in Table 1.18), etc.

It can be seen from Table 1.19 that, based on the total grading, Underfill D2, D3, D1, and H are ranked on top of the others. Thus, they are chosen for the solder bumped flip chip on board applications. However, it should be noted that these chosen underfills are not perfect. For example, Underfill D2 has the best shear force results, however, its flow rate is less than half of that of Underfill H.

1.12.8 Summary

Nine different underfills with different resin epoxy and filler content and size have been studied. Their curing conditions and material properties have also been measured using DSC, TMA, DMA, and TGA equipment. Furthermore, their flow rate and effects on the mechanical (shear) and electrical (voltage) performance in a

Table 1.19 Ranking of underfill A, B, C, D1, D2, D3, E, F, G, H, and I

Underfill	Properties grading (from measured results and Table 1.6									
	1	2	3	4	5	6	7	8	9	Sum	Rank
A	6	1	3	2	3	5	2	3	3	28	5
B	6	3	3	2	2	2	2	3	3	26	6
C	6	5	4	3	3	6	1	3	3	34	4
D1	6	6	4	5	5	4	4	5	5	44	2
D2	5	5	3	5	5	6	4	6	6	45	1
D3	6	6	4	5	5	6	4	6	3	45	1
E	1	6	2	3	2	2	2	3	3	24	7
F	1	3	1	4	1	1	2	4	3	20	8
G	1	3	3	6	4	3	2	1	1	24	7
H	3	5	5	6	6	6	6	3	4	44	2
I	4	4	6	4	4	6	5	2	2	37	3

solder-bumped functional flip chip on an organic substrate have been experimentally determined. Some important results are summarized as follows.

- A simple method has been presented for choosing (ranking) underfill materials.
- For all the underfills considered, curing time is temperature dependent, they cured faster at higher temperatures.
- For all the underfills considered, the higher the filler content the lower the TCE.
- In general, the Tg is higher for underfill materials with bis-phenol epoxy than that with cycloaliphatic epoxy, except Underfill I which is with a high Tg resin.
- For all the underfills considered, the storage modulus is temperature dependent, the higher the temperature the lower the modulus.
- For all the underfills considered, the moisture content after 20 h of steam aging is at least three times more than the dry condition.
- In general, the flow rate is strongly affected by the filler content and size, the more the filler content and the larger the size the smaller the flow rate.
- In general, the moisture content in the underfill significantly affects the mechanical shear strength of solder bumped flip chip assemblies.
- For all the underfills considered, the effects of underfill on the electrical (voltage readout) performance of the functional chip are insignificant.

1.13 Reliability–Stencil Printing of Underfill

1.13.1 New Stencil Design

In order to increase the throughput of CUF and avoid the drawbacks of MUF in post-assembly underfill, a new stencil designed of printing underfill for flip chips on organic-panel and Si-wafer assemblies is proposed in this study and shown in Figs. 1.56, 1.57, and 1.58. The following can be observed.

(a) A very small rectangular opening of the stencil is designed for each chip, and it is located on one edge of the chip.
(b) The stencil is with a dry film underneath with many rectangular openings (one for each chip), and the size is a little larger than the chip size.
(c) There is a gap between the stencil and the backside of the chips.
(d) There is a fixture to apply heat to the stencil and underfill to lower the underfill's viscosity, as shown in Fig. 1.56.

The size of the stencil opening, and the thickness of the stencil and dry film determine the underfill volume, which should be approximately equal to the space between the chip, solder joints, and the substrate plus the fillet on the four sides of the chip.

During printing [Figure 1.56b], the underfill will fill the opening of the stencil and fall into the space between one edge of the chip and the dry film. After printing (usually it takes only a few seconds), remove the assembly [Figure 1.56c] from the stencil printer and place it on a hot plate (~120 °C) for the underfill to flow between the chip, solder joints, and the substrate by the capillary action [Figure 1.56d]. Finally, cure the underfill.

The advantages of the current design are as follows.

(a) For a given chip size, solder joint size, and standoff height, the opening and thickness of the stencil and dry film can be estimated, and there will not be any waste of the underfill material.
(b) Because of the dry film underneath the stencil and the gap between the stencil and the backside of the chip, the chance of damaging the chip is very slim.
(c) With the aid of heat, the underfill is easier to be printed and the printed underfill on one edge of the chip is more uniform, and the residue of underfill on the stencil is reduced.

1.13.2 Test Vehicles

In order to demonstrate the feasibility of the new stencil design for post-assembly underfill, test vehicles are built.

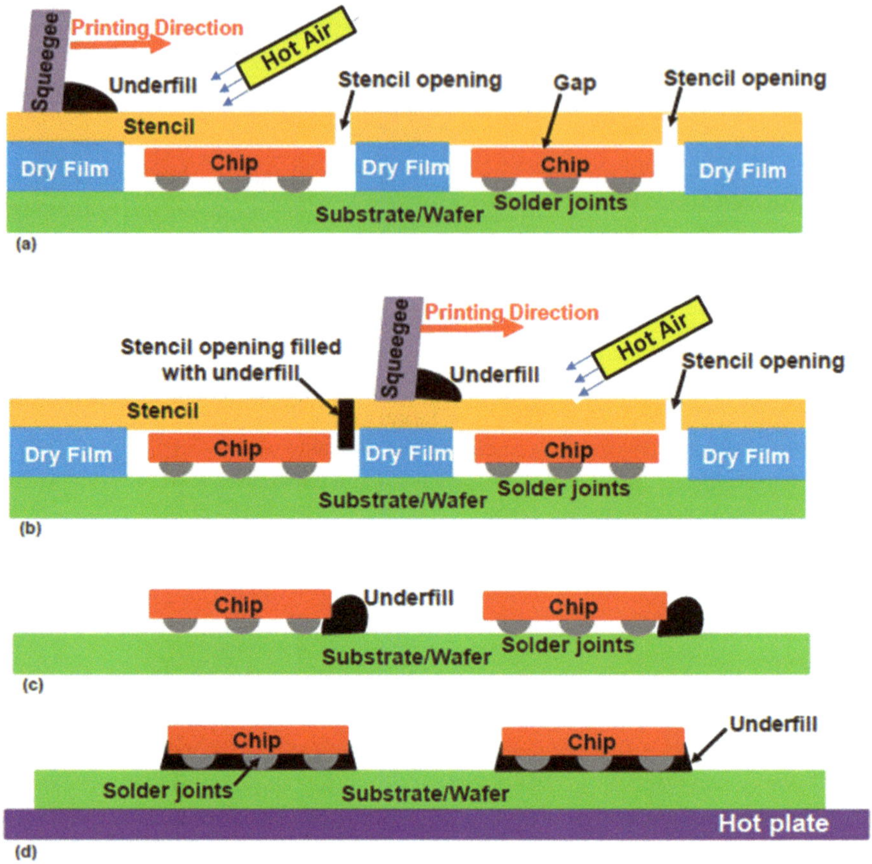

Fig. 1.56 **a** Before printing. **b** During printing. **c** After printing. **d** After capillary action. Hot air (optional)

(A) Test Chip

The test chips for both organic-panel and Si-wafer assemblies are the same and shown in Fig. 1.59. It can be observed that the chip dimensions are 5 mm × 5 mm × 150 μm and there are 31 × 31 (961) Cu-pillar + SnAg solder cap bumps, which are on 160-μm-pitch. The diameter of the Cu pillar is 40 μm, and its height is 25 μm, while the SnAg solder cap is 17 μm as shown in Fig. 1.59.

(B) Test Substrates

In this paper, the test substrates for the flip-chip assemblies are organic panel and Si wafer and are shown, respectively, in Figs. 1.60 and 1.61. It can be observed that for the organic-panel substrate (Fig. 1.60), the dimensions are 240 mm × 63 mm × 0.32 mm. There are 36 units, and the dimensions for each unit are 15.4 mm × 15.4 mm × 0.32 mm. There are pads and traces on each chip site. The diameter of the

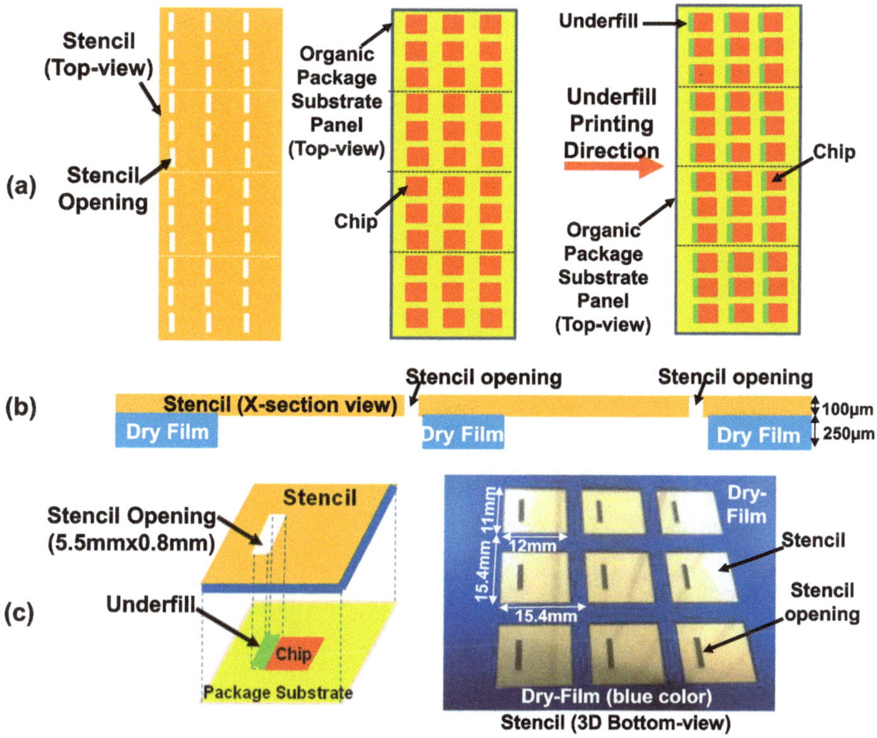

Fig. 1.57 Printing underfill for flip chips on organic substrate. **a** Top view. **b** X-section view. **c** 3-D view

organic solderability preservative (OSP) Cu pad is 80 μm and is on a 320-μm pitch. The trace's (leads) width is 25 μm and will be bump on lead (BOL).

Figure 1.61 shows the 200 mm Si-wafer substrate. The dimensions of the chip site are 5 mm × 5 mm × 760 μm. There are 961 Cu pads and are on 160-μm pitch on each chip site. The pad diameter is 60 μm. Because the street (kerf) width between the chip sites on the wafer is too narrow (<160 μm) to place the underfill, every other chip site will be used.

(C) Flip-Chip Assemblies

The flip chips (Fig. 1.59) are dipped into flux, placed on the organic-panel or Si-wafer substrates, and then massed reflowed. The flux residues are cleaned by jetting water at 60 °C. The X-ray image of the flip-chip assembly on organic-panel substrate is shown in Fig. 1.62a. It can be observed that there are two kinds of solder joints: (1) bump on pad (BOP) and (2) BOL. The pitch of BOP is 320 μm and that of BOL is 160 μm. The X-ray image of the flip-chip assembly on the Si-wafer substrate is shown in Fig. 1.62b. It can be observed that the 40-μm-diameter Cu pillar on the chip is soldered on the 60-μm-diameter Cu pad on the substrate. They are on a 160-μm pitch.

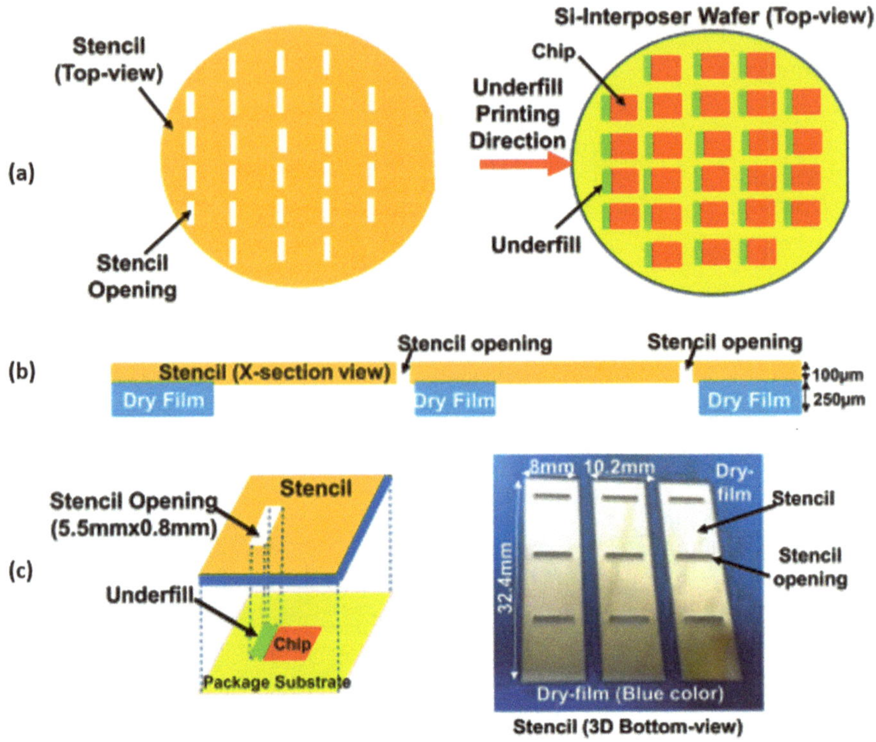

Fig. 1.58 Stencil printing of underfill for flip chips on Si substrate. **a** Top view. **b** X-section view. **c** 3-D view

(D) Stencil Designs

The stencil designs for the test vehicles are shown in Fig. 1.57 for flip chips on organic-panel substrate and in Fig. 1.58 for flip chips on Si-wafer substrate. For both cases, the stencil thickness is 100 μm and the opening is 5.5 mm × 0.8 mm. The dry film thickness underneath the stencil is 250 μm. The estimated underfill volume per chip for this stencil design is ~1.52 mm^3. For the case with organic-panel substrate, the dry film opening underneath the stencil is 11 mm × 12 mm (Fig. 1.57). For the case with Si-wafer substrate, the dry film opening underneath the stencil is 8 mm × 32.4 mm, as shown in Fig. 1.58. The dry film (photoresist) is laminated on the stencil with a heater roller (~125 °C). The opening of the dry film is by ultraviolet exposure machine and chemical solution.

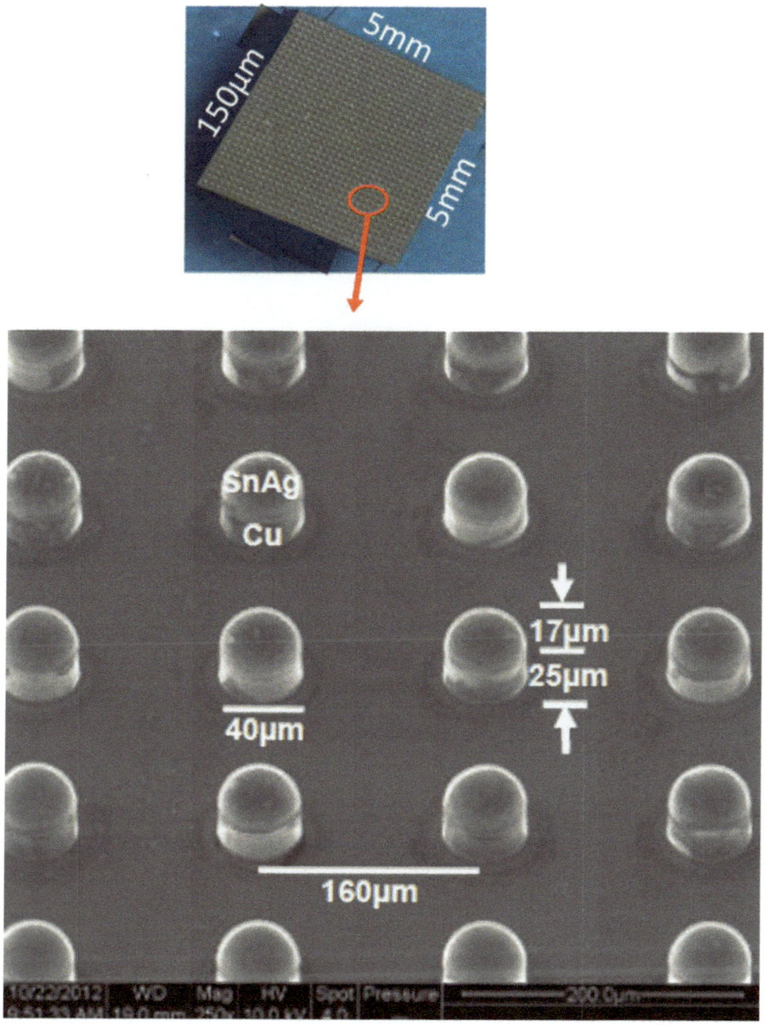

Fig. 1.59 Test chip with Cu-pillar + SnAg solder cap

1.13.3 Experiments

The test matrix, baking of assemblies, printing process, underfill capillary action, and curing will be briefly mentioned.

(A) Test Matrix

Three different underfill materials are considered in this paper. Their viscosities are: (1) material #1 = 34 Pa · s room temperature (RT); (2) material #2 = 68 Pa · s (RT); and (3) material #3 = 15 000 Pa · s (RT) and are shown in Fig. 1.63. For underfill

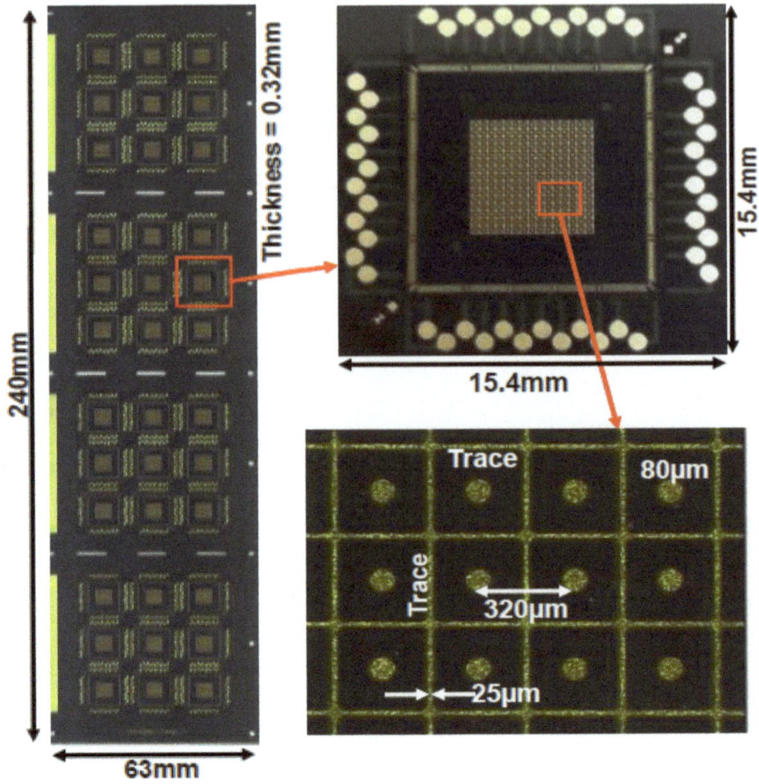

Fig. 1.60 Organic-panel substrate. The diameter of the OSP Cu pads is 80 μm on a 320-μm pitch. The trace width is 25 μm for BOL

material #3, since the viscosity is so high that the squeegee of stencil hardly moves (even with 60 °C thermal enhancement), it will not be considered for the rest of the studies. The test matrix is shown in Table 1.20. It can be observed that: (1) there are two different substrates (organic panel and Si wafer); (2) for each substrate, there are two difference underfill materials (#1 and #2); (3) for each underfill, there are two different print numbers (1 time and 10 times); and (4) for each time, there are two different temperature environments (25 °C and 45 °C).

(B) Baking Substrates

The underfill process begins by baking the organic- and Si-substrate flip-chip assemblies. The baking condition is at 120 °C for 60 min (on a hot plate). This step is critical for voidless underfill. Otherwise, the entrapped moisture in the assemblies will create voids in the underfill as shown in the C-SAM (C-mode scanning acoustic microscopy) image (Fig. 1.64).

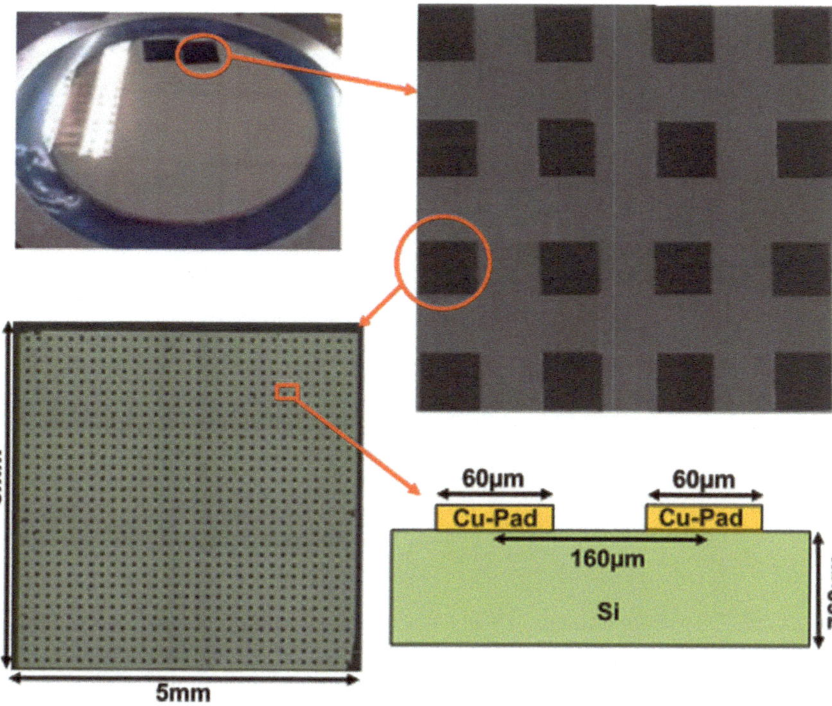

Fig. 1.61 Si-wafer substrate. The diameter of the Cu pad is 60 μm on a 160-μm pitch

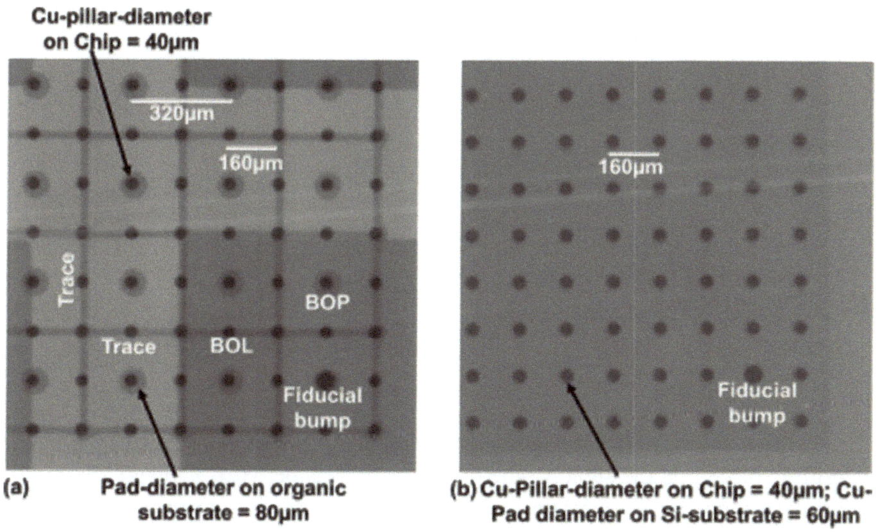

Fig. 1.62 X-ray images. **a** Flip-chip assembly on organic substrate that includes BOLs and BOPs. **b** Flip-chip assembly on Si substrate

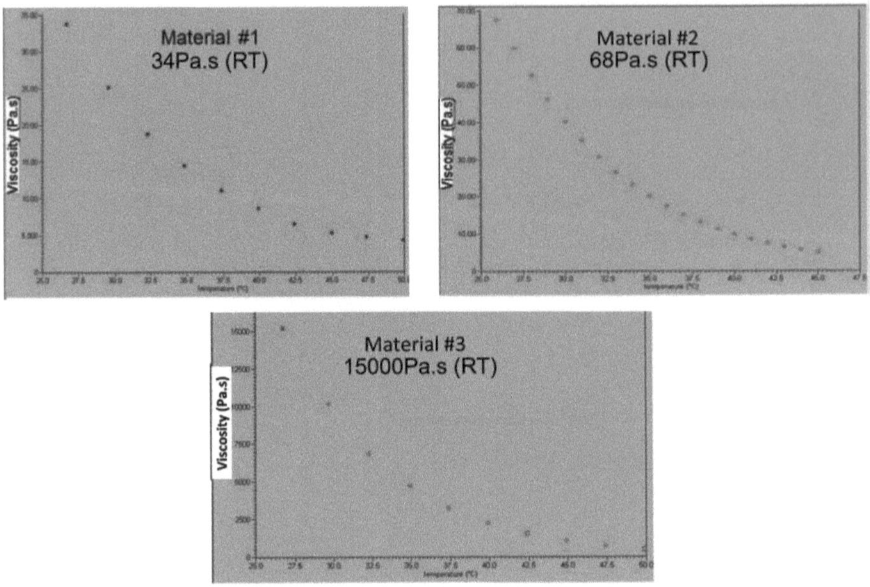

Fig. 1.63 Viscosity versus temperature for underfill materials #1–#3

Table 1.20 Test matrix for stencil printing of underfills for flip-chip assemblies

Parameters	Test structure															
	Chip on organic-panel								Chip on Si-wafer							
Underfills (Pa.s) (Viscosity at 25 °C)	34 Material #1				68 Material #2				34 Material #1				68 Material #2			
Multiple print (times)	1		10		1		10		1		10		1		10	
Thermal enhanced (°C)	25 No	45	25 No	45	25 No	45	25 No	45	25 No	45	25 No	45	25 No	45	25 No	45

(C) Printing Process

After baking, the flip-chip assemblies are loaded in a stencil printer (DEK). The underfill is placed on the stencil. The printing speed is between 150 and 290 mm/s, the printing force is 8 kg, and the snap-off height is zero. The printing patterns are quite uniform as shown in Fig. 1.65a and b for organic and Si substrates, respectively.

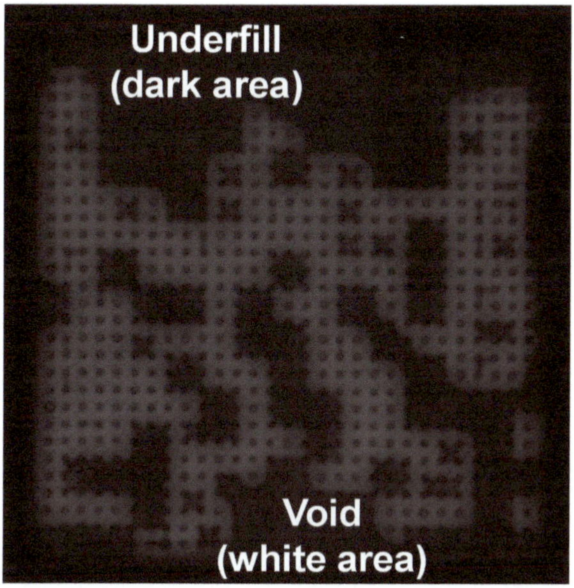

Fig. 1.64 Voids in underfill sample without baking of the assembly

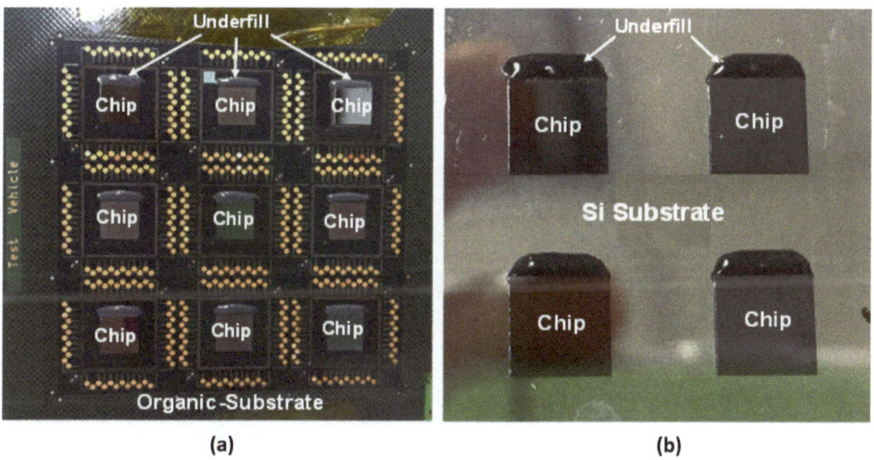

Fig. 1.65 Stencil printing of underfill on **a** organic-panel substrate and **b** Si-wafer substrate

(D) Capillary Action and Curing

The printed underfill flip-chip assemblies are placed on a hot plate at 120 °C. The underfill will flow underneath the chip and fill the space between the chip, solder joints, and substrate by the capillary action. Once the underfill flows out from the other edges of the chip, then the underfill can be cured.

1.13.4 Characterizations

The experimental samples are characterized by C-mode SAM, X-ray, cross-sectioning, shear test, and SEM (scanning electron microscope) methods.

(A) Effects of Underfill Viscosity, Thermal Enhancement, and Multiple Prints

The experimental results are shown in Table 1.21. For both underfills #1 and #2 (with and without thermal enhancement), there are only three chips with voids, which are due to the chip/substrate surface contamination and flux residues from the cleaning (Fig. 1.66). It can also be observed from Table 1.21 that there is not a significant effect on multiple prints. Typical C-mode SAM images are shown in Fig. 1.67. Figure 1.67(top) shows flip chips on Si-substrate assembly with underfill material #1, while Figure 1.67(bottom) shows flip chips on organic substrate with material #2. There are not any voids in these assemblies.

The effect of thermal enhancement for stencil printing underfill material #1 is shown in Fig. 1.68. It can be observed that: (1) there is minimal underfill residue after stencil printing underfill at 45 °C [Figure 1.68a] and (2) there is significant underfill residue after stencil printing underfill at RT (no thermal enhancement) [Figure 1.68b]. The effect of thermal enhancement for stencil printing underfill material #2 is very similar.

(B) Cross Sections

Figure 1.69 shows a typical cross section of the stencil printed underfill for flip chip on organic-panel assemblies. The following can be observed.

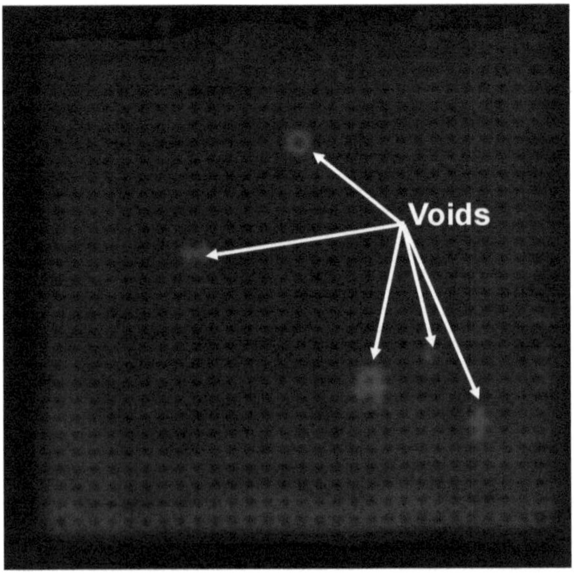

Fig. 1.66 C-mode SAM image of underfill flip-chip assembly with voids

Fig. 1.67 Typical C-mode SAM. Top: flip chip on Si-substrate assemblies with underfill material #1. Bottom: flip chip on organic substrate with underfill material #2

(a) The underfill fillets on the edges of the chip are clearly shown.
(b) The underfill between the chip, solder joints, and substrate is voidless and properly processed.
(c) The solder joints on the Cu pad (BOP) and lead (BOL) of the organic substrate look very good.

Figure 1.70 shows a typical cross section of the stencil-printed underfill for flip chip on Si-wafer assemblies. The following can be observed.

(a) The underfill fillets on the edges of the chip are clearly shown.
(b) The underfill between the chip, solder joints, and substrate is voidless and properly processed.
(c) The solder joints with 40-μm-diameter Cu pillar of the chip on 60-μm-diameter Cu pad of the Si substrate look good. The calculated underfill volume per chip is $5 \times 5 \times 0.035 - 31 \times 31 \times (\pi \times 0.02252) \times 0.035 + 4 \times 0.5 \times 0.3 \times 0.135 \times 5.5 = 1.27\ \text{mm}^3$.

(C) Underfill Filler Density

Figure 1.71 shows a flip-chip assembly with voidless underfill. It is observed that even though there is no void, there are darker spots. For example, it is slightly darker between solder joints 7 and 8 than between solder joints 8 and 9 on the cross section shown in the top portion of Figure 1.71. The SEM images show that the silica fillers between solder joints 7 and 8 are denser than those between solder joints 8 and 9.

(D) Shearing Test

The stencil-printed underfill flip-chip assemblies are subjected to a shearing test. The tip of the shear blade is placed at 30 μm from the substrate surface (near to

Table 1.21 Experimental results

		Runs	Temperature (°C)	Multiple prints	Voids (Y/N)	Die shear force (kg)	Remark
Underfill material #1	Chip to organic-panel	R1	25	1	N (0/5)	>60 kg	
		R2	25	10	N (0/5)	>60 kg	
		R3	**45**	**1**	**Y (2/4)**	>60 kg	
		R4	45	10	N (0/4)	53 kg	Die crack
	Chip to Si-wafer	R5	25	1	N (0/4)	57 kg	Die crack
		R6	25	10	N (0/4)	45 kg	Die crack
		R7	45	1	N (0/4)	43 kg	Die/sub crack
		R8	45	10	N (0/4)	>60 kg	
Underfill material #1	Chip to organic-panel	R9	25	1	N (0/4)	>60 kg	
		R10	25	10	N (0/5)	>60 kg	
		R11	45	1	N (0/4)	39 kg	Die crack
		R12	45	10	N (0/4)	>60 kg	
	Chip to Si-wafer	**R13**	**25**	**1**	**Y (1/3)**	>60 kg	
		R14	25	10	N (0/4)	26 kg	Die crack
		R15	45	1	N (0/3)	41 kg	Die/sub crack
		R16	45	10	N (0/4)	45 kg	Die crack

the bottom of the flip chip). The maximum shear force of the tester is set at 60 kg. Some of the test results are shown in Table 1.21 and Fig. 1.72. It can be observed that many samples passed 60 kg and have no failure. For the failed samples, the failure mode is chip fracture (cracking) and there is not any underfill failure. For some of the flip chips on Si-substrate assemblies, not only did the chip fracture but there was cracking of the Si substrate as shown in Fig. 1.72. This shows the toughness of the underfill.

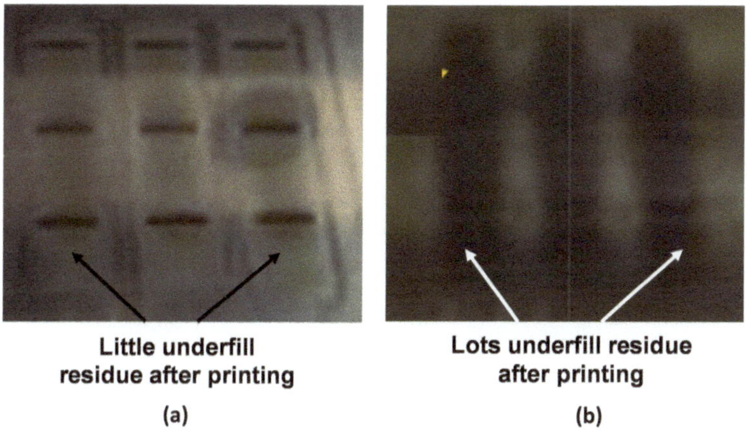

Fig. 1.68 **a** Stencil printing underfill at 45 °C. **b** Stencil printing underfill at RT

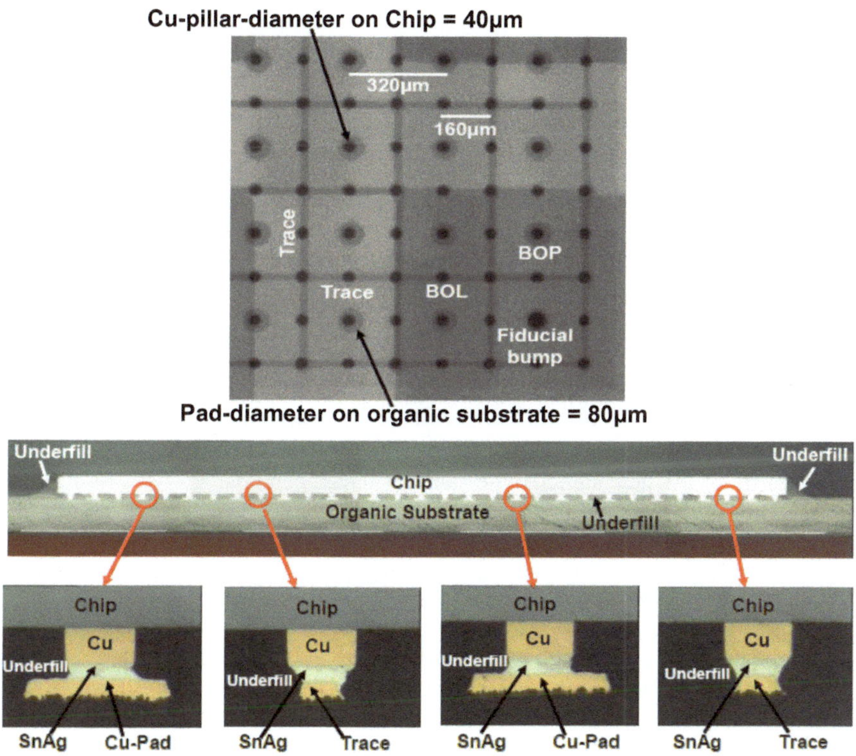

Fig. 1.69 Typical cross section of stencil-printed underfill for flip chip on organic-panel assembly

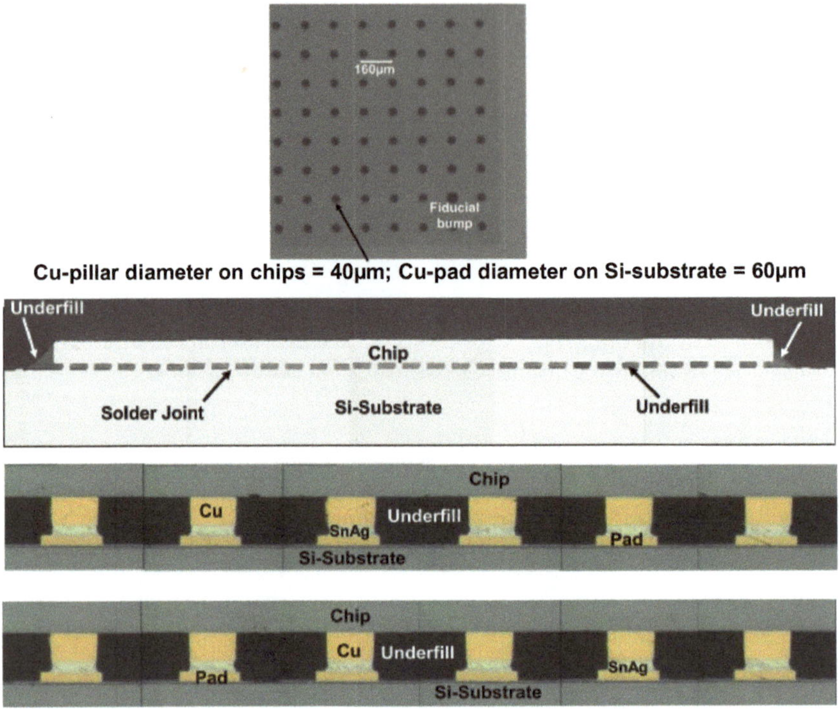

Fig. 1.70 Typical cross section of the stencil-printed underfill for flip chip on Si-wafer assembly

1.13.5 Summary

A new stencil design for printing underfill of flip chips on organic-panel and Si-wafer assemblies has been proposed. Test vehicles have been built to demonstrate the feasibility of the new stencil. The samples have been characterized by C-mode SAM, X-ray, cross sectioning, shear test, and SEM. Some important results and recommendations are summarized as follows.

- With the small opening of the stencil right in front of a chip edge, no underfill is wasted.
- With the dry film underneath the stencil and the gap between the stencil and the back of the chip, the chance of damaging the chip during underfill printing is slim.
- There is no significant difference between 1-time and 10-time underfill prints.
- For both underfill materials #1 and #2, with and without thermal enhancement, most of the underfill flip-chip assemblies are voidless.
- With the thermal enhancement, the underfill is easier to be printed and the residue on the stencil is reduced.
- The darker spots in a C-mode SAM image of the underfill represent that there are denser silica fillers.

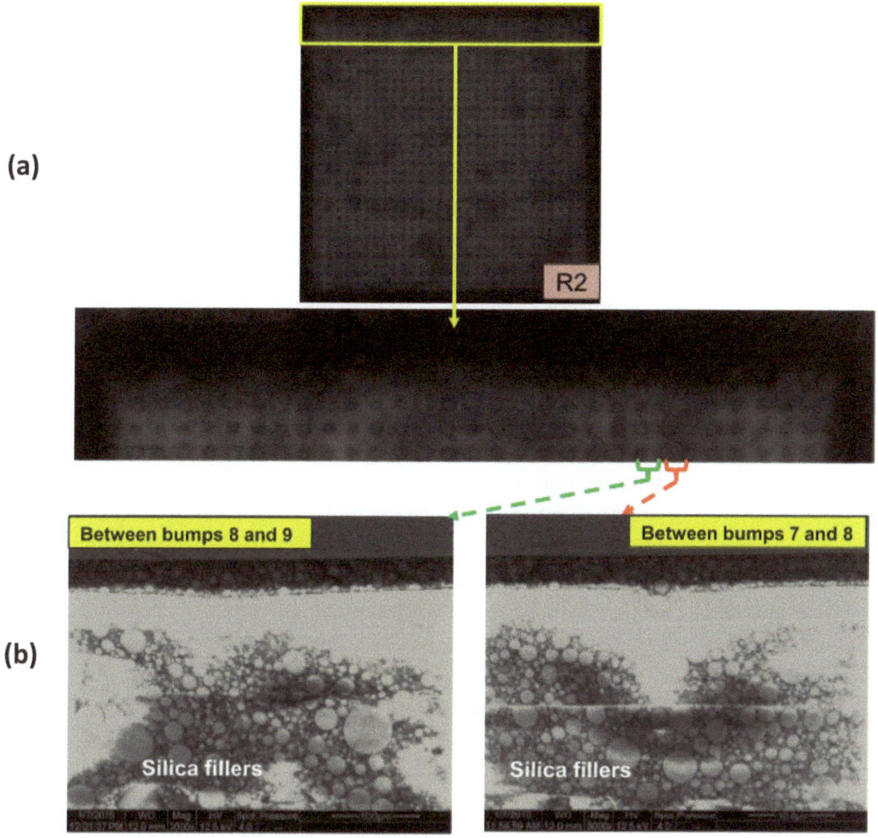

Fig. 1.71 **a** Flip-chip assembly with voidless underfill but with darker spots. **b** SEM images showing the distribution of underfill silica fillers

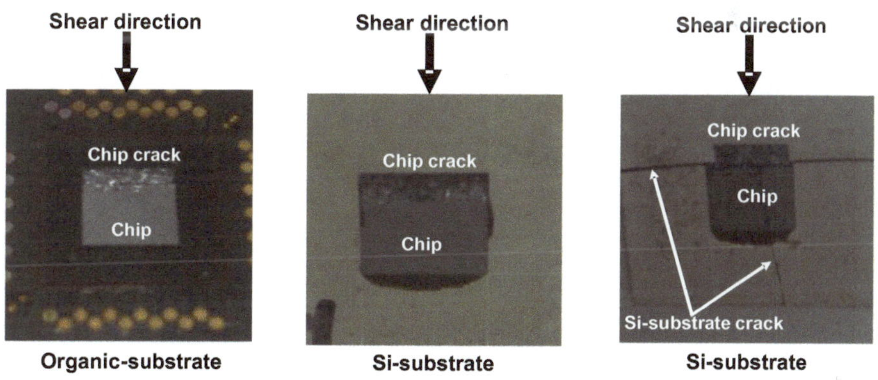

Fig. 1.72 Shearing tests and failure modes

- As is well known, the flip-chip assembly must be baked before underfilling; otherwise, there are many voids in the underfill.
- As it is well known, the chip/substrate surface contamination and flux residues must be totally cleaned; otherwise, the entrapped flux residues and contamination will create voids in the underfill.
- Because of the narrow street (kerf) widths between the interposers on a Si-wafer substrate, stencil printing of flip chips on every interposer on a Si-wafer substrate is impossible.
- Stencil printing of underfill for flip chips on organic-panel substrate is a simple, low-cost, and high-throughput process.

1.14 LPC–A High Throughput TCB Process

The increase in functionality together with miniaturization of electronic devices has driven the development of fine pitch copper pillars with solder cap flip chip packages. Both mass reflow (MR) and thermal compression bonding (TCB) have been used as chip assembly approaches. With continuously reducing the pitch size, solder volume, and also increasing tendency to package warpage caused by thin die/substrate and coefficient of thermal expansion (CTE) mismatch, there is a growing need for using TCB process in order to achieve high assembly yield and high-level of package reliability. However, low throughput is a major limitation for TCB process.

In this study, a new TCB-flux process, namely, liquid phase contact (LPC) TCB approach is proposed and studied. The significant difference of the LPC TCB process from the traditional TCB-flux process is that the solder cap on the copper pillar has been heated to a molten state before contacting the bond pad/lead on the substrate under a small force. The bonding cycle time for LPC TCB with or without BH cooling step is greatly reduced compared with the traditional TCB-flux process, therefore a high throughput assembly process could be achieved. Another advantage of LPC TCB process is that the solder thickness (stand-off height) after assembly could be precisely controlled, which is very important for the subsequent underfill process and reliability consideration.

1.14.1 Experimental Procedure

(A) Test Vehicles

In this study, two kinds of test vehicles are used for chip-on-organic substrate (CoS) and chip-on-chip (CoC) bonding, as summarized in Table 1.22. For CoS bonding, the chip size is 5.0 mm × 5.0 mm × 0.15 mm. As shown in Fig. 1.73, the chip has a 31 × 31 area array of Cu pillar with SnAg solder cap. The pillar diameter is 40 or 60 μm, the heights for Cu pillar and solder cap are 25 μm and 27 μm, or 25 μm and 17 μm, respectively. The bump pitch is 160 μm. The solder composition is 97.5%

Table 1.22 Test vehicles

	Chip-on-substrate (CoS)	Chip-on-chip (CoC)
Chip	Bump diameter = ϕ60 μm Height = 25 μm Cu + 27 μm solder Bump pitch = 160 μm	Bump diameter = ϕ40 μm Height = 25 μm Cu + 17 μm solder Bump pitch = 160 μm
	Bump diameter = ϕ40 μm Height = 25 μm Cu + 17 μm solder Bump pitch = 160 μm	
Substrate	BT metallization = CuOSP Cu lead width = 18 μm Cu pad diameter = ϕ80 μm Cu lead/pad thickness = 15 μm	Surface finish = Cu Cu pad diameter = ϕ60 μm Cu pad thickness = 10 μm
		Surface finish = SOP Cu pad diameter = ϕ40 μm Cu pad thickness = 10 μm

Sn and 2.5% Ag. The BT substrate had a Cu OSP metallization with both BOP and BOL structures. The pad diameter is 80 μm, while the lead width is 18 μm.

For CoC bonding, the top die has 40 μm pillar diameter with 25 μm Cu pillar height and 17 μm solder height. The bottom die has two surface metallizations: bare Cu and solder on pad (SOP). For the die with Cu finish, the pad diameter is 60 μm

Fig. 1.73 CoS test vehicle

and the pad thickness is 10 μm. For the die with SOP finish, the pad diameter is 40 μm and the solder height is 17 μm.

(B) LPC TCB Process

As shown in Fig. 1.74, LPC TCB process has following operation sequences:

(a) Flux is printed or sprayed on the substrate.
(b) BH picks up a die from the carrier at an elevated temperature, which is below the solder melting point.
(c) The BH heats up to a temperature higher than the solder melting point.
(d) The chip is aligned with the substrate.
(e) After the vision alignment, the chip contacts and wets on the substrate at a predetermined bonding level.
(f) After a certain bonding time, the BH could move away at the bonding temperature or cool down to a temperature below the melting point of solder.

Figure 1.75 compares the bonding steps between the conventional TCB-flux process and LPC TCB process. For the conventional TCB-flux process, the flux is applied on the die by a flux dipping method. The dipping temperature is lower than 100 °C due to the flux wicking and evaporation issues under high temperatures. A large temperature increment, such as from 100 °C to 250 °C for SnAg solder, is required after the chip contacting with the substrate. The heating time needed depends on the heating capacity/heating rate of the heater. However, for LPC TCB process, the standby temperature of the bond head is relatively high, only 20–30 °C lower than the solder melting point. The temperature heating up could occur during the bond head moving up and the vision alignment. The whole bonding cycle for

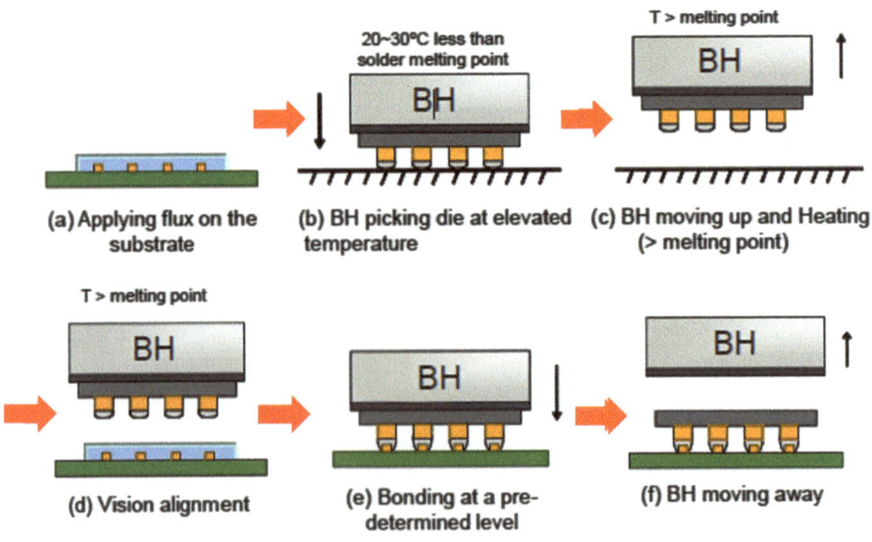

Fig. 1.74 Process flow of LPC TCB Process

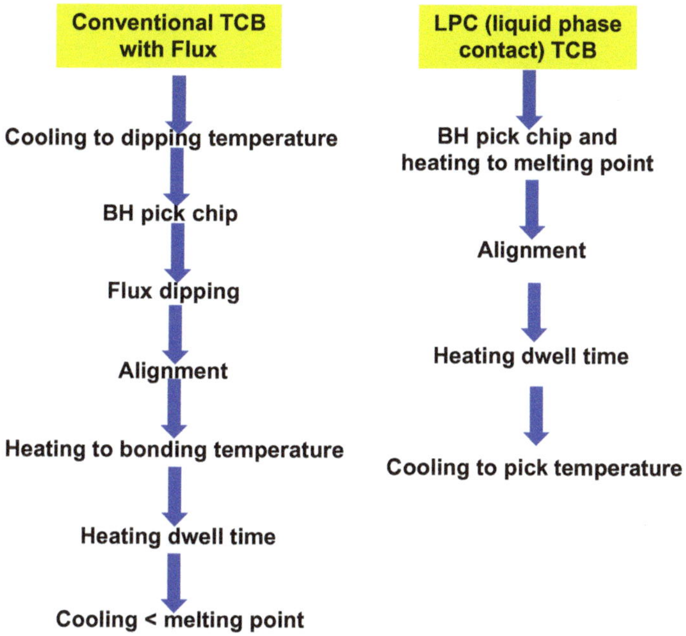

Fig. 1.75 Bonding steps of conventional TCB-flux and LPC TCB processes

LPC process could be less than 4 s. In some circumstances, if the required stand-off height is near the solder equilibrium level, the cooling step could be omitted. The bonding cycle could be further reduced to 3 s or even less. The solder on pad (SOP). UPH (units per hour) could be as high as 1.2 k for a single bond head configuration, compared with UPH of ~600 for a conventional TCB-flux process.

1.14.2 Bonding Quality Evaluation

Bonding temperature, bonding force, bonding time, bonding level, and cooling step are major process parameters for TCB LPC process, which need to be optimized. Cross sections of CoS and CoC bonded samples Fig. 1.76a and b, respectively show excellent solder wetting with a solder thickness variation less than 2 μm. It is noted higher temperatures for both bond head and work holder are required for CoC bonding. This is because silicon has a higher thermal conductivity (149 W/m·K) than organic materials (BT: 0.35 W/m·K), for CoC bonding heat from the bond head could be dissipated rapidly into the bonding stage by the bottom silicon chip. The amount of additional heat required is controlled by the thickness and size of the bottom chip. The solder joints obtained from LPC TCB process are evaluated together with the samples assembled using MR and conventional TCB-flux processes.

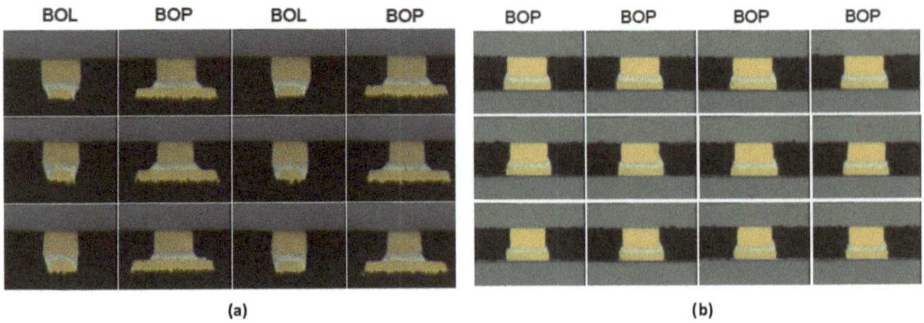

Fig. 1.76 **a** Cross section images of CoS packages. Stand-off height: 45.6 (left) → 46.2 (center) → 46.4 μm (right). (Bonding conditions: WH = 80 °C, BH = 260 °C, BT = 1.0 s). **b** Cross section images of CoC packages. Stand-off height: 38.4 (Left) → 37.6 (center) → 38.1 μm (right). (Bonding conditions: WH = 120 °C, BH = 350 °C, BT = 1.0 s)

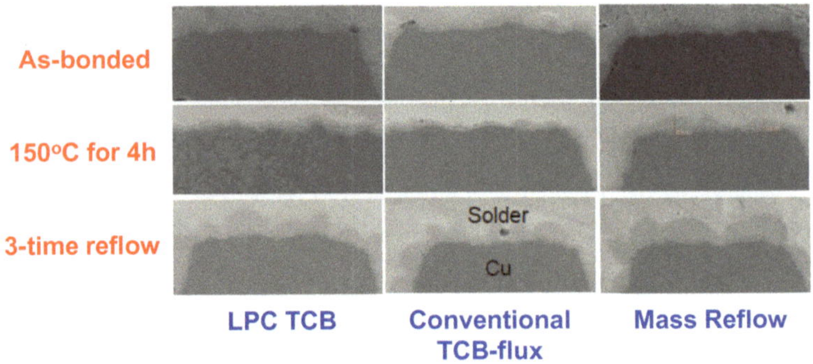

Fig. 1.77 Interfacial microstructure for LPC TCB, conventional TCB-flux and MR Processes

Figure 1.77 shows the interfacial microstructure using three bonding processes. For as-bonded condition, a very thin intermetallic compound (IMC) layer, about 1 μm, could be detected for both LPC TCB and conventional TCB-flux processes. For the MR process, IMC is relatively thick, about 2.0 μm. This is due to the longer liquidus time during MR process.

Short thermal aging and 3 times reflow tests are performed to examine the IMC growth behavior. After thermal aging at 150 °C for 4 h, the IMC growth is limited, only 0.1–0.2 μm, while significant IMC growth (3 μm for LPC TCB and traditional TCB-flux processes, and 4 μm for MR process) is observed after 3 times reflow. The scallop-like Cu_6Sn_5 could be found at the bonding interfaces for all three bonding processes.

The die shear test is also performed to evaluate the solder joint integrity. High values of shear force, ranging from 13.0 kgf to 15.0 kgf are obtained for all three processes under as-bonded conditions. The fracture occurred inside the solder layer. Compared with as-bonded condition, there is only a little change in shear force after thermal aging at 150 °C for 4 h, a relatively large drop in shear force is detected after multiple reflow (shear force after reflowed: 10.0 kg to 13.0 kg). The results on interfacial microstructure, shear strength and failure mode indicate that LPC TCB process could provide an excellent solder joint.

1.14.3 Stand-Off Height Control

Stand-off height control is important to facilitate the subsequent underfill process and to achieve excellent joint reliability. Coplanarity adjustment, precise position control and thermal management are the key controlling factors. Compared to the conventional TCB-flux process, there is no bonding tool heat-up step after the die contacted with the substrate for LPC TCB process. Thermal expansion induced by the bonding tools could be neglected. Because of the relatively short bonding time, the substrate temperature is low, hence the thermal expansion of the substrate is also less than that in the conventional TCB-flux process.

In addition, only a small bonding force (<1 kg) is required in LPC TCB process since the solder is in the molten state during bonding. The small bonding force could also minimize the elastic deformation of the substrate to maintain an accurate stand-off height.

In the following sections, effects of two bond parameters (cooling step, bonding temperature) on the solder height are reported and discussed.

(A) Effect of Cooling Step

As explained previously, after bonding the bond head could move away from the die at the bonding temperature (without a cooling step), or the bond head cools down to a temperature below the solder melting point before moving away (with a cooling step). Different solder height values could be obtained under these two conditions.

The pictures on the left in Fig. 1.78 show that without a cooling step the final solder height is almost the same for the samples bonded at different pre-determined bonding levels (from 4 μm to 20 μm), while the solder heights are consistent with the bonding levels if the BH cools down to 200 °C before releasing (the pictures in the middle), which are also in good agreement with the simulation results predicted using Surface Evolver model (the pictures on the right).

In order to understand how the restoring force driving the molten solder to the equilibrium height after BH is removed at the temperature above the solder melting point, Surface Evolver model is used to calculate the force generated by the surface tension of liquid solder. Based on the principle of virtual work, the restoring force at z direction is the rate of change of energy with respect to the solder thickness [3]. In this study, the central difference scheme is used for calculating the restoring force. Figure 1.79 shows the relationship between the restoring force and the solder thickness. When the solder is compressed, the force is repulsive, while if the solder is elongated it is attractive. There is one equilibrium point, at which the net force is zero. This point is the equilibrium solder thickness without any external force exertion.

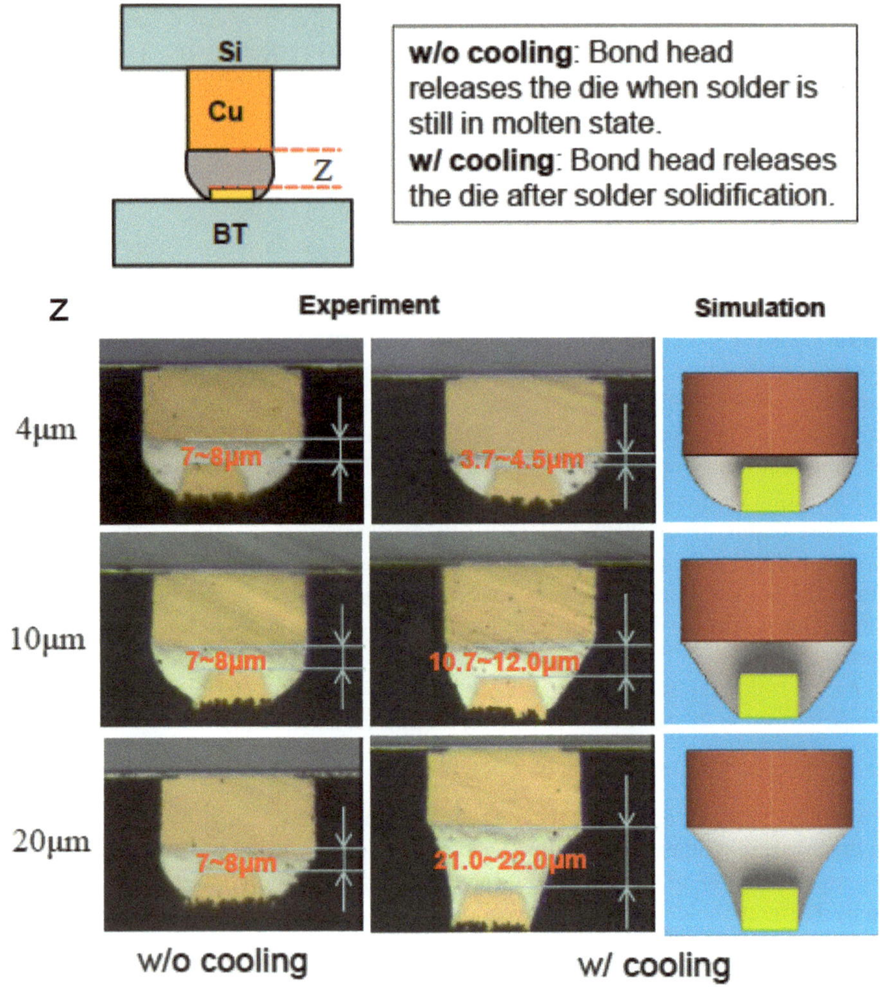

Fig. 1.78 Cooling effect on solder height at different bonding levels for CoS packages

If the pre-determined bonding level is 4 μm, the chip moved up to the equilibrium height driven by the repulsive force after the bond head moving away from the die top. While for the samples bonded at bonding levels of 10 and 20 μm, the chip moves down to the equilibrium height by the attractive force. That is why the solder height maintains the same regardless of the bonding level.

(B) Effect of Bonding Temperature

It should be noted that in aforementioned cases the bonding temperature is high enough that no solder solidification occurred during the bonding process. If the bonding temperature is relatively low, the molten solder might solidify, or partially

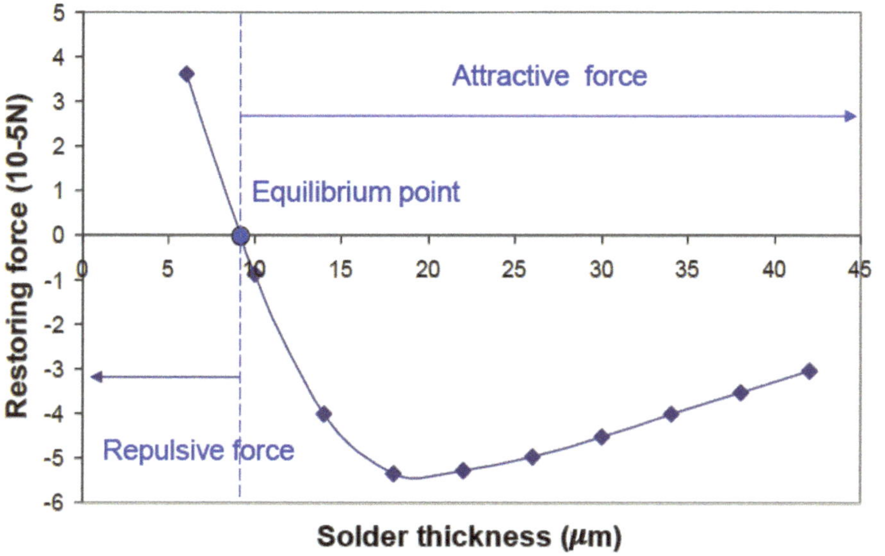

Fig. 1.79 Restoring force versus solder thickness

solidify, as the BH contacting with the substrate. The final stand-off height at the low bonding temperature could be higher than that at the high bonding temperature. As shown in the Fig. 1.80, 8 μm and 15 μm solder heights are obtained for bonding temperatures at 280 °C and 250 °C, respectively. Similar phenomenon is also observed for CoC bonding (Fig. 1.81).

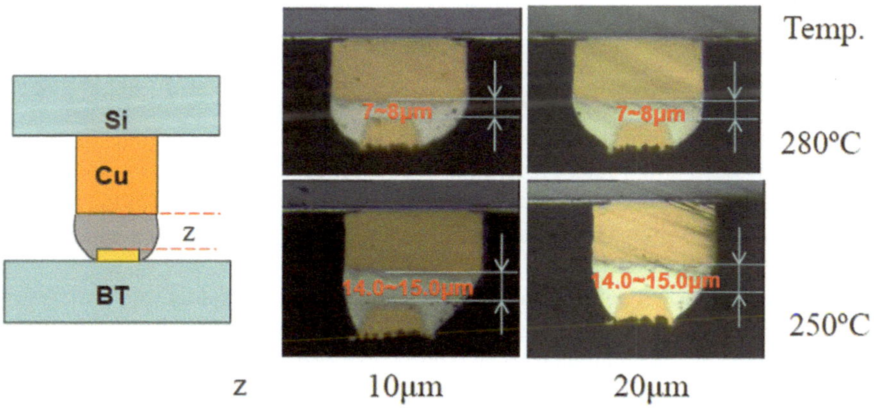

Fig. 1.80 Temperature effect on solder height for CoS packages (w/o cooling)

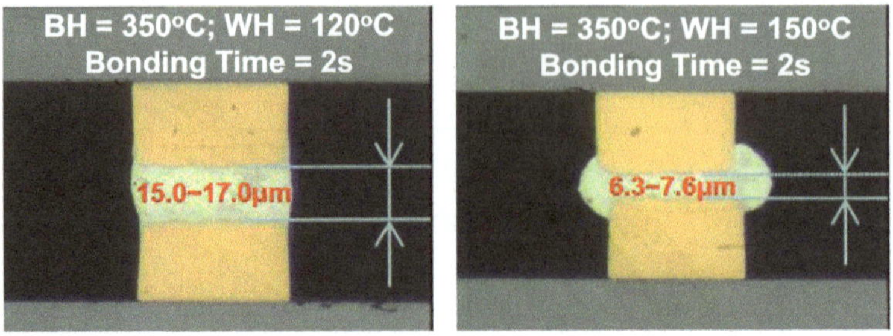

Fig. 1.81 Temperature effect on solder height for CoC packages (w/o cooling)

1.14.4 Thermal Fatigue

Bump geometry is the dominant factor that affects solder joint reliability. As shown in Fig. 1.82, the flip chip assembly is modeled as a three-dimensional slice that captured the construction along a diagonal path from the geometric center to corner. In this model, there are two kinds of bumps: BOP and BOL. The joint geometries with 10 μm solder thickness are first simulated by Surface Evolver. The simulated geometries are then imported into the commercial finite element analysis (FEA) tools ANSYS 14.0 for simulation. The critically affected bumps are meshed finely, while the others are meshed coarsely. Thermal cycle loads recommended by IPC-970, are applied on all the nodes. The temperature ranged from −55 °C ↔ 125 °C with 12 min ramp-up time and 3 min dwell time. Totally three cycles are simulated in the current model. The material properties used were listed in Table 1.23.

As shown in the below equation, a double power law model [139] is used to study the evolution and accumulation of irreversible strains and energies of solder bumps.˙

$$\epsilon_{cr}^{'} = A_1 \exp\left(\frac{-H_1}{kT}\right)\left(\frac{\sigma}{\sigma_n}\right)^{n1} + A_2 \exp\left(\frac{-H_2}{kT}\right)\left(\frac{\sigma}{\sigma_n}\right)^{n2}$$

where $\epsilon_{cr}^{'}$ is creep strain rate, A_1 and A_2 are power law multipliers, H_1 and H_2 are activation energies, σ is applied stress, n_1 and n_2 are stress exponents, T is the absolute temperature, k is Boltzmann's constant. The values of all these parameters for SnAg solder are listed in Table 1.24.

The accumulated creep strain contours of the outmost BOP solder joint for the first cycle, second cycle, and third cycle are, respectively shown in Fig. 1.83a, b and c. It can be seen that the accumulated creep strain was stabilized at the second cycle. The maximum creep strain occurs at the interface between the circumferential solder joint and the OSP Cu-pad. The accumulated creep strain contours of the outmost BOL solder joint for the first cycle, second cycle, and third cycle are, respectively shown in Fig. 1.84a, b and c. Again, it can be seen that the accumulated creep strain is stabilized at the second cycle. The maximum creep strain occurs at the interface between the solder joint and the Cu pillar and at a localized area.

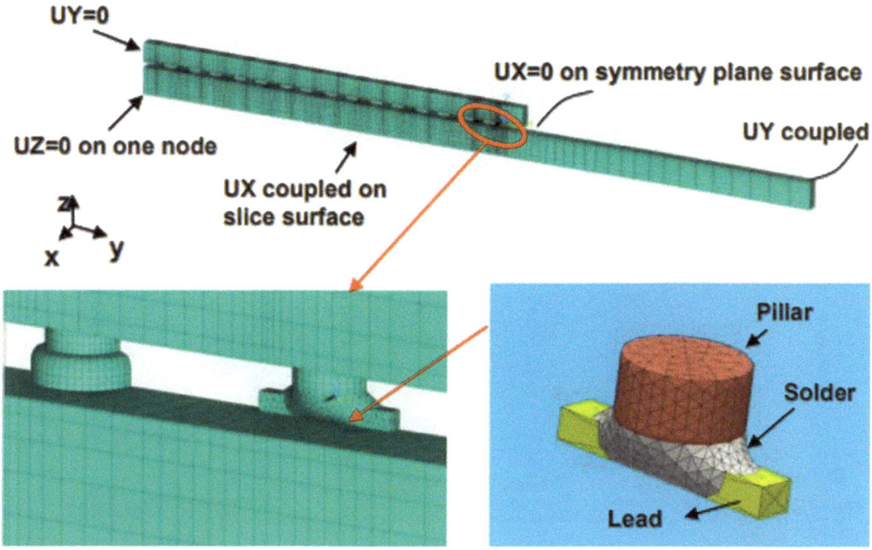

Fig. 1.82 Slice model of a CoS flip chip package. (Underfill is not shown so the BOP and BOL solder joints can be seen.)

Table 1.23 Material properties

Materials	CTE (ppm/K)	Young's modulus (GPa)	Poisson's ratio
Copper [1]	16.3	121	0.34
BT substrate [2]	16	25.5	0.37
Silicon chip [3]	2.7	131	0.28
Solder: SnAg [1]	21.5	49	0.40
Underfill [1]	29	7.6	0.32

[1] J. Jhou, et al., "Thermal stresses and deformations of Cu pillar flip cihp BGA package: Analyses and measurements." in Proc. Of IMPACT., Oct. 20–22, 2010, pp. 1–4.5
[2] D. Chong, et al, "Reliability assessment of a high performance flip-chip BGA package (organic substrate bases) using finite element analysis." in Proc. IEEE Electronic Components and Technol. Conf. (ECTC), May 2730, 2003, pp. 207–213
[3] V. Kripesh, et al., "Design and development of a multi-die embedded micro wafer level package." in Proc. IEEE Electronic Components and Technol. Conf. (ECTC), Lake Buena Vista, FL, May 27–30, 2008, pp. 15,441,549

Table 1.24 Creep data of SnAg solder

$A_1 = 4\text{E}{-7}\ \text{s}^{-1}$; $H_1/k = 3223$; $n = 3.0$
$A_2 = 1\text{E}{-12}\ \text{s}^{-1}$; $H_2/k = 7348$; $n_2 = 12$
$\sigma_n = 1$ MPa

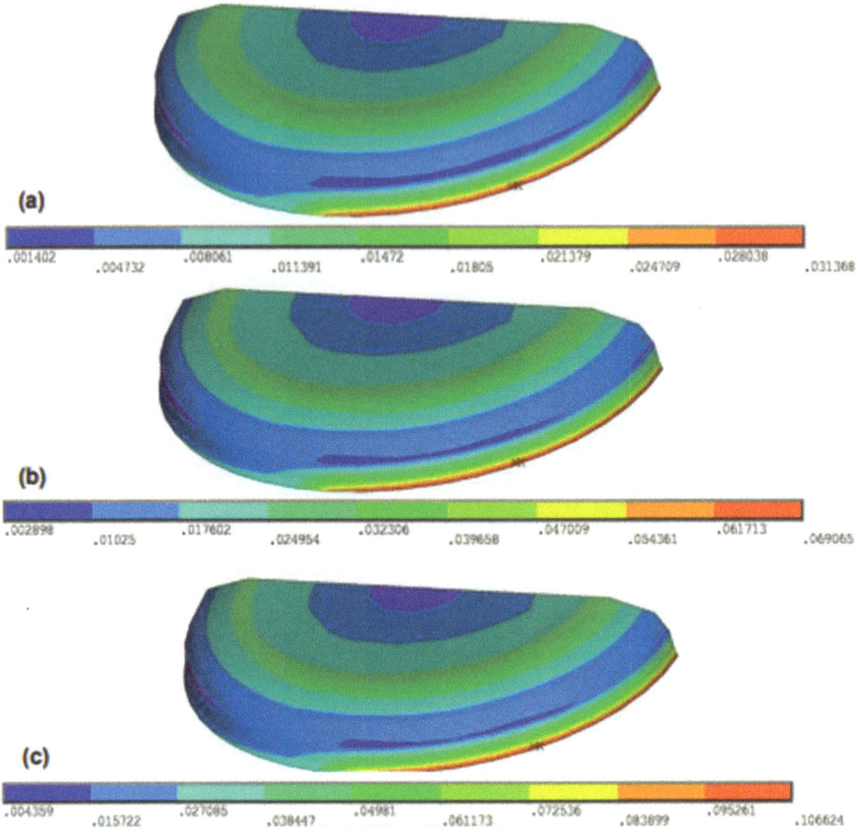

Fig. 1.83 Creep strain contours for the utmost BOP solder joint. **a** 1st cycle. **b** 2nd cycle. **c** 3rd cycle

The crack paths for estimating the average thermal fatigue life of the outmost solder joint are shown in Fig. 1.85a for the BOP case and Fig. 1.85b for the BOL case. It can be seen that for the BOP solder joint the crack path is assumed to be along the circumference between the solder joint and the OSP Cu-pad. On the other hand, the crack path for the BOL solder joint is assumed to be along the circumference between the solder joint and the Cu-pillar. The calculated average accumulated creep strain vs. time is shown in Fig. 1.86. It can be seen that the average accumulated creep strain for the outmost BOP solder joint (0.0148) is smaller than that for the outmost BOL solder joint (0.0167) and thus the thermal fatigue life of the BOP solder joint is longer than that of the BOL solder joint. This could be due to the larger distance-to-neutral point and less solder volume to resist thermal fatigue of the BOL solder joint.

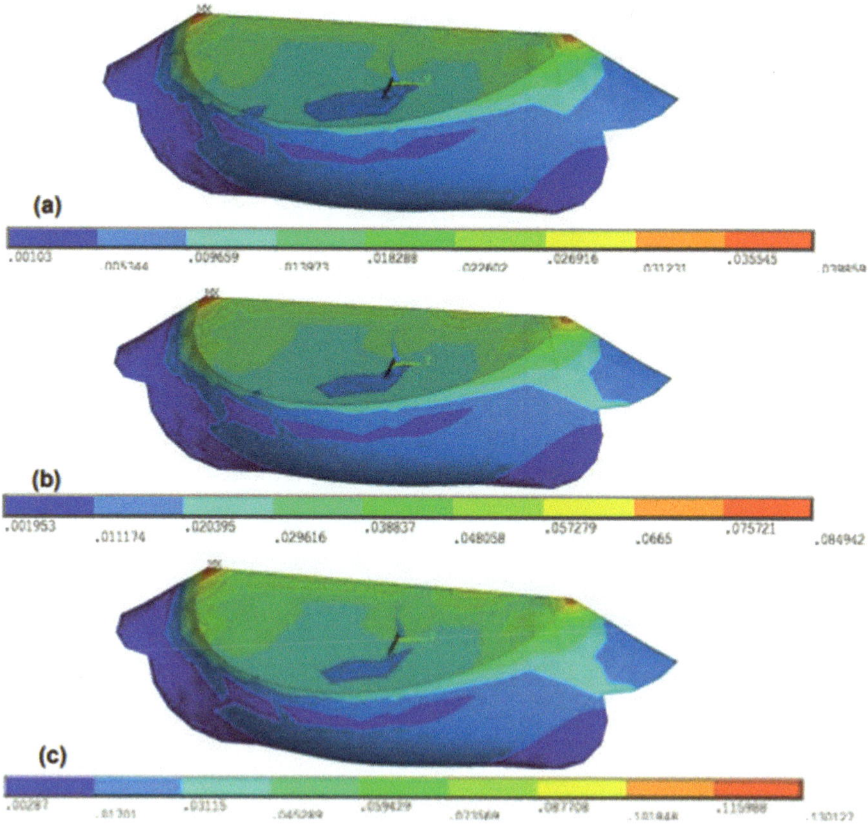

Fig. 1.84 Creep strain contours for the utmost BOL solder joint. **a** 1st cycle. **b** 2nd cycle. **c** 3rd cycle

Besides the 10 μm solder joint height, 20 μm solder joint height is also considered. The simulation results are shown in Fig. 1.86. It can be seen that when the standoff height of the solder joint increased (from 10 μm to 20 μm): (1) the average accumulated creep strain decreased (from 0.0148 to 0.0105 for the BOP case and 0.0167 to 0.0151 for the BOL case).

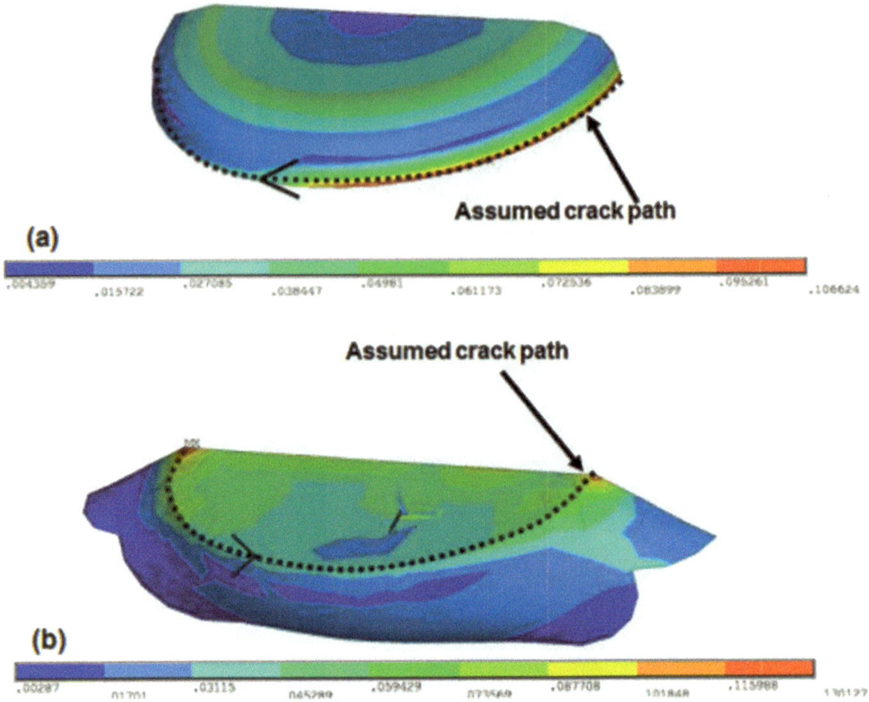

Fig. 1.85 Crack paths for calculating the average accumulated creep strain per cycle. **a** BOP solder joint. **b** BOL solder joint

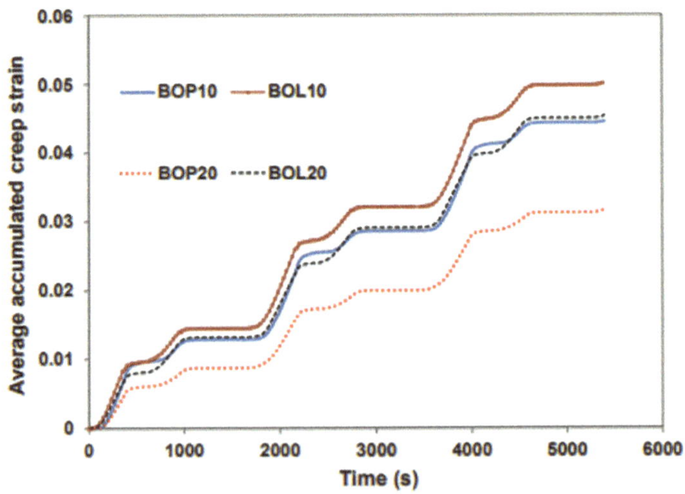

Fig. 1.86 Average accumulated creep strain vs. time

1.14.5 Summary

A LPC TCB process has been developed for the assembly of fine-pitch copper pillar with solder cap flip chip packages. Some important results are summarized in the follows.

- LPC TCB process is a high throughput method; UPH of 1.2 k could be achieved, compared with UPH of ~600 for a conventional TC-flux process.
- The results on cross section examination, interfacial microstructure, shear strength and failure mode demonstrated that excellent wetting and robust solder joint have been achieved using LPC TCB process.
- With a cooling step in the bonding profile, a precise solder height can be obtained based on the predetermined bonding level.
- Without a BH cooling step, the solder height remained at a constant value (the equilibrium level) regardless of the bonding level.
- Solder height can be influenced by the maximum bonding temperature. The higher the bonding temperature the lower the solder height.
- FEA simulation and simple life prediction results indicated that for both BOP and BOL solder joints, the thermal fatigue life increased with increased solder height. Also, their average values are greater than 1,000 cycles. Thus, these solder joints are reliable for most of the operating and environmental conditions.

1.15 Summary

Some important results are summarized in the follows.

- Flip chip is a very important assembly technology for advanced packaging which is shown in Fig. 1.87 for their electrical performance and interconnect density ranking and Fig. 1.88 for their groups such as 2D, 2.1D, 2.3D, 2.5D, and 3D IC integration.
- The manufacturing roadmap (in the next three years) for flip-chip assembly on the organic substrate is given in Fig. 1.89. It can be seen that: (a) for flip chip with mass reflow of C4 or C2 bumps with CUF, the pad-pitch can be as small as 60 μm, (b) for flip chip TCB with small force and reflow of C4 or C2 bumps with CUF, the pad-pitch could be as small as 50 μm, and (c) for flip chip TCB with large force and reflow of C2 bumps with NCP or NCF, the pad-pitch could be as small as 30 μm.
- The manufacturing roadmap (in the next three years) for the pitch of flip chip embedded in fan-out EMC/RDL could be as small as 50 μm
- The manufacturing roadmap (in the next three years) for flip-chip assembly on the silicon chip or substrate is given in Fig. 1.90. It can be seen that: (a) for flip chip with mass reflow of C4 or C2 bumps with CUF, the pad-pitch could be as small as 50 μm, (b) for flip chip TCB with small force and reflow of C4 or C2 bumps with CUF, the pad-pitch could be as small as 40 μm, (c) for flip chip TCB

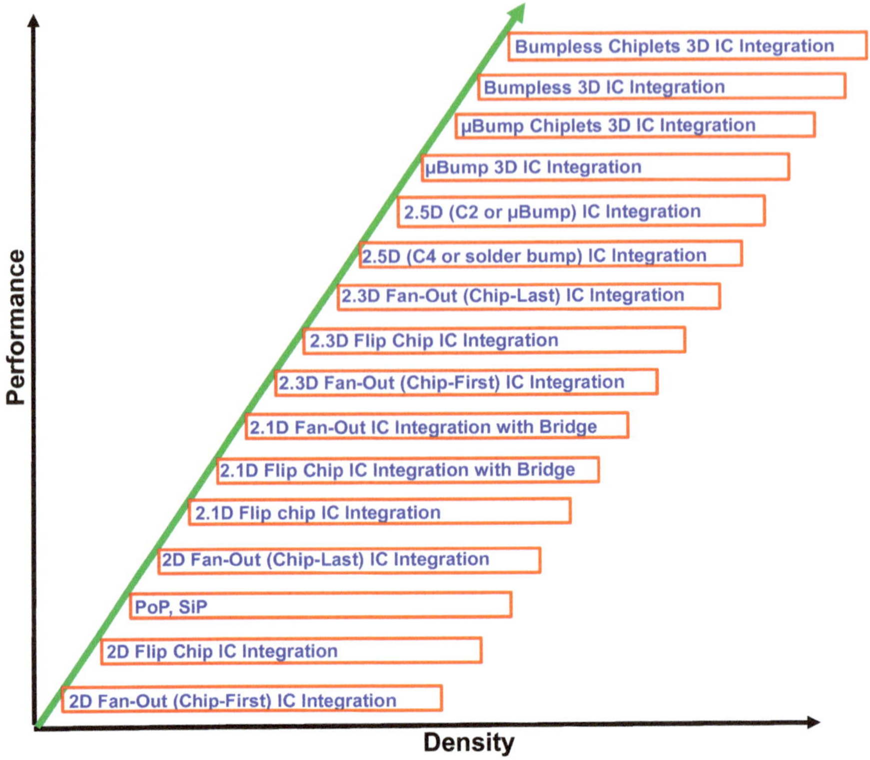

Fig. 1.87 Advanced packaging ranking according to their density and performance

with large force and reflow of C2 bumps with NCP or NCF, the pad-pitch could be as small as 20 μm, and (d) for Bumpless flip chip Cu-Cu hybrid bonding, the pad-pitch could be as small as 4 μm in high-volume manufacturing.

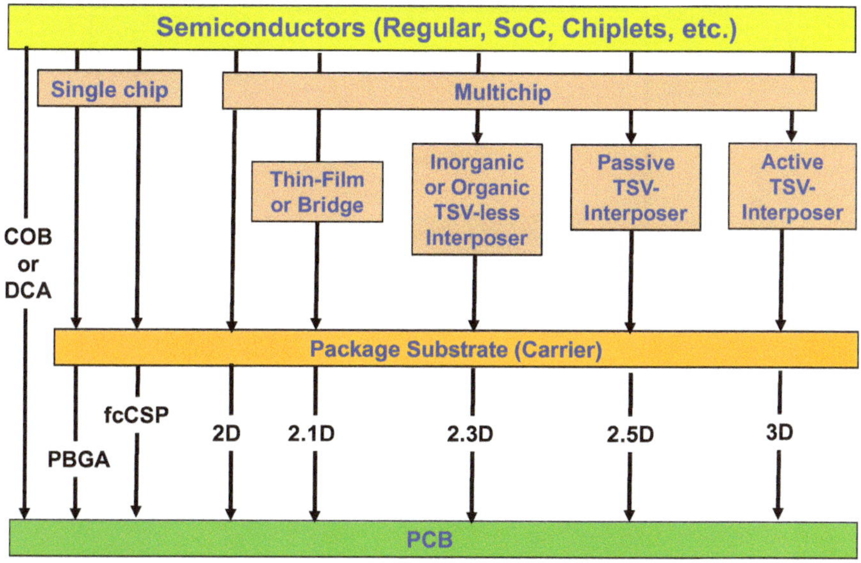

Fig. 1.88 Groups of advanced packaging: 2D, 2.1D, 2.3D, 2.5D, and 3D IC integration

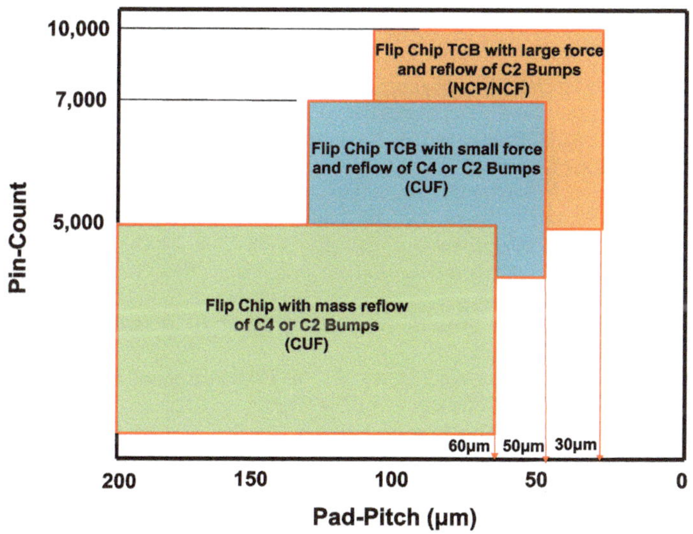

Fig. 1.89 Roadmap (in the next three years) for chip on organic build-up package substrates

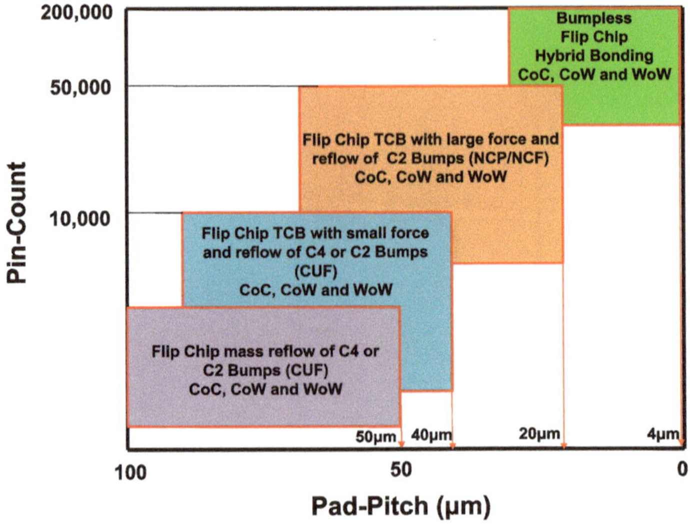

Fig. 1.90 Roadmap (in the next three years) for chip on chip, chip on TSV interposers, chip on silicon wafer, and silicon wafer on silicon wafer

References

1. Lau, J.H. 1996. *Flip chip technologies*, editor. New York: McGraw-Hill Book Company.
2. Lau, J.H. 2000. *Low cost flip chip technologies*. New York: McGraw-Hill Book Company.
3. Lau, J.H., C. Wong, J. Prince, and W. Nakayama. 1998. *Electronic packaging*. New York: McGraw-Hill Book Company.
4. Lau, J.H., N. Lee, C. Wong, and R. Lee. 2003. *Electronic manufacturing*. New York: McGraw-Hill Book Company.
5. Lau, J.H. 2016. *3D IC integration and packaging*. New York: McGraw-Hill Book Company.
6. Lau, J.H. 2013. *Through-silicon vias for 3D integration*. New York: McGraw-Hill Book Company.
7. Lau, J.H. 2011. *Reliability of RoHS-compliant 2D and 3D IC interconnects*. New York: McGraw-Hill Book Company.
8. Lau, J.H., C. Lee, C. Premachandran, and A. Yu. 2010. *Advanced MEMS packaging*. New York: McGraw-Hill Book Company.
9. Lim, S., V. Rao, W. Hnin, W. Ching, V. Kripesh, C. Lee, J.H. Lau, J. Milla, and A. Fenner. 2010. Process development and reliability of microbumps. *IEEE Transactions on CPMT* 33 (4): 747–753.
10. Sutanto, J. 2012. POSSUM^{TM} die design as a low cost 3D packaging alternative. *3D Packaging* 25: 16–18.
11. Davis, E., W. Harding, R. Schwartz, and J. Corning. 1964. Solid logic technology: Versatile, high performance microelectronics. *IBM Journal of Research and Development* 102–114.
12. Totta, P., and R. Sopher. 1969. SLT device metallurgy and its monolithic extension. *IBM Journal of Research and Development* 226–238.
13. Lau, J.H., and Y. Pao. 1997. *Solder joint reliability of BGA, CSP, and flip chip assemblies*. New York: McGraw-Hill Book Company.
14. Lau, J.H., and C. Chang. 2000. Taguchi design of experiment for wafer bumping by stencil printing. *IEEE Transactions on Electronics Packaging Manufacturing* 21 (3): 219–225.

15. Elenius, P., and H. Yang. 1998. The ultra CSP wafer-scale package. *High-Density Interconnect* 1 (6): 36–40.
16. Elenius, P., J. Leal, J. Ney, D. Stepniak, and S. Yeh. 1999. Recent advances in flip chip wafer bumping using solder paste technology. In *Proceedings of IEEE ECTC*, 260–265.
17. Li, L., S. Wiegele, P. Thompson, and R. Lee. 1998. Stencil printing process development for low cost flip chip interconnect. In *Proceedings of IEEE ECTC*, 421–426.
18. Wiegele, S., P. Thompson, R. Lee, and E. Ramsland. 1998. Reliability and process characterization of electroless nickel-gold/solder flip chip interconnect technology. In *Proceeding of ECTC*, 861–866.
19. Kloeser, J., K. Heinricht, E. Jung, L. Lauter, A. Ostmann, R. Aschenbrenner, and H. Reichl. 1998. Low cost bumping by stencil printing: Process qualification for 200mm pitch. In *Proceedings of International Symposium on Microelectronics*, 288–297.
20. Cho, M., S. Kang, Y. Kwon, D. Jang, and N. Kim. 1999. Flip-chip bonding on PCB with electroless Ni-Au and stencil printing solder bump. In *Proceedings of SMTA International Conference*, 159–164.
21. Fowlkes, W., and C. Creveling. 1995. *Engineering methods for robust product design*. Reading, MA: Addison-Wesley.
22. Love, D., L. Moresco, W. Chou, D. Horine, C. Wong, and S. Eilin. 1994. Wire interconnect structures for connecting an integrated circuit to a substrate. *US Patent no. 5,334,804.*
23. Tung, F. 2003. Pillar connections for semiconductor chips and method of manufacture. *US patent* 6 (578): 754.
24. Tung, F. 2003. Pillar connections for semiconductor chips and method of manufacture. *US Patent no. 6,592,019.*
25. Tung, F. 2004. Pillar connections for semiconductor chips and method of manufacture. *US Patent no. 6,681,982.*
26. Tsukada, Y., S. Tsuchida, and Y. Mashimoto. 1992. Surface laminar circuit packaging. In *IEEE Proceedings of Electronic and Components Technology Conference*, 22–27.
27. Tsukada, Y., and S. Tsuchida. 1992. Surface laminar circuit, a low cost high density printed circuit board. In *Proceedings of Surface Mount International Conference*, 537–542.
28. Tsukada, Y. 1994. Solder bumped flip chip attach on SLC board and multichip module. In *Chip on board technologies for multichip modules*, ed. Lau, J. H. Van Nostrand Reinhold, 410–443.
29. Lau, J.H., and S.W.R. Lee. 2001, *Microvias for low cost, high density interconnects.* New York: McGraw-Hill Book Company.
30. Koide, M., K. Fukuzono, H. Yoshimura, T. Sato, K. Abe, and H. Fujisaki. 2006. High-performance flip-chip BGA technology based on thin-core and coreless package substrate. In *IEEE Proceedings of Electronic and Components Technology Conference*, 1869–1873.
31. Sung, R., K. Chiang, Y. Wang, and C. Hsiao. 2007. Comparative analysis of electrical performance on coreless and standard flip-chip substrate. In *IEEE Proceedings of Electronic and Components Technology Conference*, 1921–1924.
32. Lin, E., D. Chang, D. Jiang, Y. Wang, and C. Hsiao. 2007. Advantage and challenge of coreless flip-chip BGA. In *Proceedings of IEEE/IMPACT*, 346–349.
33. Chang, D., Y. Wang, and C. Hsiao. 2007. High performance coreless flip-chip BGA packaging technology. In *IEEE Proceedings of Electronic and Components Technology Conference*,1765–1768.
34. Savic, J., P. Aria, J. Priest, N. Dugbartey, R. Pomerleau, B. Shanker, M. Nagar, J. Lim, S. Teng, L. Li, and J. Xue. 2009. Electrical performance assessment of advanced substrate technologies for high speed networking applications. In *IEEE Proceedings of Electronic and Components Technology Conference*, 1193–1199.
35. Kurashina, M., D. Mizutani, M. Koide, and N. Itoh. 2009. Precision improvement study of thermal warpage prediction technology for LSI packages. In *IEEE Proceedings of Electronic and Components Technology Conference*, 529–534.
36. Wang, J., Y. Ding, L. Liao, P. Yang, Y. Lai, and A. Tseng. 2010. Coreless substrate for high performance flip chip packaging. In *Proceedings of IEEE 11th International Conference on Electronic Packaging Technology & High Density Packaging*, 819–823.

37. Kimura, M. 2011. Shinko officially announces volume production of coreless substrate. *Nikkei Electronics*.
38. Fujimoto, D., K. Yamada, N. Ogawa, H. Murai, H. Fukai, Y. Kaneko, and M. Kato. 2011. New fine line fabrication technology on glass-cloth prepreg without insulation films for package substrate. In *IEEE Proceedings of Electronic and Components Technology Conference*, 387–391.
39. Kim, G., S. Lee, J. Yu, G. Jung, J. Kim, N. Karim, H. Yoo, and C. Lee. 2011. Advanced coreless flip-chip BGA package with high dielectric constant thin film embedded decoupling capacitor. In *IEEE Proceedings of Electronic and Components Technology Conference*, 595–600.
40. Nickerson, R., R. Olmedo, R. Mortensen, C. Chee, S. Goyal, A. Low, and C. Gealer. 2012. Application of coreless substrate to package on package architectures. In *IEEE Proceedings of Electronic and Components Technology Conference*, 1368–1371.
41. Kim, G., J. Yu, C. Park, S. Hong, J. Kim, G. Rinne, and C. Lee. 2012. Evaluation and verification of enhanced electrical performance of advanced coreless flip-chip BGA package with warpage measurement data. In *IEEE Proceedings of Electronic and Components Technology Conference*, 897–903.
42. Nishitani, Y. 2012. Coreless packaging technology for high-performance application. *IEEE/ECT/CMPT Seminar on Advanced Coreless Package Substrate and Material Technologies*, 1–23.
43. Kurashina, M., D. Mizutani, M. Koide, M. Watanabe, K. Fukuzono, and H. Suzuki. 2012. Low warpage coreless substrate for large-size LSI packages. In *IEEE Proceedings of Electronic and Components Technology Conference*, 1378–1383.
44. Kurashina, M., D. Mizutani, M. Koide, M. Watanabe, K. Fukuzono, N. Itoh, and H. Suzuki. 2012. Low warpage coreless substrate for IC packages. *Transactions of the Japan Institute of Electronics Packaging* 5 (1): 55–62.
45. Manusharow, M., S. Muthukumar, E. Zheng, A. Sadiq, and C. Lee. 2012. Coreless substrate technology investigation for ultra-thin CPU BGA packaging. In *IEEE Proceedings of Electronic and Components Technology Conference*, 892–896.
46. Kim, J., S. Lee, J. Lee, S. Jung, and C. Ryu. 2012. Warpage issues and assembly challenges using coreless package substrate. In *IPC APEX EXPO Proceedings*, 1–13.
47. Sakuma, K., E. Blackshear, K. Tunga, C. Lian, S. Li, M. Interrante, O. Mantilla, and J. Nah. 2013. Flip chip assembly method employing differential heating/cooling for large dies with coreless substrates. In *IEEE Proceedings of Electronic and Components Technology Conference*, 667–673.
48. Shimizu, N., W. Kaneda, H. Arisaka, N. Koizumi, S. Sunohara, A. Rokugawa, and T. Koyama. 2013. Development of organic multi chip package for high performance application. In *IMAPS Proceedings of International Symposium on Microelectronics*, 414–419.
49. Baloglu, B., W. Lin, K. Stratton, M. Jimarez, and D. Brady. 2013. Wargape characterization and improvements for IC packages with coreless substrate. In *IMAPS Proceedings of International Symposium on Microelectronics*, 260–284.
50. Sun, Y., X. He, Z. Yu, and L. Wan. 2013. Development of ultra-thin low warpage coreless substrate. In *IEEE Proceedings of Electronic and Components Technology Conference*, 1846–1849.
51. Lin, W., B. Baloglu, and K. Stratton. 2014. Coreless substrate with asymmetric design to improve package warpage. In *IEEE Proceedings of Electronic and Components Technology Conference*, 1401–1406.
52. Liu, W., G. Xiaa, T. Liang, T. Li, X. Wang, J. Xie, S. Chen, and D. Yu. 2015. Development of high yield, reliable fine pitch flip chip interconnects with copper pillar bumps and thin coreless substrate. In *IEEE Proceedings of Electronic and Components Technology Conference*, 1713–1717.
53. Pendse, R., K. Kim, K. Kim, O. Kim, and K. Lee. 2006. Bond-on-lead: a novel flip chip interconnection technology for fine effective pitch and high I/O density. In *IEEE Proceedings of Electronic and Components Technology Conference*, 16–23.

54. Pendse, R. 2008. Bump-on-lead flip chip interconnection. US Patent #7,368,817, filed on Nov. 10, 2004.
55. Ouyang, E., M. Chae, S. Chow, R. Emigh, M. Joshi, R. Martin, and R. Pendse. 2010. Improvement of ELK reliability in flip chip packages using bond-on-lead (BOL) interconnect structure. In *Proceedings of IMAPS International Symposium on Microelectronics*, 197–203.
56. Pendse, R. 2011. Bump-on-lead flip chip interconnection. US Patent #7,901,983, filed on May 26, 2009.
57. Pendse, R., C. Cho, M. Joshi, K. Kim, P. Kim, S. Kim, S. Kim, H. Lee, K. Lee, R. Martin, A. Murphy, V. Pandey, and C. Palar. 2010. Low cost flip chip (LCFC): An innovative approach for breakthrough reduction in flip chip package cost, 1–9.
58. Movva, S., S. Bezuk, O. Bchir, M. Shah, M. Joshi, R. Pendse, E. Ouyang, Y. Kim, S. Park, H. Lee, S. Kim, H. Bae, G. Na, and K. Lee. 2011. CuBOL (Cu-Column on BOL) technology: A low cost flip chip solution scalable to high I/O density, fine bump pitch and advanced Si-nodes. In *IEEE Proceedings of Electronic and Components Technology Conference*, 601–607.
59. Lan, A., C. Hsiao, J.H. Lau, E. So, and B. Ma. 2012. Cu pillar exposed-die molded FCCSP for mobile devices. In *IEEE Proceedings of Electronic and Components Technology Conference*, 886–891.
60. Kuo, F., J. Lee, F. Chien, R. Lee, C. Mao, and J.H. Lau. 2014. Electromigration Performance of Cu pillar bump for flip chip packaging with bump on trace by using thermal compression bonding. In *IEEE Proceedings of Electronic and Components Technology Conference*, 56–61.
61. Li, M., D. Tian, Y. Cheung, L. Yang, and J.H. Lau. 2015. A high throughput and reliable thermal compression bonding process for advanced interconnections. In *Proceedings of IEEE/ECTC*, 603–608.
62. F. Liu, C. Nair, V. Sundaram, and R. Tummala. 2015. Advances in embedded traces for 1.5μm RDL on 2.5D glass interposers. In *IEEE Proceedings of Electronic and Components Technology Conference*, 1736–1741.
63. Chen, C., M. Lin, G. Liao, Y. Ding, and W. Cheng. 2015. Balanced embedded trace substrate design for warpage control. In *IEEE Proceedings of Electronic and Components Technology Conference*, 193–199.
64. Zhang, L., and G. Joseph. 2015. Low cost high performance bare die pop with embedded trace coreless technology and coreless cored build up substrate manufacture process. In *IEEE Proceedings of Electronic Components and Technology Conference*, 882–887.
65. Lu, M. 2014. Challenges and opportunities in advanced IC packaging. *Chip Scale Review* 18 (2): 5–8.
66. Lee, K., S. Cha, and P. Shim. 2016. Form Factor and Cost Driven Advanced Package Substrates for Mobile and IoT Applications. *Proceedings of CSTIC*, 1–4.
67. Oi, K., S. Otake, N. Shimizu, S. Watanabe, Y. Kunimoto, T. Kurihara, T. Koyama, M. Tanaka, L. Aryasomayajula, and Z. Kutlu. 2014. Development of new 2.5D package with novel integrated organic interposer substrate with ultra-fine wiring and high density bumps. *IEEE Proceedings of Electronic and Components Technology Conference*, 348–353.
68. Lau, J.H. 2015. Patent issues of fan-out wafer/panel-level packaging. *Chip Scale Review* 19 (6): 42–46.
69. Lau, J.H., N. Fan, and M. Li. 2016. Design, material, process, and equipment of embedded fan-out wafer/panel-level packaging. *Chip Scale Review* 20 (3): 38–44.
70. Selvanayagam, C., J.H. Lau, X. Zhang, S. Seah, K. Vaidyanathan, and T. Chai. 2009. Nonlinear thermal stress/strain analyses of copper filled TSV (Through Silicon Via) and their flip-chip microbumps. *IEEE Transactions on Advanced Packaging* 32 (4): 720–728.
71. Yu, A., N. Khan, G. Archit, D. Pinjala, K. Toh, V. Kripesh, S. Yoon, and J.H. Lau. 2009. Fabrication of silicon carriers with TSV electrical interconnections and embedded thermal solutions for high power 3-D packages. *IEEE Transactions on CPMT* 32 (3): 566–571.
72. Chai, T., C. Zhang, J.H. Lau, C. Selvanayagam, D. Pinjala, Y. Hoe, Y. Ong, V. Rao, E. Wai, H. Li, E. Liao, N. Ranganathan, V. Kripesh, S. Liu, J. Sun, M. Ravi, C. Vath, and Y. Tsutsumi. 2011. Development of large Die Fine-Pitch Cu/low-*k* FCBGA package with through silicon via (TSV) interposer. *IEEE Transactions on CPMT* 1 (5): 660–672.

73. Yu, A., J.H. Lau, S. Ho, A. Kumar, W. Hnin, W. Lee, M. Jong, V. Sekhar, V. Kripesh, D. Pinjala, S. Chen, C. Chan, C. Chao, C. Chiu, C. Huang, and C. Chen. 2011. Fabrication of high aspect ratio TSV and assembly with fine-pitch low-cost solder microbump for Si interposer technology with high-density interconnects. *IEEE Transactions on CPMT* 1 (9): 1336–1344.
74. Lau, J.H. 2011. Overview and outlook of TSV and 3D integrations. *Journal of Microelectronics International* 28 (2): 8–22.
75. Lau, J.H., and G.Y. Tang. 2012. Effects of TSVs (through-silicon vias) on thermal performances of 3D IC integration system-in-package (SiP). *Journal of Microelectronics Reliability* 52 (11): 2660–2669.
76. Lau, J.H. 2014. Overview and outlook of 3D IC packaging, 3D IC integration, and 3D Si integration. *ASME Transactions Journal of Electronic Packaging* 136 (4): 1–15.
77. Lau, J.H. 2014. The future of interposer for semiconductor IC packaging. *Chip Scale Review* 18 (1): 32–36.
78. Lau, J.H., S. Lee, M. Yuen, J. Wu, J. Lo, H. Fan, and H. Chen. 2013. Apparatus having thermal-enhanced and cost-effective 3D IC integration structure with through silicon via interposer. US Patent No: 8,604,603.
79. Banijamali, B., S. Ramalingam, K. Nagarajan, and R. Chaware. 2011. Advanced reliability study of TSV interposers and interconnects for the 28nm technology FPGA. *IEEE Proceedings of Electronic and Components Technology Conference*, 285–290.
80. Chaware, R., K. Nagarajan, and S. Ramalingam. 2012. Assembly and reliability challenges in 3D integration of 28nm FPGA die on a large high density 65nm passive interposer. In *IEEE Proceedings of Electronic and Components Technology Conference*, 279–283.
81. Banijamali, B., S. Ramalingam, H. Liu, and M. Kim. 2012. Outstanding and innovative reliability study of 3D TSV interposer and fine pitch solder micro-bumps. In *IEEE Proceedings of Electronic and Components Technology Conference*, 309–314.
82. Banijamali, B., C. Chiu, C. Hsieh, T. Lin, C. Hu, S. Hou, S. Ramalingam, S. Jeng, L. Madden, and D. Yu. 2013. Reliability evaluation of a CoWoS-enabled 3D IC package. In *Proceedings of ECTC*, 35–40.
83. Coudrain, P., J. Charbonnier, A. Garnier, P. Vivet, R. Vélard, A. Vinci, F. Ponthenier, A. Farcy, R. Segaud, P. Chausse, L. Arnaud, D. Lattard, E. Guthmuller, G. Romano, A. Gueugnot, F. Berger, J. Beltritti, T. Mourier, M. Gottardi, S. Minoret, C. Ribière, G. Romero, P.-E. Philip, Y. Exbrayat, D. Scevola, D. Campos, M. Argoud, N. Allouti, R. Eleouet, C. Fuguet Tortolero, C. Aumont, D. Dutoit, C. Legalland, J. Michailos, S. Chéramy, and G. Simon. Active interposer technology for chiplet-based advanced 3D system architectures. In *IEEE/ECTC Proceedings*, May 2019, 569–578.
84. Chiu, C., Z. Qian, and M. Manusharow. 2014. Bridge interconnect with air gap in package assembly. *US Patent Application No.* 8,872,349.
85. Sikka, K., R. Bonam, Y. Liu, P. Andry, D. Parekh, A. Jain, M. Bergendahl, et al. 2021. Direct bonded heterogeneous integration (DBHi) Si bridge. In *IEEE/ECTC Proceedings*, June 2021, 136–147.
86. You, J., J. Li, D. Ho, J. Li, M. Zhuang, D. Lai, et al. 2021. Electrical performances of fan-out embedded bridge. In *Proceedings of IEEE/ECTC*, June 2021, 2030–2034.
87. Lee, L., Y. Chang, S. Huang, J. On, E. Lin, and O. Yang. 2021. Advanced HDFO packaging solutions for chiplets integration in HPC application. *Proceedings of IEEE/ECTC*, June 2021, 8–13.
88. Lee, J., G. Yong, M. Jeong, J. Jeon, D. Han, and M. Lee, et al. 2021. S-connect fan-out interposer for next-gen heterogeneous integration. In *Proceedings of IEEE/ECTC*, June 2021, 96–100.
89. Chong, C., T. Lim, D. Ho, H. Yong, C. Choong, and S. Lim, et al. 2021. Heterogeneous integration with embedded fine interconnect. In *Proceedings of IEEE/ECTC*, June 2021, 2216–2221.
90. Hu, Y., Y. Liang, H. Hu, C. Tan, C. Shen, C. Lee, et al. 2023. CoWoS architecture evolution for next-generation HPC on 2.5D system in package. In *IEEE/ECTC Proceedings*, June 2023, 1022–1026.

91. Hou, S., C. Lee, T. Wang, and H. Hu. 2023. Supercarrier redistribution layers to realize ultra-large 2.5D wafer-scale packaging by CoWoS. In *IEEE/ECTC Proceedings*, June 2023, 510–514.
92. Su, P., D. Ho, J. Pu, and Y. Wang. 2023. FO-EB-T package solution for high-performance computing. In *IEEE/ECTC Proceedings*, June 2023, 1038–1042.
93. Lin, V., T. Shih, A. Jang, and Y. Wang. 2023. Fan-out embedded bridge with TSV (FO-EB-T) package characterization and evaluation. In *IEEE/ECTC Proceedings*, June 2023, 1344–1347.
94. Chen, K., S. Lee, P. Andry, C. Tsang, A. Topop, Y. Lin, Y.J. Lu, A. Young, M. Ieong, and W. Haensch. 2006. Structure, design and process control for Cu bonded interconnects in 3D integrated circuits. In *IEEE International Electron Devices Meeting*, 367–370.
95. Liu, F., R. Yu, A. Young, J. Doyle, X. Wang, L. Shi, K. Chen, X. Li, D. Dipaola, D. Brown, C. Ryan, J. Hagan, K. Wong, M. Lu, X. Gu, N. Klymko, E. Perfecto, A. Merryman, K. Kelly, S. Purushothaman, S. Koester, R. Wisnieff, and W. Haensch. 2008. A 300-mm wafer-level three-dimensional integration scheme using tungsten through-silicon via and hybrid Cu-adhesive bonding. In *IEEE International Electron Devices Meeting*, 1–4.
96. Yu, R., F. Liu, R. Polastre, K. Chen, X. Liu, L. Shi, E. Perfecto, N. Klymko, M. Chace, T. Shaw, D. Dimilia, E. Kinser, A. Young, S. Purushothaman, S. Koester, and W. Haensch. 2009. Reliability of a 300-mm-compatible 3DI technology based on hybrid Cu-adhesive wafer bonding. In *Symposium on VLSI Technology*, 170–171.
97. Eitan, A., and K. Jing. 2015. Thermocompression bonding for fine-pitch copper-pillar flip chip interconnect–tool features as enablers of unique technology. In *Proceedings of IEEE/ECTC*, 460–464.
98. Lau, J.H., and C. Chang. 2002. Characteristics and reliability of fast-flow, snap-cure, and reworkable underfills for solder bumped flip chip on low-cost substrates. *IEEE Transactions on Electronics Packaging Manufacturing* 25 (3): 231–230.
99. Lau, J.H., and R. Lee. 2000. Fracture mechanics analysis of low cost solder bumped flip chip assemblies with imperfect underfills. *ASME Transactions Journal of Electronic Packaging* 222 (4): 306–310.
100. Lau, J.H., R. Lee, and C. Chang. 2000. Effects of underfill material properties on the reliability of solder bumped flip chip on board with imperfect underfill encapsulants. *IEEE Transactions on CPMT* 23 (2): 323–333.
101. Lau, J.H., and S.W. Lee. 2000. Effects of underfill delamination and chip size on the reliability of solder bumped flip chip on board. *International Journal of Microcircuit & Electronic Packaging, IMAPS Transactions*, 23 (1), First Quarter, 33–39.
102. Lau, J.H., C. Chang, and C. Chen. 1999. Characteristics and reliability of no-flow underfills for solder bumped flip chip assemblies. *International Journal of Microcircuit & Electronic Packaging, IMAPS Transactions* 22 (4): 370–381.
103. Lau, J.H., and C. Chang. 1999. How to select underfill materials for solder bumped flip chips on low cost substrates? *International Journal of Microelectronics & Electronic Packaging, IMAPS Transactions* 22 (1): 20–28.
104. Lau, J.H., and C. Chang. 1999. Characterization of underfill materials for functional solder bumped flip chips on board applications. *IEEE Transactions CPMT, Part A* 22 (1): 111–119.
105. Lau, J.H., C. Chang, and C. Ouyang. 1998. SMT compatible no-flow underfill for solder bumped flip chip on low-cost substrates. *Journal of Electronics Manufacturing* 8 (3 & 4): 151–164.
106. Lau, J.H., C. Chang, and R. Chen. 1997. Effects of underfill encapsulant on the mechanical and electrical performance of a functional flip chip device. *Journal of Electronics Manufacturing* 7 (4): 269–277.
107. Lau, J.H., and B. Wun. 1995. Characterization and evaluation of the underfill encapsulants for flip chip assembly. *Journal of the Institute of Interconnection Technology* 21 (1): 25–27.
108. Lau, J.H., E. Schneider, and T. Baker. 1996. Shock and vibration of solder bumped flip chip on organic coated copper boards. *Journal of Electronic Packaging, Transactions of ASME* 118 (2): 101–104.

109. Lau, J.H. 1998. Flip chip on PCBs with anisotropic conductive film. *Advanced Packaging, July/August* 1998: 44–48.
110. Lau, J.H., C. Chang, and R. Lee. 2000. Failure analysis of solder bumped flip chip on low-cost substrates. *IEEE Transactions on Electronics Packaging Manufacturing* 23 (1): 19–27.
111. Lau, J.H., R. Lee, S. Pan, and C. Chang. 2002. Nonlinear time-dependent analysis of micro via-in-pad substrates for solder bumped flip chip applications. *ASME Transactions, Journal of Electronic Packaging* 124 (3): 205–211.
112. Lau, J.H., Q. Zhang, M. Li, K. Yeung, Y. Cheung, N. Fan, Y. Wong, M. Zahn, and M. Koh. 2015. Stencil printing of underfill for flip chips on organic-panel and Si-wafer assemblies. In *Proceedings of IEEE/ECTC*, 168–174.
113. Nakano, F., T. Soga, and S. Amagi. 1987. Resin-insertion effect on thermal cycle resistivity of flip-chip mounted LSI devices. In *ISHM Proceedings of the International Symposium on Microelectronics*, 536–541.
114. Lau, J.H., T. Krulevitch, W. Schar, M. Heydinger, S. Erasmus, and J. Gleason. 1993. Experimental and analytical studies of encapsulated flip chip solder bumps on surface laminar circuit boards. *Journal of the Institute of Interconnection Technology* 19 (3): 18–24.
115. Wun, B., and and. Lau, J. H. 1995. Characterization and Evaluation of the Underfill Encapsulants for Flip Chip Assembly. *Circuit World* 21 (3): 25–32.
116. Lai, Y.M., C.K. Chee, E. Then, C.H. Ng, and M.F. Low. 2007. Capillary underfill and mold encapsulantion method and apparatus. *US Patent #7,262,077*.
117. Cremaldi, J., M. Gaynes, P. Brofman, N. Pesika, and E. Lewandowski. 2014. Time, temperature, and mechanical fatigue dependence on underfill adhesion. In *Proceedings on IEEE/ECTC*, 255–262.
118. Gilleo, K., B. Cotterman, and I.A. Chen. 2000. Molded underfill for flip chip in package. In *Proceedings of High Density Interconnects*, 28–31.
119. Rector, L.P., S. Gong, T.R. Miles, and K. Gaffney. 2000. Transfer molding encapsulation of flip chip array packages. *The International Journal of Microcircuits and Electronic Packaging* 23 (4): 401–406.
120. Lee, J.Y., K.S. Oh, C.H. Hwang, C.H. Lee, and R.D. St. Amand. 2009. Molded underfill development for flipstack CSP. In *Proceedings of IEEE/ECTC*, 954–959.
121. Joshi, M., R. Pendse, V. Pandey, T.K. Lee, I.S. Yoon, J.S. Yun, Y.C. Kim, and H.R. Lee. 2010. Molded underfill (MUF) technology for flip chip packages in mobile applications. In *Proceedings of IEEE/ECTC*, 1250–1257.
122. Ferrandon1, C., A. Jouve, S. Joblot, Y. Lamy, A. Schreiner, P. Montmeat, M. Pellat, M. Argoud, F. Fournel, G. Simon, and S. Cheramy. 2013. Innovative wafer-level encapsulation & underfill material for silicon interposer application. In *Proceedings of IEEE/ECTC*, 761–767.
123. Lau, J.H., Q. Zhang, M. Li, K. Yeung, Y. Cheung, N. Fan, Y. Wong, M. Zahn, and M. Koh. 2015. Stencil printing of underfill for flip chips on organic-panel and Si-wafer substrates. *IEEE Trans. CPMT* 5 (7): 1027–1035.
124. Wong, C.P., D. Baldwin, M.B. Vincent, B. Fennell, L.J. Wang, and S.H. Shi. 1998. Characterization of a no-flow underfill encapsulant during the solder reflow process. In *Proceedings of IEEE/ECTC*, 1253–1259.
125. Lee, M., M. Yoo, J. Cho, S. Lee, J. Kim, C. Lee, D. Kang, C. Zwenger, and R. Lanzone. 2009. Study of interconnection process for fine pitch flip chip. In *Proceedings of IEEE/ECTC*, 720–723.
126. Okayama, Y., M. Nakasato, K. Saitou, Y. Yanase, H. Kobayashi, T. Yamamoto, R. Usui, and Y. Inoue. 2010. Fine pitch connection and thermal stress analysis of a novel wafer level packaging technology using laminating process. In *Proceedings of IEEE/ECTC*, 287–292.
127. Honda, K., T. Enomoto, A. Nagai, and N. Takano. 2010. NCF for wafer lamination process in higher density electronic packages. In *Proceedings of IEEE/ECTC*, 1853–1860.
128. Honda, K., A. Nagai, M. Satou, S. Hagiwara, S. Tuchida, and H. Abe. 2012. NCF for pre-applied process in higher density electronic package including 3D-package. In *Proceedings of IEEE/ECTC*, 385–392.

129. Fukushimal, T., Y. Ohara, J. Bea, M. Murugesan, K.-W. Lee, T. Tanaka, and M. Koyanagi. 2012. Non-conductive film and compression molding technology for self-assembly-based 3D integration. In *Proceedings of IEEE/ECTC*, 393–398.
130. Itol, Y., M. Murugesan, H. Kino, T. Fukushima, K. Lee, K. Choki, T. Tanaka, and M. Koyanagi. 2015. Development of highly-reliable microbump bonding technology using self-assembly of NCF-covered KGDs and multi-layer 3D stacking challenges. In *Proceedings of IEEE/ECTC*, 336–341.
131. Choubey, A., E. Anzures, D. Fleming, A. Dhoble, L. Herong, R. Barr, J. Calvert, and J. Oh. 2014. Non-conductive film (NCF) underfill for flip chip assembly and high reliability. In *Proceedings of International Wafer Level Packaging Conference*, 1–7.
132. Lee, D., K. Kim, K. Kim, H. Kim, J. Kim, Y. Park, J. Kim, D. Kim, H. Park, J. Shin, J. Cho, K. Kwon, M. Kim, J. Lee, K. Park, B. Chung, and S. Hong. 2014. A 1.2V 8Gb 8-channel 128GB/s high-bandwidth memory (HBM) stacked DRAM with effective microbump I/O test methods using 29nm process and TSV. In *Proceedings of IEEE/ISSCC*, 433–435.
133. Shin, J., Y. Kim, H. Lee, U. Kang, S. Seo, and K. Paik. 2015. Effects of thermo-compression bonding parameters on joint formation of micro-bumps in non-conductive film (NCF). In *Proceedings of IEEE/ECTC*, 910–915.
134. Lee, H., Y. Choi, J. Shin, and K. Paik. 2015. Wafer level packages (WLPs) using B-stage non-conductive films (NCFs) for highly reliable 3D-TSV micro-bump Interconnection. In *Proceedings of IEEE/ECTC*, 331–335.
135. Hiner, D., D. Kim, S. Ahn, K. Kim, H. Kim, M. Lee, D. Kang, M. Kelly, R. Huemoeller, R. Radojcic, and S. Gu. 2015. Multi-die chip on wafer thermo-compression bonding using non-conductive film. In *Proceedings of IEEE/ECTC*, 17–21.
136. Nonaka, T., Kobayashi, Y., Asahi, N., Niizeki, S., and Fujimaru, K. 2014. High Throughput Thermal Compression NCF Bonding. *Proceedings of Electronic Components & Technol. Conference*, 913–918.
137. Matsumura, K., M. Tomikawa, Y. Sakabe, and Y. Shiba. 2015. New non conductive film for high productivity process. In *IEEE Proceedings of CPMT Symposium Japan*, 19–20.
138. Asahi, N., Y. Miyamoto, M. Nimura, Y. Mizutani, and Y. Arai. 2015. High productivity thermal compression bonding for 3D-IC. In *Proceedings of IEEE International 3D Systems Integration Conference*, TS7.3.1–TS7.3.5.
139. Syed, A. 2004. Accumulated creep strain and energy density based thermal fatigue life prediction models for SnAgCu solder joints. In *Proceedings of Electronic Components & Technol. Conference*, Las Vegas, Nevada, June 1–4, 737–746.

Chapter 2
Cu-Cu Hybrid Bonding

2.1 Introduction

Hybrid bonding has achieved quite a bit of traction recently. For example, the 2023 IEEE Electronic Components and Technology Conference (ECTC) featured more than 80 papers related to hybrid bonding. In [1, 2], it was mentioned that Cu-Cu bumpless hybrid bonding so called direct bond interconnect (DBI) was invented by the Research Triangle Institute (RTI) in US [3–6] around 2000. Actually, at about the same time, Suga and his colleagues in Japan proposed a similar technology called room-temperature bumpless direct bonding [7–16]. The DBI has been in very high-volume manufacturing since 2016 by Sony on their CMOS (complementary metal–oxide–semiconductor) image sensors (CIS) [17–22], and since 2021 by Yangtze Memory Technologies Company Ltd. (YMTC) on their 3-D NAND [23].

The brief fundamentals and recent advances and trends of Cu-Cu bumpless hybrid bonding from the IEEE/ECTC 2022 have been reported in [1, 2]. The state-of-the-arts Cu-Cu bumpless hybrid bonding from the IEEE/ECTC 2023 will be discussed in this chapter.

2.2 Samsung's Hybrid Bonding

During IEEE/ECTC 2023, Samsung published at least 6 papers on hybrid bonding [24–29]. In [24], they presented a multi-stack hybrid Cu bonding (HCB) technology development using ultra-thin chips. As mentioned in Sect. 1.11.2 of this book, Samsung have been in production for 3D IC integration of their high bandwidth memory (HBM) by flip chip high-bonding force TCB (thermocompression bonding) of the C2 (Cu-pillar with solder cap or μbump) TSV (through-silicon via) [30–32] chips with non-conductive film (NCF). Figure 2.1a schematically shows the 16H multi-stack by flip chip TCB solder reflow method. It can be seen that there are joint

J. H. Lau, *Flip Chip, Hybrid Bonding, Fan-In, and Fan-Out Technology*,
https://doi.org/10.1007/978-981-97-2140-5_2

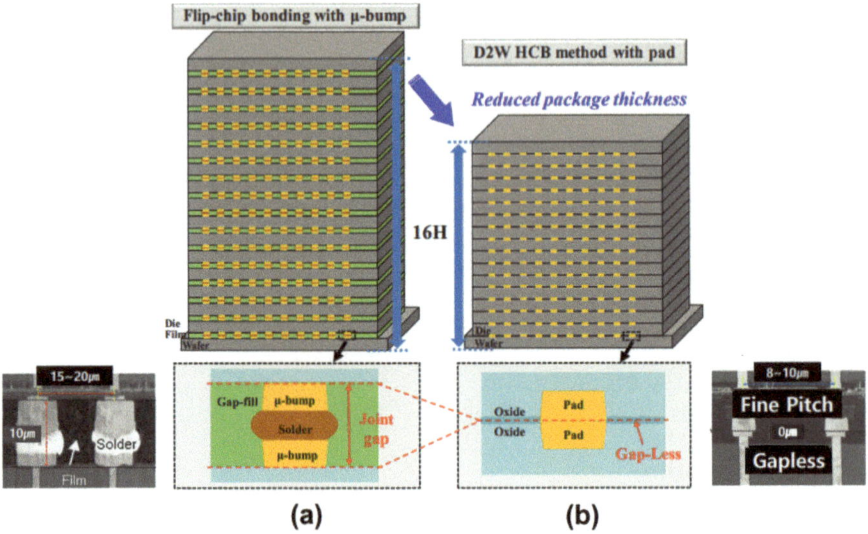

Fig. 2.1 16H HBM structure. **a** Flip chip with microbump solder reflow with joint gap. **b** D2W Cu-Cu hybrid bonding with gapless joint

gaps between the chips. However, with HCB, the chips are bumpless and the Cu-pad to Cu-pad are bonded with D2W (die-to-wafer) and D2D (die-to-die) method [1, 2]. Figure 2.1b schematically shows the 16H multi-stack by HCB method. It can be seen that the total package thickness is smaller than the one with TCB solder bumped flip chip and there is no gap (gap-less) between the chips.

The HCB mechanism is shown in Fig. 2.2a. CMP (chemical–mechanical polishing) is conducted for the bonding surface of the core memory die after dicing and the buffer wafer. Plasma processing on the passivation surface creates an activation state and generates the hydroxyl group (OH) when deionized (DI) water is sprayed onto the surface. Then, the chip is picked and aligned with the wafer (D2W), and the pressure applied for joining generates bindings among the oxides. These methods are repeated to stack several core memory dies as die-to-die (D2D) layers. After pre-bonding is completed by stacking the layers, the bond between oxides is strengthened by high-temperature annealing; the Cu pads of the top and bottom plates expand and meet, electrically connecting as Cu diffusion occurs. Figure 2.2b shows the scanning electron microscope (SEM) image of HCB 16H stacking HBM package.

One of the challenges of HBM is thermal management. Because there are no gaps between the chips, the vertical thermal resistant of the 16H HBM with the HCB is 20% lower than that by the flip chip with gaps for the μbumps and underfill (NCF) as shown in Fig. 2.3 [24].

At ECTC 2023, Samsung published another paper on thermal improvement of HBM with joint thermal resistance reduction for scaling 12 stacks and beyond [25].

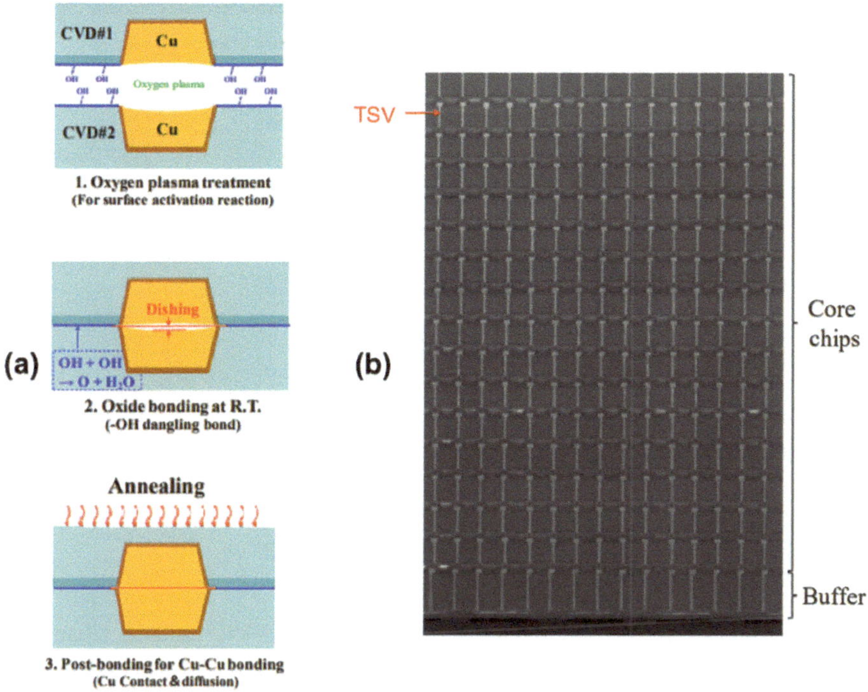

Fig. 2.2 **a** Key hybrid bonding process. **b** 16H HBM structure

Fig. 2.3 Thermal resistance in 16H HBM structure. Conventional flip chip versus hybrid Cu bonding (HCB)

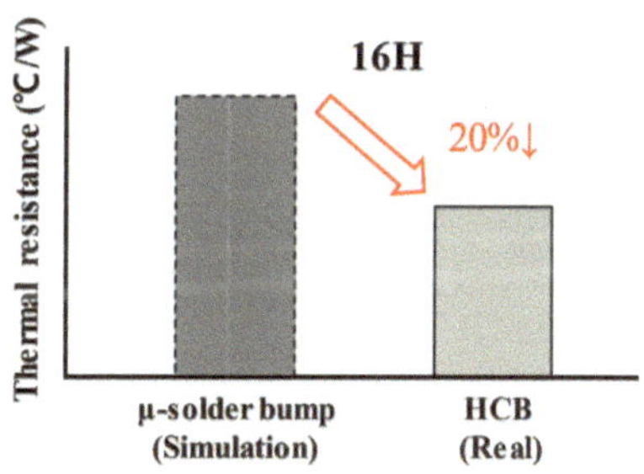

They stated that there are two factors affecting the HBM temperature: (a) internally (HBM itself), thermal resistance, power magnitude and map (distribution), and temperature sensor location, and (b) externally (system), cooling performance, graphics processing unit (GPU), application-specific integrated circuit (ASIC), central processing unit (CPU) power, and system-in-package (SiP) structure. In this paper their focus is on the inside of HBM and dealing with thermal resistance improvement method such as HCB.

Figure 2.4 shows the simulation results on joint thermal resistance with arbitrary unit (A.U.) versus joint gap thickness for mass reflow (MR)-molded underfill (MUF),

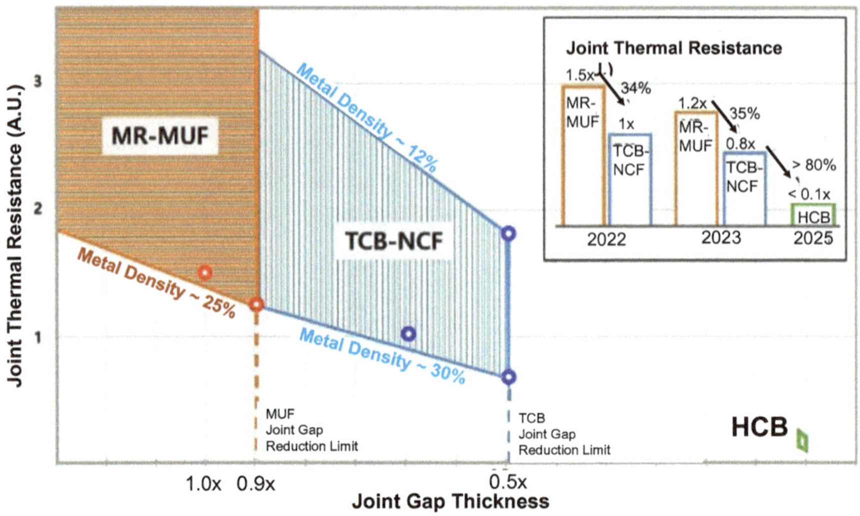

Fig. 2.4 Joint thermal resistance of MR-MUF, TCB-NCF, and HCB

TCB-NCF, and HCB. MR-MUF and TCB-NCF have been discussed in Chap. 1 of this book. It can be seen from Fig. 2.4 that: (a) the joint thermal resistance is the lowest for HCB, (b) the joint thermal resistance is highest for MR-MUF, (c) the reduction of joint thermal resistance from MR-MUF to TCB-NCF is 35%, and (d) the reduction of joint thermal resistance from MR-MUF to HCB is more than 80%. Again, this is because of thinner packaging by the HCB and there are not the Cu-pillar with solder cap and underfill/NCF.

Figure 2.5 shows the simulation and measurement results of the 16H HBM package with the HCB and TCB-NCF methods. It can be seen that (a) the simulation and measurement results correlate very well (more than 95% matched), and (b) the stack thermal resistance fabricated with the HCB is 15~30% less than that with the TCB-NCF.

At ECTC 2023, there is another paper, a study on the surface activation of Cu and oxide for hybrid bonding joint interface [26], published by Samsung. In this paper, they looked at the effects of three different plasma treatments on the bonding strength and adhesion force between the oxide interfaces. In particular, a simple test method is proposed to evaluate oxide interfaces. The surface roughness of these three plasma treatments is shown in Fig. 2.6. It can be seen that Plasma C leads to the smallest Cu-pad and oxide surface roughness.

To evaluate the bonding strength between the oxide interfaces, the first top chip is bonded on the bottom wafer. Then, the bonding is performed by shift bonding the second top chip to make a place for uniform load as shown in Fig. 2.7. The fabricated samples are evaluated by the modified single cantilever beam (SCB) test. By measuring the crack length of ‘a’ from load–displacement graph, the bonding energy *Gc* can be calculated using the equation shown in Fig. 2.7.

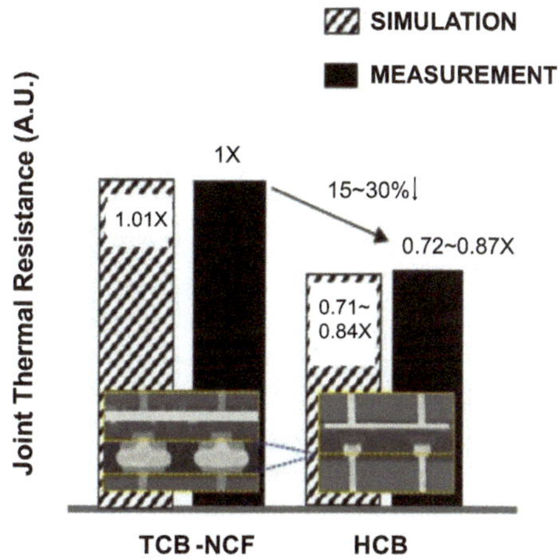

Fig. 2.5 Measured and simulated stack thermal resistance of HBM with TCB and HCB

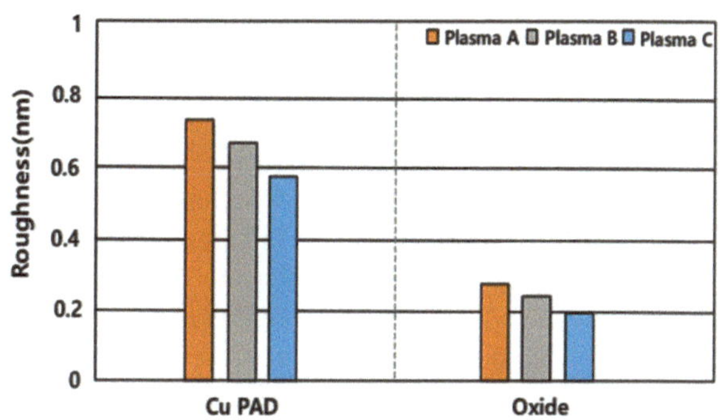

Fig. 2.6 Surface roughness (nm) of Cu pads and oxide according to Plasma A, B, and C

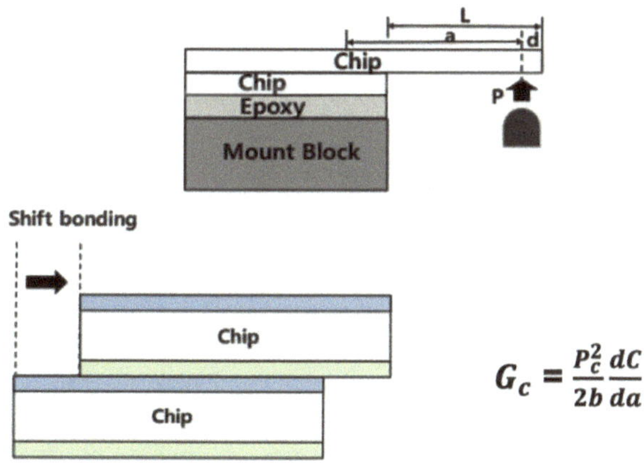

Fig. 2.7 Single cantilever beam (SCB) method for measuring bonding strength. Where *Gc* is the bonding energy, *Pc* is the critical load that is applied at the end of cantilever, '*a*' is the crack length, '*b*' is the width of the beam, and '*C*' is the compliance of the beam

Where *Pc* is the critical load that was applied at the end of cantilever, '*a*' is the crack length, '*b*' is the width of the beam, and '*C*' is the compliance of the beam.

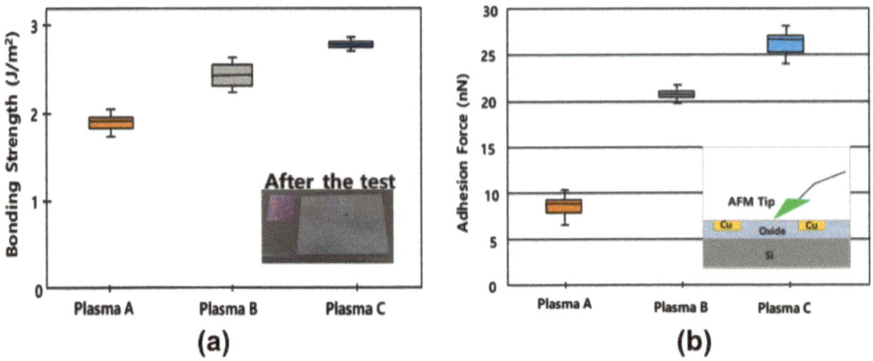

Fig. 2.8 **a** The bonding strength (inset image is the exfoliated top chip and the bottom wafer after the SCB method). **b** The adhesion force according to various Plasma A, B, and C (inset image is the measuring structure for adhesion force)

Figure 2.8a shows the bonding strength according to Plasma A, Plasma B, and Plasma C treatments. The bonding strength is highest (average 2.8 J/m^2) when Plasma C condition (which yields the smallest surface roughness of the Cu-pad and oxide surface) is applied. The top chip and the bottom chip are cleanly separated after measuring the bonding strength with the SCB method, Fig. 2.7.

Figure 2.8b shows the adhesion force measured by an atomic force microscopy (AFM) according to Plasma A, Plasma B, and Plasma C treatments. It can be seen that the highest adhesion force is achieved with the Plasma C treatment. On the other hand, the adhesion force according to the Plasma A treatment is the lowest because this treatment yields the largest surface roughness of the Cu-pad and oxide surface.

Figure 2.9a shows the SEM image of the 16H stacked HBM package structure assembled with Plasma C treatment and the process shown Fig. 2.2a. After annealing, the upper Cu pads and the lower Cu pads contact together through Cu diffusion at the interface, Fig. 2.9b. It can be seen that (a) there is not any void at the interface and (b) the Cu diffusion and grain growth at triple points has been observed at the Cu interface.

Samsung published another paper, die to wafer hybrid Cu bonding for fine pitch 3D-IC applications [27] on the HBM with HCB. In this paper they only considered D2W bonding (the dynamic random-access memory (DRAM) die to the buffer logic wafer). Figure 2.10 shows the HCB process. It can be seen that for both the DRAM die and buffer wafer, they are CMP first to get the controlled dishing of Cu pad and erosion and roughness of dielectrics. For die sawing of the top wafer, care must be taken on minimizing particles. Then, the top dies and the bottom wafers are individually pretreated with oxygen plasma that activates surface reaction of the dielectrics. Finally, the top dies are bonded on the bottom wafer. First, dielectric bonding at the bonding interface at room temperature, and then Cu-to-Cu diffusion bonding at an annealing temperature. Figure 2.11 shows the cross-sectional SEM

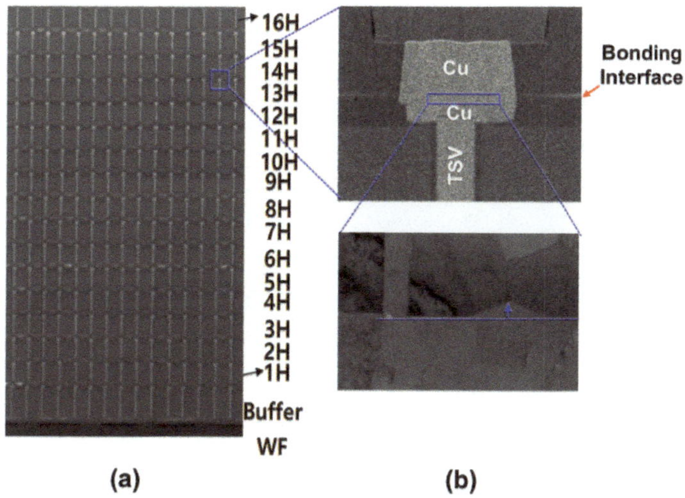

Fig. 2.9 **a** A typical cross sectional SEM image of the 16H HBM structure. **b** SEM image of the Cu-Cu interface showing recrystallized Cu grains across the bonding interface

Fig. 2.10 Key steps in D2W HCB process

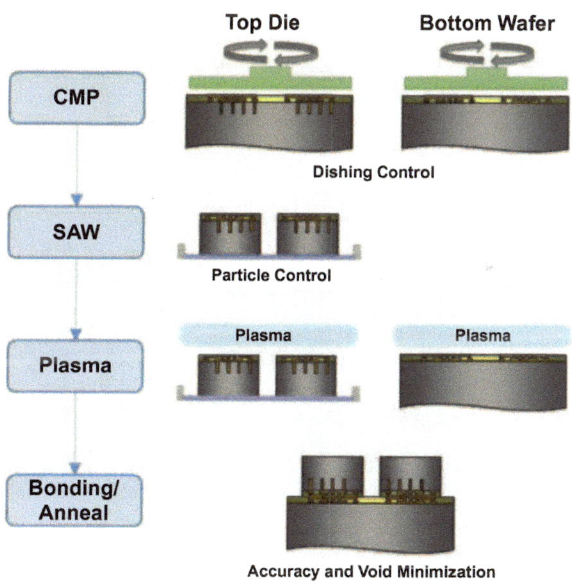

image of the HCB interface. The wafer image of HCB bonding chip is shown in Fig. 2.12a, and the cross-section image of the 3D IC package is shown in Fig. 2.12b.

Figure 2.13 shows electrical connectivity results according to three key factors, namely, dishing, bonding accuracy, and void. Comparing the resistance measurement

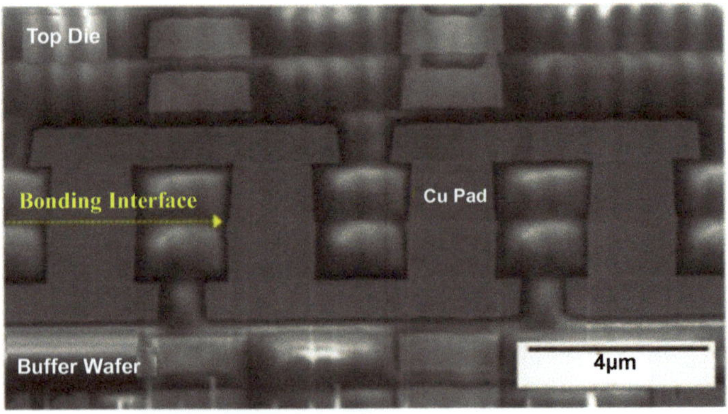

Fig. 2.11 The cross-sectional SEM image of HCB interface

(a)

(b)

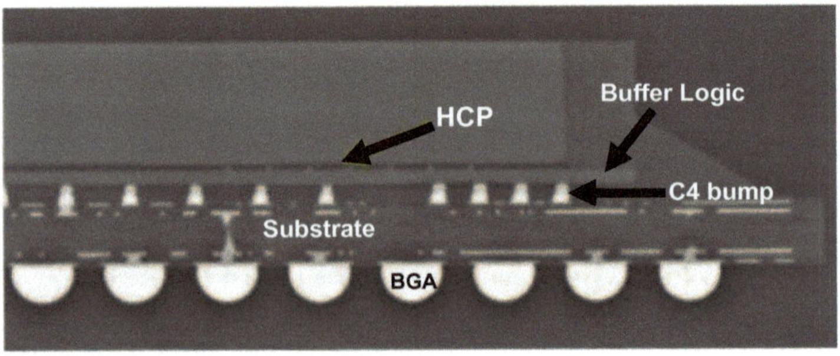

Fig. 2.12 **a** The wafer image of bonding chips. **b** The cross-sectional image of 3D IC package fabricated by D2W HCB

results of Test1 and Test2, the electrical connectivity of the daisy chains is dramatically enhanced when the dishing targeting is satisfied. In the case of Test3, that has reflected the improvements on bonding accuracy and minimization of void defects, the electrical connectivity was 98%.

Key Factors	Test1	Test2	Test3
Dishing targeting	x	0	0
Bonding accuracy	Low	Low	High
Void generation	Many	Many	Few

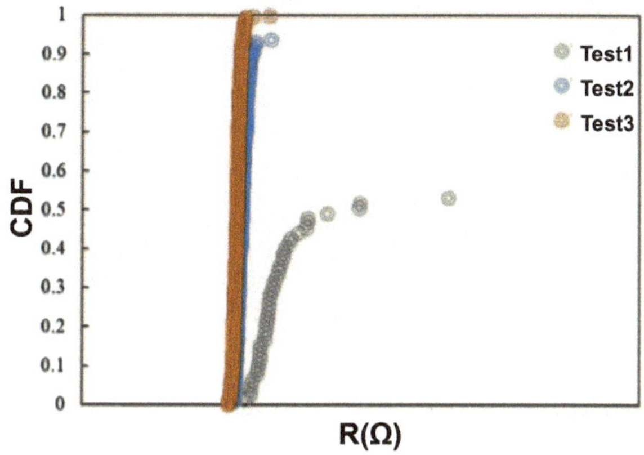

Fig. 2.13 The effects of dishing, bonding accuracy, and void in electrical connectivity of Test1, Test2, and Test3

Another hybrid bonding paper published by Samsung at ECTC2023 is contamination-free Cu/SiCN hybrid bonding process development for sub-μm pitch devices with enhanced bonding characteristics [28]. It is well-known that as the Cu-pad pitch is getting smaller (submicron), there is a concern in the surface activation step. During the plasma treatment process there is a chance for the re-sputtering of Cu caused by ion bombardment from the plasma source as shown in Fig. 2.14. For very small pad-pitch, the re-sputtered atoms may create an electrical path over the dielectric layer and shorten the circuits. In [28], both N_2 and O_2 plasmas have been studied and the material properties of both have also been given.

Figure 2.15 shows the plasma-surface interactions analysis results with molecular dynamics simulations. Figure 2.15a shows the O_2 plasma on SiCN and Fig. 2.15b shows the N_2 plasma on SiCN, where the values in the subset indicate the number of bonds/nm^2. Comparing the O_2 plasma (Fig. 2.5a) and N_2 plasma (Fig. 2.5b), it can be seen that for the N_2 plasma treatment case, little deviation in the number of bonds is observed [28]. Figure 2.15c shows the evaluation of surface inactive and active bond species.

Figure 2.16 shows the TEM (transmission electron microscopy) images of the SiCN-SiCN bonding surfaces with Fig. 2.16a for the O_2 plasma and Fig. 2.16b for the N_2 plasma. It can be seen that the O_2 plasma treatment introduces more oxidation compared to that of N_2 plasma but the dielectric surface oxidation does not necessary correlate to the bonding strength as reported in [28] (the bonding strength of the O_2 plasma is 1.65 J/m^2 and that of the N_2 plasma is 1.80 J/m^2).

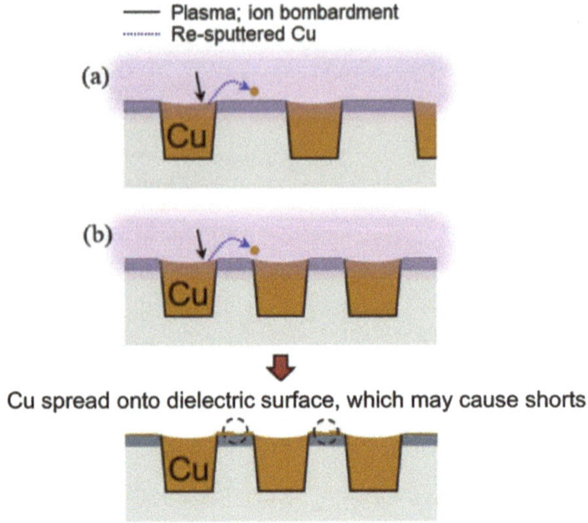

Fig. 2.14 An illustration of surface activation process by plasma causing a re-sputtering of Cu pads. **a** With sufficient distance (large-pitch pads). **b** with fine pitch interconnects which can potentially be problematic due to the electrical pathway over the dielectric surface

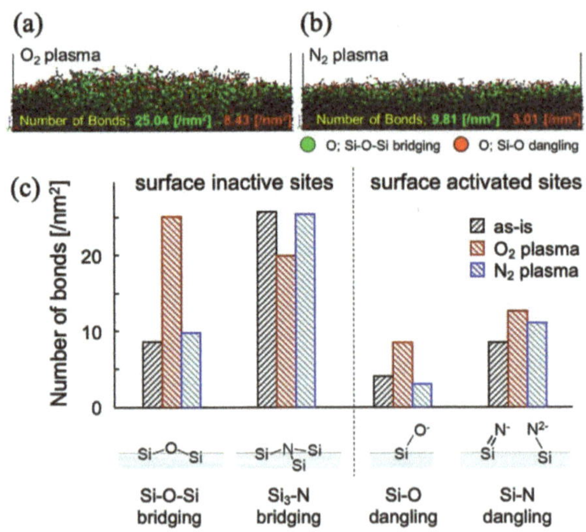

Fig. 2.15 Plasma-surface interactions investigated with molecular dynamics simulations. **a** O_2 plasma on SiCN. **b** N_2 plasma on SiCN where the values in the subset indicate the number of bonds/nm^2 **c** Evaluation of surface inactive and active bond species

Figure 2.17a and b show the TEM images of Cu-Cu bonding interface with Fig. 2.17a for the O_2 plasma treatment and Fig. 2.17b for the N_2 plasma treatment. It can be seen that the top and bottom Cu pads are reasonably in good contact and well-aligned. However, the EDS (energy-dispersive x-ray spectroscopy) images shown in Fig. 2.17c show that O_2 plasma treatment induces an oxidation layer at the Cu-Cu interface. On the other hand, the EDS images shown in Fig. 2.17d show that there is not any oxidation layer at the Cu-Cu interface for the N_2 plasma treatment.

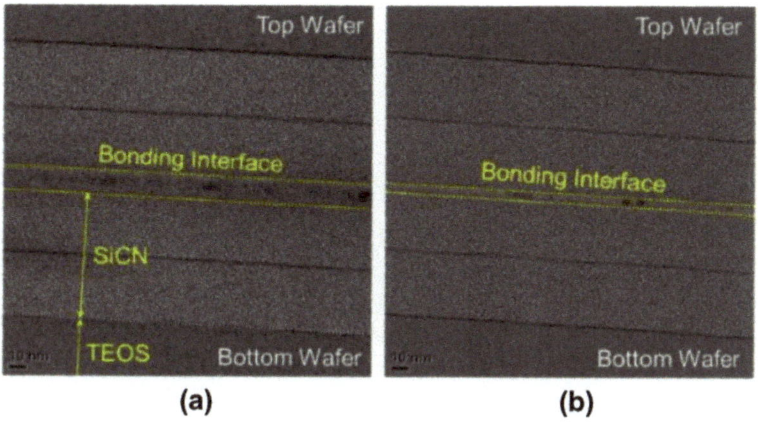

Fig. 2.16 TEM images of SiCN-SiCN bonding interfaces. **a** O_2 reference plasma. **b** N_2 reference plasma

2.3 SK Hynix's Hybrid Bonding

During IEEE/ECTC 2023, SK Hynix presented at least a paper on C2W hybrid bonding interconnect technology for higher density and better thermal dissipation of high bandwidth memory [33]. Their focus is on the demonstration of 8Hi-HBI (hybrid bonding interconnect) with chip-to-wafer (C2W) multi stack HBM as shown in Fig. 2.18. It can be seen that comparing with the solder bumped mass reflow method (Chap. 1 of this book), the thickness of the 8Hi-HBM package can be reduced by 15% with the Cu-Cu bumpless HBI method.

Plasma treatment is a key process step for preparing the bonding surface which makes oxide bonding at room temperature with CMP in HBI. They consider three different kinds of plasma conditions, namely, A, B, and C. Figure 2.19 shows the bonding strength and scanning acoustic microscopy (SAM) images of samples under plasma conditions A, B, and C after end-of-line and reliability test. It can be seen that plasma condition B yields the highest bonding strength which is measured by the Maszara test in wafer-to-wafer base. The annealing temperature is 200 °C.

The thermal resistance of MR + MUF (8Hi) and HBI (8Hi) structures with different bump (pad) density is shown in Fig. 2.20. It can be seen that: (a) for the same bump (pad) density (20%), the thermal resistance of the HBI (8Hi) is 22% lower than that of the MR + MUF (8Hi), and (b) the thermal resistance of the HBI (8Hi) (with 8% pad density) is 13% lower than that of the MR + MUF (8Hi) (with 20% bump density).

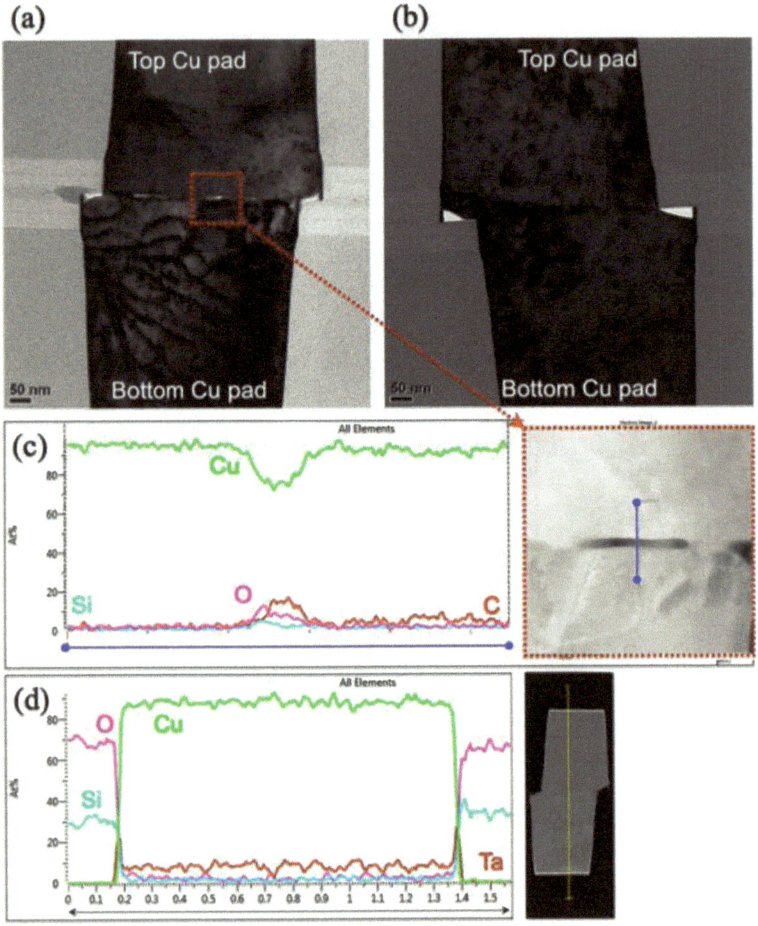

Fig. 2.17 TEM images of Cu-Cu bonding interfaces. **a** O_2 reference plasma and **b** N_2 reference plasma. EDS analysis denotes that O_2 plasma induces an oxidation layer at the Cu-Cu interface **c**, while there is no oxidation layer at the Cu-Cu interface for the N_2 plasma treatment **d**

2.4 Micron's Hybrid Bonding

During ECTC2023, Micron published at least one paper on critical challenges with copper hybrid bonding for chip-to-wafer memory stacking [34]. Figure 2.21a shows a conventional stacking of TSV-dies with μbumps solder reflow method for HBM. The μbump pitch can go down to 20 μm. However, with hybrid bonding, the Cu pad pitch can easily go down to 10 μm and there is no bump, i.e., bumpless as shown in Fig. 2.21b. Figure 2.21c shows the electron backscatter diffraction (EBSD) across the bonding interface where two Cu pads are joined together. Due to Cu-to-Cu diffusion, these two Cu pads become one with the Cu crystal growth occurring

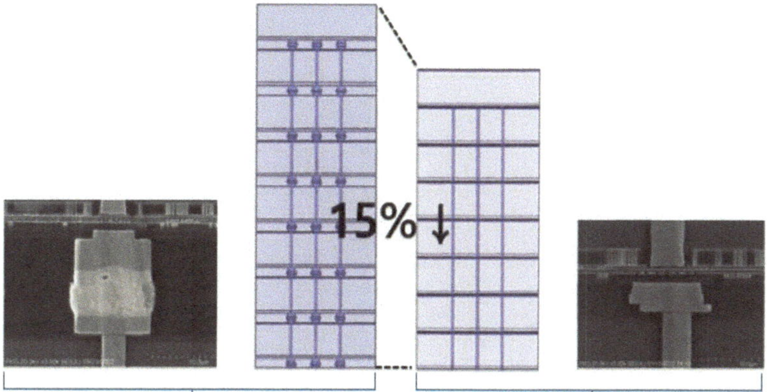

Fig. 2.18 Structural comparison between solder bumped flip chip and hybrid bonding interconnect. (Chip thickness = 58 μm)

Plasma Conditions		A	B	C
Bond Strength		1.0J	2.5J	2.0J
SAM	End of Line @ TV			
	After Reliability @ Real device		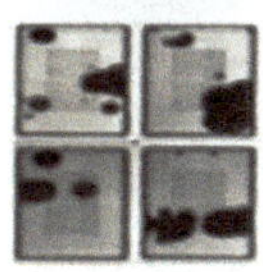	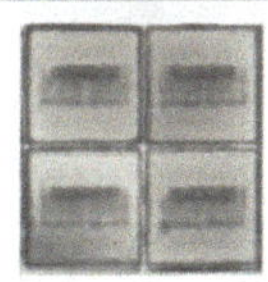

Fig. 2.19 Bonding strength and SAM images of samples under plasma conditions A, B, and C after end-of-line and reliability test

across the bonding interface. However, there are some challenges of hybrid bonding for HBM as pointed out by Micron [34].

(A) Particles Generated by Conventional Wafer Dicing.

Figure 2.22 shows the particle counts before and after a laser stealth dicing, classifications of dicing particles, and signatures of particles. Micron find that laser stealth dicing, Fig. 2.22b has generated far more particles than that allowed by a Cu hybrid

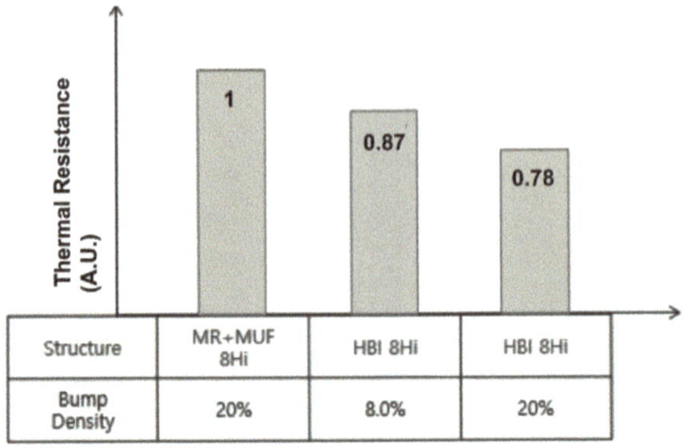

Fig. 2.20 Thermal resistance comparison according to different structures (MUF versus HBI)

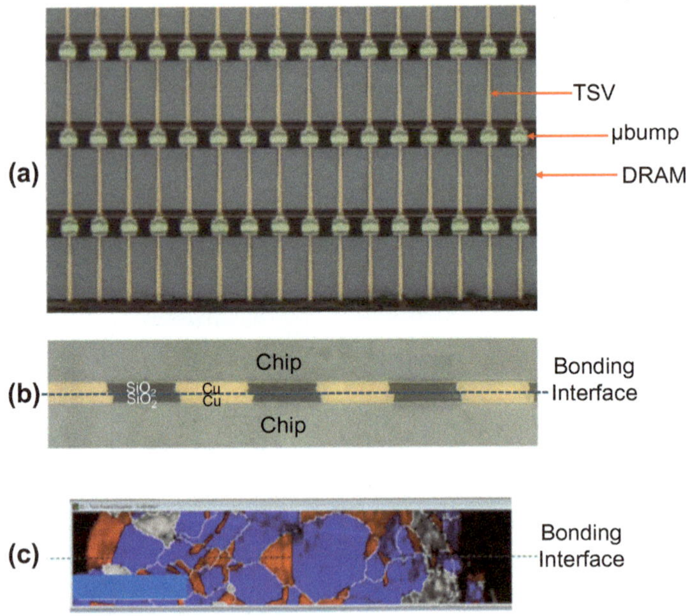

Fig. 2.21 **a** HBM with flip chip solder reflow (20 μm pitch). **b** HBM with hybrid bonding. **c** Cu crystal growth across bonding interface

bonding process, Fig. 2.22a—an increase by 7x. The dicing particles can be classified as inorganics such as silicon dust, Fig. 2.22c, and organics such as tape residues, Fig. 2.22d. Both of them have different impacts on void formulation at bonding interface. The signature of particle impacts on voids are circular shaped as shown in Fig. 2.22e and circular shape attached with a comet tail due to the bonding wave influence as shown in Fig. 2.22f [34].

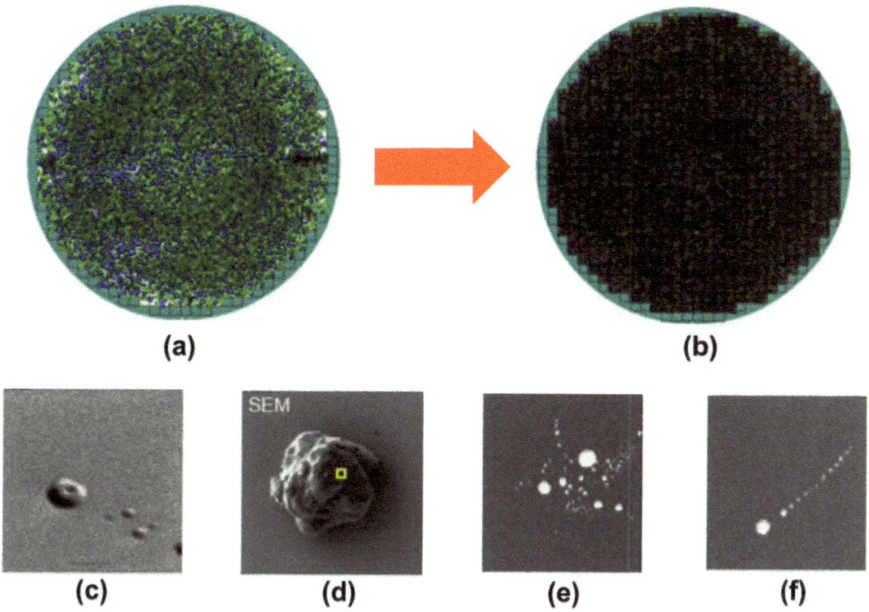

Fig. 2.22 **a** After post-debond clean (particle counts = X). **b** After laser stealth dicing, GPTC wet clean (particle counts = 7X). Classifications of dicing particles: **c** Inorganics and **d** organics. Signatures of particle induced voids: **e** Clusters and **f** comet tails

(B) Particle Mitigation by Combination Laser Grooving and Plasma Dicing.

Figure 2.23a shows a typical memory scribe design with five dielectric films. Micron propose a combination of laser grooving plus plasma etching to achieve an optimum balance between the dicing quality and a reasonable throughput. By doing this an innovative cleaning method to successfully remove the debris and dust (due to lasering) before plasma etching has been developed and the effects are shown in Fig. 2.23b for before cleaning and Fig. 2.23c for post cleaning. The post wafer dicing and cleaning images at saw streets are shown in Fig. 2.23d at wafer center and Fig. 2.23e at wafer edge. The overall particle counts are shown in Fig. 2.23f. It shows that the saw streets are clean, and the particle counts are significantly reduced.

(C) Limitation of Conventional Wafer Support System (WSS).

Figure 2.24a shows the process of a typical conventional WSS. The bottom interface (IF) wafer is usually attached on a temporary carrier wafer with a thick organic glue. Because of the limitation of the thermal budget of the organic glue, the whole process flow must run at a low temperature range (<200 °C). Figure 2.24b shows the severe voids with two different organic glues.

Figure 2.24c shows the effect of low annealing temperatures of less than 200 °C. Although the Cu-Cu pad joint looks good, there is a porous layer formed inside the bonding interface between the chip and the IF wafer dielectric films.

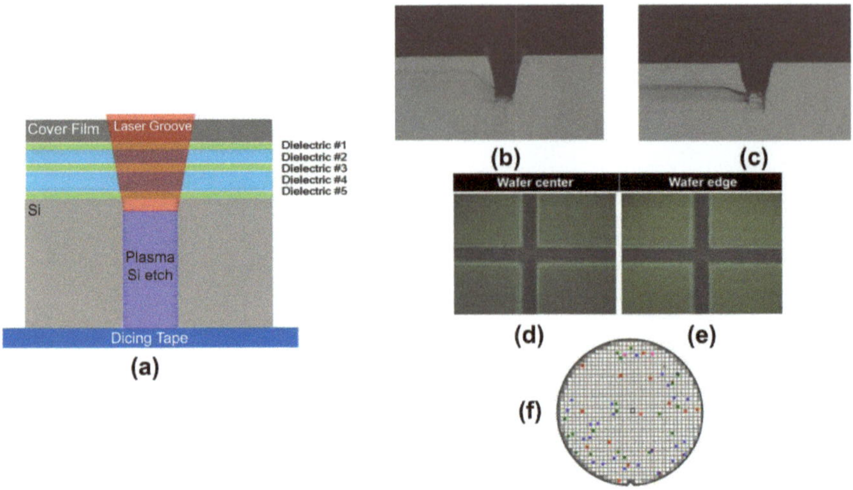

Fig. 2.23 **a** Wafer dicing by laser grooving and plasma etching. Innovative cleaning to remove the laser grooving induced edge burrs and Si dust: **b** before and **c** post cleaning. Post wafer dicing images at saw streets at **d** wafer center and **e** wafer edge. **f** Overall particle counts

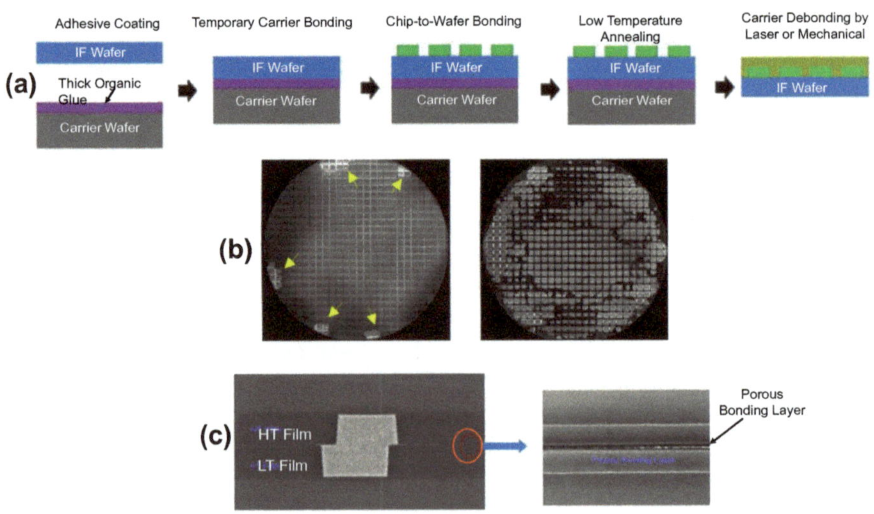

Fig. 2.24 **a** Application of conventional WSS in the chip-to-wafer stacking process flow. **b** Voids (left) and delamination observed in conventional WSS with two different types of organic glues (right). **c** Issues of low temperature annealing with Cu creeping (left) and porous layer between the bonding interface (right)

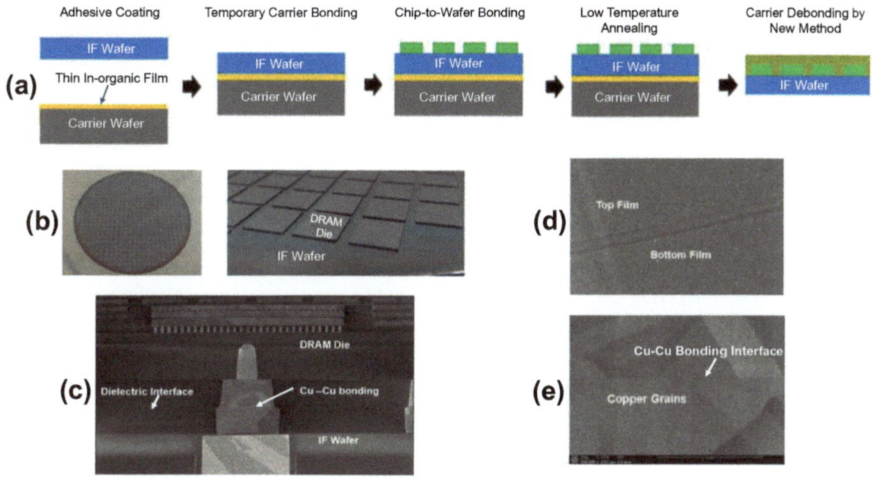

Fig. 2.25 **a** Application of a new WSS in the chip-to-wafer stacking process flow. **b** Finished chip-to-wafer hybrid bonding for a fine-pitch HBM device with pictures of full wafer (left) and local (right) image. **c** Coress-section of copper hybrid bonding between a DRAM die and IF wafer. Magnified images of the bonding interface between **d** top and bottom dielectric films and **e** Cu-Cu joints

(D) Innovative New WSS.

Figure 2.25a shows the new WSS process, which is with a thin inorganic film to connect the IF wafer and the temporary carrier. This film has a much higher temperature budget which allows the annealing temperature to 350 °C. Also, there is no limitation on the types of bonding dielectric films on the backside of DRAMs or IF wafer surface. Furthermore, the Cu pad dishing by CMP is more stable and a thinner package can be achieved. Figure 2.25b shows the finished C2W hybrid bonding for a fine-pitch HBM device. Figure 2.25c shows the SEM cross-section of Cu hybrid bonding between a DRAM die and the IF wafer. The bonding quality of the interface between the top and bottom dielectric films and the Cu grain structure are shown in Fig. 2.25d and e, respectively. It can be seen that both the dielectric-to-dielectric bonding and the copper pad-to-copper pad bonding between the DRAM die, and the IF wafer are properly done.

2.5 Mitsui's Hybrid Bonding

During ECTC 2023, Mitsui published at least two papers on hybrid bonding [35, 36]. In [35], they proposed a novel polymer adhesive for Cu/polymer-dielectric hybrid bonding of multi-chip stacking process as shown in Fig. 2.26. The adhesive can be temporarily bonded to SiO_2, SiN, SiCN and itself at room temperature (RT) even

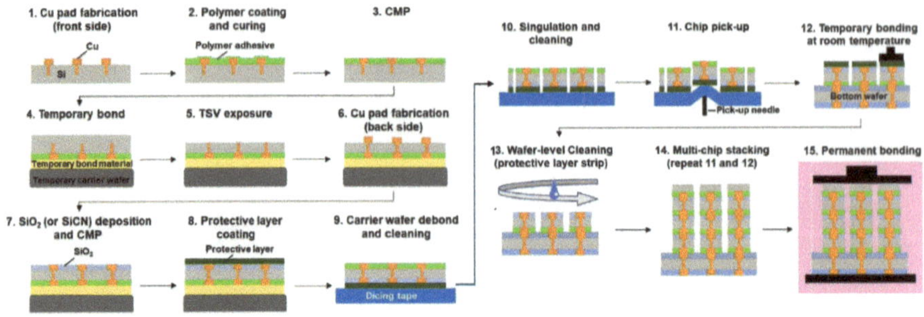

Fig. 2.26 Schematic images of process steps for the proposed multi-chip stacking in this work

after curing and CMP process. Strong permanent bonding strength can be obtained by further annealing.

The fabrication process of the test wafer with Cu pads on both sides is shown in Fig. 2.27a. It can be seen that the Cu pad is 10 μm-diameter, 3 μm-height, and 40 μm-pitch. The polymer is spin-coated and cured, and then CMP. It is followed by thinning the wafer to 200 μm-thick and attaching the wafer to a temporary glass wafer. Then, Cu pads are formed on the back side of the wafer, and it is followed by SiO_2 (about 4 μm) deposition, and the Cu surface is exposed by CMP. Finally, a positive-type photoresist is spin-coated on the Cu/SiO_2 surface as a protective layer and UV-exposed [35]. Figure 2.27b shows the stacking of 4-layer of chips (a total of 8 chips) on a silicon wafer. Figure 2.27c shows the C-SAM images of the interface between the bottom wafer and the first-layer chip and Fig. 2.27d shows that between the third-layer and fourth-layer chips. It can be seen that because of the polymer adhesive there is no void even blade dicing is used.

The process of multi-chip stacking is shown in Fig. 2.28a. After temporary mounting 8 chips on the bottom wafer at RT, TCB is performed at 350 °C, 5N/mm^2, and for 90 min in N_2. Figure 2.28b shows the SEM cross-section images after permanent bonding. It can be seen that there are no voids at any of the bonding interfaces, and the Cu-Cu bonding is without any gaps.

Mitsui published another paper on a new adhesive for CoW (chip-on-wafer) Cu-Cu hybrid bonding with high throughput and room temperature pre-bonding [36]. In this paper, they proposed a unique adhesive for high throughput and low-temperature CoW hybrid bonding. Their process is shown in Fig. 2.29 and MA is the name of the new adhesive. The effects of MA with plasma to SiO_2, MA without plasma to SiO_2, and SiO_2 to SiO_2 without plasma on the surface energy are shown in Fig. 2.30a. It can be seen that the surface energy (bonding strength) is the largest for MA with plasma treatment to SiO_2.

There are two MAs, namely MA-0200 and MA-1200. The coefficient of thermal expansion at 30–150 °C, elastic modulus at RT, 5% weight loss temperature, and Glass transition temperature are respectively 100×10^{-6}/°K, 6GPa, > 400 °C, and 170 °C for MA-0200 and 60×10^{-6}/°K, 6GPa, > 400 °C, and > 400 °C for MA-1200. It can be seen by Fig. 2.30b that for all the baking temperatures the thickness

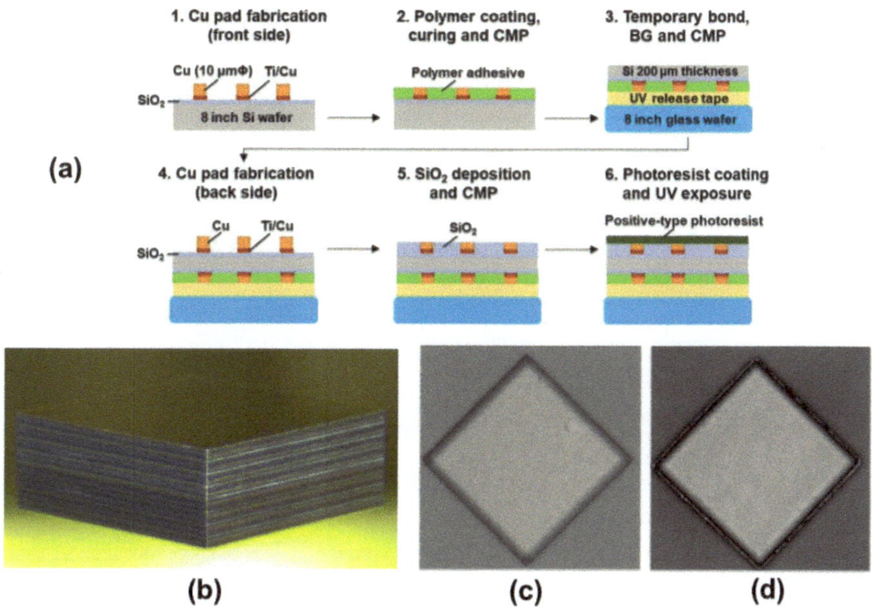

Fig. 2.27 **a** The fabrication process of test wafers with Cu pads on both sides. **b** Optical microscope image of the four-layer chips stacked wafer. C-SAM images of the interface between **c** the bottom wafer and the first layer chip and **d** the third- and fourth-layer chips

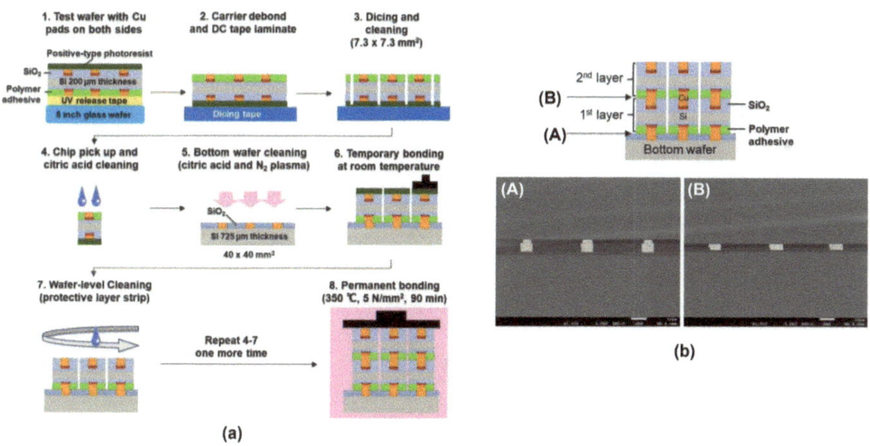

Fig. 2.28 **a** Schematic diagram of process steps for multi-chip stacking using test wafers with Cu pads on both sides. **b** Cross-sectional SEM images of a chip after thermal compression bonding

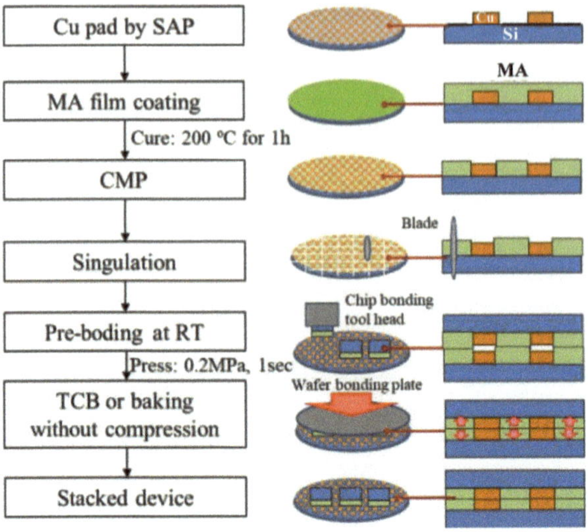

Fig. 2.29 Chip on wafer standard process by using MA. At the TCB or baking process MA shrinks slightly then the two Cu contacts and are compressed

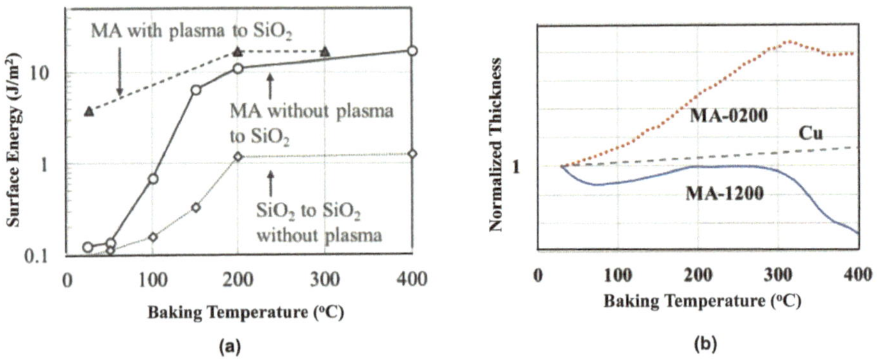

Fig. 2.30 **a** Surface energy as a function of baking temperature. The connecting lines are a guide for the eye. **b** Measured thickness of MA (200 °C cured) as a function of baking temperature. The Cu data are calculated data

of MA-0200 is thicker than that of the Cu pad (10 μm-diameter and 430 nm-thick) and the thickness of Cu is thicker than that of the MA-1200. Figure 2.31 shows the top view and cross-sectional view of a hybrid bonding with compression (5 MPa, 400 °C for 30 min) sample with the MA-0200 adhesive. There is no void or seam at the MA to MA bonding interface. Figure 2.32a and b show the cross-section SEM images of a hybrid bonding without compression (350 °C for 90 min) sample with the MA-1200 adhesive and Fig. 2.32c shows the EBSD Cu grain map of the same sample. It can be seen that the bonding interface is reasonably well done.

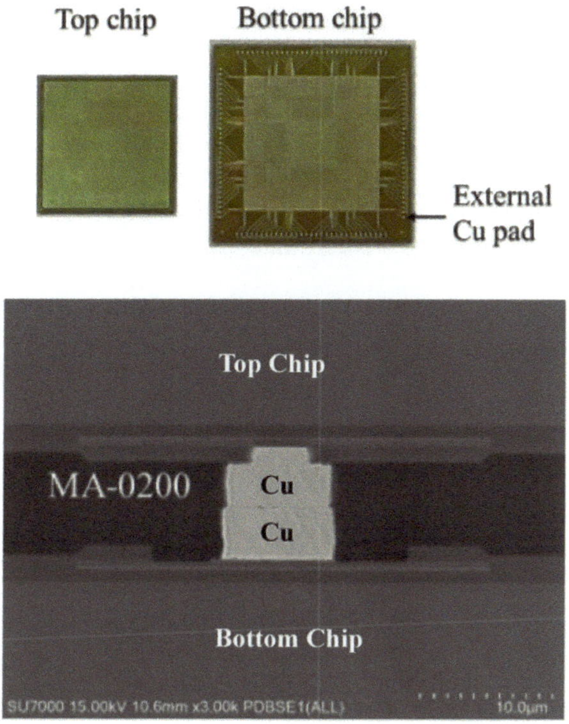

Fig. 2.31 **a** Optical microscope images of top (left) and paired bottom (right) chips with daisy chains. **b** Cross sectional SEM images of the Cu-Cu hybrid bonding with MA-0200

2.6 Sony's Hybrid Bonding

During ECTC2023, Sony published at least four hybrid bonding papers [37–40]. In [37] Sony presented a Cu-Cu wiring: the novel structure of Cu-Cu hybrid bonding. Figure 2.33a shows the conventional Cu-Cu hybrid bonding structure. Figure 2.33b shows the novel Cu-Cu wiring structure, where only Cu patterns at boning interface are shown and the vias are omitted. In the conventional structure, the Cu-Cu connection pads at bonding interface are used as just electrical contacts. On the new structure, besides the Cu-Cu pad connections, the Cu-Cu wirings at bonding interface are used as normal interconnects as shown in Fig. 2.33b.

The basic structure of the Cu-Cu hybrid bonding wiring is shown in Fig. 2.34a. The line width (L1) is 0.7 μm and spacing is 0.7 μm and the minimum pitch is 1.4 μm. The length of Cu (L2) is variable as shown in Fig. 2.34a. The process flow of the Cu-Cu hybrid bonding wiring structure is shown Fig. 2.34b. It can be seen that after the Cu connections are fabricated at the top layer of each wafer, then upper wafer and lower wafer are bonded. If there are three wafers stacked, the first two wafers are face-to-face bonded, and the last wafer is face-to-back bonded. Figure 2.34c shows the Cu connections before hybrid bonding and after bonding. Figure 2.35 shows an example of Cu-Cu wiring interconnects by hybrid bonding (which effectively used as low resistance wiring to prevent infrared (IR) drop) for CIS applications. It can

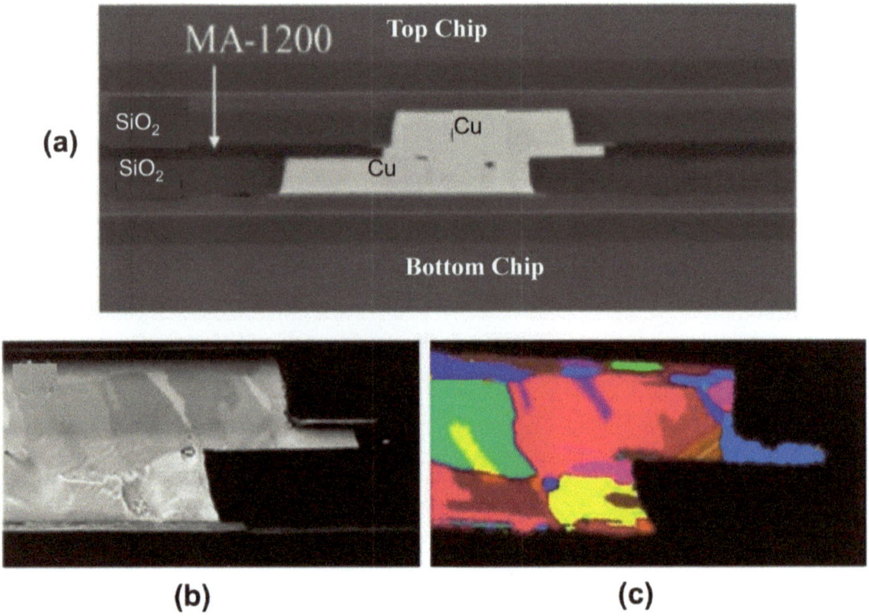

Fig. 2.32 **a** and **b** Cross-sectional SEM images of the Cu-Cu hybrid bonding layer. **c** EBSD Cu grain map of the same sample. The bonding condition is 350 °C for 90 min without pressing. The thickness of MA-1200 is 0.5 μm and the top chip's SiO_2 thickness is 1.5 μm, and the bottom chip's SiO_2 thickness is 2 μm

be seen that there is no void along the Cu-Cu wiring interconnect bonding interface [37].

Sony published another paper at ECTC2023 on development of high reliability 6 μm-pitch Cu-Cu connections using over-400mm^2-large chip on wafer bonding process [38]. Figure 2.36 shows the difference between CoW and WoW hybrid bonding process [38]. It can be seen that for CoW process it includes the additional dicing step and special pretreatment step, and then chip stacking step. Developing a pretreatment process for high-cleanness chip surface is utmost important to remove dust from the dicing process. Because the residue of dust causes voids after CoW bonding as shown in Fig. 2.37a and even larger voids after CoW bonding anneal (due to outgassing) as shown in Fig. 2.37b. This effect is compounded with larger chip size (area) as shown in Fig. 2.37c. This is due to the longer bonding wave propagation distance and the importance of controlling the propagation.

Figure 2.38a shows CoW bonding voids before optimization (left) and after optimization (right) [38]. It can be seen that there are many voids before the optimization of the CoW process while there is no void after optimization. Sony's optimization involves the optimal pretreatment process which focuses on removal of the dust from the dicing process and the optimal bonding process which focuses on the control of the bonding wave propagation for large chips.

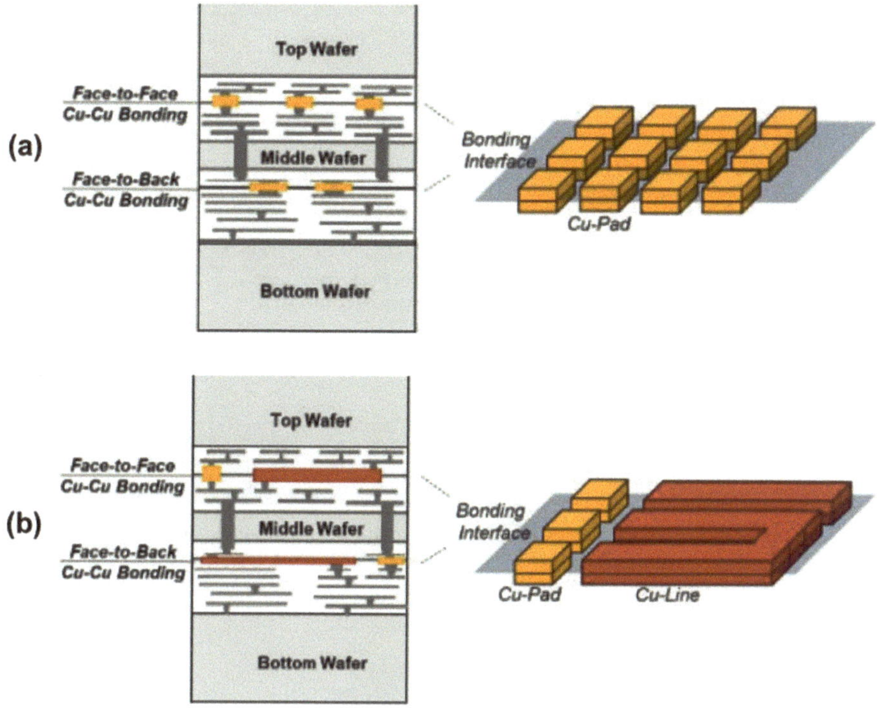

Fig. 2.33 Cross-section images and schematic diagrams of **a** conventional Cu-Cu hybrid bonding structure and **b** novel Cu-Cu wiring structure. In the schematic diagrams, only Cu patterns at bonding interface are shown (vias are omitted)

(A) Optimal Pretreatment Process

Sony's pretreatment process of CoW is tried to have the chip surface for CoW bonding would be as high cleanness as wafer surface for WoW bonding. This could be achieved with a protective film before dicing and plasma cleaning after dicing. Figure 2.38b shows the chips before optimization (there are voids after annealing) and after optimization (void-free after annealing).

(B) Optimal Bonding Process

The bonding substrate shapes between CoW (rectangular) and WoW (round) are different. Chip bonding voids could occur if the bonding wave propagation is not controlled. Thus, an optimal bonding process in controlling of the bonding wave propagations is very important. This is especially for large chips. Figure 2.38c shows the CoW bonding voids without and with optimal bonding process.

Figure 2.39a shows the cross-section SEM images of CoW hybrid bonding of a lager chip over-400 mm^2 with optimal pretreatment process and optimal bonding process. Figure 2.39b shows the unique Cu grain map image by ASTAR TEM analysis showing recrystallized Cu grains across the bonding interface.

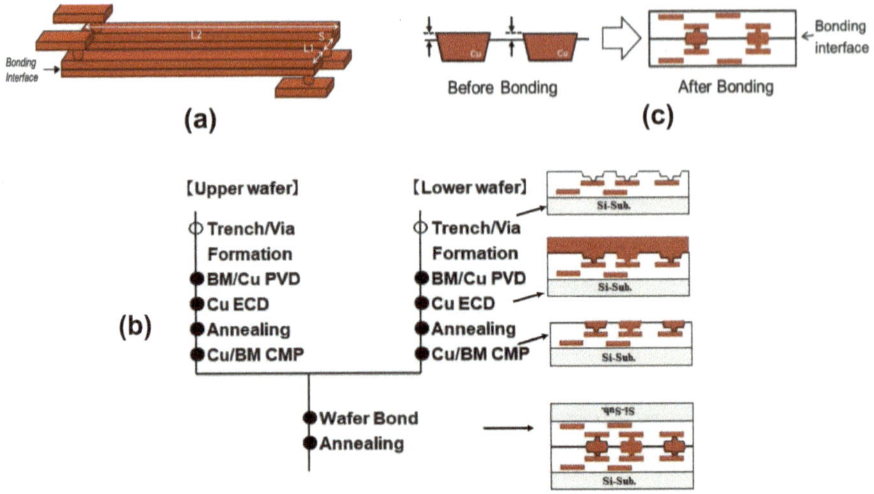

Fig. 2.34 **a** Basic Cu-Cu wiring structure. L1, L2 and S are line width, line length and space respectively. **b** Process flow of Cu-Cu hybrid bonding (left). **c** Cu connection pad structure for ultra-fine pitch Cu-Cu hybrid bonding

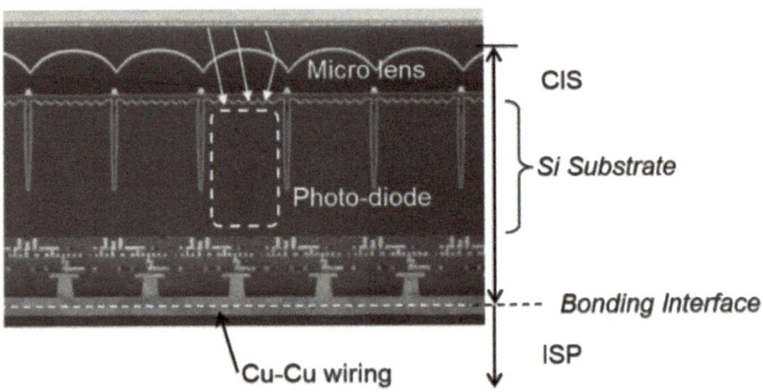

Fig. 2.35 Cross section of stacked CIS with Cu-Cu wiring. CIS chip and Cu-Cu wiring are only shown in this view. White arrows are images of incidental light

2.7 CEA-Leti's Hybrid Bonding

CEA-Leti published at least two papers on hybrid bonding at ECTC 2023 [41, 42]. Demonstration of a wafer level face-to-back (F2B) fine pitch Cu-Cu hybrid bonding with high density TSV for 3D integration Applications has been presented in [41].

CEA-Leti has been working on a Cu-Cu hybrid bonding based on a technology combined with high-density TSVs in order to achieve a 3-layer functional device as shown in Fig. 2.40a and with its 3-layer cross section shown in Fig. 2.40b. However, in [41], they decided to start with a more simplified F2B test structure as an intermediate

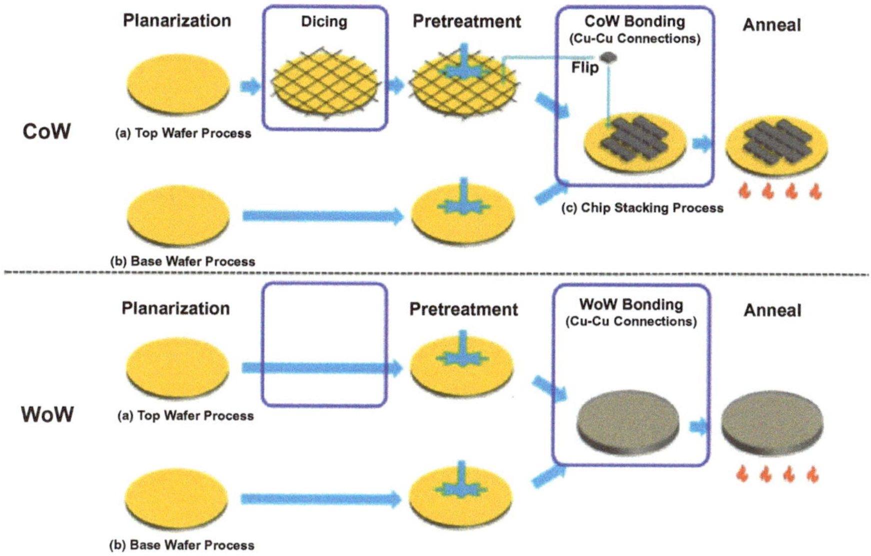

Fig. 2.36 CoW bonding process flow versus WoW bonding process flow. **a** top wafer process, **b** base wafer process, and **c** chip stacking process

Fig. 2.37 CoW bonding process challenges. **a** voids after CoW bonding, **b** voids after CoW anneal bonding, and **c** difficulty of large chip area bonding

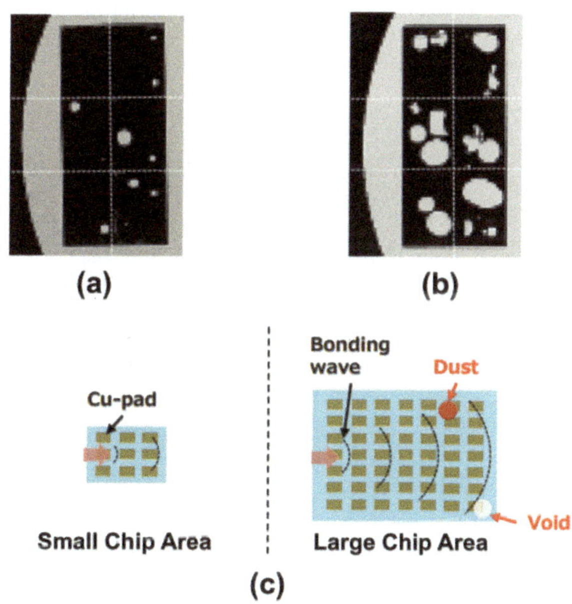

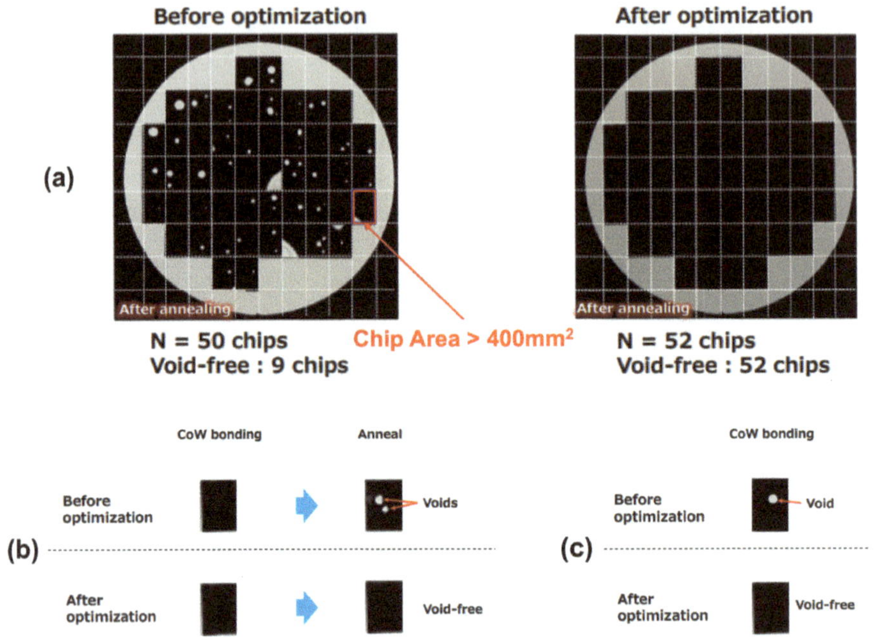

Fig. 2.38 **a** CoW bonding voids without (left) and with optimal pretreatment (right) process and bonding process. **b** CoW bonding voids without and with optimal annealing process. **c** CoW bonding voids without and with optimal pretreatment process

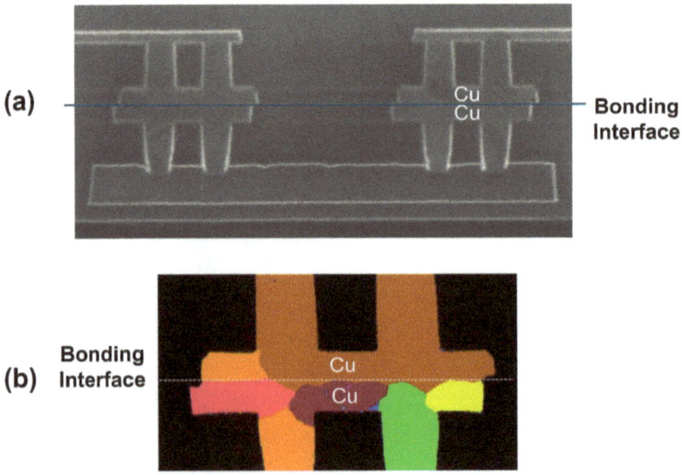

Fig. 2.39 Unique copper grain map image by ASTAR TEM analysis showing recrystallized copper grains across the bonding interface

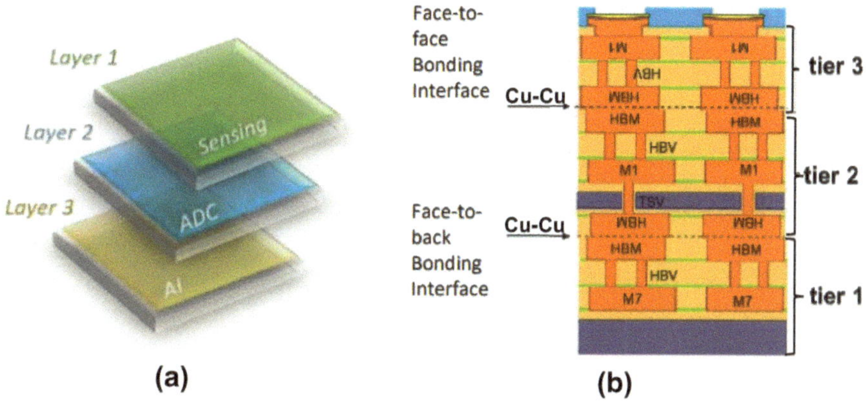

Fig. 2.40 **a** 3-Layers final device scheme. **b** 3-Layers test vehicle cross-section

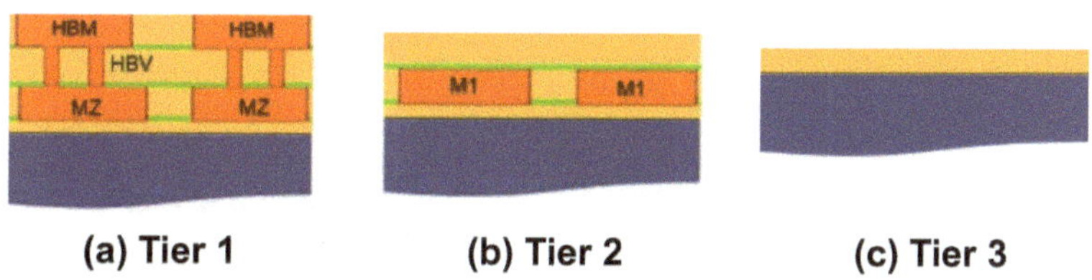

Fig. 2.41 Cross-section of tier 1 **a**, tier2 **b** and tier 3 **c**

step to the 3-layer structure. Figure 2.41 shows the three basic tiers of the F2B structure. Tier 1 is formed by three levels: a line level (MZ), a hybrid bonding via (HBV) level, and a hybrid bonding metal (HBM) level. Tier 2 is formed by a metal line (M1) level encapsulated by the dielectric layer. Tier 3 is just a thin oxide layer.

Figure 2.42 shows the process flow of the F2B structure. It can be seen that, first Tier 2 and Tier 3 direct oxide/oxide bonding, Fig. 2.42a, and then thin down the Tier 2 to 10 μm, Fig. 2.42b. It is followed by TSV (10 μm-diameter) fabrication, Fig. 2.42c, and hybrid bonding (HB) Cu-pad fabrication, Fig. 2.42d. F2B hybrid bonding is made between Tier 1 and Tier 2 or Tier 3, Fig. 2.42e. Then, the silicon carrier is removed from Tier 2 or Tier 3, Fig. 2.42f. Finally, the passivation is opened for electrical testing, Fig. 2.42g. The focused ion beam (FIB)-SEM image of the final F2B test structure is shown in Fig. 2.43a. The F2B stack consists of the M1 (metal line level), the TSV level, two HBM (hybrid bonding metal) levels, the HBV (hybrid bonding via) level and the final MZ (metal level). The hybrid bonding interface between the HBM and the bottom of the F2B test structure are shown in Fig. 2.43b. Figure 2.43c and d show the FIB-SEM images of the top TSV and metal line level (M1), Fig. 2.43c, and the bottom TSV and HBM level, Fig. 2.43d.

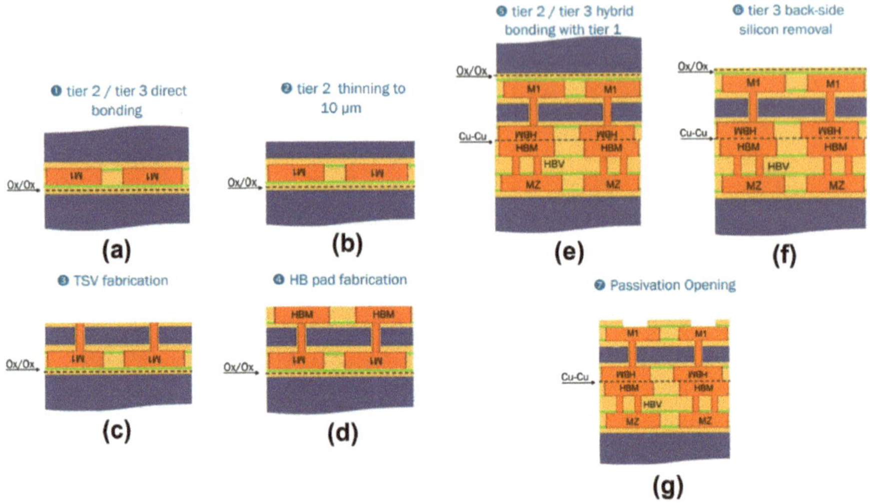

Fig. 2.42 Different stages of F2B test vehicle fabrication process

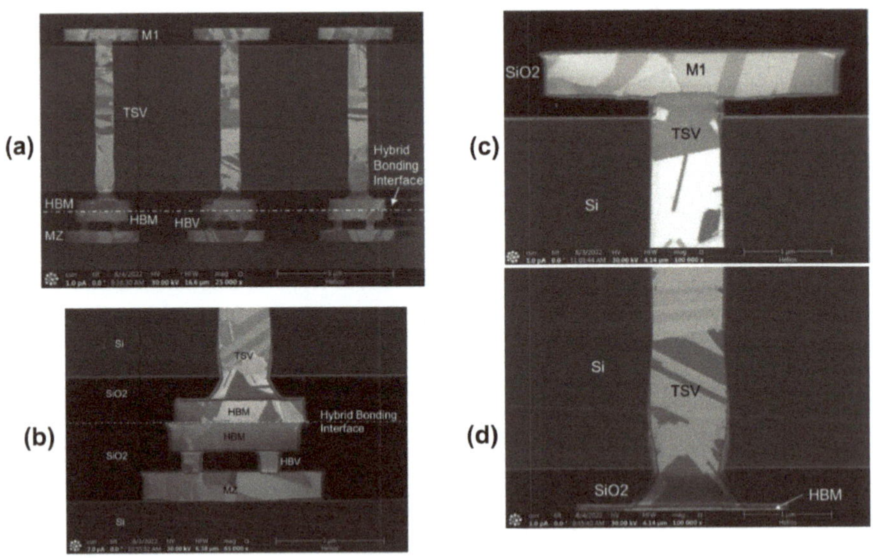

Fig. 2.43 **a** FIB-SEM cross section image of the F2B TV. **b** Hybrid bonding interface after Cu-Cu bonding. FIB-SEM images of **c** top and **d** bottom of a TSV

2.8 TEL's Hybrid Bonding

During ECTC 2023, TEL presented at least four papers on hybrid bonding [43–46]. In [43], they presented electrical analysis of W2W copper hybrid bonding at submicron pitches with two different schemes to make the test structures, namely single damascene (SD) and dual damascene (DD). Figure 2.44a shows the Cu-Cu W2W hybrid bonding cross-sectional SEM image of the 1 μm pitch Cu-pads by SD method and the probability of failures versus average resistance per via for different via chain. The Cu recess depth is 5 nm by CMP. It can be seen that there is a gap at the dielectric bonding interface regardless of if the chain is electrical connect (left) or disconnected (right). During annealing the Cu can bridge the gap to enable electrical contact in a small percentage of the shorter via chains with a quantity under 1 k vias, the bottom of Fig. 2.44a.

(A) Single Damascene Integration Scheme

Figure 2.44b shows the Cu-Cu W2W hybrid bonding cross-sectional SEM image of the 0.5 μm pitch Cu-pads by SD method and the cumulative probability of failures versus resistance for isolated via and dense via. It can be seen that the isolated SD kelvin via structure with no gaps at the dielectric bonding interface, the left-side of Fig. 2.44b. However, when a dense via pattern surrounds the SD kelvin via structure a recessed region is created during CMP which causes a gap at the bonding interface

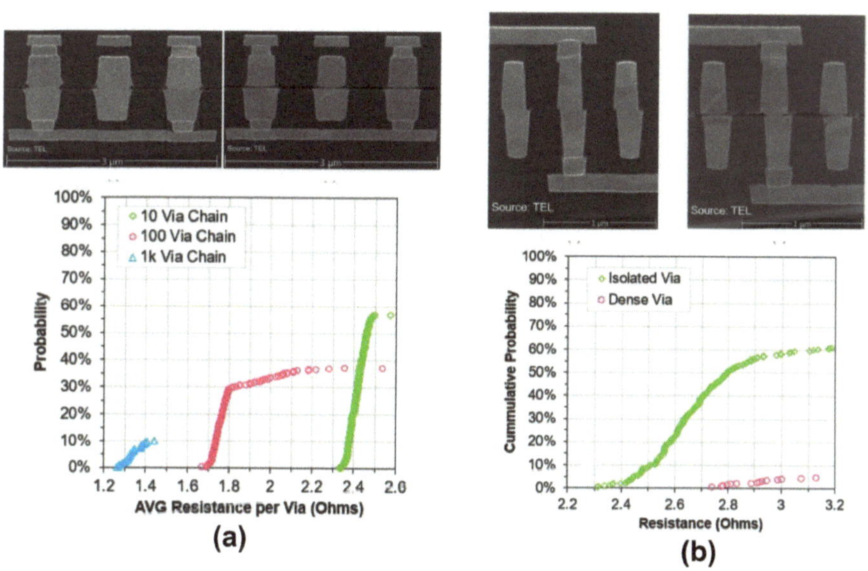

Fig. 2.44 **a** SEM images of 1 μm pitch electrically connected (top-left) and disconnected SD via chain (top-right). Average resistance per via plot of 1 μm pitch SD via-chain macro (bottom). **b** SEM images of 0.5 μm pitch isolated (top-left) and dense SD kelvin via (top-right). Electrical resistance results of corresponding macros (bottom)

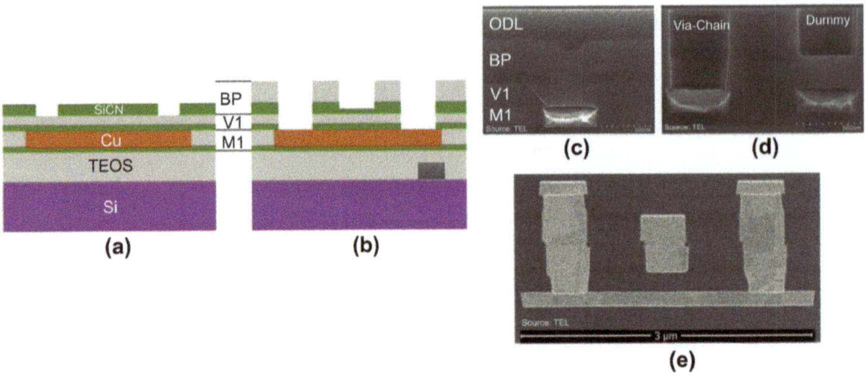

Fig. 2.45 V1 and BP layer DD schematic **a** post V1 SiCN open and **b** post final BP etch. SEM Image of 1 um pitch DD structure at **c** via memorization post BP oxide and ODL deposition, **d** post optimized BP etch with via chain and dummy patterns and **e** bonded DD via-chain with no gap remaining at the dielectric interface

to be formed, the right-side of Fig. 2.44b. Electrical resistance measurements, the bottom of Fig. 2.44b, correlate with the SEM images, the isolated via macro has much higher yield than the dense via macro.

(B) Dual Damascene Integration Scheme

Figure 2.45a shows a dual damascene process to etch the via pattern into a thick SiCN etch stop layer (ESL). When the final bond pad (BP) etch is completed the open via pattern enables contact with the underlying M1 line as shown in Fig. 2.45b. Figure 2.45c shows the SEM images of 1 µm pitch DD hybrid bonding structure at via memorization post BP oxide and ODL deposition. Figure 2.45d shows that at post optimized BP etch with via chain and dummy patterns. Figure 2.45e shows that at bonded DD via-chain with no gap remaining at the dielectric interface. Figure 2.46 shows the white light interferometry (WLI) measurement of the line scan step height at edge of the via-chain macros. It can be seen that the DD step height is smaller than the SD step height. The cumulative probability versus resistance of the 1 µm-pitch kelvin via test structure with SD and DD integration is shown in Fig. 2.47. It can be seen that the DD kelvin via resistance is lower than the SD kelvin via resistance.

2.9 Applied Materials' Hybrid Bonding

During ECTC2023 Applied Materials published at least three papers on hybrid bonding [47–49]. In [47], they presented 0.5 µm pitch wafer-to-wafer hybrid bonding with SiCN bonding interface for advanced memory. Figure 2.48a shows Applied Materials' Cu-Cu W2W hybrid bonding integration schematics. It can be seen that SiCN is used as the bonding layer on both wafers. Before bonding, the top and

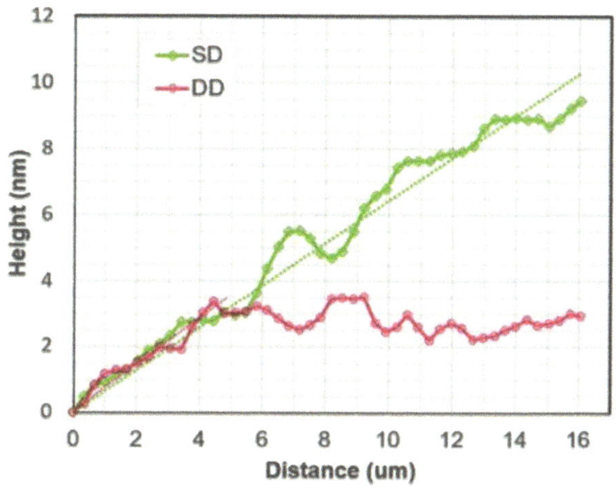

Fig. 2.46 Line scan step height at edge of the via-chain macros with SD and DD integration

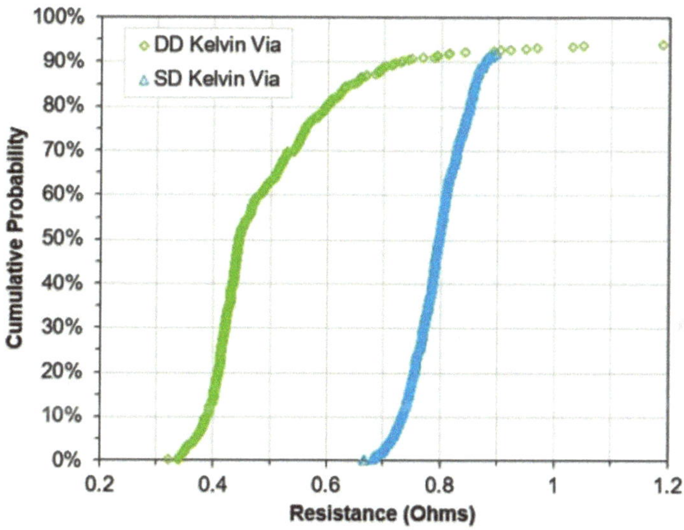

Fig. 2.47 Electrical resistance plot of 1um pitch kelvin via test structure with SD and DD integration

bottom wafers each has two metal layers. Figure 2.48b shows the post-bonding strength versus annealing temperatures for various SiCN and a reference TEOS (tetraethyl orthosilicate). The SiCN bonding strength is dependent on its composition, which could be tuned by adjusting the plasma enhanced chemical vapor deposition (PECVD) process conditions. SiCN#2 has been selected for this paper because it has the highest bonding strength at the highest annealing temperature.

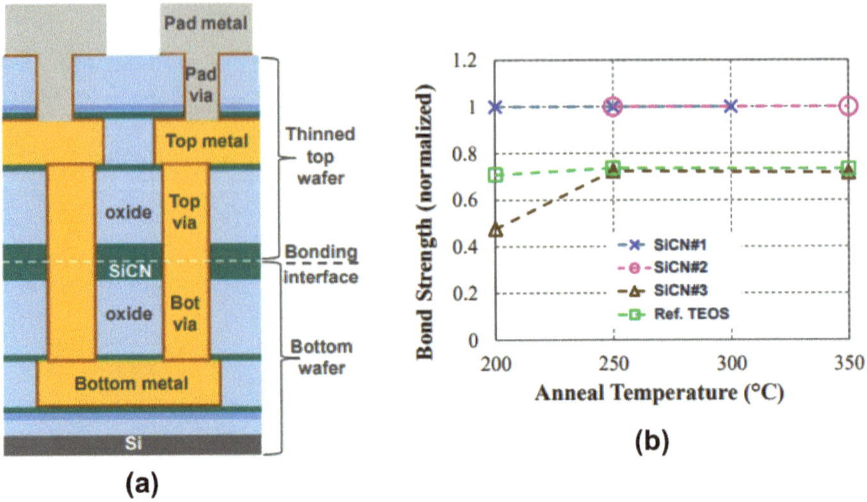

Fig. 2.48 **a** Wafer-to-wafer hybrid bonding integration flow illustration with SiCN as the bonding layer. The top and bottom vias may or may not have the same dimensions. **b** Bond strength of three different SiCN films and a reference TEOS film vs post-bonding anneal temperature

By using Applied Materials' Reflextion® LK Prime® CMP tool, less than 2-3 nm Cu dishing can be achieved within 0.5 nm accuracy. Their W2W hybrid bonding is performed on an EVG Gemini FB XT platform using a SmartView NT3 bond aligner.

Figure 2.49 shows the TEM images and electron energy loss spectroscopy (EELS) maps and scans of a SiCN-SiCN bonded interface. Figure 2.49a shows the TEM image of the bonding interface where a 4-5 nm thick SiO_x layer is formed between SiCN. Figure 2.49b shows the EELS elemental (Si, C, N, O) maps of the bonding interface. Figure 2.49c shows the EELS scan of the bonding interface. Figures 2.49a–c confirm the SiCN-to-SiCN bonding interface converted to a ~ nm-thick SiO_x.

Figure 2.50 summarizes the post bonding results (TEM images of via chain bonding) of a variety of feature sizes [47]. Figure 2.50a shows a 0.5 μm pitch, top and bottom via critical dimension (CD) are both 200 nm, top and bottom wafer Cu densities are both 16%. Figure 2.50b shows a 0.5 μm pitch, top via CD is 250 nm, bottom via CD is 200 nm, top wafer Cu density is 25% and bottom wafer density is 16%. Figure 2.50c shows a 0.8 μm pitch, top and bottom via CD are both 300 nm, top and bottom wafer Cu densities are both 14%. Figure 2.50d shows 1 μm pitch, top and bottom via CD are both 400 nm, top and bottom wafer Cu densities are both 16%. Figure 2.50e shows a 1.2 μm pitch, top via CD is 200 nm, bottom via CD is 400 nm, top wafer Cu density is 3% and bottom wafer density is 11%. Figure 2.50f shows 1.2 μm pitch, top via CD is 300 nm, bottom via CD is 600 nm, top wafer Cu density is 6% and bottom wafer density is 25%. Figure 2.50g shows a 1.2 μm pitch, top via CD is 400 nm, bottom via CD is 600 nm, top wafer Cu density is 11% and bottom wafer density is 25%.

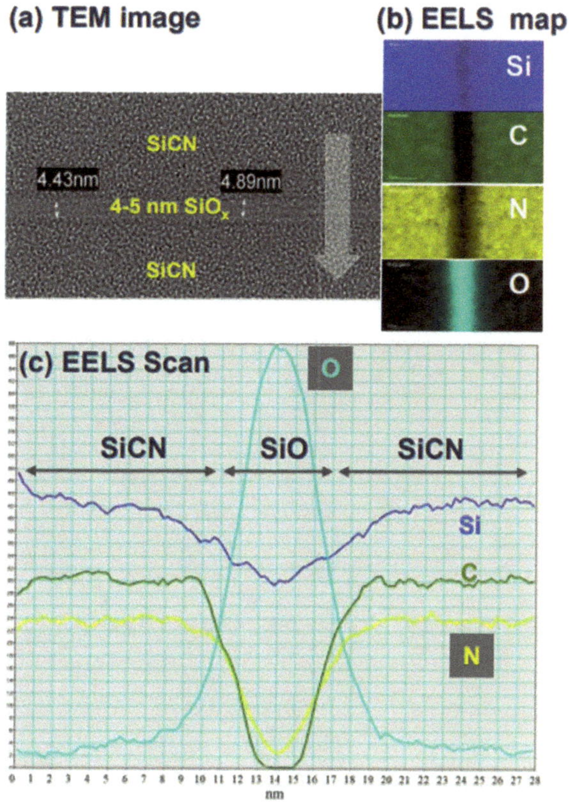

Fig. 2.49 TEM and EELS of a SiCN-to-SiCN bonded interface. **a** TEM image of the bonding interface. A 4–5 nm thick SiOx layer is formed between SiCN. **b** EELS elemental maps of the bonding interface. **c** EELS scan of the bonding interface

2.10 Intel's Hybrid Bonding

During ECTC 2023, Intel published at least three papers on hybrid bonding [50–52]. In [50], they presented the characterization of 300 mm low temperature SiCN physical vapor deposition (PVD) films for hybrid bonding applications. Table 2.1 shows the chemical composition such as Si, C, N, O, Ar, and H (normalized by Si atomic %) and density of SiCN films. Plasma-enhanced chemical vapor deposition (PECVD) film is hydrogen rich, as shown in Table 2.2, and it has much less density than of all of the PVD SiCN films [50].

Figure 2.51a shows the refractive index (at 633 nm wavelength) from ellipsometry measurement. It can be seen that the higher the amount of N_2 (PVD SiCN#1) in the process atmosphere, the lower the refractive index (n). Also, n remains constant over a wide range of thickness, it implies that film quality does not change across thickness and the PVD process is able to deposit a 1–2 μm uniform film.

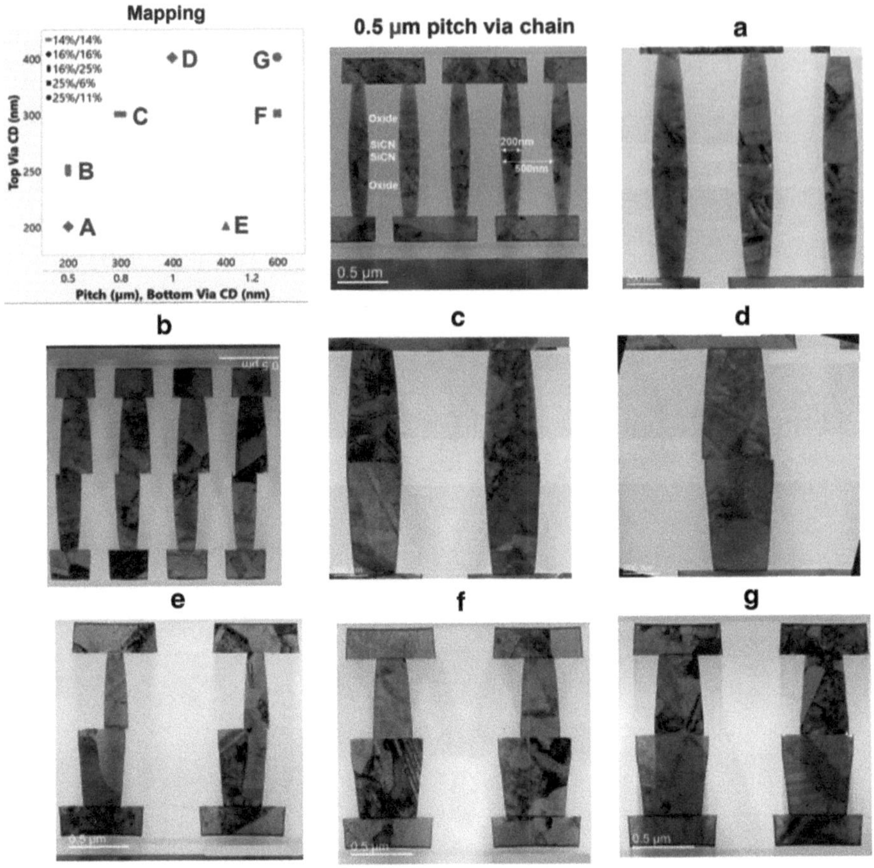

Fig. 2.50 Hybrid bonding feature scalability is established for a variety of bond-via pitches, CD, and Cu density. **a–g** are TEM images of via chain bonding with different feature sizes

Table 2.1 Chemical composition and density of SiCN films

	Composition						Density (g/cm^3)
	Si	C	N	O	Ar	H	
PVD SiCN #1 (Ar/N_2 1:1)	1.00	1.01	1.37	0.01	0.06	0.01	2.81
PVD SiCN #2 (Ar/N_2 2:1)	1.00	1.02	1.10	0.01	0.13	0.01	2.77
PVD SiCN #3 (Ar/N_2 3:1)	1.00	1.03	0.88	0.01	0.16	0.01	2.65
PECVD SiCN	1.00	1.05	0.73	0.00	0.00	1.87	2.00

Figure 2.51b shows the with-in-wafer thickness uniformity (1σ %) of the films at varying mean thicknesses. All three films have good uniformity (< 4%) of the films at varying mean thicknesses. Films deposited with higher N2 flow (PVD SiCN#1) seems to have better uniformity.

Table 2.2 Comparison dangling bond density

Sample	Activation	DB density (10^6 g^{-1} cm^2)[a]
PVD SiCN #3 (Ar/N_2 3:1)	None	3.194
PVD SiCN #3 (Ar/N_2 3:1)	N_2 Plasma	3.864
PECVD SiCN	None	0.413
PECVD SiCN	N_2 plasma	0.505

a is the overall DB signal intensity normalized by the corresponding film density and post-CMP thickness

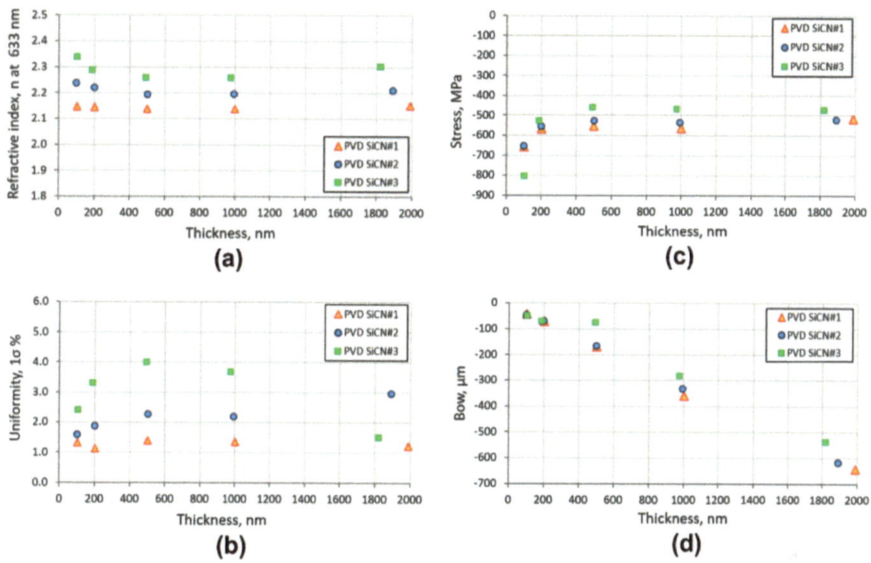

Fig. 2.51 **a** Refractive index (at 633 nm) of PVD films with thickness. **b** PVD SiCN film uniformity (1σ%) with thickness. **c** Estimated film stress from ellipsometry measurement versus thickness. **d** Wafer bow and PVD film thickness

Figure 2.51c shows the measured stress versus the film thickness. All films exhibit a compressive stress. Initially stress decreases with increase in thickness and eventually (>500 nm) reaches a plateau.

Figure 2.51d shows the bow of the wafer recorded after film deposition. All the bows increase almost linearly with the thickness. Also, wafer with 500 nm-thick films does not exceed bow of 200 μm in any cases.

Figure 2.52 shows the impact of plasma activation on SiCN films—X-band (~9.628 GHz) electron spin resonance (ESR) spectra measured at 298 K. Figures 2.52a and c show the spectra of PECVD SiCN, respectively before activation and after N_2 activation. Figure 2.52b and d show the spectra of PVD SiCN#3, respectively before activation and after N_2 activation. It can be seen that there are multiple overlapping signals in PECVD SiCN case, while in PVD SiCN case, there is only single signal.

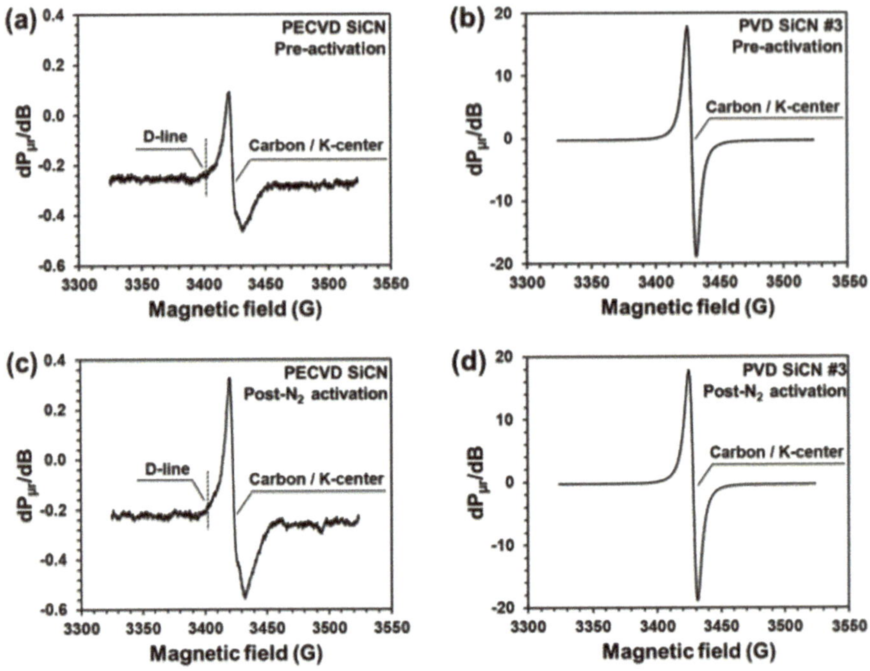

Fig. 2.52 X-band (~9.628 GHz) ESR spectra measured at 298 K on unbonded PECVD SiCN **a** before activation and **c** after N_2 plasma activation, and PVD SiCN #3 **b** before activation and **d** after N_2 plasma activation

Figure 2.53 shows the bonding strength (energy) of SicN-SiCN bonding interfaces with two annealing temperatures (200 and 250 °C) and various bonding pairs (N_2 activation has been used for all cases). It can be seen that the pair of PECVD SiCN—PECVD SiCN at 250 °C annealing temperature leads to the highest bonding energy.

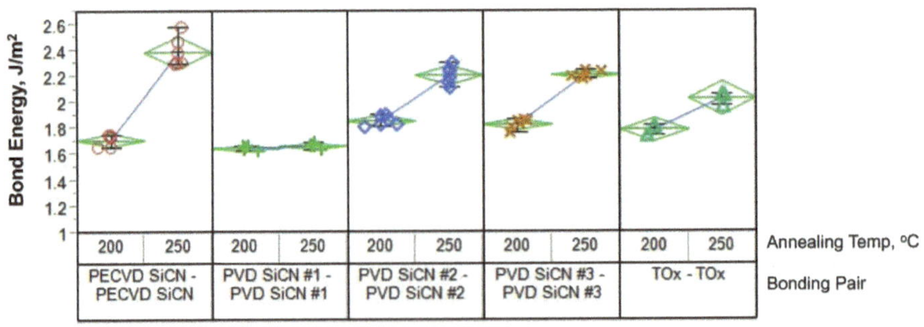

Fig. 2.53 Bond Energy (strength) of SiCN-SiCN interfaces with two annealing temperatures, 200 and 250 °C; N_2 activation was used for all cases

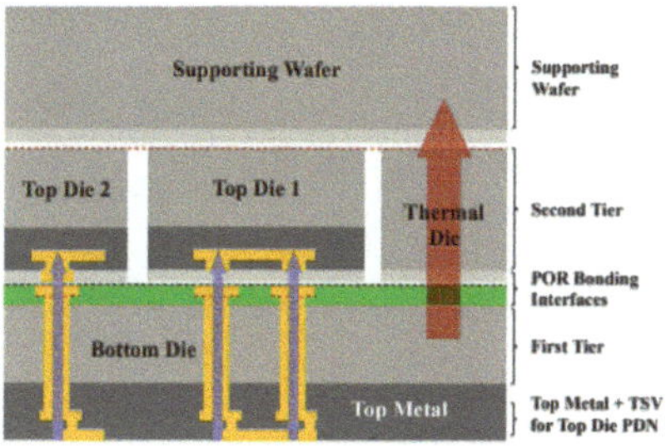

(a)

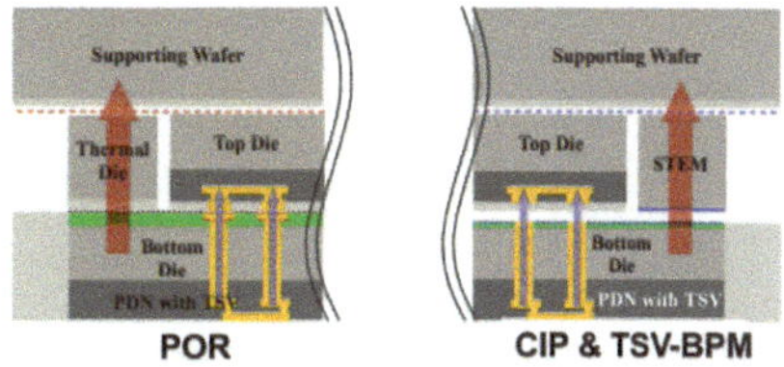

Co-Optimization	POR	CIP & TSV-BPM Bonding	
Bonding Interface Thickness in STEM Area	1x	< 0.20x	< 0.10x
Thermal Resistance	1x	0.79x	0.78x
TSV-BPM Parasitic R	1x	0.80x	0.78x
TSV-BPM Parasitic C	1x	0.98x	1.05x
EEP	1x	1.31x	1.17x

(b)

Fig. 2.54 **a** PND and thermal dissipation path in a SoIC. **b** Electrical-thermal co-optimization for backside bonding interface optimization

2.11 TSMC's Bumpless Cu-Cu Bonding

During ECTC 2023, TSMC published at least two papers on hybrid bonding [53, 54]. However, they don't call it hybrid bonding. During the TSMC 2023 North America Technology Symposium at Santa Clara, CA (April 26, 2023) TSMC announced that there are two 3D silicon stacking system on integrated circuits (SoIC), namely SoIC-X and SoIC-P. SoIC-P is dealing with μbumps (18-25 μm pitch) by CoW method. SoIC-X is dealing with bumpless (Cu-Cu) pads (4.5-9 μm pitch) by CoW method and with bumpless (Cu-Cu) pads (<3 μm pitch) by WoW method.

In [53], TSMC presented a thermally friendly bonding scheme for 3D system integration—SoIC-X. They proposed: (a) to replace the bonding pad-to-bonding pad bonding to a TSV-BPM (bonding pad metallization) bonding to avoid any unnecessary contact/boundary resistance for power delivery, (b) to thin down the bonding film which implies a much lower thermal resistance, and (c) a silicon thermal enhancement module (STEM) with an ultra-thin bonding film is integrated to further enhance the heat dissipation form the bottom die as shown in Fig. 2.54. Figure 2.54a shows the schematic power delivery network (PDN) and thermal dissipation path in a SoIC and Fig. 2.54b shows the electrical-thermal co-optimization for backside bonding interface optimization.

Figure 2.55 shows some of the results. Figure 2.55a shows the cross-sectional TEM image of a TSV-BPM bonding interface (left) and the schematic drawing of the bonding interface (right). A local barrier for metal contamination prevention is also integrated into the TSV-BPM bonding scheme [53]. Figure 2.55b shows the ultra-thin film bonding assessment on STEM with different die sizes and aspect ratios. For both cases, the void area ratio is less than 1.5%. Figure 2.55c shows the TSV-BPM alignment accuracy using TSV alignment marks. It can be seen that the TSV-BPM alignment accuracy is less than 0.5 μm for the worst die corner.

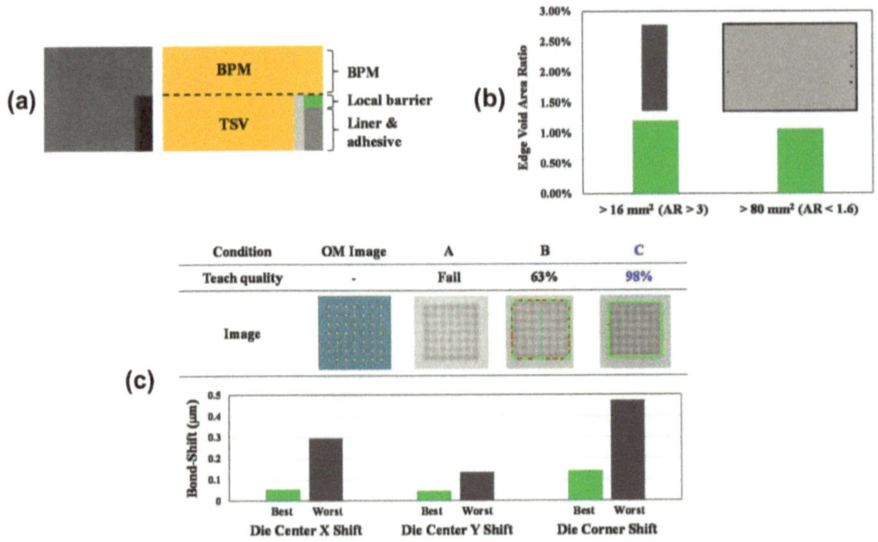

Fig. 2.55 **a** Cross-sectional TEM image of a TSV-BPM bonding interface (left) and schematic drawing of the bonding interface (right). **b** Ultra-thin film bonding assessment on STEM with different die sizes and aspect ratios. **c** TSV-BPM alignment accuracy using TSV alignment marks

2.12 MKS' Hybrid Bonding

During ECTC 2023, Atotech of MKS published at least one paper on hybrid bonding [55]. They presented the optimization of the Cu microstructure to improve copper-to-copper direct bonding for 3D Integration. Figure 2.56 shows the FIB SEM cross-section of Cu microstructures and various Cu-to-Cu bonding formations. Figure 2.56a shows the coarse-grained Cu microstructure while Fig. 2.56b shows the metastable Cu microstructure. Figure 2.56c shows Cu-Cu bonding of two stable, coarse-grained microstructures. There is not any further grain growth and a clearly visible interface is expected. Figure 2.56d shows Cu-Cu bonding of two metastable Cu layers. Grain growth through the bonding interface may be obtained. Finally, Fig. 2.56e shows Cu-Cu bonding of stable coarse-grained Cu on metastable Cu. One of the advantages of this case is all time-consuming process steps, e.g., dicing, could be performed with the stable microstructure, which could eventually be bonded to a metastable deposit [55].

Figure 2.57a shows the FIB SEM images of the coarse-grained Cu, while Fig. 2.57b of the metastable Cu. Figure 2.57c shows a consecutively deposited layer of coarse-grained Cu and metastable Cu after annealing for one hour at 200 °C. It can be seen that there is not visible interface and a proper growth of grains through the bonding interface [55].

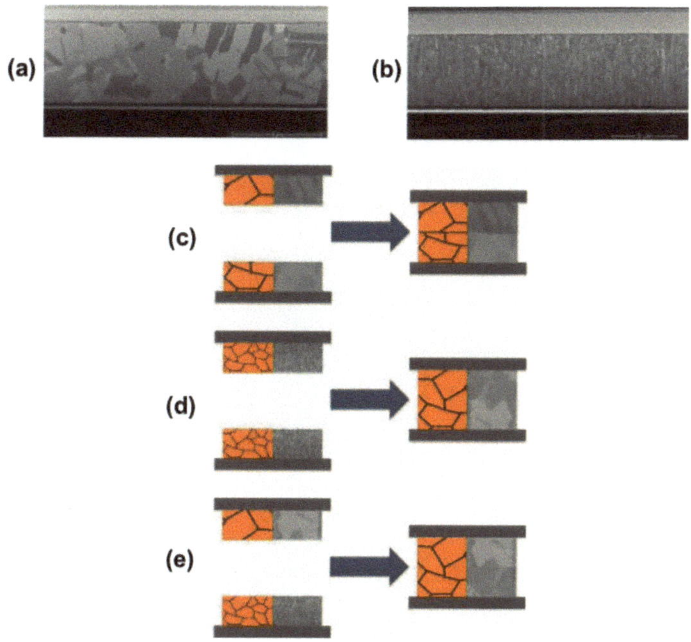

Fig. 2.56 FIB SEM cross-sections: **a** coarse-grained and **b** metastable Cu microstructures. Cu-Cu bond formations: **c** bonding of two stable, coarse-grained microstructures, **d** bonding of two metastable microstructures, and **e** bonding of a coarse-grained and metastable microstructures

2.13 Resonac's Hybrid Bonding

At ECTC 2023 Resonac published at least one paper on hybrid bonding [56]. They presented the hybrid bonding utilizing molding compound and dielectric systems. Instead of the front-end technology such as dry vacuum processes and Si-based inorganic materials, they recommend a novel hybrid bonding system utilizing organic materials focusing on epoxy molding compound (EMC) and photoimageable dielectric (PID).

Figure 2.58a shows the process in making the EMC chips. First, compression bonding the EMC (200 μm) on a silicon-chip wafer. After post-mold curing, the EMC is backgrinded to 50 μm, Fig. 2.58b, then CMP to 20 μm, Fig. 2.58c. Finally, singulate the wafer, Fig. 2.58d, into individual EMC chip, Fig. 2.58d.

Figure 2.59a shows the process in making the PID wafer. First, coat the PID varnish on a 300 mm-wafer. The PID thickness on the wafer is 10 μm after curing at 230 °C. CMP is applied to make the PID surface more planar. Figure 2.59c shows the EM image of the PID wafer after CMP.

Figure 2.60 shows the thermal bonding process (CoW) using EMC chips and PID wafer. First, two kinds of surface treatments, namely, O_2 plasma ashing and UV modification are conducted on the surface of the PID wafer, Fig. 2.60a. It is followed

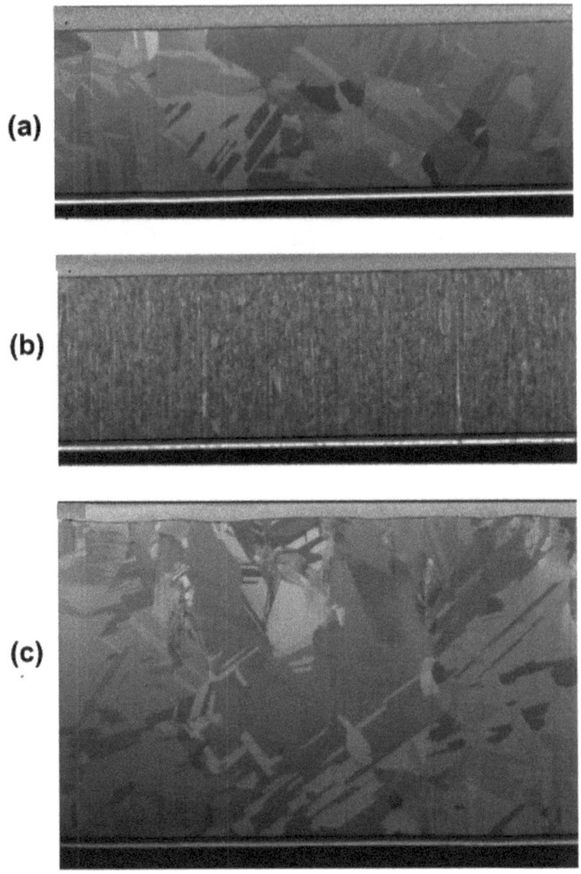

Fig. 2.57 FIB SEM images of coarse-grained copper **a** metastable Cu **b** and **c** a consecutively deposited layers of coarse-grained and metastable copper after annealing for 1 h at 200 °C

by pick and place the EMC chips on the PID wafer and TCB (5 MPa by 300 °C for 5 min by gang bonder) under a nitrogen atmosphere, Fig. 2.60c. Finally, annealing is carried out under a nitrogen atmosphere at 250 °C for two hours by an inert gas oven, Fig. 2.60d.

Figure 2.61a shows the SEM images of EMC and PID after O_2 plasma ashing and UV modification at each treatment time, respectively 0. 60, 120, and 300 s. It can be seen that (a) the surface morphologies of EMC and PID change as processing times increase by using an O_2 plasma ashing, and (b) the surface morphologies have little changes for UV modification. Figure 2.61b shows a photo of an assembled test sample using the EMC and PID system after pre-bonding, Fig. 2.60c. Figure 2.51c shows the cross-section SEM image of an TCB test sample using the EMC and PID system.

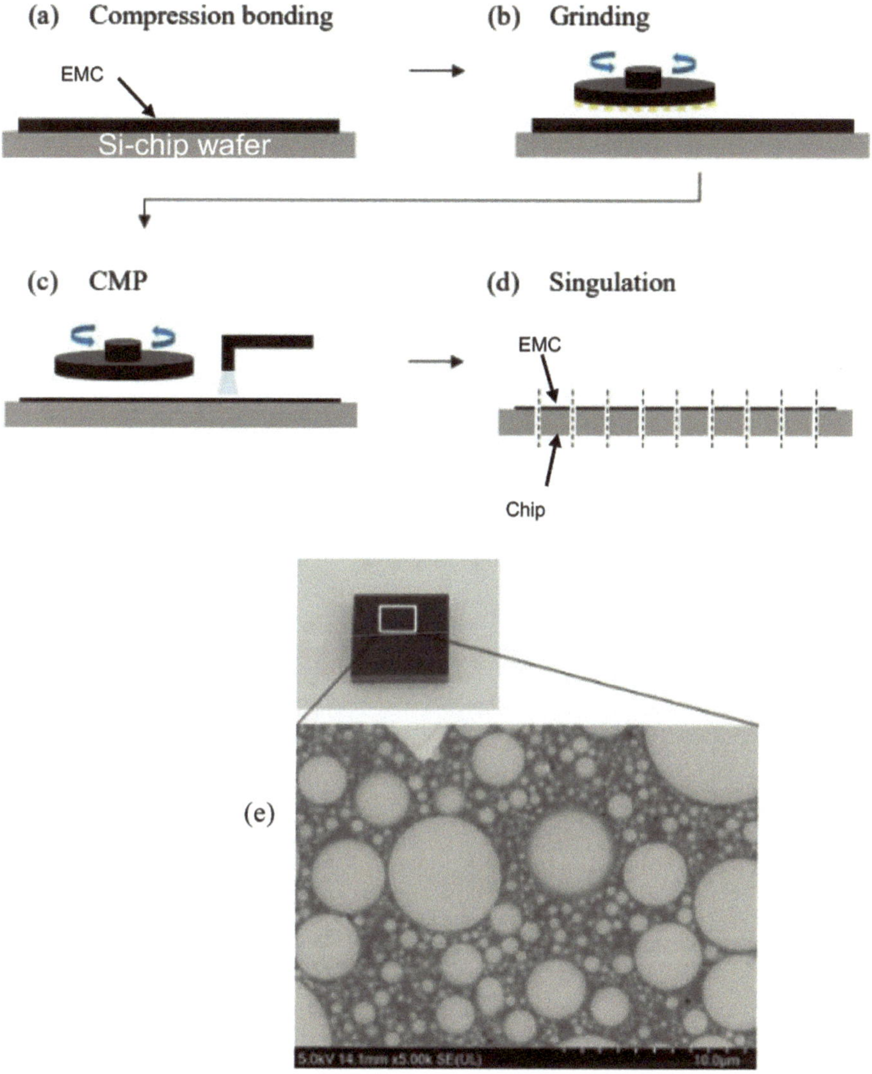

Fig. 2.58 **a–d** Fabrication process of EMC chips. **e** The SEM images of the EMC chip

2.14 Adeia's Hybrid Bonding

Adeia published at least one paper on hybrid bonding at ECTC2023 [57], in which they presented fine pitch die-to-wafer hybrid bonding. Figure 2.62a shows the D2W hybrid bonding test vehicle which is design to evaluate the effect of overall x–y misalignment, rotation alignment, and edge effects on electrical yield in a 0.5 million interconnect device. Figure 6.2b shows the typical cross-sectional SEM image of a

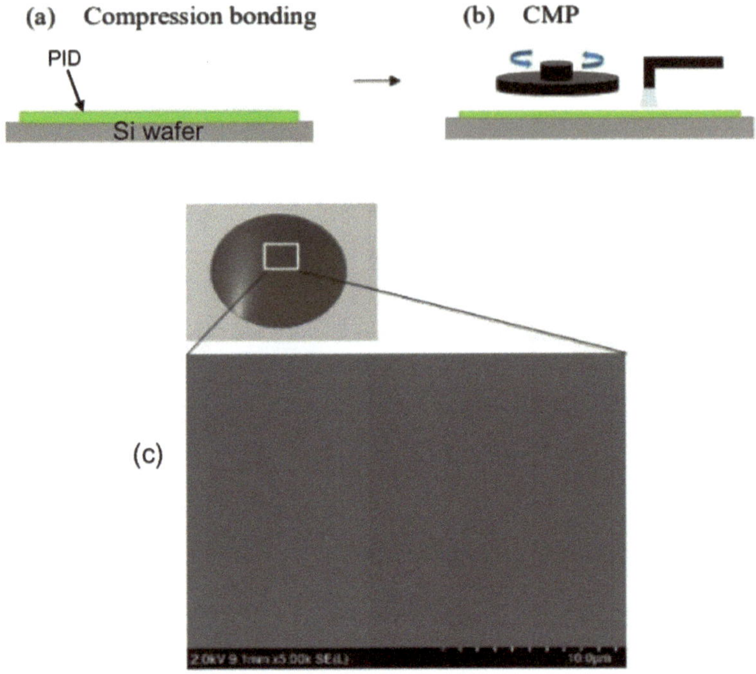

Fig. 2.59 **a–b** Fabrication process of PID wafer. **c** The SEM images of the PID wafer after CMP

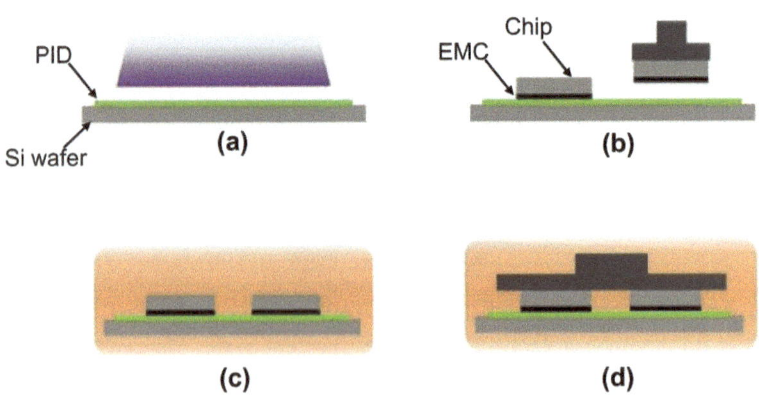

Fig. 2.60 **a–d** Thermal bonding process using EMC and PID

single die stack with dimensions of the isolation, redistribution-layer (RDL), via, and bond-pad (DBI) layers.

Figure 2.63a shows the post-bond alignment accuracy for the four separate bonded lots. It can be seen the distribution of die stretch, rotation, worst corner offset, and die center offset relative to the substrate for all four lots. Dotted lines are the global

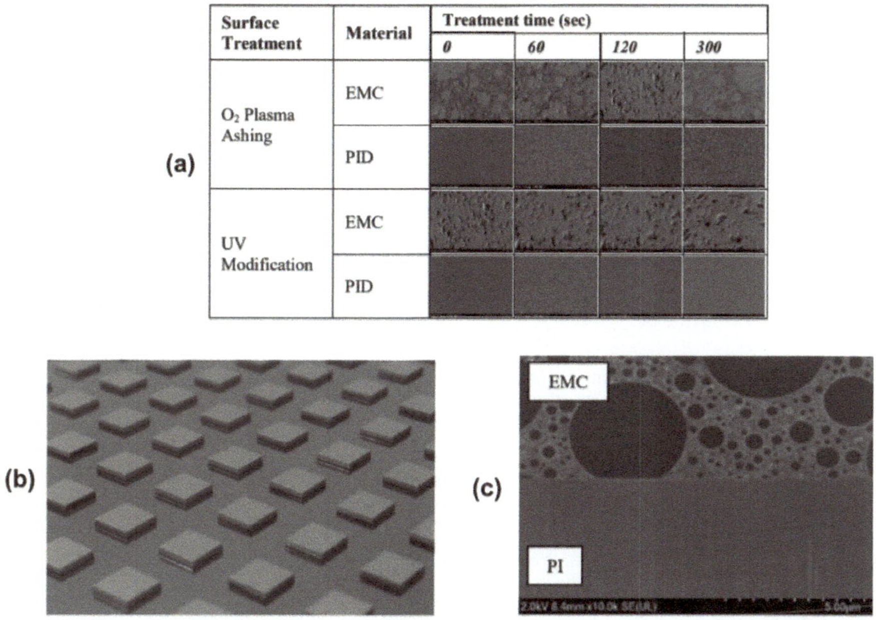

Fig. 2.61 **a** The SEM images of EMC and PID after O_2 plasma ashing and UV modification. **b** A photo of an assembled coupon using the EMC/PID system after pre-bonding. **c** The cross-sectional SEM image of a thermal bonded coupon using EMC and PID

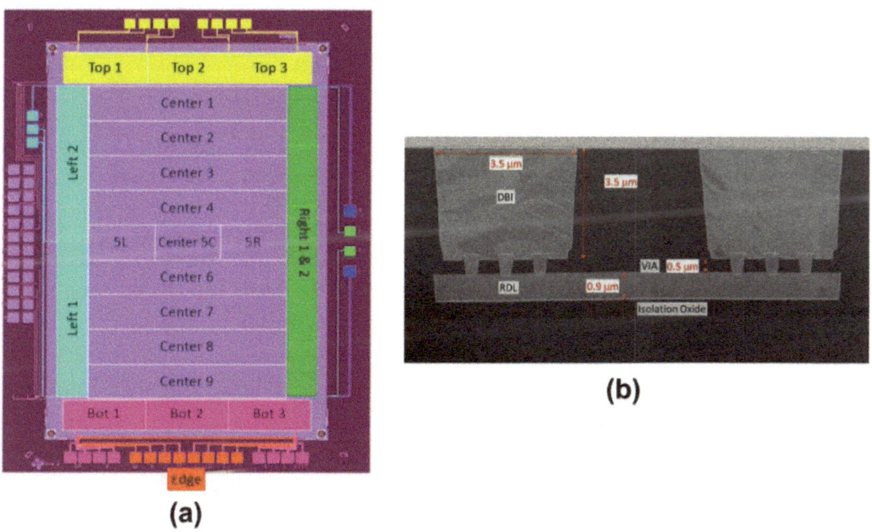

Fig. 2.62 **a** Test chip layout showing the six main daisy chain arrays and their sub-chain divisions. **b** Cross-sectional SEM image of a single die stack with the dimensions of the isolation, RDL, via, and bond pad layers

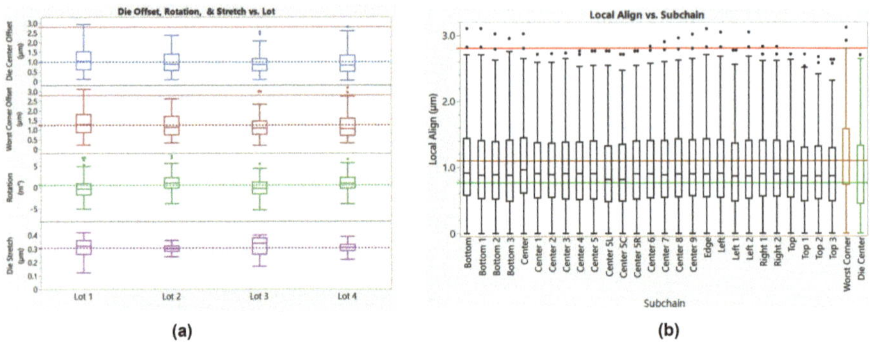

Fig. 2.63 **a** Postbond alignment accuracy for the four bonded lots. **b** The subchain local offset distribution for each die within one lot

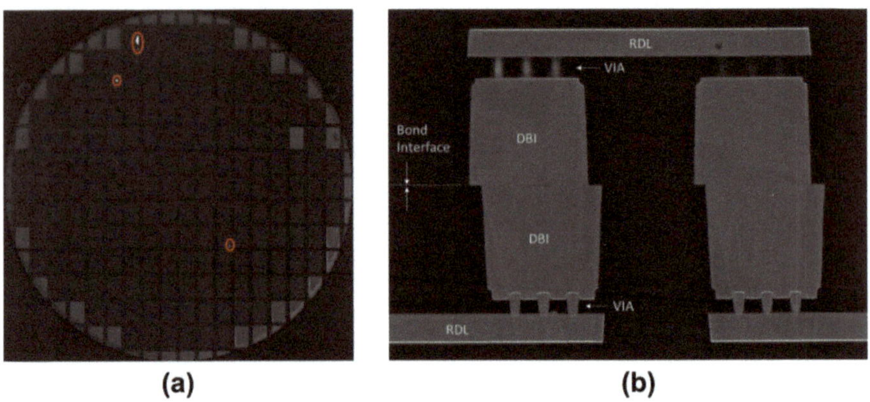

Fig. 2.64 **a** The CSAM image of a bonded wafer with bond interface voids on only 1.7% of the bonded die. **b** Cross-sectional SEM image of a daisy chain link within the main array

average and red lines at 2.8 μm offset indicate the maximum offset expected to yield electrically [57]. It can be seen that the die alignment improves slightly with each lot as the programmable offset is fine-tuned. Figure 2.63b shows the sub-chain local distribution for each die within one lot. It can be seen that test chains farther from the center of the die such as Edge have larger offset than those near the center such as Center 5C [57].

Figure 2.64a shows the CSAM image of a bonded interface voids on only 1.7% of the bonded die from Lot 3. In general, a low level of voiding (1–4%) is measured in the fine pitch lots.

Figure 2.64b shows the cross-sectional SEM image of a daisy chain link within the main array. It can be seen that good alignment and high quality dielectric-dielectric and metal–metal bonds are obtained.

Figure 2.65 shows the Cu recess within-wafer and within-die variation on a 200 mm wafer. Figure 2.65a shows the map of die measured for Cu recess.

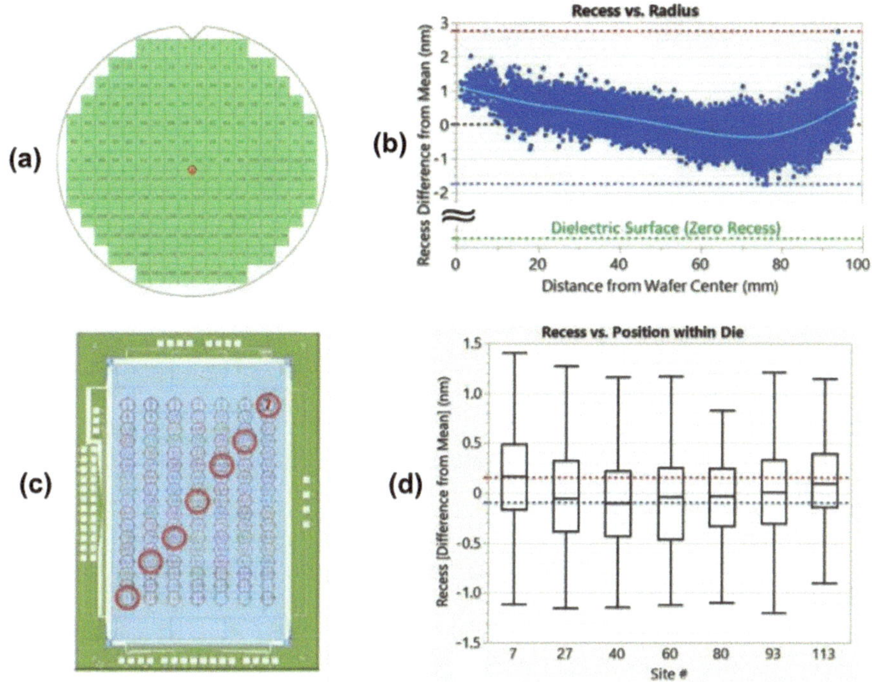

Fig. 2.65 **a** Map of die measured for Cu recess. **b** Cu recess versus distance from the wafer center (within wafer variation **c** Mapping of sites measured within each die to determine within die variation. **d** Cu recess variation across a die diagonal

Figure 2.65b shows the Cu recess versus distance from the wafer center (within wafer variation) showing a 4.5 nm difference between maximum (red line) and minimum (blue line) recess. Figure 2.65c shows the map of sites measured within each die to determine within die variation. Figure 2.65d shows the Cu recess variation across a die diagonal, showing a 0.3 nm difference between edge mean (red) and center mean recess (red). All these data show excellent uniformity of Cu recess over the entire wafer and within each die, enabled by the carefully engineered CMP process [57].

2.15 Technische Universitaet Dresden's Hybrid Bonding

During ECTC 2023, Technische Universitaet Dresden published at least one paper on hybrid bonding [58], where they presented intermetallic growth study of ultra-thin copper and tin bilayer for hybrid bonding applications. Figure 2.66 shows the process flow of their electroplating Cu using damascene method in patterned oxide and then electroless Sn deposition on Cu. Three different pairs of Cu and Sn thickness

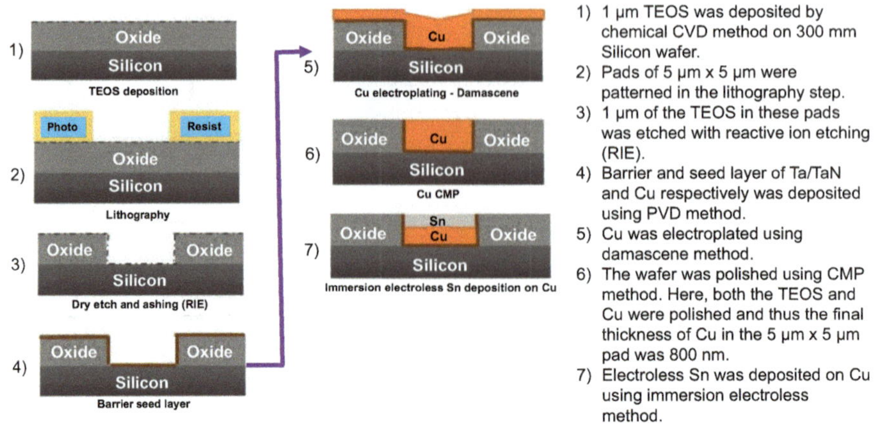

Fig. 2.66 Process flow of electroplating copper using damascene method in patterned oxide and then electroless Sn deposition on Cu

have been studied, namely, Cu(570 nm)–Sn(230 nm), Cu(530 nm)–Sn(270 nm), and Cu(480 nm)–Sn(320 nm).

Figure 2.67a shows the intermetallic growth of ultra-thin Cu and Sn bilayers at RT. It can be seen that pads with 320 nm Sn show relatively less intermetallic compound (IMC) growth area as compared to pads with 230 nm and 270 nm Sn. Thus, pads with thicker Sn layer could be beneficial for hybrid bonding.

Figure 2.67b and c show the FIB cross section of one of the pads with 320 nm Sn and 480 nm Cu after RT. Figure 2.67b shows the presence of Sn layer on Cu with a few areas within Sn layer and at the interface of Sn and Cu layers being converted into Cu_6Sn_5 IMC at Day 0. Here, Cu/(Sn + IMC) average stack thickness of (409.9 nm)/(394.8 nm) is calculated [58]. Figure 2.67c shows the growth in Cu_6Sn_5 IMC area within Sn layer is observed at Day 84 at RT. However, areas with Sn are also present. Here, Cu/(Sn + IMC) average stack thickness of (381.3 nm)/(418.1 nm) is calculated [58].

Figure 2.68a shows the intermetallic growth of ultra-thin Cu and Sn at different annealing temperatures. It can be seen that, just like those shown in RT, the pads with 320 nm show relatively less average IMC growth area as compared to pads with 230 and 270 nm.

Figure 2.68b shows the FIB cross section of one of the pads with 320 nm Sn and 480 nm Cu after annealing at 80 °C for 10 s. Here, Cu/(Sn + IMC) average stack thickness of (399.6 nm)/(401.9 nm) is calculated. Figure 2.68c shows the FIB cross section of one of the pads with 320 nm Sn and 480 nm Cu after annealing at 250 °C for 10 s. Here, Cu/(Sn + IMC) average stack thickness of (165.5 nm)/(635.9 nm) is calculated. Moreover, Kirkendall voids are observed at the Cu_3Sn/Cu interface. The intrinsic diffusion coefficient of Cu atoms is known to be larger than Sn atoms for the Cu_3Sn layer (with a ratio of around 30 at 200 °C). This diffusion coefficient

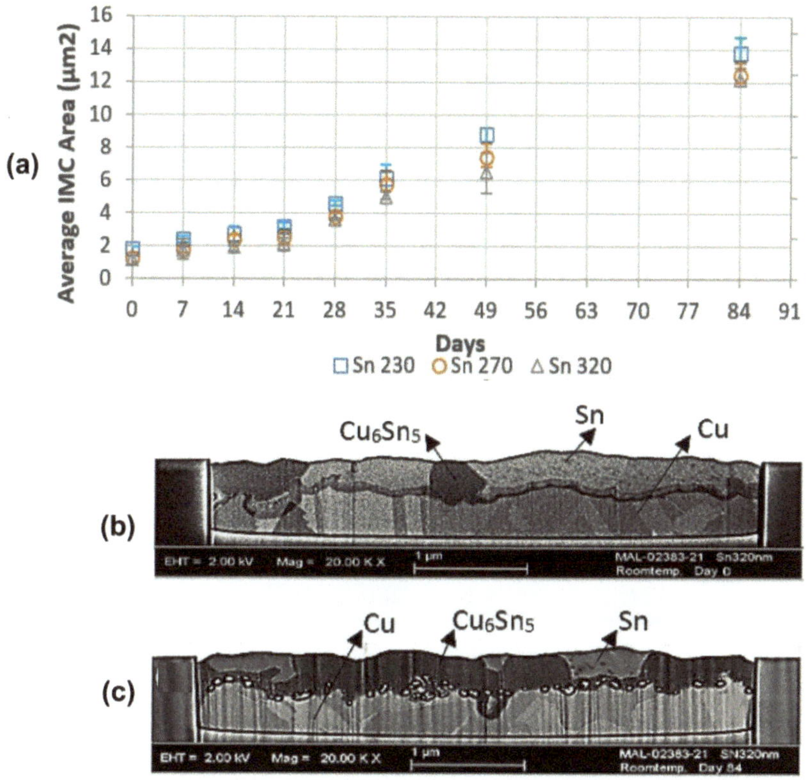

Fig. 2.67 **a** Intermetallic growth of ultra-thin Copper and Tin bilayers at room temperature. **b** Cross section of pad with 320 nm Sn and 480 nm Cu after RT day 0. **c** After RT day 84

difference thus indicates the presence of these Kirkendall voids at the interface of Cu_3Sn/Cu [58].

2.16 EVG's Hybrid Bonding

At ECTC 2023, EVG published at least one hybrid bonding paper [59], in which they presented a novel IR laser cleaving for ultra-thin layer transfer and 3D stacked devices. Glass carrier wafer has been used for 3D integration [31, 32]. However, for hybrid bonding silicon carrier wafer is preferred because hybrid bonding is mostly suitable for silicon-to-silicon assemblies such as D2D, D2W, and W2W. Figure 2.69 shows EVD's reconstructed D2W integration combining die placement and processing and reconstruction of ultra-thin chiplets with high bandwidth W2W hybrid bonding. The new layer release technology for silicon, called NanoCleave™,

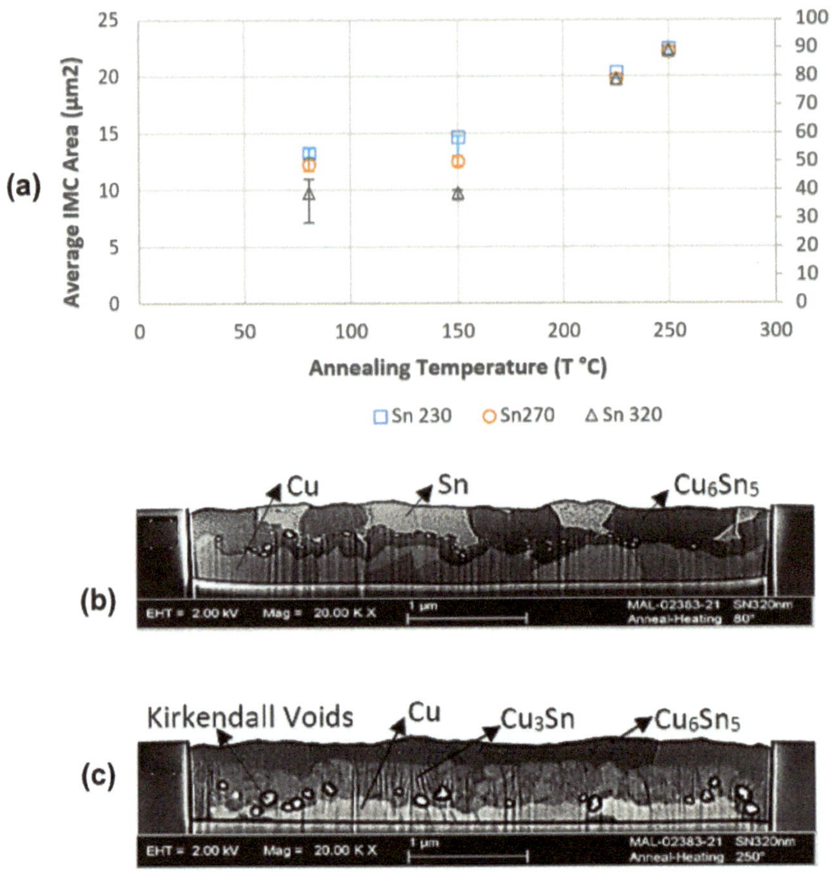

Fig. 2.68 **a** Intermetallic growth of ultra-thin Copper and Tin bilayers at different annealing temperatures. **b** Cross section of pad with 320 nm Sn and 480 nm Cu after 10 s anneal at 80 °C. **c** After 10 s anneal at 250 °C

utilizes an IR laser and specially formulated inorganic release materials to enable laser deboning on silicon with nanometer precision [59].

Figure 2.70a shows the face-to-back stacking of ultra-thin wafers on front-end compatible silicon carriers enabled by NanoCleave™ process. While Fig. 2.70b shows the process of the presented experiment for ultra-thin layer transfer. In the experiment, EVG start with a 300 mm silicon wafer enabling a Nanocleave (NC) layer to trigger later IR laser release and about 300 nm bondable TEOS oxide on top. A second wafer has been implanted using He + /H + ion implantation to a depth of about 200 nm and a bondable oxide with a thickness of 50 nm TEOS. The wafers are then bonded using a standard plasma activated wafer bond such as EVG850LT. The He + /H + layer is activated dependent on the annealing dose and distribution between 300 °C and 380 °C in order to trigger blistering, while final debonding is done using mechanical debond initiation and lift off [59].

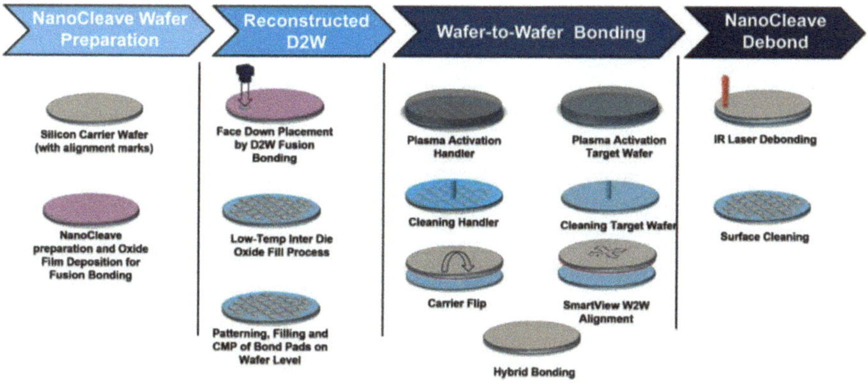

Fig. 2.69 Reconstructed D2W integration combining die placement and processing and reconstruction of ultra-thin chiplets with high bandwidth W2W hybrid bonding

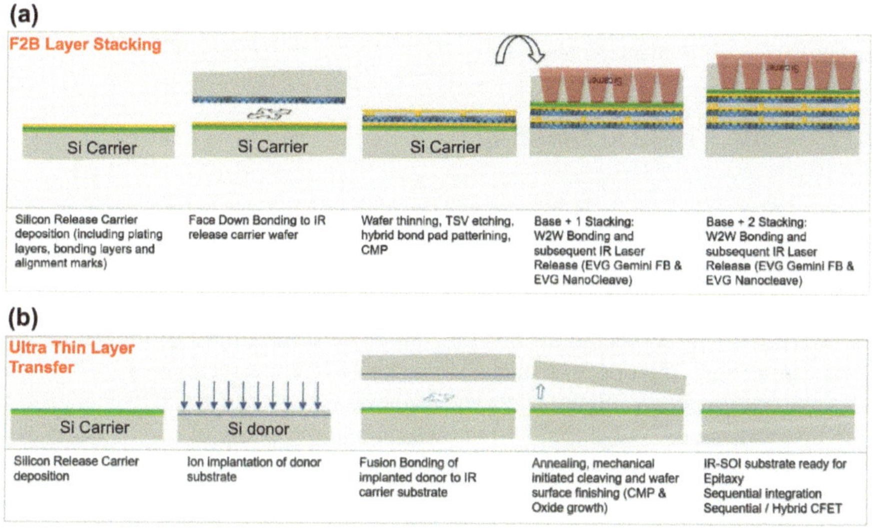

Fig. 2.70 **a** Face to back stacking of ultra-thin wafers on front-end compatible silicon carriers enabled by NanoCleave process. **b** Schematic process flow of the presented experiment for ultrathin layer transfer

Figure 2.71a shows the X-ray secondary-emission microscopy (XSEM) image of the release layer after layer transfer onto a final carrier substrate and Fig. 2.71b shows the photo of the full 300 mm wafer after layer release.

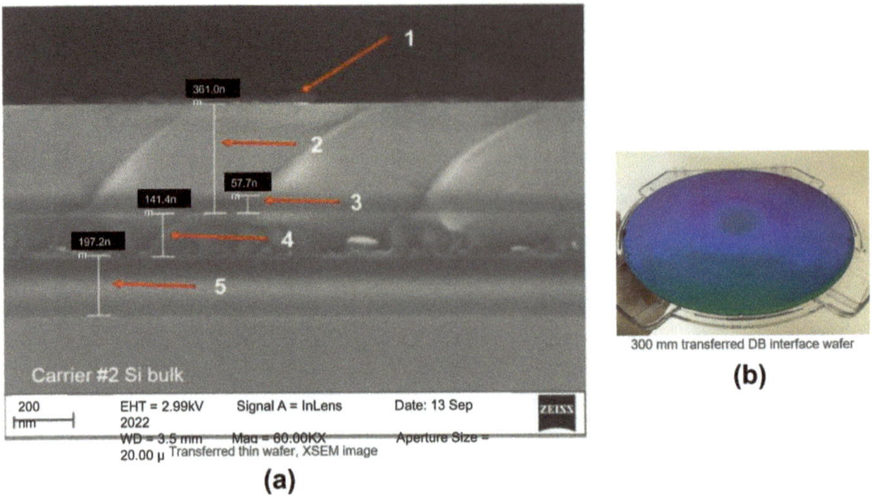

Fig. 2.71 XSEM of the release layer after layer transfer onto a final carrier substrate and photo of the full 300 mm wafer after layer released

2.17 Iemc's Hybrid Bonding

During ECTC 2023, imec published at least three papers on hybrid bonding [60–62]. In [60], they presented a new Cu "bulge-out" mechanism supporting sub-micron scaling of hybrid wafer-to-wafer bonding. In sub-micron assembly of hybrid bonding CMP of a tight control of surface topography is utmost important. The post CMP Cu pad recess with respect to the bonding dielectric surface must be limited to only a few nanometers. Figures 2.72a and b show the TEM cross sections of 200/400 nm pad combination arrays. Figure 2.72a shows the defective (due to CMP) connections with up to 91 nm of "bulge-out" of bottom pad into the cavity of corroded top pad. While Fig. 2.72b shows excellent connections in (a 700 nm pitch array) achieved due to the same bulge-out mechanics, as evidenced by the raised Cu-Cu interface. Figures 2.72c and d show the test vehicle for bulge-out measurement over large number of pads. Figure 2.72c shows the bottom wafer before bonding and Fig. 2.72d shows the bottom wafer after bonding and partial removal of top wafer. They found that "bulge-out" mechanism occurs by Cu surface diffusion at post-bond annealing temperature as low as 230 °C. Also, they found that the ideal ratio of large to small pad is 2:1 [60].

Figure 2.73a shows that, in ultra-shallow pads, bulge-out formation is observed which generates a large void at the pad sidewall. Figure 2.73b shows a creep finite element model and result has been built and calibrated to match the observed bulge-out values, Fig. 2.73c. It shows that the physical mechanism underlying the bulge-out phenomenon is not due to creep deformation but Cu surface diffusion [60].

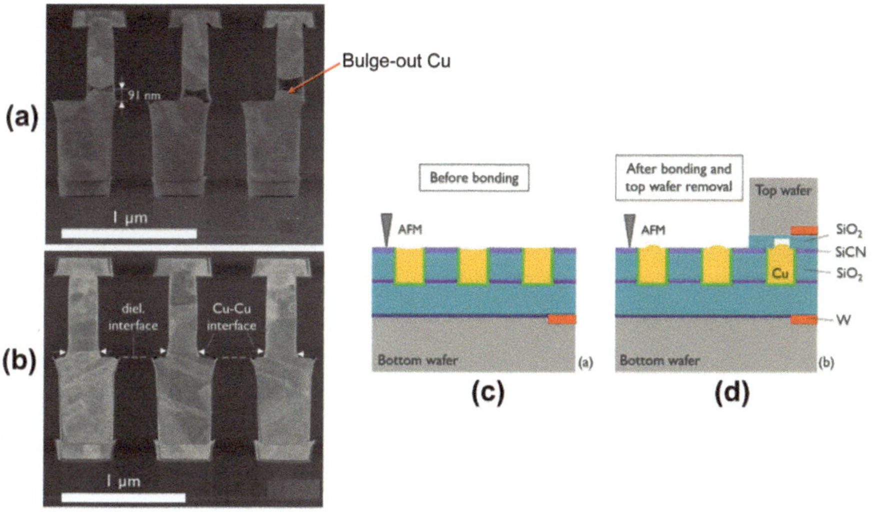

Fig. 2.72 TEM cross sections of 200 and 400 nm pad combination arrays: **a** Defective connections. **b** Excellent connection by raising the Cu-Cu interface. TV for bulge-out measurements: **c** Bottom wafer before bonding. **d** Bottom wafer after bonding and partial removal of the top wafer

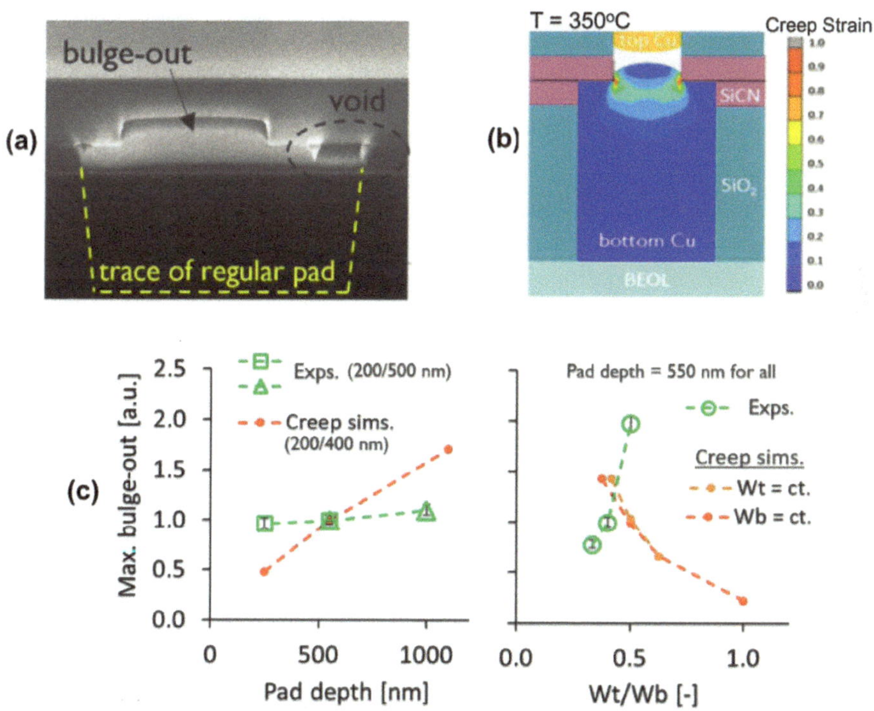

Fig. 2.73 **a** FIB-SEM cross section image of ultra-shallow pad. **b** Creep strain around the bonding interface. **c** Comparison of bulge-out experiment and simulation results

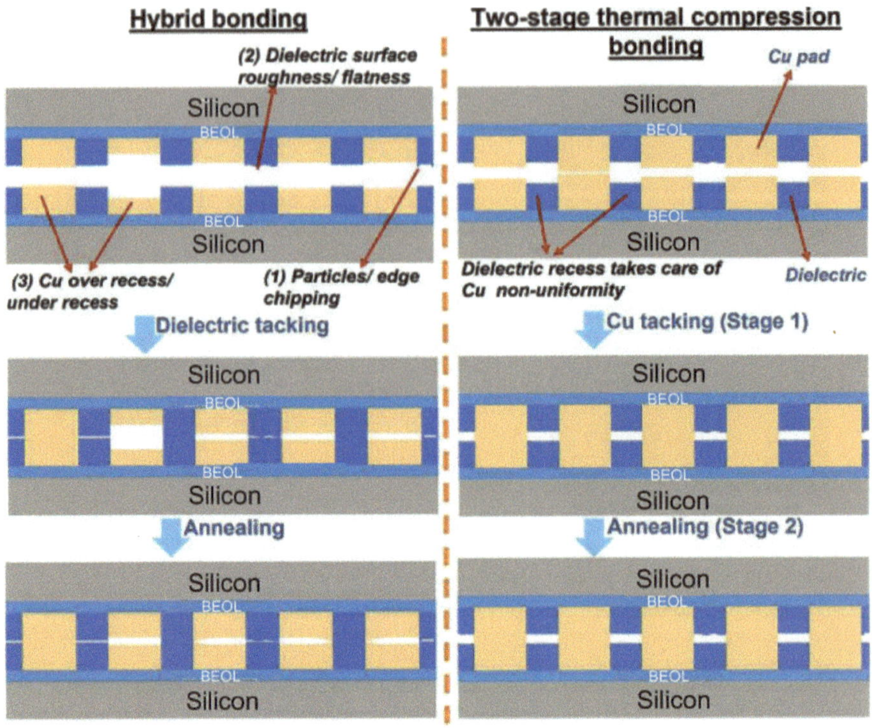

Fig. 2.74 Comparison between D2W hybrid bonding and D2W two-stage Cu-Cu TCB

2.18 UCLA's Hybrid Bonding

At ECTC2023, UCLA published at least one paper on hybrid bonding [63], where they presented a high throughput two-stage die-to-wafer thermal compression bonding scheme for heterogeneous integration. Figure 2.74 shows the difference between the conventional hybrid bonding and the proposed two-stage TCB hybrid bonding. It can be seen that a D2W-TCB process is much simpler than the conventional process. In TCB, bondable pads/pillars are recessed to achieve metal-to-metal contact. Surface asperities on Cu pads are flattened by temperature and pressure during TCB. Since there is no dielectric bonding, dielectric roughness requirements are not important. The optimized two-stage direct Cu-Cu TCB process is shown Table 2.3, It can be seen that the die cleaning time and die placement time are, respectively 3 s and 5 s. The tacking pressure is 50 MPa. The annealing temperature and annealing pressure are, respectively 300 °C and 100 MPa for low temperature variant and 400 °C and 50 MPa for high temperature variant.

Figure 2.75 shows the SEM image of a typical cross section of a two-stage D2W-TCB hybrid bonding. It can be seen that there are no voids visible along the bonding

Table 2.3 Optimized two-stage direcr CU-CU tab proce

Parameter name	Unit	Low temp. (LT) variant	High temp. (HT) variant
Die cleaning time	s	3	3
Die placement time	s	5	5
Chuck temperature	°C	120	120
Tacking pressure	MPa	50	50
Annealing temperature	°C	300	400
Annealing pressure	MPa	100	50

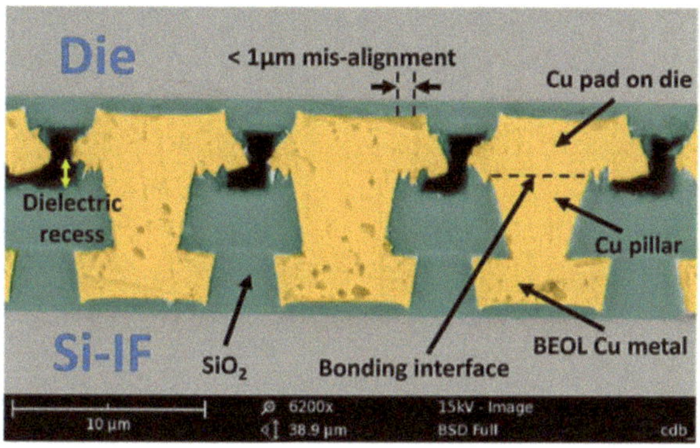

Fig. 2.75 Cross section SEM image of two-stage Cu-Cu TCB

interface. Dielectric recess enables direct metal-to-metal contact without dielectric involvement. Only Cu pillar-recess for bonding is also feasible [63].

The cumulative distribution of resistance measured across 40 daisy chains subjected to the unbiased highly accelerated stress test (UHAST) reliability measurement is presented in Fig. 2.76. It can be seen that (a) the average resistance change after UHAST is <10% for the low temperature variant, and (b) the average resistance does not change much after UHAST for the high temperature variant.

2.19 Summary

Some important results and recommendations are summarized as follows.

- The thickness, weight, thermal resistance, and electrical performance of HBMs with hybrid bonding are, respectively thinner, lighter, smaller, and better than that with the conventional flip chip solder reflow.
- All the producers of HBMs, SK Hynix, Samsung, and Micron are working on hybrid bonding, e.g., [24–34]. It is expected that high volume manufacturing of HBMs with hybrid bonding could be happening in 2025.

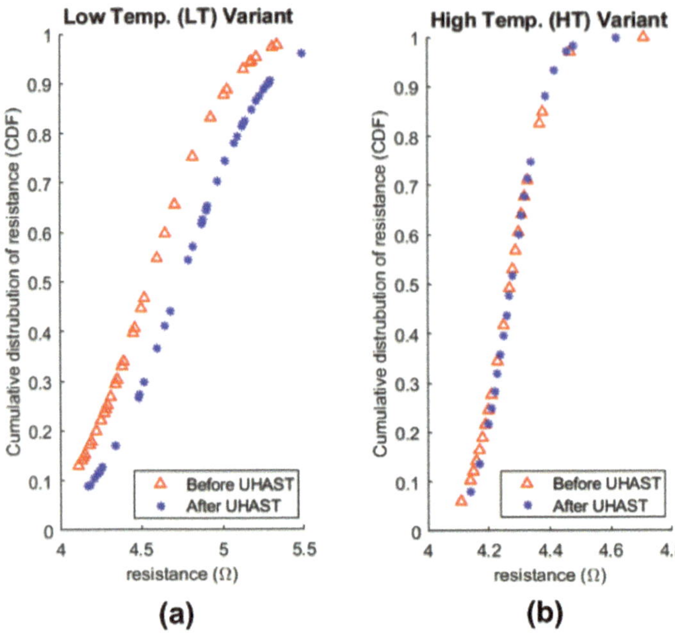

Fig. 2.76 Cumulative distribution of resistance measured across 40 daisy chains subjected to UHAST

- Most of the HBMs with hybrid bonding are by C2W method. One of the most challenges is the flatness and cleanness of the oxide surface of the chips so after they are stacked, the oxide-to-oxide of all the chips can be bonded at RT with minimum-to-no voids/seams before annealing all the chips at once.
- HBMs with hybrid bonding is a 3D IC integration packaging method. 3D NAND flash memory is a 3D monolithic manufacturing method within the chip. As mentioned in [1, 2, 23], Yangtze Memory Technologies Company Ltd. (YMTC) have been manufacturing their 3D monolithic NAND by hybrid bonding method within the chip (not the package) since they licensed the Xperi (now Adeia)'s Cu-Cu hybrid bonding technology on October 12, 2021.
- Just like the 3D NAND, 3D DRAM is a 3D monolithic chip manufacturing method which is a very hot topic today. On the research and development results, such as patents (as late as October 2022) on 3D monolithic DRAMs, Micron (30) is leading Samsung (15) and SK Hynix (10). Recently, Samsung and SK Hynix have been working very hard on 3D DRAM.
- Just like the 3D NAND by YMTC, hybrid bonding can also be combined with the 3D DRAM monolithic manufacturing within a chip. In order to enforce this area, Micron licensed the Cu-Cu hybrid bonding technology from Adeia on February 23, 2022. 3D DRAM with hybrid bonding within a chip product could be shipped in 2030.

- The works by Samsung, SK Hynix, Micron, Mitsui, Sony, CEA-Leti, TEL, Applied Materials, Intel, TSMC, MKS, Resonac, Adeia, Technische Universitaet Dresden, EVG, imec, and UCLA have been briefly mentioned and discussed in this report. Some other ECTC 2023 hybrid bonding papers [64–83] should also be noted.
- As of today, Cu-Cu hybrid bonding is applied to silicon chip to silicon chip or substrate. Since the announcement of Intel's glass interposer (substrate) for supporting their one trillion transistors on a processor chip (scheduled to be shipped by 2030), the research and development of silicon chip on glass substrate, such as the preparation of dielectric for oxide-to-oxide RT bonding and Cu dishing for Cu diffusion during annealing bonding, should begin now.

References

1. Lau, J.H. 2023. Recent advances and trends in Cu-Cu hybrid bonding. *IEEE Transactions on CPMT* 13 (3): 399–425.
2. Lau, J.H. 2023. *Chiplet design and heterogeneous integration packaging*. New York: Springer.
3. Tong, Q., G. Fountain, and P. Enquist. 2000a. Method for low temperature bonding and bonded structure. U.S. Patent 6 902 987, Patented on Feb. 16, 2000.
4. Tong, Q., Fountain, G., Enquist, P. 2000b. Method for low temperature bonding and bonded structure. U.S. Patent 7 387 944, Patented on Feb. 16, 2000.
5. Tong, Q., G. Fountain, and P. Enquist. 2000c. Method for low temperature bonding and bonded structure. U.S. Patent 8 053 329, Patented on Feb. 16, 2000.
6. Tong, Q., P. Enquist, A. Rose. 2003. Method for room temperature metal direct bonding. U.S. Patent 6 962 835, Patented on Feb. 7, 2003.
7. Suga, T. 2000. Feasibility of surface activated bonding for ultra-fine pitch interconnection – a new concept of bump-less direct boding for system level packaging. In *Proceedings of IEEE/ECTC*, 702–705, May 2000.
8. Suga, T., and K. Otsuka. 2001. Bump-less interconnect for next generation system packaging. In *Proceedings of IEEE/ECTC*, 1003–1008, May 2001.
9. Shigetou, A., Hosoda, N., Itoh, T., and Suga, T. 2001. Room-temperature direct bonding of CMP-Cu film for bumpless interconnection. In *Proceedings of IEEE/ECTC*, 755–760, May 2001.
10. Shigetou, A., T. Itoh, M. Matsuo, N. Hayasaka, K. Okumura, and T. Suga. 2006. Bumpless interconnect through ultrafine Cu electrodes by mans of surface-activated bonding (SAB) method. *IEEE Transaction on Advanced Packaging* 29(2):226.
11. Shigetou, A., T. Itoh, K. Sawada, and T. Suga. 2008. Bumpless interconnect of 6-um pitch Cu electrodes at room temperature. In *Proceedings of IEEE/ECTC*, 1405–1409, May 2008.
12. Wang, C., and T. Suga. 2009. A novel moire fringe assisted method for nanoprecision alignment in wafer bonding. In *Proceedings of IEEE/ECTC*, 872–878, May 2009.
13. Wang, C., and T. Suga. 2009. Moire method for nanoprecision wafer-to-wafer alignment: theory, simulation and application. In *IEEE Proceedings of International Conference on Electronic Packaging Technology & High Density Packaging*, 219–224, August 2009.
14. Higurashi, E., D. Chino, T. Suga, R. Sawada. 2009. Au-Au surface-activated bonding and its application to optical microsensors with 3-D structure. *IEEE Journal of Selected Topic in Quantum Electronics* 15(5):1500–1505.
15. Tsukamoto, K., E. Higurashi, and T. Suga. 2010. Evaluation of surface microroughness for surface activated bonding. In *Proceedings of IEEE CPMT Symposium Japan*, 147–150, August 2010.

16. Kondou, R., C. Wang, and T. Suga. 2010. Room-temperature Si-Si and Si-SiN wafer bonding. In *Proceedings of IEEE CPMT Symposium Japan*, 161–164, August 2010.
17. Kagawa, Y., N. Fujii, K. Aoyagi, Y. Kobayashi, S. Nishi, and N. Todaka. 2016. Novel stacked CMOS image sensor with advanced Cu2Cu hybrid bonding. In *Proceedings of IEEE/IEDM*, 8.4.1–4, Dec. 2016.
18. Kagawa, Y., N. Fujii, K. Aoyagi, Y. Kobayashi, S. Nishi, N. Todaka, S. Takeshita, J. Taura, H. Takahashi, and Y. Nishimura. An advanced CuCu hybrid bonding for novel stack CMOS image sensor. In *IEEE/EDTM Proceedings*, 1–3, March 2018.
19. Kagawa, Y., H. Hashiguchi, T. Kamibayashi, M. Haneda, N. Fujii, S. Furuse, T. Hirano, and H. Iwamoto. Impacts of misalignment on 1μm pitch Cu-Cu hybrid bonding. In *Proceeding of IEEE international interconnect technology conference (IITC)*, 148–150, Oct. 5–9, 2020.
20. Oike, Y. 2022. Evolution of image sensor architectures with stacked device technologies. *IEEE Transactions on Electron Devices* 69 (6): 2757–2765.
21. Kagawa, Y., T. Kamibayashi, Y. Yamano, K. Nishio, A. Sakamoto, T. Yamada, K. Shimizu, T. Hirano, and H. Iwamoto. 2022. Development of face-to-face and face-to-back ultra-fine pitch Cu-Cu hybrid bonding. In *Proceedings of IEEE/ECTC*, 306–311, May 2022.
22. Furuse, S., N. Fujii, K. Kotoo, N. Ogawa, S. Saito, T. Yamada, T. Hirano, Y. Hagimoto, and H. Iwamoto. 2022. Behavior of bonding strength on wafter-to-wafer Cu-Cu hybrid bonding. In *Proceedings of IEEE/ECTC*, 591–594, May 2022.
23. Lau, J.H. (2024). State of the art of Cu-Cu hybrid bonding. *IEEE Transactions on CPMT* 14(3): 376–396.
24. Kim, M., H. Lee, A. Jang, W. Lee, S. Back, I. Kim, J. Park, Y. Jee, I. Lee, K. Moon, U. Kang, and D. Kim. 2023. Multi-stack hybrid cu bonding technology development using ultra-thin chips. In *Proceedings of IEEE/ECTC*, 1672–1676, May 2023.
25. Taehwan, K., J. Lee, Y. Kim, H. Park, H. Hwang, J. Kim, H. Jung, and D. Kim. Thermal improvement of HBM with joint thermal resistance reduction for scaling 12 stacks and beyond. In *Proceedings of IEEE/ECTC*, 767–771, May 2023.
26. Hwang, B., S. Kim, J. Lee, S. Lee, Y. Jee, S. Park, G. Jo, K. Kim, S. Han, J. Park, H. Jung, D. Kang, and U. Kang. 2023. A study on the surface activation of Cu and oxide for hybrid bonding joint interface. In *Proceedings of IEEE/ECTC*, 78–82, May 2023.
27. Kim, Y., J. Kim, H. Kim, H. Lee, D. Kim, S. Seo, C. Jo, and D. Kim. 2023. Die to wafer hybrid Cu bonding for fine pitch 3D-IC applications. In *Proceedings of IEEE/ECTC*, 1043–1047, May 2023.
28. Hahn, S., W. Kim, D. Shin, Y. Lee, S. Kim, W. Choi, K. Lim, B. Moon, and M. Rhee. Contamination-free Cu/SiCN hybrid bonding process development for sub-μm pitch devices with enhanced bonding characteristics. In *Proceedings of IEEE/ECTC*, 1390–1396, May 2023.
29. Baek, K., J. Kim, M. Han, K. Lim, and M. Rhee. 2023. Chip level evaluation of wafer-to-wafer direct bonding strength with bending test. In *Proceedings of IEEE/ECTC*, 310–317, May 2023.
30. Lau, J.H. 2021. *Semiconductor advanced packaging*. New York: Springer.
31. Lau, J.H. 2016. *3D IC integration and packaging*. New York: McGraw-Hill.
32. Lau, J.H. 2013. *Through-silicon via (TSV) for 3D integration*. New York: McGraw-Hill.
33. Kim, K., S. Lim, D. Jung, J. Choi, S. Na, J. Yeom, M. Lee, J. Kim, J. Kwon, K. Moon, G. Lee, and K. Lee. 2023. C2W hybrid bonding interconnect technology for higher density and better thermal dissipation of high bandwidth memory. In *Proceedings of IEEE/ECTC*, 1048–1052, May 2023.
34. Zhou, W., M. Kwon, Y. Chiu, H. Guo, B. Bhushan, B. Street, K. Parekh, and A. Singh. 2023. Critical challenges with copper hybrid bonding for chip-to-wafer memory stacking. In *Proceedings of IEEE/ECTC*, 336–341, May 2023.
35. Nakamura, Y., K. Tamura, Y. Kayaba, W. Okada, T. Shikama, and S. Inada. 2023. A study on multi-chip stacking process by novel dielectric polymer adhesive for Cu-Cu hybrid bonding. In *Proceedings of IEEE/ECTC*, 1129–1134, May 2023.
36. Kayaba, Y., Y. Nakamura, W. Okada, T. Shikama, K. Tamura, and S. Inada. 2023. A new adhesive for CoW Cu-Cu hybrid bonding with high throughput and room temperature pre-bonding. In *Proceedings of IEEE/ECTC*, 571–578, May 2023.

37. Kagawa, Y., T. Kamibayashi, N. Fujii, S. Furuse, T. Yamada, T. Hirano, and H. Iwamoto. (2023). Cu-Cu wiring: the novel structure of Cu-Cu hybrid bonding. In *Proceedings of IEEE/ECTC*, 103–108, May 2023.
38. Kamei, T., H. Yoshioka, T. Hiratsuka, A. Sakamoto, K. Shimizu, and H. Iwamoto. 2023. Development of high reliability 6 μm-pitch Cu-Cu connections using over-400mm^2-large chip on wafer bonding process. In *Proceedings of IEEE/ECTC*, 1285–1290, May 2023.
39. Hirano, T., T. Yamada, S. Kobayashi, Y. Hagimoto, and H. Iwamoto. 2023. Simulation of device structure impacts on bonding wave and strain in wafer-to-wafer Cu-Cu hybrid bonding. In *Proceedings of IEEE/ECTC*, 1314–1318, May 2023.
40. Yoshioka, H., N. Fujii, T. Shigetoshi, T. Kamei, K. Kotoo, N. Ogawa, S. Furuse, T. Horikiri, S. Saito, S. Saito, Y. Hagimoto, and H. Iwamoto. 2023. Influence of H2O in bonding interfaces on bonding strength of plasma-activated bonded silicon oxide. In *Proceedings of IEEE/ECTC*, 1119–1123, May 2023.
41. Berru, J., S. Nicolas, N. Bresson, M. Assous, and S. Borel. 2023. Demonstration of a wafer level face-to-back (F2B) fine pitch Cu-Cu hybrid bonding with high density TSV for 3D integration applications. In *Proceedings of IEEE/ECTC*, 97–102, May 2023.
42. Bourjot, E., A. Bond, N. Nadi, T. Enot, L. Sanchez, P. Montmeat, B. Martin, A. Campo, F. Fournel, F. Eid, and J. Swan. 2023. Integration and process challenges of self assembly applied to die-to-wafer hybrid bonding. In *Proceedings of IEEE/ECTC*, 1397–1402, May 2023.
43. Ryan, K., C. Netzband, A. Gildea, Y. Mimura, S. Hoshino, S. LeFevre, A. Tuchman, I. Son, H. Aizawa, and K. Maekawa. 2023. Electrical analysis of wafer-to-wafer copper hybrid bonding at sub-micron pitches. In *Proceedings of IEEE/ECTC*, 114–117, May 2023.
44. Netzband, C., K. Ryan, Y. Mimura, S. Ilseok, H. Aizawa, N. Ip, X. Chen, H. Fukushima, and S. Tan. 2023. 0.5μm pitch next generation hybrid bonding with high alignment accuracy for 3D integration. In *Proceedings of IEEE/ECTC*, 1100–1104, May 2023.
45. Kim, H., Y. Tsai, S. Hoshino, I. Son, M. Kaoru, P. Biolsi, and S. Arkalgud. Atomistic simulation analysis of plasma surface activation in wafer-to-wafer oxide fusion bonding. In *Proceedings of IEEE/ECTC*, 2034–2039, May 2023.
46. Netzband, C., D. Burns, K. Tapily, S. Ilseok, and C. Wajda. 2023. Impact of plasma activation on copper surface layer for low temperature hybrid bonding. In *Proceedings of IEEE/ECTC*, 350–355, May 2023.
47. Ma, K., N. Bekiaris, S. Ramaswami, T. Ding, G. Probst, J. Burggraf, and T. Uhrmann. 2023. 0.5 μm pitch wafer-to-wafer hybrid bonding with SiCN bonding interface for advanced memory. In *Proceedings of IEEE/ECTC*, pp. 1110–1114, May 2023.
48. Sherwood, T., R. Sreenivasan, J. Appell, R. Patlolla, K. Li, K. Ahn, J. Salfelder, R. Ley, M. Gorchichko, T. Kasbauer, G. Probst, and T. Uhrmann. 2023. Aggressive pitch scaling (sub-0.5 μm) of W2W hybrid bonding through process innovations. In *Proceedings of IEEE/ECTC*, 13–18, May 2023.
49. Dag, S., M. Liu, L. Jiang, A. Kiaee, G. See, P. Lianto, B. Ayyagari-Sangamalli, and E. Bazizi. 2023. Development of copper thermal coefficient for low temperature hybrid bonding. In *Proceedings of IEEE/ECTC*, 342–349, May 2023.
50. Niazi, H., B. Zhang, X. Brun, S. Krishnan, and J. Chuang. 2023. Critical dimension scatterometry as a scalable solution for hybrid bonding pad recess metrology. In *Proceedings of IEEE/ECTC*, 1403–1409, May 2023.
51. Brun, X., M. Hasan, Y. Yang, P. Cara, P.C. Drechsel, and E. Strolz. 2023. Characterization of 300 mm low temperature SiCN PVD films for hybrid bonding application. In *Proceedings of IEEE/ECTC*, 548–555, May 2023.
52. Ajayi, T., A. Gatimu, X. Brun, C. Woods, K. Song, A. Bhat, C. Galande, P. Le, and Z. Karim. 2023. Impact of thermal annealing and other process parameters on hybrid bonding performance for 3D advanced assembly technology. In *Proceedings of IEEE/ECTC*, pp. 1426–1432, May 2023.
53. Kuo, G., C. Chen, C. Hsieh, C. Lee, J. Wu, J. Cui, Y. Kuo, C. Tung, T. Ku, C. Tsai, K. Yee, and D. Yu. 2023. A thermally friendly bonding scheme for 3D system integration. In *Proceedings of IEEE/ECTC*, 1973–1976, May 2023.

54. Chia, H., S. Tai, C. Wang, C. Tung, K. Yee, and D. Yu. 2023. Ultra high density low temperature SoIC with Sub-0.5 μm bond pitch. In *Proceedings of IEEE/ECTC*, 1–4, May 2023.
55. Schmidt, R., and C. Schwarz. 2023. Optimization of the Cu microstructure to improve copper-to-copper direct bonding for 3D integration. In *Proceedings of IEEE/ECTC*, 2024–2028, May 2023.
56. Imazu, Y., and K. Mitsukura. 2023. Hybrid bonding utilizing molding compound and dielectric systems. In *Proceedings of IEEE/ECTC*, 1649–1654, May 2023.
57. Mirkarimi, L., T. Workman, J. Theil, G. Fountain, K. Bang, O. Zhao, B. Lee, C. Uzoh, D. Suwito, and G. Gao. 2023. Fine pitch die-to-wafer hybrid bonding. In *Proceedings of IEEE/ECTC*, 83–90, May 2023.
58. Khurana, G., and I. Panchenko. Intermetallic growth study of ultra-thin copper and tin bilayer for hybrid bonding applications. In *Proceedings of IEEE/ECTC*, 866–871, May 2023.
59. Uhrmann, T., P. Urban, B. Povazay, M. Gruber, J. Bravin, D, Burgstaller, M. Wimplinger, and B. Thallner. 2023. Novel IR laser cleaving for ultra-thin layer transfer and 3D stacked devices. In *Proceedings of IEEE/ECTC*, 1749–1753, May 2023.
60. Messemaeker, J., L. Witters, B. Zhang, Y. Tsau, F. Fodor, J. Vos, G. Beyer, K. Croes, and E. Beyne. 2023. New Cu "Bulge-Out" mechanism supporting sub-micron scaling of hybrid wafer-to-wafer bonding. In *Proceedings of IEEE/ECTC*, 109–113, May 2023.
61. Noda, H., A. Hiro, N. Nishiguchi, J. Derakhshandeh, J. Slabbekoorn, and E. Beyne. 2023. An investigation on particle embedding capability of wafer level spin-on polymer underfill enabling low temperature bonding of hybrid bonding system. In *Proceedings of IEEE/ECTC*, 1677–1681, May 2023.
62. Iacovo, S., K. D'havé, O. Okudur, J. Vos, T. Uhrmann, T. Plach, T. Conard, J, Meersschaut, P. Bex, S. Brems, A. Phommahaxay, M. Gonzalez, L. Witters, G. Beyer, and E. Beyne. 2023. A study of SiCN wafer-to-wafer bonding and impact of wafer warpage. In *Proceedings of IEEE/ECTC*, 1410–1417, May 2023.
63. Sahoo, K., H. Ren, and S. Iyer. 2023. A high throughput two-stage die-to-wafer thermal compression bonding scheme for heterogeneous integration. In *Proceedings of IEEE/ECTC*, 362–366, May 2023.
64. Cheemalamarri, H., B. Varghese, S. Jaibir, L. Hongyu, C. Rao N. Singh, V. Rao, K. Chui. 2023. CMOS-compatible fine pitch Al-Al bonding. In *Proceedings of IEEE/ECTC*, 1725–1729, May 2023.
65. Vasarla, N., D. Mishra, P. Lianto, S. Chong, and S. Vempati. 2023. Cu damascene process on temporary bonded wafers for thin chip stacking using Cu-Cu hybrid bonding. In *Proceedings of IEEE/ECTC*, 564–570, May 2023.
66. Chong, S., K. Jason, S. Nagendra, C. Ismael, and S. Vempati. 2023. Development of 4 die stack module using hybrid bonding approach. In *Proceedings of IEEE/ECTC*, 788–794, May 2023.
67. Sakuma, K., R. Yu, N. Polomoff, L. Darling, P. Chowdhurry, S. Raghavan, K. Seifert, J. Knickerbocker, D. McHerron, M. Li, S. So, S. Kwok, C. Fan, and S. Lau. 2023. Voids-free die-level Cu/ILD hybrid bonding. In *Proceedings of IEEE/ECTC*, 800–805, May 2023.
68. Chiu, W., O. Lee, T. Kuo, J. Lo, C. Shih, H. Chiu, and H. Chang. 2023. Low temperature and fine pitch nanocrystalline Cu/SiCN wafer-to-wafer hybrid bonding. In *Proceedings of IEEE/ECTC*, 1105–1109, May 2023.
69. Lin, Y., T. Yang, O. Lee, C. Lee, K. Ko, Y. Chen, H. Chang, M. Gallagher, C. Gilmore, P. Chuang, Y. Tseng, P. Tsai, P. Huang, and C. Lee. 2023. A novel polymer-based ultra-high density bonding interconnection. In *Proceedings of IEEE/ECTC*, 1779–1784, May 2023.
70. Murugesan, M., M. Sawa, E. Sone, T. Miazaki, M. Koyanagi, and T. Fukushima. 2023. Chip-to-chip hybrid bonding with larger-oriented Cu grains for μ-joints beyond 100 K. In *Proceedings of IEEE/ECTC*, 1713–1718, May 2023.
71. Mitsuishi, H., H. Maeda, M. Ushijima, A. Kamashita, M. Okada, M. Aramata, T. Shiomi, S. Sakamoto, K. Takahata, T. Chiba, M. Fukuda, M. Kanbayashi, T. Shimoda, and I. Sugaya. 2023. 50 nm overlay accuracy for wafer-to-wafer bonding by high-precision alignment technologies. In *Proceedings of IEEE/ECTC*, 1664–1671, May 2023.

72. Choi, K., J. Joo, G. Choi, C. Lee, K. Jang, J. Oh, J. Shin, H. Yun, S. Moon, Y. Moon, and Y. Eom. 2023. Laser-assisted bonding with compression (LABC) based tiling bonding technology, enabling technology for chiplet integration. In *Proceedings of IEEE/ECTC*, 1385–1389, May 2023.
73. Inoue, F., S. Teranishi, T. Iwata, K. Onishi, N. Yamamoto, A. Kawai, S. Aoki, T. Hare, and A. Uedono. 2023. Inorganic temporary direct bonding for collective die to wafer hybrid bonding. In *Proceedings of IEEE/ECTC*, 556–563, May 2023.
74. Dubey, V., D. Wünsch, K. Gottfried, M. Wiemer, T. Fischer, S. Schermer, N. Dittmar, C. Helke, M. Hasse, S. Ghosal, A. Hanisch, J. Bonitz, J. Luo-Hofmann, L. Hofmann, M. Lykova, F. Stoll, K. Vogel, and S. Schulz. 2023. Impact of dielectric and copper via design on wafer-to-wafer hybrid bonding. In *Proceedings of IEEE/ECTC*, 795–799, May 2023.
75. Lee, S., G. Oh, J. Choi, Y. Kim, S. Park, and S. Eunkyung Kim. 2023. Investigation of Cu-to-Cu and oxide-to-oxide bonding. In *Proceedings of IEEE/ECTC*, 1519–1523, May 2023.
76. Fan, C., P. Ng, S. So, N. Man, C. Lee, M. Li, T. Uhrmann, J. Burggraf, and M. Pires. 2023. Direct die to wafer cu hybrid bonding for volume production. In *Proceedings of IEEE/ECTC*, 91–96, May 2023.
77. Li, A., J. Jiang, Q. Wang, N. Jing, Z. Dong, S. Ji, X. Cheng, and Y. Zhao. 2023. A method to improve 3D interconnections resource utilization and reliability in hybrid bonding process considering the effects on signal integrity. In *Proceedings of IEEE/ECTC*, 2131–2136, May 2023.
78. Zheng, D., M. Zhou, S. Liu, C. Ke, X. Zhang. 2023. An intensive study of effects of orientations of Cu bumps on Cu–Cu direct bonding for 3D integration by molecular dynamics simulation. In *Proceedings of IEEE/ECTC*, 1760–1766, May 2023.
79. Nakagawa, T., D. Furuyama, S. Nakagawa, Y. Mori, K. Iwata, K. Nakaya, and T. Katase. 2023. Investigation of Cu-Cu direct bonding process utilized by high porosity and nanocrystal structure. In *Proceedings of IEEE/ECTC*, 356–361, May 2023.
80. Ip, N., N. Nejadsadeghi, N. Kohama, H. Tanoue, and K. Motoda. 2023. Distortion simulation for direct wafer-to-wafer bonding process. In *Proceedings of IEEE/ECTC*, 694–698, May 2023.
81. Yang, S., and C. Chen. 2023. High temperature storage of Cu-Cu joints fabricated by highly (111)-oriented nanotwinned Cu. In *Proceedings of IEEE/ECTC*, 2075–2078, May 2023.
82. Ong, J., D. Tran, W. Chiu, Y. Chen, O. Lee, H. Chang, and C. Chen. Surface modification on hydrophilicity enhancement using NH4OH, NaOH, and KOH on fine-pitch low-temperature Cu/SiO_2 hybrid bonding. In *Proceedings of IEEE/ECTC*, 1549–1552, May 2023.
83. Kwak, B., J. Lee, S. Yoon, and B. Yoo. 2023. Copper-to-copper direct bonding process using current-induced enhancement method. In *Proceedings of IEEE/ECTC*, 2120–2124, May 2023.

Chapter 3
Fan-In Technology

3.1 Introduction

The wafer-level chip-scale package (WLCSP) era began in 2001, when Amkor led the outsourced semiconductor assembly and tests and foundries to license the UltraCSP [1–3] from Flip Chip Technologies (now Flip Chip International, a subsidiary of Huatian Technology in China). In the past 20 years [3–39], WLCSPs have been used extensively for low-cost, low-end, low-profile, low-pin count, small-form factor, and high-volume applications such as mobile (smartwatches, smartphones, and tablets) and portable (digital cameras and notebooks) products. Specially, it is used for housing the semiconductor devices with low pin-counts ($\leq$200), small die size ($\leq$6 mm x 6 mm), low-cost, low-end, low-profile, and high-volume applications.

The packages made from the fan-in wafer-level packaging technology are called wafer-level chip scale packages (WLCSPs). The unique feature of most WLCSPs is the use of a metal (usually Cu) RDL (redistribution-Lyer) to redistribute the fine-pitch peripheral-arrayed pads on the chips of a wafer to much larger-pitch area-arrayed pads with much taller solder joints as shown in Figs. 3.1, 3.2, 3.3, 3.4 and 3.5. Thus, with WLCSPs, the demand on the PCB is relaxed and the underfill may not be necessary. From the system houses' point of view, WLCSPs is just like another "solder-bumped flip chip" surface-mount component, except for the followings: (a) the solder bumps of WLCSP are much taller and bigger; (b) the PCB assembly of WLCSP is more robust; and (c) may not have to struggle with the underfill encapsulant. Also, with WLCSP, the electrical performance could be enhanced by: (1) distributing most of the power and ground toward the center of the die in an area-array format to minimize voltage drop, and (2) adding more power and ground solder bumps surround the sensitive I/O signal solder bumps to minimize the simultaneous switching noise and ground bounce.

J. H. Lau, *Flip Chip, Hybrid Bonding, Fan-In, and Fan-Out Technology*,
https://doi.org/10.1007/978-981-97-2140-5_3

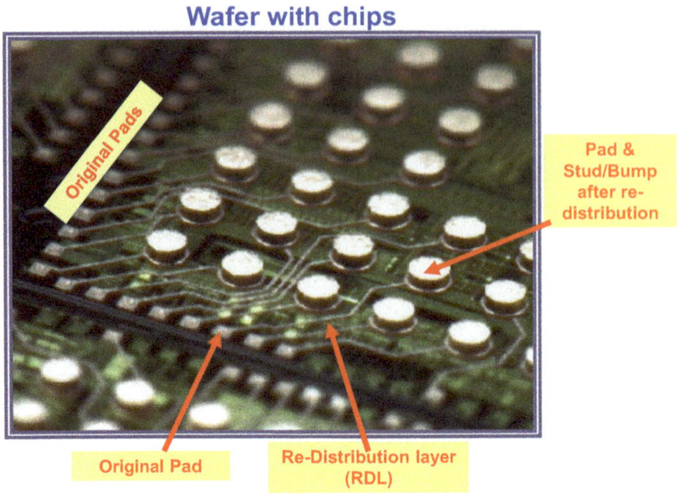

Fig. 3.1 Wafer level chip scale package (WLCSP)

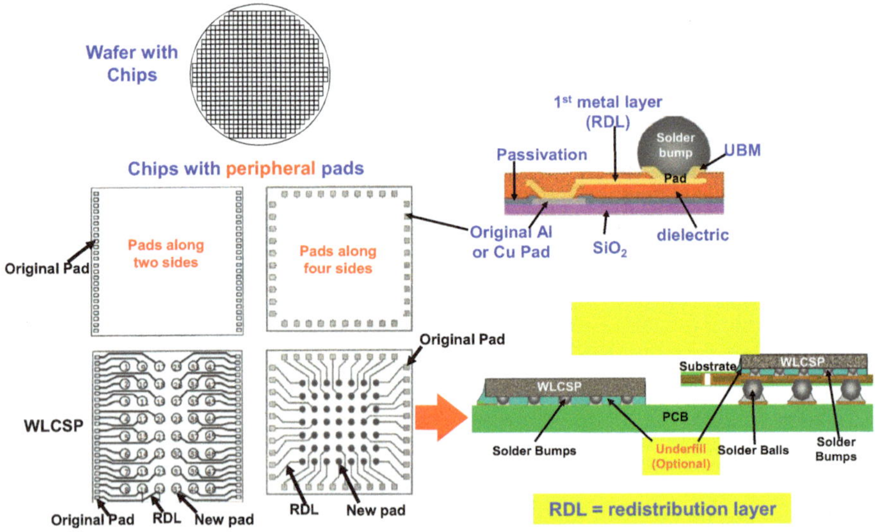

Fig. 3.2 Key elements of WLCSP. Fan-in from peripheral pads

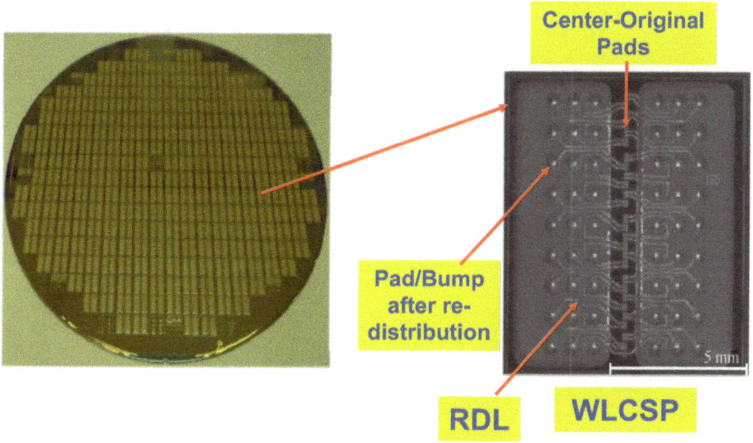

Fig. 3.3 WLCSP. Fan-in from the center line pads

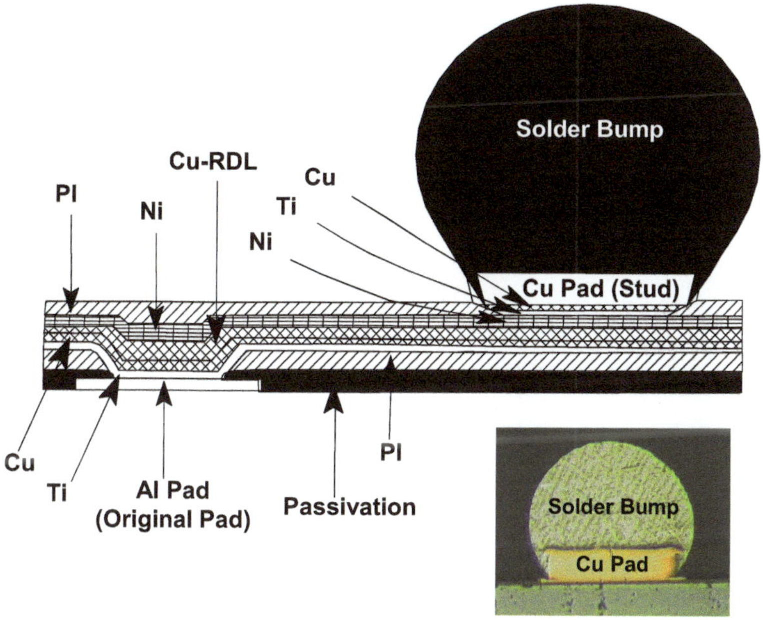

Fig. 3.4 Fan-in from the original peripheral pads

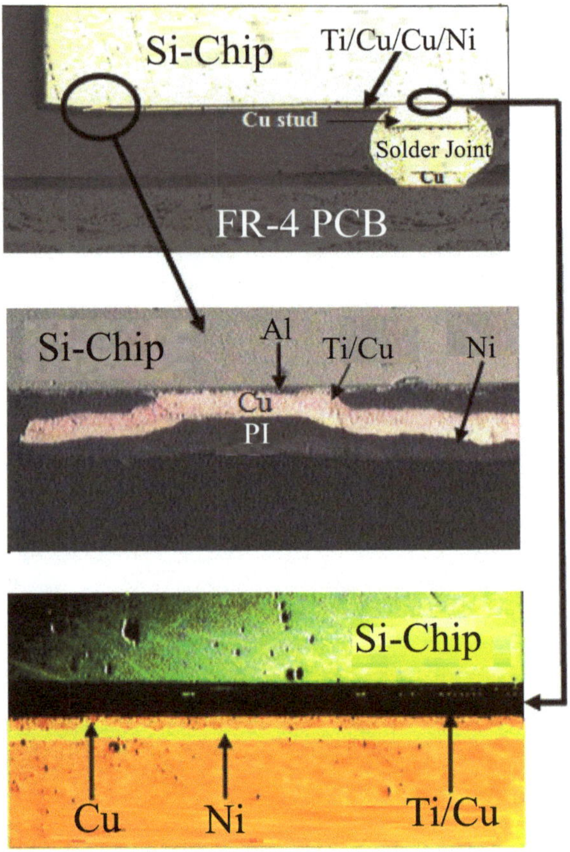

Fig. 3.5 Details of WLCSP showing the original pad and RDL

3.2 Creep Analysis of a Lead-Free WLCSP

3.2.1 The Structure

In this study, two lead-free solder alloys are considered, 96.5wt%Sn-3.5wt%Ag and 100wt%In [39]. The melting point for 96.6Sn-3.5Ag and 100In solder alloys are 221 °C and 156.6 °C, respectively, and thus, each of them has some unique applications. The 62Sn-36Pb-2Ag solder alloy is also considered to establish a baseline. From the cost point of view, 100In is the most expensive, and of course, 62Sn-36Pb-2Ag is the cheapest.

For solder bumped flip chip on low-cost substrate applications, even with the wafer-level pad re-distribution (from peripheral-array to area-array) to relax the pressure on PCB, however, in most of the cases, one to two, or even three to four build-up layers with microvias are needed to fan-out the circuitry. In this study, the effects of microvia build-up circuits on solder joint reliability of the WLCSP are presented.

Since there is no underfill for most of the WLCSP assemblies and the thermal expansion mismatch between the silicon chip and the FR-4 epoxy glass PCB is very large, thus solder joint reliability of the WLCSP assemblies becomes a critical issue. In this study, the focus is not on manufacturability but on the solder joint reliability with these lead-free solder alloys in WLCSP on PCB assemblies. Only thermal cycling loadings are considered. Emphasis is placed on the creep behaviors of these lead-free solder joints in WLCSP on PCB assemblies and is not on the life prediction of these solder joints.

Figure 3.6 schematically shows the silicon chip under consideration. The dimensions of the chip are: 9 × 9 × 0.51 mm. It has 121 peripheral pads (0.06 × 0.06 mm) with a spacing of 0.1 mm. After wafer-level redistribution, the pads (0.3 mm in diameter) are in an area-arrayed format with 0.75 mm pitch.

Figure 3.7 shows the PCB assembly of the lead-free solder bumped WLCSP. It can be seen that the PCB is 1.575 mm thick and is made of FR-4 Epoxy glass. The copper pad thickness is 0.018 mm as shown in Fig. 3.7. The solder joint height is 0.1524 mm and is made of lead-free and eutectic solder alloys.

3.2.2 Material Properties

Figure 3.8 shows a typical finite element model for creep analysis of the lead-free solder bumped WLCSP on PCB assemblies. Due to symmetry, only one-half of the structure is modeled and it is a 2-D analysis. Since the focus is on the corner solder joint, finer meshes are used to model it. The finite element code used in this study is ANSYS. It can solve boundary-value problems with Garofalo-Arrhenius steady-state creep constitutive equation expressed by [40]:

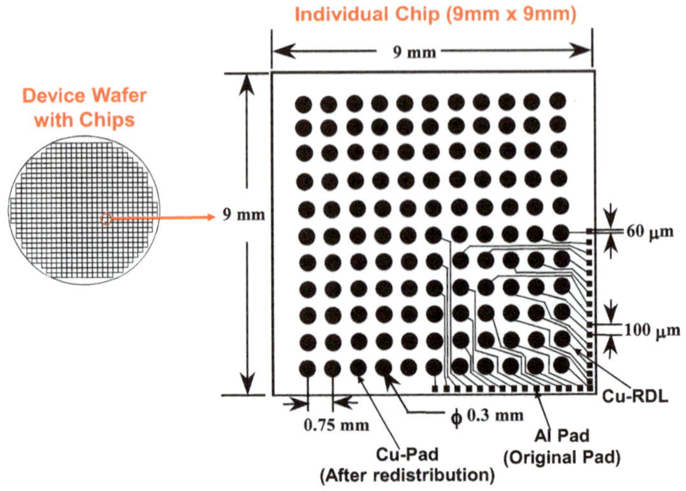

Fig. 3.6 Schematic of a chip with pad re-distribution

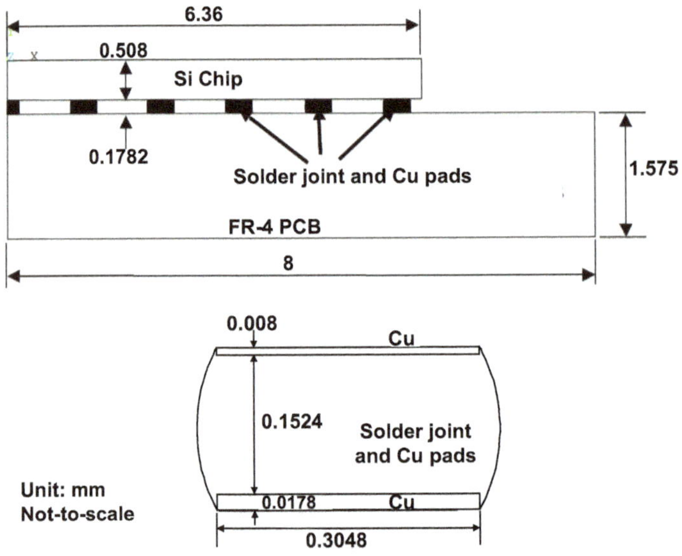

Fig. 3.7 Schematic of one-half of the WLCSP assembly and dimensions of the solder joint

$$\frac{d\gamma}{dt} = C\left(\frac{G}{\Theta}\right)\left[\sinh\left(\omega\frac{\tau}{G}\right)\right]^n \exp\left(-\frac{Q}{k\Theta}\right) \tag{3.1}$$

where γ is the steady-state creep shear strain, $d\gamma/dt$ is the steady-state creep shear strain rate, t is the time, C is a material constant, G is the temperature-dependent

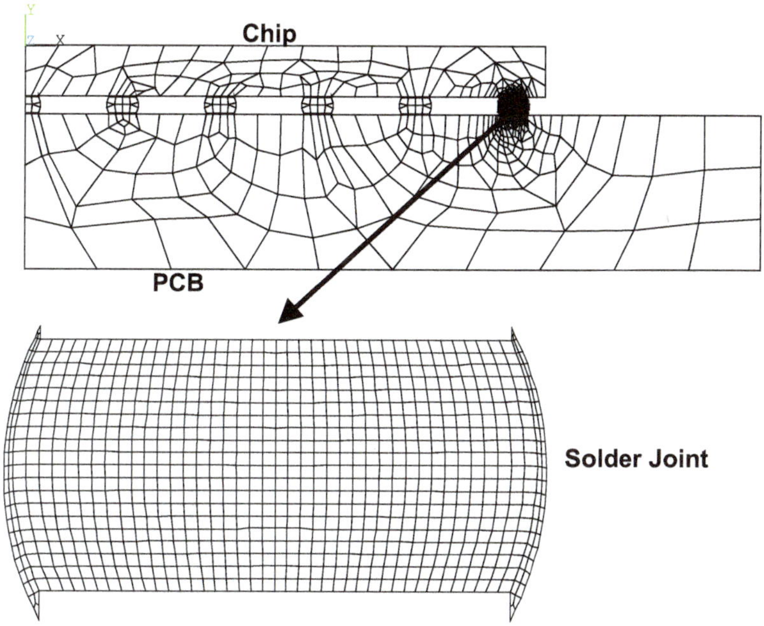

Fig. 3.8 Finite element model of one-half of the WLCSP assembly

shear modulus, Θ is the absolute temperature (°K), ω defines the stress level at which the power law stress dependence breaks down, τ is the shear stress, n is the stress exponent, Q is the activation energy for a specific diffusion mechanism (for example, dislocation diffusion, solute diffusion, lattice self-diffusion, and grain boundary diffusion), and k is the Boltzmann's constant (8.617×10^{-5} eV/°K). For 62Sn36Pb2Ag, 96.5Sn3.5Ag, and 100In solder alloys, the material constants of Eq. (3.1) have been experimentally determined by Darveaux and Banerji [40] with a single hyperbolic sine function given by [40]

$$\frac{d\gamma}{dt} = A\left[\sinh\left(\frac{\tau}{B}\right)\right]^n \exp\left(-\frac{Q}{kT}\right) \tag{3.2}$$

Table 3.1 shows the values of constant n, Q, A, and B.

Table 3.1 Fitted equations from shear creep test results

Solder alloys	n	Q (eV)	A (1/sec)	B (psi)
62Sn36Pb2Ag	3.3	0.548	801(508-T)/T	3163-6.23 T
96.5Sn3.5Ag	5.5	0.5	31(553-T)/T	3687-6.67 T
100In	5	0.72	70400(593-T)/T	158-0.27 T

Note T is absolute temperature in °K

Equations (3.2) for these three solder alloys at 50 and 75 °C are plotted in Fig. 3.9. It can be seen that: (1) for all the solder alloys and for all the temperatures, the higher the stress the higher the steady state creep strain rate; (2) for all the solder alloys and for all the shear stress, the higher the temperature the higher the steady state creep strain rate; (3) for all temperatures and for all the shear stress levels, the steady state creep strain rate of the 100In is much larger than that of the 62Sn36Pb2Ag and 96.6Sn3.5Ag; (4) for all stress levels, the steady state creep strain rate of the 100In is much larger than that of the 62Sn36Pb2Ag and 96.6Sn3.5Ag; and (5) for all operating temperatures and for most of the operating stresses, the steady state creep strain rate of 62Sn36Pb2Ag is larger than that of 96.5Sn3.5Ag, especially for higher temperatures and lower stresses.

If the solder materials obey the von Mises criterion [42], then Eq. (3.2) can be written as [41]

$$\frac{d\varepsilon}{dt} = C_1[\sinh(C_2\sigma)]^{C3}\exp\left(-\frac{C_4}{T}\right) \tag{3.3}$$

where C_1, C_2, C_3, and C_4 are given in Table 3.2 for the three solder alloys under consideration. It should be noted that Eq. (3.3) is exactly the same form of input for implicit creep model (TBOPT = 8) of ANSYS [43, 44]. In Eqs. (3.2) and (3.3), σ is the uniaxial stress, $d\varepsilon/dt$ is the uniaxial steady-state creep strain rate. The unit for σ and τ is in lb/in^2 (psi). The material properties of the silicon chip, FR-4 PCB, copper, and underfill are shown in Table 3.3.

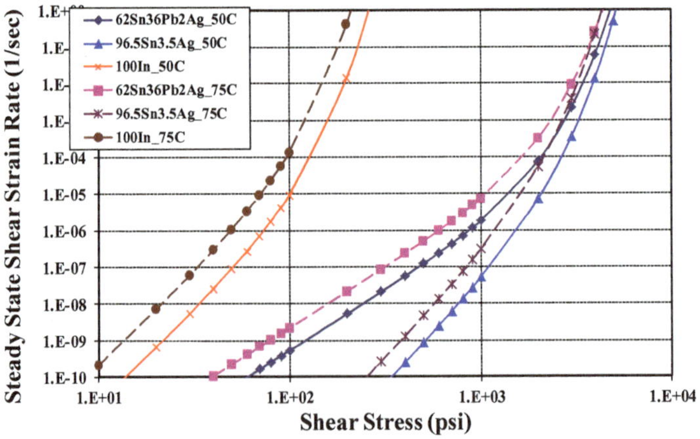

Fig. 3.9 Creep relations of some lead-free solders

Table 3.2 ANSYS input for implicit creep analysis

$\frac{d\varepsilon}{dt} = C_1[\sinh(C_2\sigma)]^{C_3} \exp\left(-\frac{C_4}{T}\right)$

Solder alloys	C_1 (1/sec)	C_2 (1/psi)	C_3	C_4 (°K)
62Sn36Pb2Ag	462(508-T)/T	1/(5478-10.79T)	3.3	6360
96.5Sn3.5Ag	18(553-T)/T	1/(6368-11.55T)	5.5	5802
100In	40647(593-T)/T	1/(274-0.47T)	5	8356

Note T is absolute temperature in °K

Table 3.3 Material properties of flip chip assemblies

Material properties	Young's modulus (MPa)	Poisson's ratio (ν)	CTE (α) ppm/°C
62Sn36Pb2Ag	34,441-152 T	0.35	24.5
96.5Sn3.5Ag	52,708-67.14 T $-0.0587T^2$	0.4	21.85+0.02039 T
100In	2,200	0.4	32.1
Underfill	9,292-35.4 T	0.35	31.0395+0.0923 T
Si	131,000	0.3	2.8
FR4	22,000	0.28	18
Copper	76,000	0.35	17
Build-up resin	20,000	0.3	50
ABF	7,000	0.3	7

Note T is temperature (°C). The electroplated copper is assumed to be electic-plastic

3.2.3 Boundary Conditions

The temperature loading imposed on the solder bumped WLCSP on PCB assemblies is shown in Fig. 3.10. It can be seen that for each cycle (60 min) the temperature is between –20 ↔ + 110 °C, with 15 min ramp, 20 min hold at hot, and 10 min hold at cold. There are two reasons for choosing this temperature profile: (1) the glass transition temperature of the FR-4 PCB is 120 °C and we do not want to introduce additional failure mechanisms of the solder joint due to the degradation of the PCB; and (2) the behavior of solder below −20 °C is not very well understood. Five full cycles are executed (Fig. 3.10).

3.2.4 Deformed Shapes, Strains, and Stresses

Figure 3.11a–e show the deformed shapes (50×) of the 62Sn2Ag36Pb solder bumped WLCSP on PCB assembly at the instants of 588, 4188, 7788, 11,388, and 14,988 s. It can be seen and expected that, due to the thermal expansion mismatch between the silicon (2.8 ppm/°C) chip and the FR-4 epoxy glass (18 ppm/°C) PCB, the solder

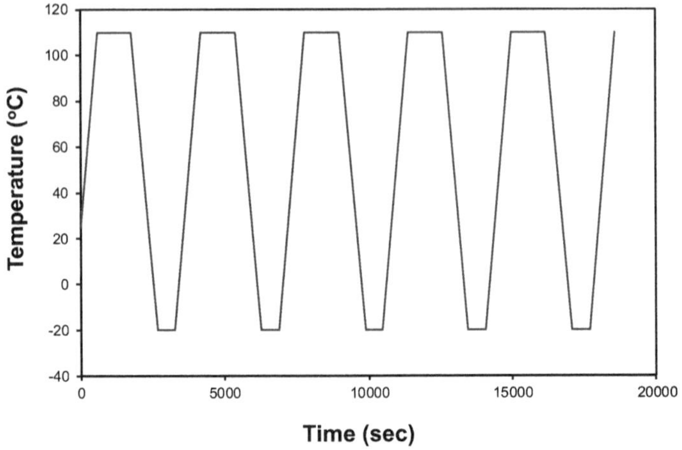

Fig. 3.10 Thermal cycling condition

joints are subjected to very large shear deformation (especially the corner solder joint). Also, the whole structure is deformed into a concave shape at these time instants (temperature ~110 °C). Figures 3.12a–e and 3.13a–e show, respectively, the effective creep strain contours and the effective stress contours in the corner solder joint of the 62Sn2Ag36Pb solder bumped WLCSP on PCB assembly at the instants of 588, 4188, 7788, 11,388, and 14,988 s.

It is interesting to note that at 588 s, the maximum deflection of the assembly is 0.028 mm and at that instant the maximum effective creep strain and stress in the corner solder joint are 0.05 and 15.01 MPa, respectively. However, at 4188 s, the maximum deflection of the assembly is reduced (by 21%) to 0.022 and at that instant the maximum effective creep strain in the corner solder joint is increased (by 78%) to 0.088 and the maximum effective stress is decreased (by 17.5%) to 12.39 MPa. This is because of the increasing deformation of the solder joints which makes the chip and the PCB less bonding together.

At 7788 s, the maximum deflection of the assembly is reduced (by 5%) to 0.021 and at that instant the maximum effective creep strain in the corner solder joint is increased (by 11%) to 0.098 and the maximum effective stress is decreased (by 3%) to 11.97 MPa.

At 11,388 s, the maximum deflection of the assembly is reduced (by 4.7%) to 0.020 and at that instant the maximum effective creep strain in the corner solder joint is increased (by 7%) to 0.105 and the maximum effective stress is decreased (by 1.8%) to 11.76 MPa.

At 14,988 s, the maximum deflection of the assembly is reduced (by 2%) to 0.0196 and at that instant the maximum effective creep strain in the corner solder joint is increased (by 4.7%) to 0.11 and the maximum effective stress is decreased (by 0.03%) to 11.76 MPa. Based on the foregoing results, it can be seen that: (1) the stress and creep strain are quite stabilized after the first thermal cycle, and (2) the

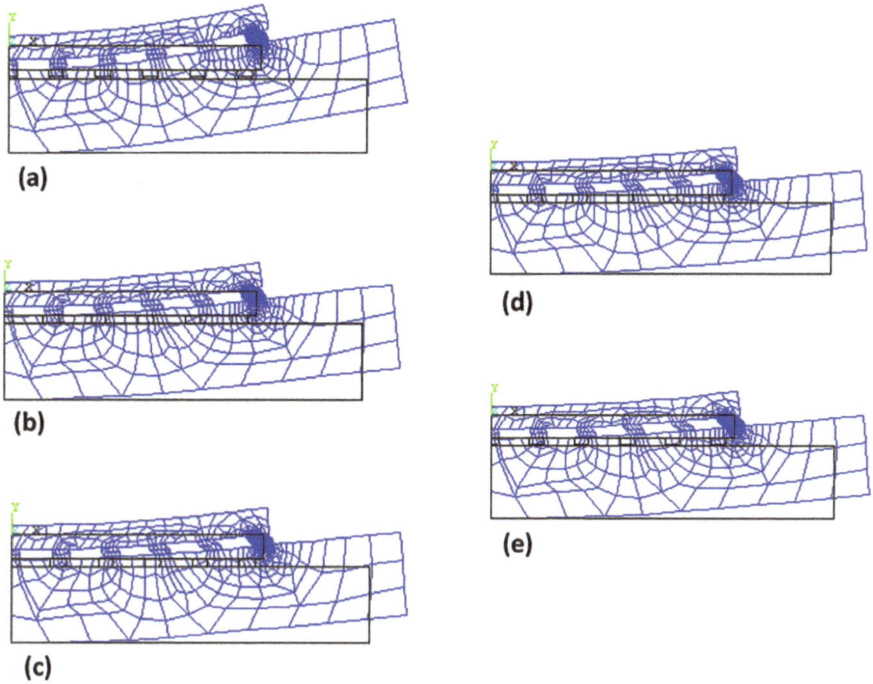

Fig. 3.11 Deformed shapes at **a** 588 s, **b** 4188 s, **c** 7788 s, **d** 11,388 s, and **e** 14,988 s

present analysis is converged after the third thermal cycles. Similar results for the WLCSP PCB assemblies with 96.5Sn3.5Ag, and 100In lead free solder joints.

3.2.5 Hysteresis Loops

For creep analysis, it is important to study the responses for multiple cycles until the hysteresis loops become stabilized. Figures 3.14, 3.15, and 3.16, respectively, show the shear stress and shear creep strain hysteresis loops for multiple cycles at the center of solder joints made by 62Sn36Pb2Ag, 96.5Sn3.5Ag, and 100In solder alloys. It can be seen that: (1) the creep shear strain is quite stabilized after the first cycle; (2) the creep shear strain range of the 100In (6.2%) is larger than that of the 62Sn36Pb2Ag (1%) and 96.5Sn3.5Ag (0.6%); and (3) the shear stress range of the 96.5Sn3.5Ag (31 MPa) is larger than that of the 62Sn36Pb2Ag (28 MPa) and 100In (3 MPa). This is expected, since 100In is the softest material. Also, 62Sn36Pb2Ag is softer than 96.5Sn3.5Ag.

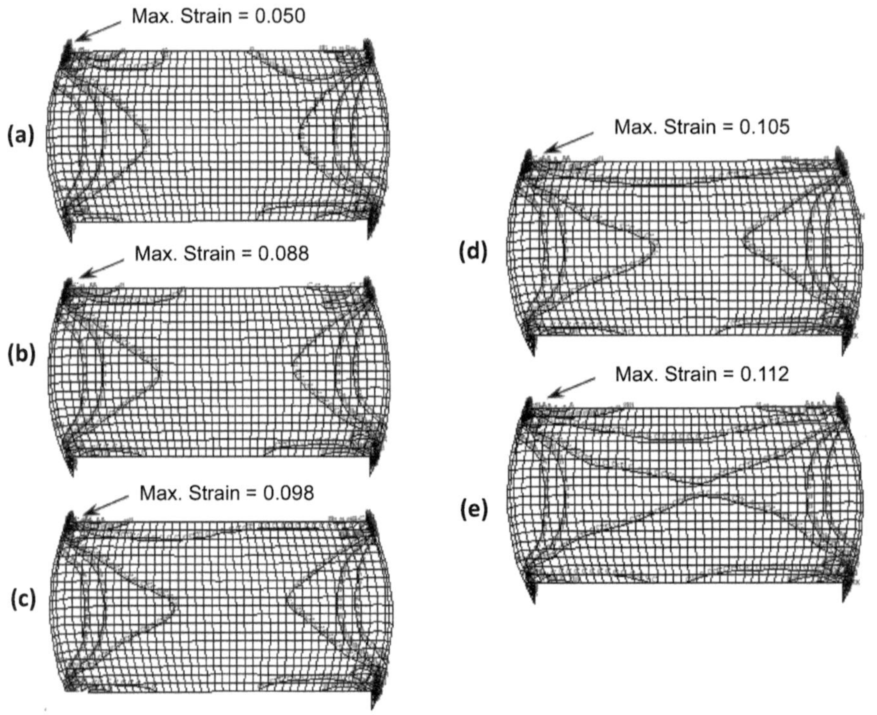

Fig. 3.12 Effective creep strain contours in the corner solder joint at **a** 588 s, **b** 4188 s, **c** 7788 s, **d** 11,388 s, and **e** 14,988 s

3.2.6 Time-Dependent Shear Stress

The shear-stress history at the center of solder joints made by the 62Sn36Pb2Ag, 96.5Sn3.5Ag, and 100In solder alloys is shown in Fig. 3.17. It can be seen that: (1) at this location, the shear-stress history of all these solder alloys follows the imposed thermal cycling condition; (2) the shear stress range of 100In is much smaller than that of 62Sn36Pb2Ag and 96.5Sn3.5Ag; and (3) the shear stress range of 96.5Sn3.5Ag is slight higher than that of 62Sn36Pb2Ag.

3.2.7 Time-Dependent Shear Creep Strain

The shear-creep-strain history at the center of solder joints made by the 62Sn36Pb2Ag, 96.5Sn3.5Ag, and 100In solder alloys is shown in Fig. 3.18. It can be seen that: (1) at this location, just like the shear-stress history, the shear-creep-strain

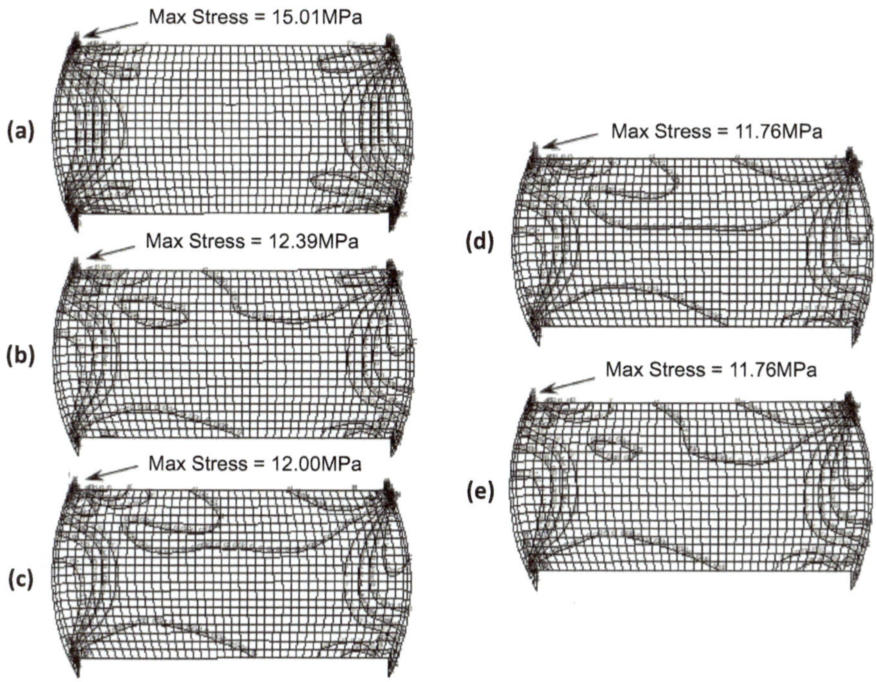

Fig. 3.13 Effective stress contours in the corner solder joint at **a** 588 s, **b** 4188 s, **c** 7788 s, **d** 11,388 s, and **e** 14,988 s

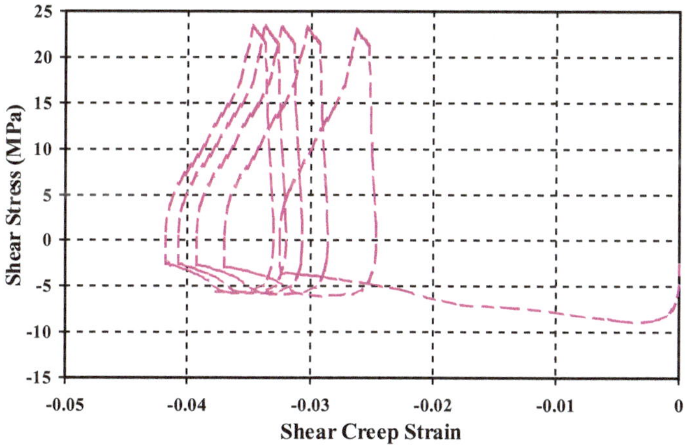

Fig. 3.14 Shear stress and shear creep strain hysteresis loops for 62Sn2Ag36Pb corner solder joint

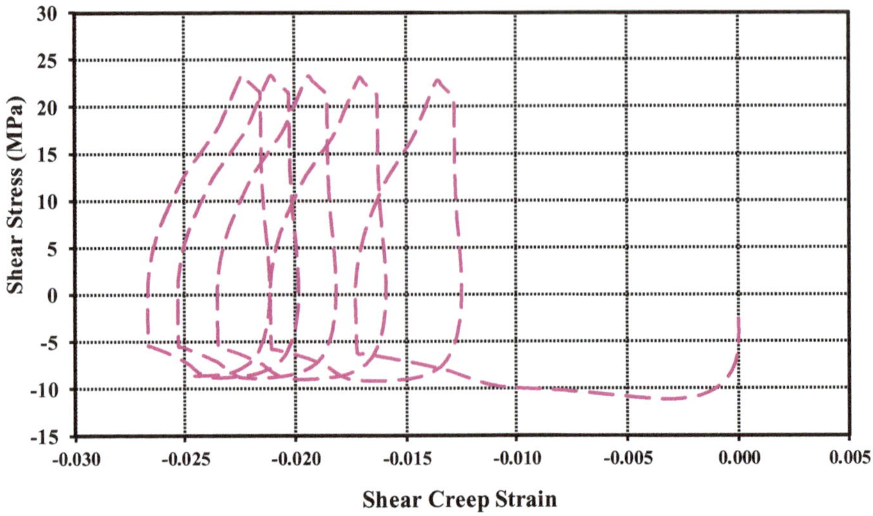

Fig. 3.15 Shear stress and shear creep strain hysteresis loops for 96.5Sn3.5Ag corner solder joint

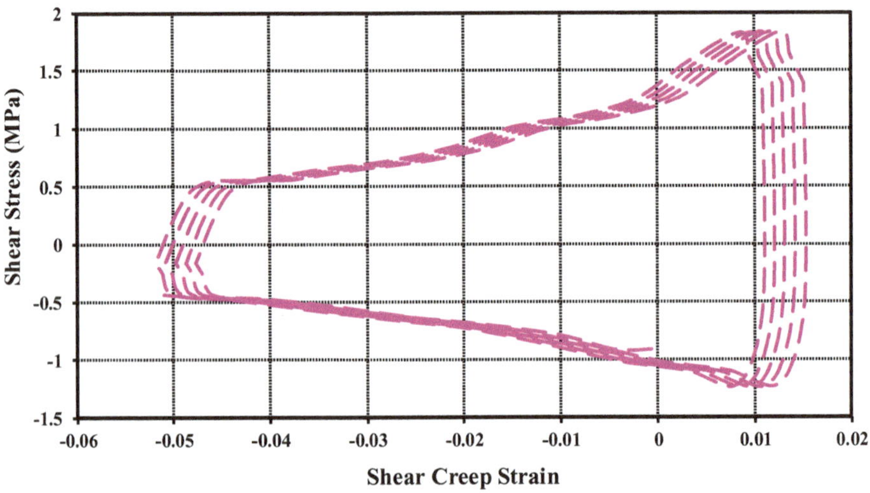

Fig. 3.16 Shear stress and shear creep strain hysteresis loops for 100In corner solder joint

history of all these solder alloys follows the imposed thermal cycling condition; (2) the shear-creep-strain range of 100In is much larger than that of 62Sn36Pb2Ag and 96.5Sn3.5Ag; and (3) the shear-creep-strain range of 96.5Sn3.5Ag is slight smaller than that of 62Sn36Pb2Ag.

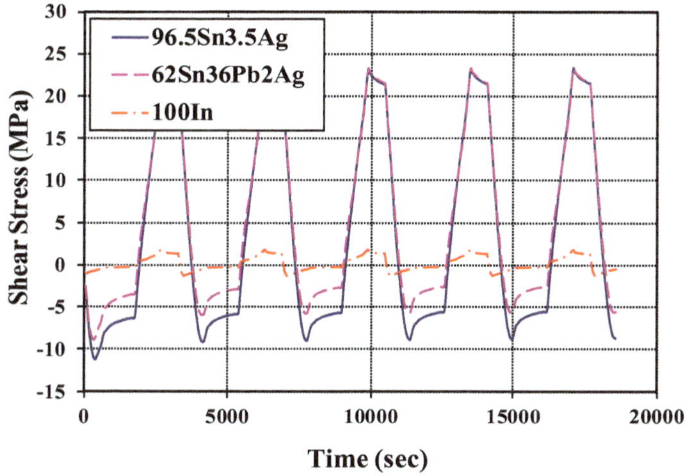

Fig. 3.17 Shear stress time-history for 62Sn2Ag36Pb, 96.5Sn3.5Ag, and 100In corner solder joints

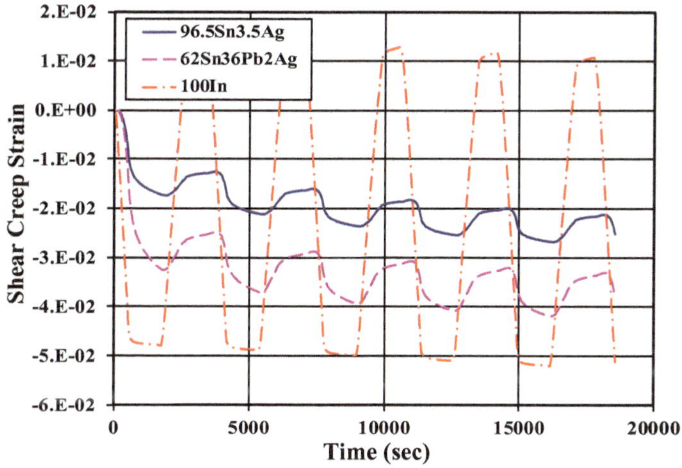

Fig. 3.18 Shear creep strain time-history for 62Sn2Ag36Pb, 96.5Sn3.5Ag, and 100In corner solder joints

3.3 Creep Analysis of a Lead-Free WLCSP on PCB with Microvia

3.3.1 Microvia Build-Up PCB

Figure 3.19 shows the cross sections of the WLCSP on a one-layer build-up with microvia PCB [39]. The dimensions of the chip, solder joints, copper, and PCB are exactly the same as those shown in Figs. 3.6 and 3.7. The build-up microvia dimensions are shown in Fig. 3.19. It can be seen that the diameters of the microvia are varying from 0.1 to 0.15 mm. The thickness of the electroplated copper microvia is 0.025 mm. The thickness of the build-up resin is about 0.125 mm. The Young's modulus, Poisson's ratio, thermal expansion of coefficient of the resin are, respectively, 20 GPa, 0.3, and 50 × 10^{-6}/°C. The electroplated copper is considered as an elastic–plastic material (yield stress = 54 MPa, yield strain = 0.007, Young's modulus = 76Gpa, and tangent modulus of plastic curve is 7.6GPa). The thermal expansion of coefficient of the electroplated copper is 17 × 10^{-6}/°C.

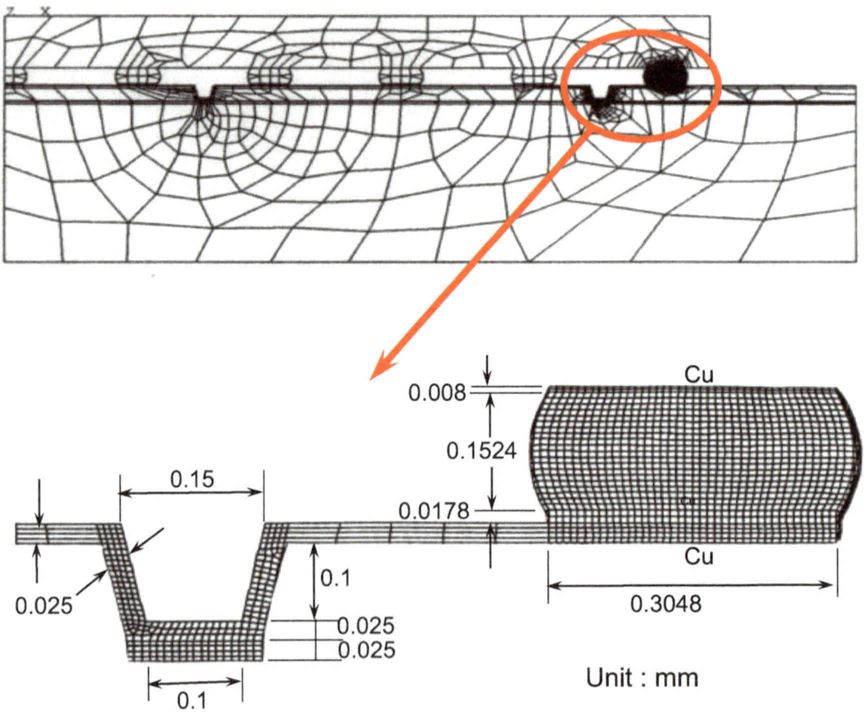

Fig. 3.19 Schematic of WLCSP on PCB with microvia build-up circuits

3.3.2 Deformed Shapes

The build-up PCB assembly shown in Fig. 3.20 is subjected to the same thermal loading condition (Fig. 3.10). Figure 3.20a–e show the deformed shapes (50×) of the WLCSP on the microvia build-up PCB assembly at the instants of 588, 4188, 7788, 11,388, and 14,988 s. It can be seen and expected that, due to the thermal expansion mismatch between the silicon (2.8 ppm/°C) chip, the FR-4 epoxy glass (18 ppm/°C) PCB, and the build-up resin (50 ppm/°C), the solder joints are subjected to very large shear deformation (especially the corner solder joint).

At 588 s, the maximum deflection of the assembly is 0.017 mm which is (65%) smaller than that of the PCB without the build-up layer (Fig. 3.7a). This is because of: (1) the global thermal expansion mismatch between the silicon chip and the build-up PCB which wants the whole assembly to deform into a concave shape; (2) the local thermal expansion mismatch between the build-up resin and the FR-4 PCB which wants the build-up PCB to deform into a convex shape (just opposite to the global deformation); and (3) the local thermal expansion mismatch between the silicon chip and the build-up resin which introduces considerable amount of shear creep strains in the solder joints.

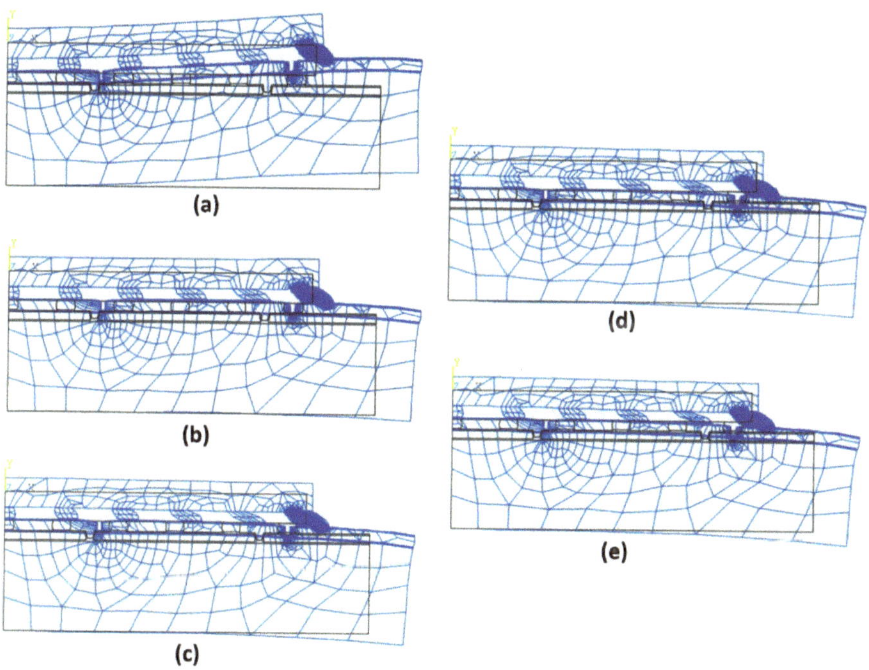

Fig. 3.20 Deformed shapes at of WLCSP on PCB with microvia build-up circuits at **a** 588 s, **b** 4188 s, **c** 7788 s, **d** 11,388 s, and **e** 14,988 s

However, at 4188 s, the deformed shape of the whole assembly is no longer in a concave shape but expands in the horizontal direction only. The reasons are basically those three mentioned above except that the cumulated shear creep strains in the solder joints are much larger than before, i.e., the structure is less bonded between the silicon chip and the build-up PCB. Consequently, the local thermal expansion mismatch between the build-up resin and the PCB became larger than the global thermal expansion mismatch between the chip and the build-up PCB.

At 7788, 11,388, and 14,988 s, the whole assembly (except the chip) is becoming to bend into a convex shape. This is because of: (1) the bonding between the chip and the build-up PCB is much less, and (2) the local mismatch is much more than the global mismatch.

3.3.3 Hysteresis Loops, Stress, and Creep Strain

Figures 3.21, 3.22, and 3.23 show, respectively, the shear stress and shear creep strain hysteresis loops, shear stress history, and shear creep strain history of the WLCSP on microvia build-in PCB. It can be seen from Fig. 3.21 that the creep responses converged at the third thermal cycle. Also, it can be seen that the stress range (Fig. 3.22) in the corner solder joint is 35 MPa, which is larger than that (28.5 MPa) of the case without the build-up layer. Similar, the shear creep strain range (Fig. 3.23) is 0.023 which is much larger than that (0.01) of the case without the microvia build-up layer. Again, this is because of the very larger thermal expansion of coefficient of the build-up resin.

3.4 Creep Analysis of a Lead-Free WLCSP on PCB with Two-Layers of Microvia

In this study, the effect of a two-layer build-up PCB with staggered microvias on the WLCSP solder joints is reported [44].

3.4.1 The Structure

Figure 3.24 shows the schematic diagram of the WLCSP under consideration. The dimensions of the chip are: 9 × 9 × 0.5 mm. It has 121 (0.06 × 0.06 mm) peripheral pads with a spacing of 0.1 mm. After the wafer-level redistribution, the pads are 0.3 mm in diameter and are in an area-arrayed format with 0.75 mm pitch (Fig. 3.6).

Figure 3.24 shows the cross-section of a 2-layer microvia build-up PCB assembly with a solder-bumped WLCSP. The PCB is a conventional FR-4 laminate with a

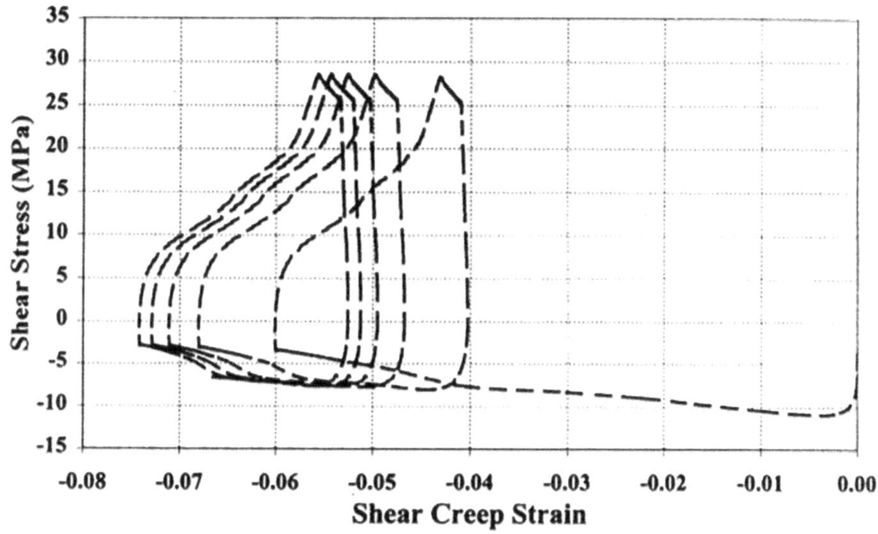

Fig. 3.21 Shear stress and shear creep strain hysteresis loops for 62Sn2Ag36Pb corner solder joint on build-up PCB

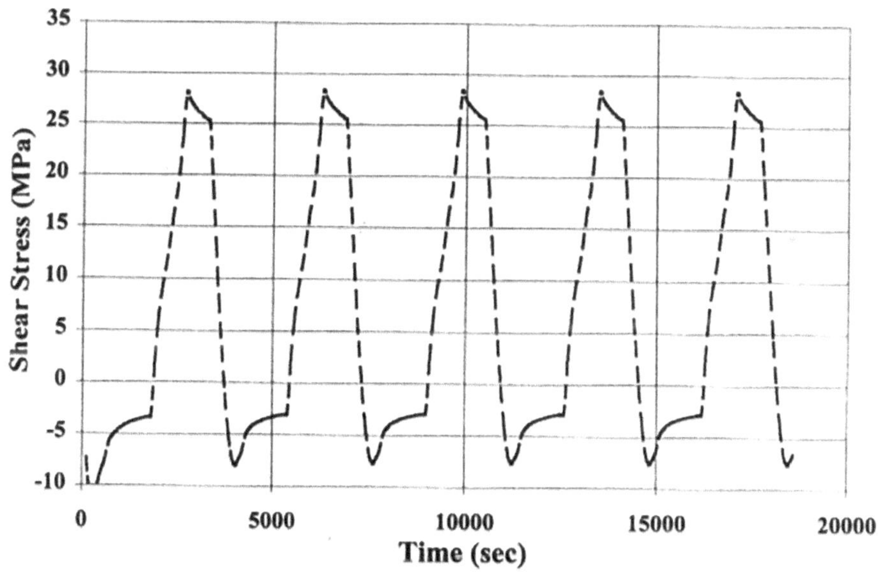

Fig. 3.22 Shear stress time-history for 62Sn2Ag36Pb corner solder joint on build-up PCB

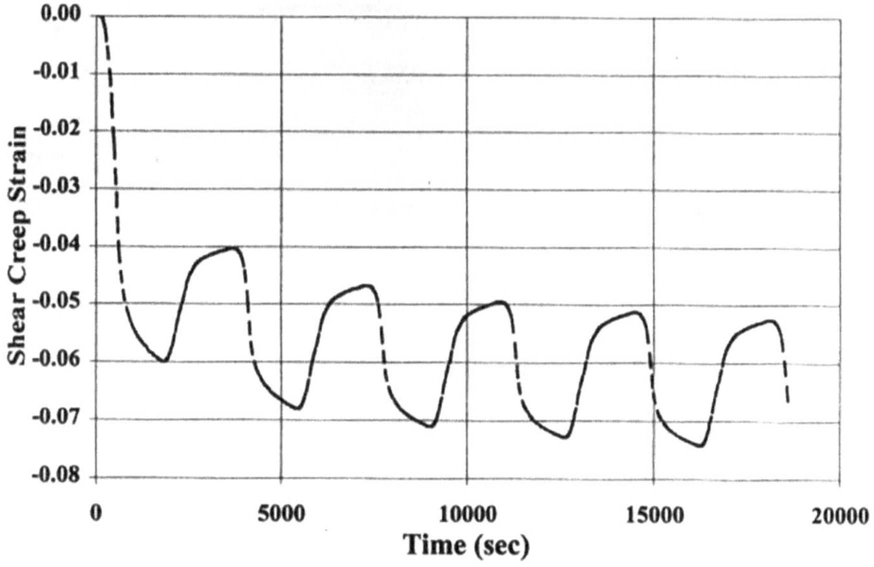

Fig. 3.23 Shear creep strain time-history for 62Sn2Ag36Pb corner solder joint on build-up PCB

thickness of 1.575 mm. The copper pad thickness is 0.0178 mm as shown in Fig. 3.24. The diameter of microvias tapers from 0.1 to 0.15 mm. The thickness of the microvia copper is 0.025 mm. The thickness of the build-up resin is about 0.125 mm.

3.4.2 Elastic Analysis and Results

Figure 3.25a shows the elastic deformation (50X) of the solder bumped WLCSP on the microvia build-up PCB. It can be seen that the whole assembly is deformed (with a maximum deflection = 0.0249 mm) into a concave shape under a temperature raise of 130 °C. This is because of the global thermal expansion mismatch between the silicon chip (2.8×10^{-6}/°C) and the FR-4 epoxy glass PCB (18×10^{-6}/°C). Also, it can be seen from Fig. 3.25a that the two microvia build-up layers and the upper-portion of the PCB tend to bend downward due to the local thermal expansion mismatch between the build-up resin (50×10^{-6}/°C) and the PCB. That is why the present global maximum deflection is slightly less than that (0.0255 mm) for the case with only one microvia build-up layer [23].

The von Mises stress contours in the second level microvia are shown in Fig. 3.25b. It can be seen that the maximum von Mises stress is 356 MPa, which is much larger than the ultimate strength of electroplated copper. This unreasonable stress result is due to the limitations of linear elastic analysis.

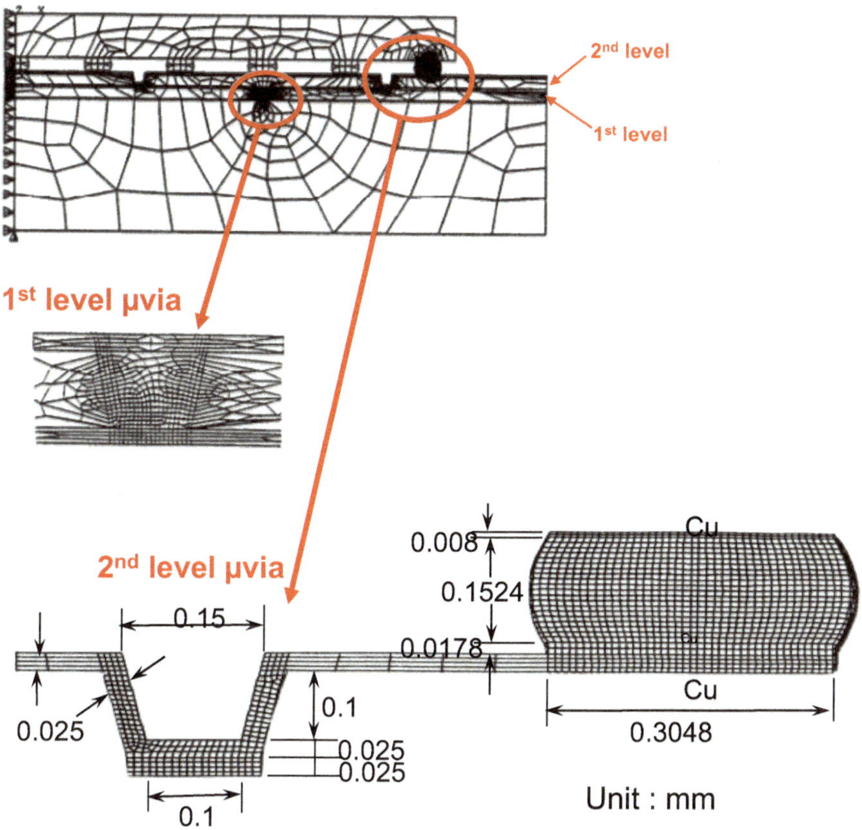

Fig. 3.24 Schematic diagram and finite element model of a WLCSP with two build-up layers PCB assemble

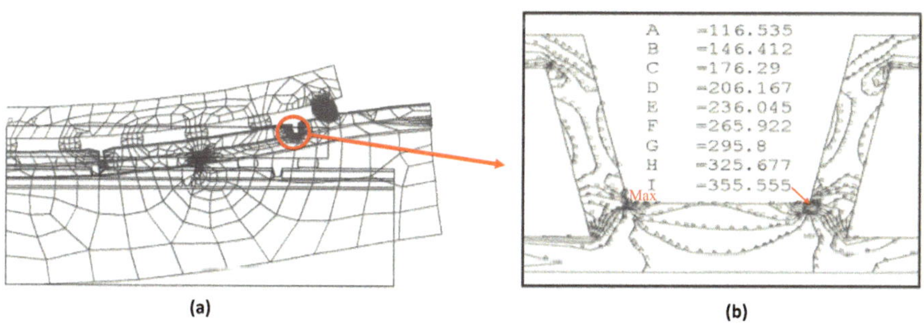

Fig. 3.25 **a** Deformation of WLCSP with two build-up layers of Microvias (elastic analysis). **b** von Mises stress contours in the 2nd Level Microvia (elastic analysis)

3.4.3 Elastic–Plastic Analysis and Results

In this analysis, the microvias and the solder joints are assumed to be elastic–plastic materials. Figure 3.26a shows the deformation (50×) of the whole assembly under a temperature change from –20 to 110 °C. It can be seen that the maximum deflection (0.0210 mm) is (19%) smaller than that (0.0249 mm) of the elastic analysis. This is due to the plastic deformation of the solder joints, which relaxes the bonding between the chip and the microvia build-up PCB.

The von Mises stress contours in the second level microvia are shown in Fig. 3.26b. It can be seen that the maximum von Mises stress (102 MPa) is much smaller than that (356 MPa) from the linear elastic analysis. By comparing the present results with those of the case with one-layer microvia build-up PCB (Table 3.4) [45], it can be seen that the maximum von Mises stress (63 MPa) and the maximum effective strain (0.0299) of the cases with one-layer microvia build-up PCB are smaller.

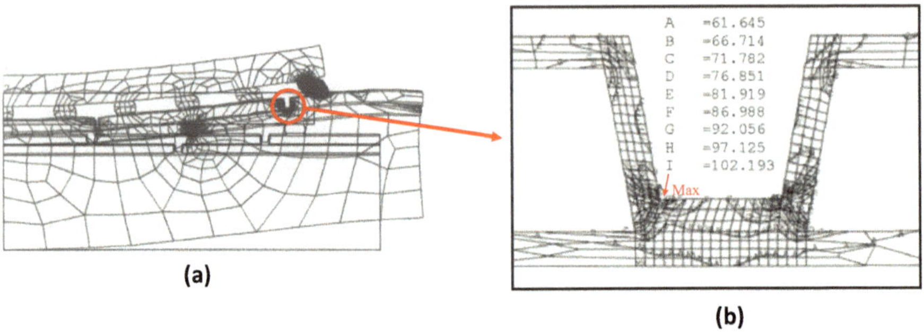

Fig. 3.26 **a** Deformation of WLCSP with two build-up layers of Microvias (elastic–plastic analysis). **b** von Mises stress contours in the 2nd level Microvia (elastic–plastic analysis)

Table 3.4 Comparison among various cases

Items	Type	1-layer μ-via Build-up PCB	2-layer μ-via Build-up PCB
Stress (MPa)	Elastic	163	198
	Plastic	63	92
	Creep	35	43
Strain (%)	Elastic	1.24	1.51
	Plastic	1.75	2.93
	Creep	2.30	4.60

3.4.4 *Creep Analysis and Results*

(A) Deformed Shapes

Figure 3.27a–e show the deformed shapes (50×) of the WLCSP on the two-layer microvia build-up PCB assembly at the instants of 588, 4188, 7788, and 14,988 s (temperature = 110 °C). It can be seen and expected that, due to the thermal expansion mismatch among the silicon chip, the FR-4 PCB, and the build-up resin, the solder joints are subjected to very large shear deformation (especially the corner solder joint). By comparing the present results with those reported in [23] with one-layer microvia build-up PCB, it can be seen that the deformed shapes are very different. In the present case: (a) the global thermal expansion mismatch between the silicon chip and the organic build-up PCB is larger; (b) the local thermal expansion mismatch between the build-up resin and the FR-4 PCB is larger; and (c) the solder joint, especially the corner one, is subjected to larger deformation.

(B) Stress and Strain in the Microvia

The von Mises stress contours in the 1st-level and 2nd-level microvias at 14,988 s are shown in Fig. 3.28a and b, respectively. (It should be pointed out that at high

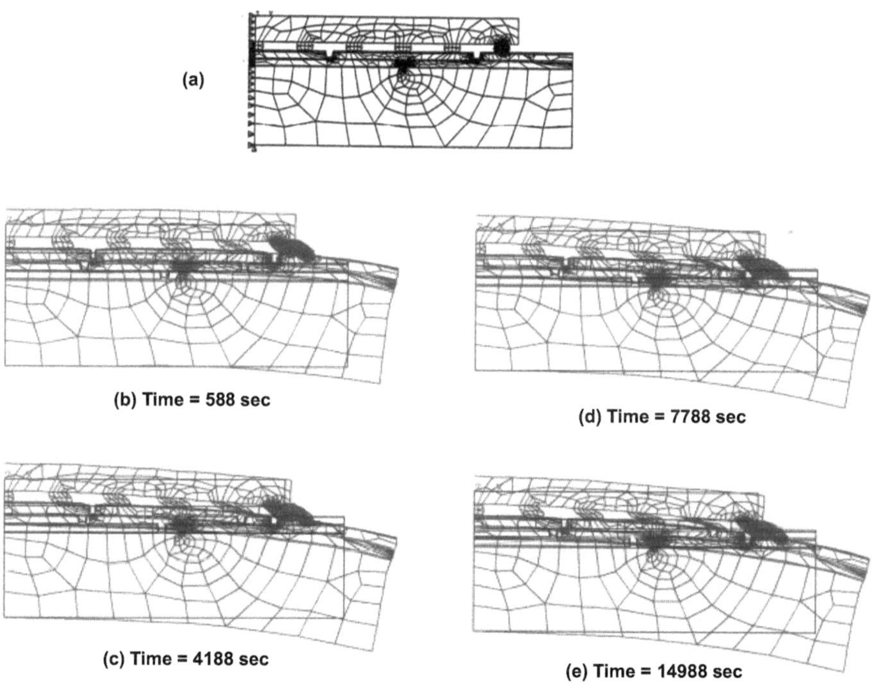

Fig. 3.27 Deformation of WLCSP-PCB assembly with two build-up layers of microvias (all at 110 °C)

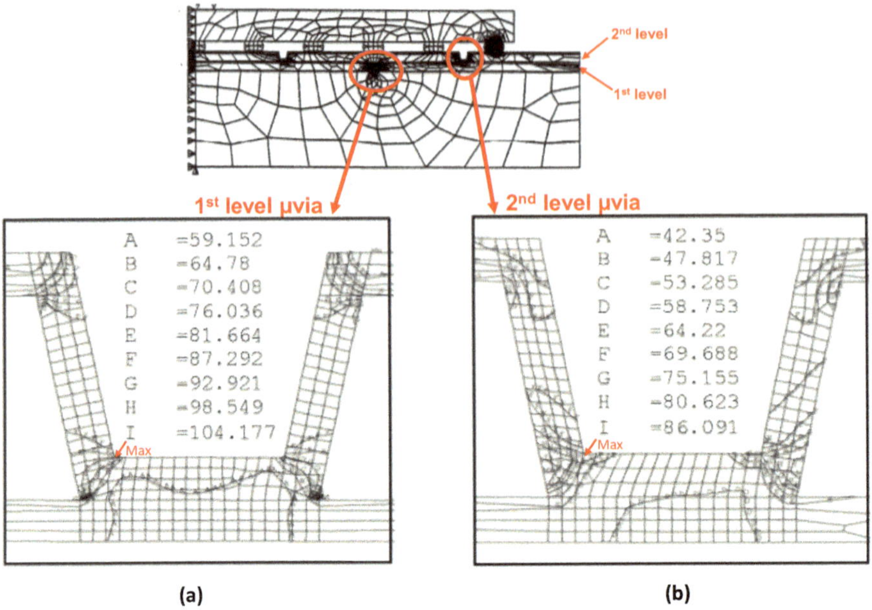

Fig. 3.28 Vvon Mises stress contours in the Microvias (Time = 14,988 s.). **a** 1st Level Microvia, **b** 2nd level Microvia

temperature, 110 °C, the stress does not change much from thermal cycle to cycle.) It can be seen that: (1) the stress in the 1st-level microvia is larger than that in the 2nd-level microvia; (2) the maximum value (104 MPa) is smaller than those from elastic and elasticplastic analyses. Again, this is due to the creep deformation of the solder joints that relaxes the mechanical bonding between the chip and the PCB. Since the maximum stress in the microvias is much smaller than the ultimate strength of the electroplated copper so the microvias should be reliable for most of the operation conditions.

3.5 Hitachi's WLCSP

Figure 3.29a shows a Hitachi's wafer with memory chips. The original pads on the memory chips are very fine-pitch and distributed along the centerline of the chip as shown in Fig. 3.29b.

A metal RDL is used to redistribute the fine-pitch original pads along the centerline of the chips on a wafer to much larger pitch area-arrayed pads all over the area within the chips as shown in Fig. 3.29b. In order to apply to larger chips (e.g., 10 mm × 10 mm) without underfill, Hitachi introduced [46] a stress relaxation layer between the Si chip and the solder bumps as shown in Fig. 3.30.

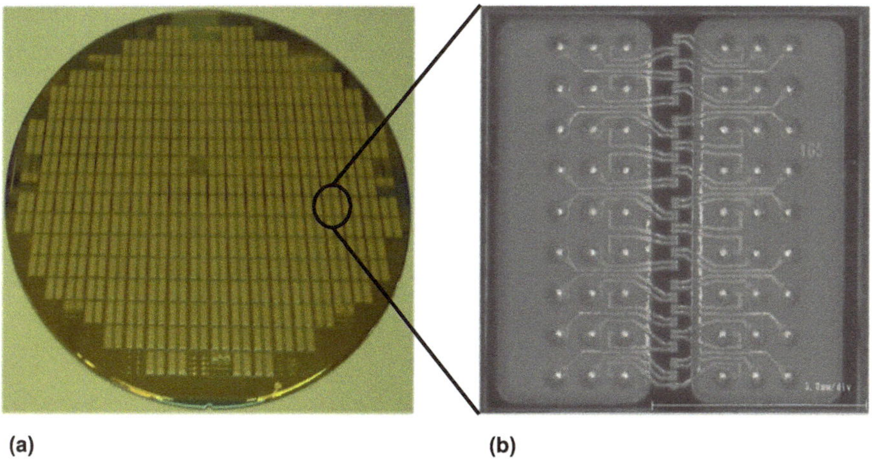

Fig. 3.29 **a** Hitachi's memory chips on a 200 mm-wafer. **b** Hitachi's WLCSP with original pads along the centerline and redistributed pads area array within the chip area

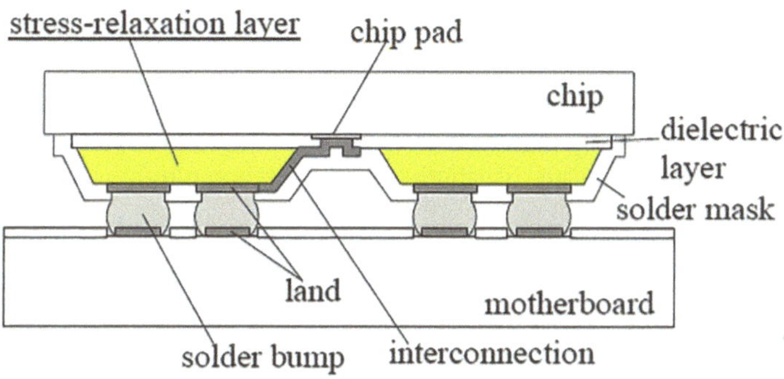

Fig. 3.30 Cross section of Hitachi's WLCSP mounted on a PCB

The key process steps of Hitachi's WLCSP with a stress relaxation layer are shown in Fig. 3.31. First, the wafer is spin-coated with a photosensitive polyimide (P–PI), and the chip pads and dicing streets are opened up by photolithography. Then, a stress relaxation layer is formed by printing liquid resin on the wafer through a stencil mask with no opening on the pads and dicing streets. The resin will flow through the open edges and form smooth slopes.

To add a layer of metal on the wafer, Hitachi sputters the seed metals (Cr and Cu) on the whole wafer, spin-costs a photoresist layer, and opens up the shapes of interconnections and new pads (in area array format) by photolithography. Then, the interconnections and pads are formed by electroplating the Cu and Ni. The photoresist and seed metals are removed by etching.

Now, the wafer is spin-coating with the P–PI again and the new chip pads and dicing streets are opened up by photolithography. This time, the P–PI acts like a

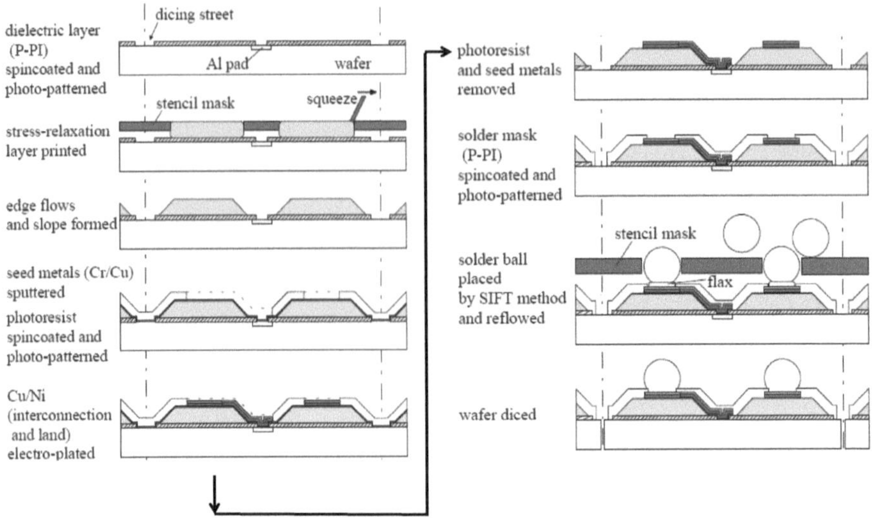

Fig. 3.31 Process steps in making Hitachi's WLCSP

solder mask. After fluxing, the SnAgCu solder bumps are placed on the new land pads on the wafer through a stencil mask with a solder ball mounter. Finally, the wafer is reflowed, cleaned, and diced.

Figure 3.32 shows the life distribution of the temperature-cycling test ($-55 \leftrightarrow 125$ °C) of the Hitachi's (10 mm × 10 mm) WLCSP (without an underfill) with a stress relaxation layer on PCB. It can be seen that the characterization life of the WLCSP with a stress relaxation layer is more than 3000 cycles. On the other hand, the characterization life is less than 200 cycles for the same chip size without stress relaxation layer and underfill.

3.6 TSMC's WLCSP

In order to save some process steps, reduce the cost, and lower package profile, TSMC [4] eliminate the under-bump metallurgy (UBM) in their WLCSP, and called it UFI (UBM-free integration) fan-in WLCSP.

Figure 3.33a shows the conventional WLCSP and Fig. 3.33b shows the TSMC's UFI WLCSP. It can be seen that for the UFI WLCSP, the polymer-2 and UBM are gone. Instead, a polymer composite protection layer (PL) is introduced to secure the solder bumps.

In TSMC's UFI WLCSP, the solder balls are directly mounted on the Cu RDL which routes electrical signals from the original pads of the chip to the desired area array pads. Then, it is followed by the PL deposition to secure the solder bumps.

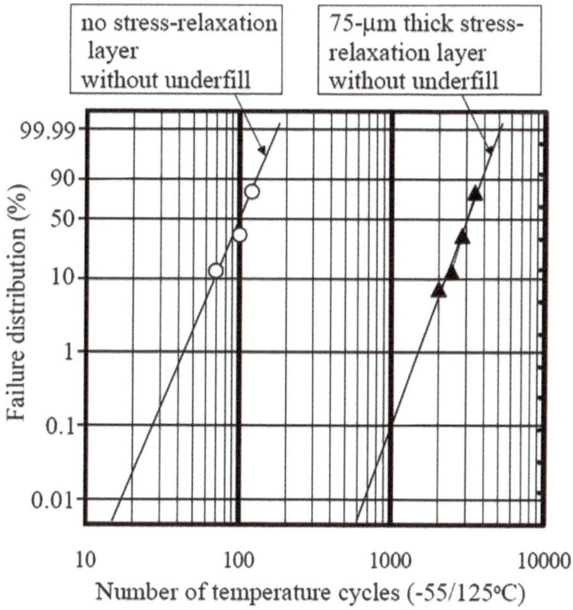

Fig. 3.32 Temperature-cycling test of Hitachi's WLCSP with a stress relaxation layer

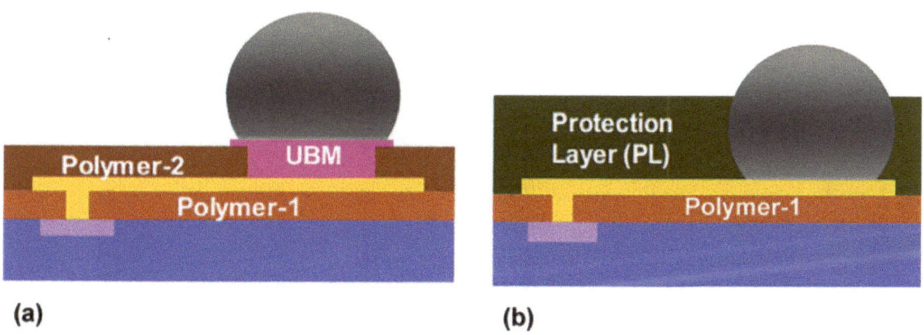

Fig. 3.33 **a** Conventional WLCSP. **b** TSMC's UFI WLCSP

Figure 3.34 shows the life distribution of the thermal cycling results of TSMC's UFI WLCSP on PCB. TSMC [4] finds that because of: (1) the mechanical support of the PL, (2) the controlling of the maximum strain location, and (3) the material optimization, the UFI WLCSP solder joint is reliable (the characteristic life is more than 800 cycles and the thermal cycling takes place between the temperatures of $-40 \leftrightarrow +125$ °C, at one cycle per hour) even for a chip size up to 10.3 mm × 10.3 mm without underfill.

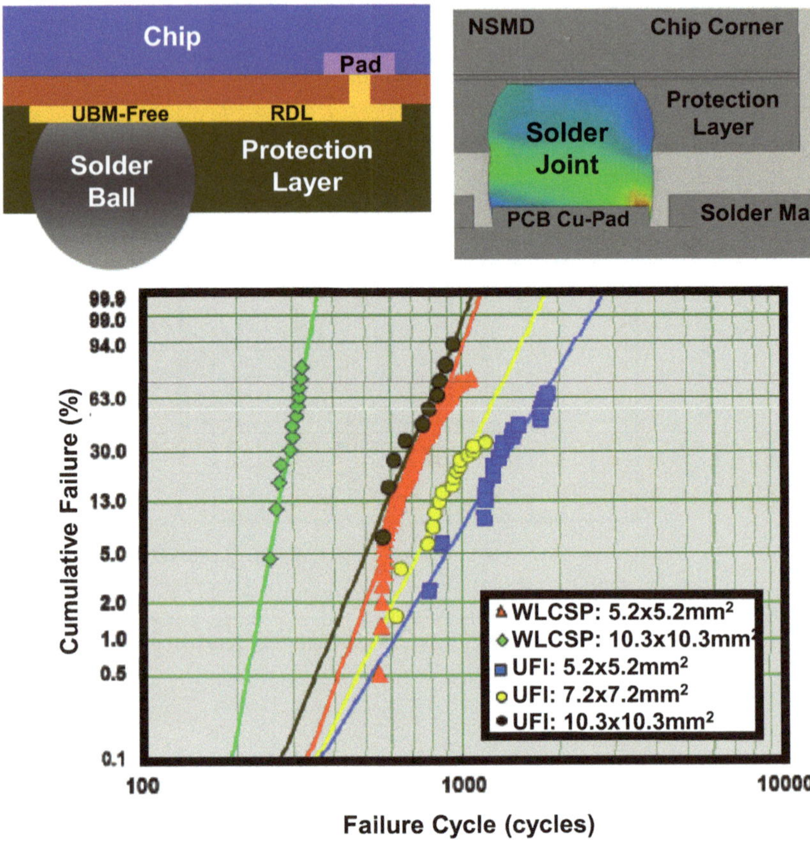

Fig. 3.34 Reliability of TSMC's UFI WLCSP with polymer composite protection layer

3.7 Material, Process, and Fabrication of Six-Sides Molded PLCSP

3.7.1 Introduction

Recently, because (1) of the delamination of the dielectric layer in the front side of the WLCSP, which is particularly true for advanced node (<14 nm process technology) products with fragile polyimides, (2) of the backside chipping and sidewall cracking due to the mechanical blade dicing the wafer, (3) of the concern of the handling and surface mount technology (SMT) pick and place to damage the chip, and (4) in automotive electronics, the trends are driving lead-free solder joint reliability for new functions such as the advanced driver-assistance systems, under-the-hood operations often require high sustained heating/cooling temperatures, a higher cost five-side or six-side molded WLCSPs are getting tractions [47–57].

The five-side molded WLCSP was first patented (e.g., US8456002) [47] in 2011 and published [48–51] by STATS ChipPAC and is called encapsulated WLCSP (eWLCSP). The backside and the four sidewalls of the chip are protected by molding. They test and then dice the original device wafer into known good dice (KGDs), pick and place the individual KGDs on a temporary wafer carrier, and then compression mold an epoxy molding compound (EMC) on the reconstituted wafer. It is followed by debonding the temporary carrier, fabricating the redistribution layer (RDL), solder ball mounting, and dicing reconstituted wafer into individual eWLCSPs. Basically, this is the embedded wafer-level ball grid array or chip-first (face-down) fan-out wafer-level packaging method [58, 59].

In 2015, UTAC published an article on six-side protected WLCSP [52]. The four sidewalls and the front side of the chip are protected by molding, and the backside of the chip is protected by a conventional method such as the lamination of a tape (film). Their method in making the five-side molded WLCSP is very different from that of STATS ChipPAC. They use the original device wafer (without dicing and P&P of the KGDs on a temporary wafer) to make the molding and solder ball mounting. In 2016, SPIL published a similar article on five-side protected (four sidewalls and the front side of the chip) and called it molded WLCSP [53]. They also proposed a six-side protection by laminating an epoxy film at the backside of the chip.

Since 2018, Huatian Technology in Kunshan, China, have been publishing a series of articles on six-side or six-side molded WLCSP [54–56]. They actually make the six-side molded WLCSP. Their process is very similar to that of UTAC and SPIL, except they mold the backside of the chip instead of using the (conventional) lamination of a film.

In 2020, MediaTek and SPIL published an article [57] on five-side and six-side molded WLCSP. Their process is very similar to that of STATS ChipPAC, but it is molded on all six sides.

In the mentioned papers [47–57], the WLCSPs are fabricated on a wafer. In this study, the design, materials, process, assembly, and reliability of a six-side molded panel-level chip scale package (PLCSP) are presented [60, 61]. Emphasis is placed on the fabrication of the RDLs of the PLCSP on a large temporary panel with multiple device wafers. After the fabrication of RDLs, the wafers from the printed circuited board (PCB) panel are debonded, and the six-side protection (molding) of the PLCSP is fabricated. Reliability assessments of the PLCSP are performed by a drop test and a thermal cycling simulation. Some recommendations are also provided (Fig. 3.35).

3.7.2 *Test Chip and Test Package*

The test chip (5 $\times$ 5 mm) and its cross section of this study are the same as those in [62] (Fig. 3.36). There are two rows of peripheral pads; the inner rows are with 160 pads (70 $\times$ 70 μm) on a 100-mm pitch, and the outer rows are with 88 pads (130 $\times$ 130 μm) on a 180-μm pitch. In this study, only the outer rows are considered. The under-bump metallurgy, Ti/Cu (0.1/0.2 μm) μbump pad size is 110 μm in diameter,

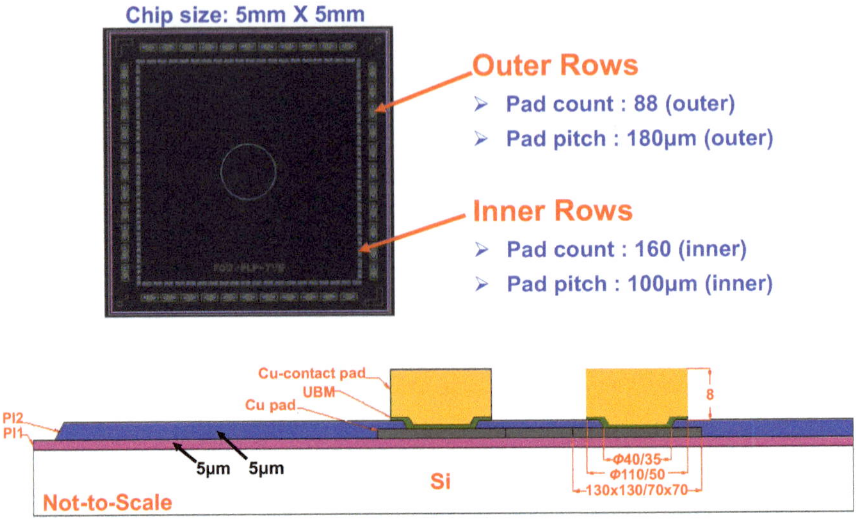

Fig. 3.35 Test chip and its cross section

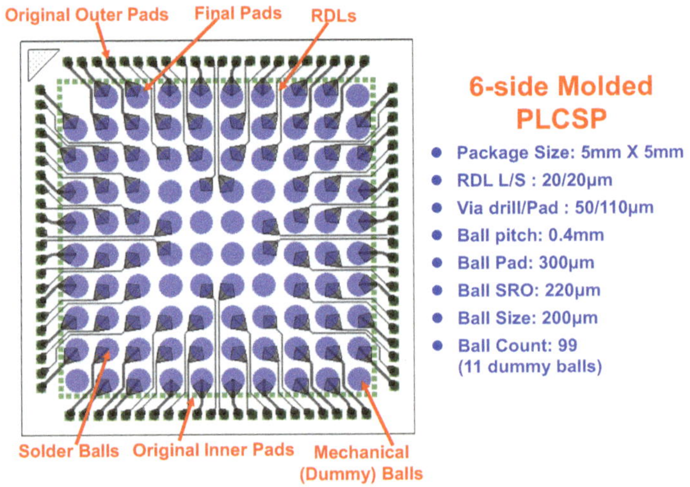

Fig. 3.36 Test package (top view)

the passivation (PI2) pad opening is 40 μm, the Cu-contact pad size is 110 μm in diameter and 8 μm in thickness. The Cu pads are daisy-chained with a 3-μm-thick copper. The diameter of the test wafer is 300 mm, and the width of the saw street is 300 μm.

The top-view and cross-sectional view of the test package are shown in Figs. 3.36 and 3.37, respectively. It can be seen from Fig. 3.36 that the peripheral pads are redistributed (fan-in) to the interior of the chip, and the new (final) pad size is 300 μm

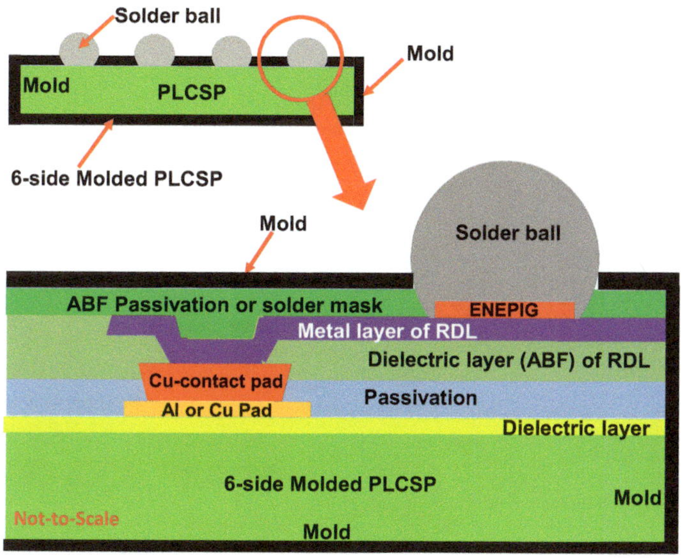

Fig. 3.37 Cross-sectional view of the six-side molded PLCSP

in diameter with a solder mask opening of 220 μm diameter. There are 99 pads and solder balls (11 are dummies). The solder ball size is 200 μm and is on a 0.4-mm pitch.

The PLCSP is protected (molded) in all six sides: front side, backside, and four sidewalls of the chip (Fig. 3.37). There is one RDL which consists of the dielectric layer and the metal layer, and their thickness is 20 μm. The metal line width and spacing of the RDL are 20/20 μm. The via opening of the dielectric layer of the RDL is 50 μm.

3.7.3 Six-Side Molded PLCSP Process Flow

Before making the RDL and the six-side molded PLCSP, a temporary panel carrier and the diced and undiced device wafers have to be prepared.

3.7.4 Wafer and Panel Preparation

In this study, a 508 mm × 508 mm panel is selected as the temporary carrier because the equipment for making the RDLs is ready. The 300-mm device wafer (780 mm thick) is diced into half and one quarter so 2.25 300-mm wafer can be put into the temporary panel as shown in Figs. 3.38 and 3.39.

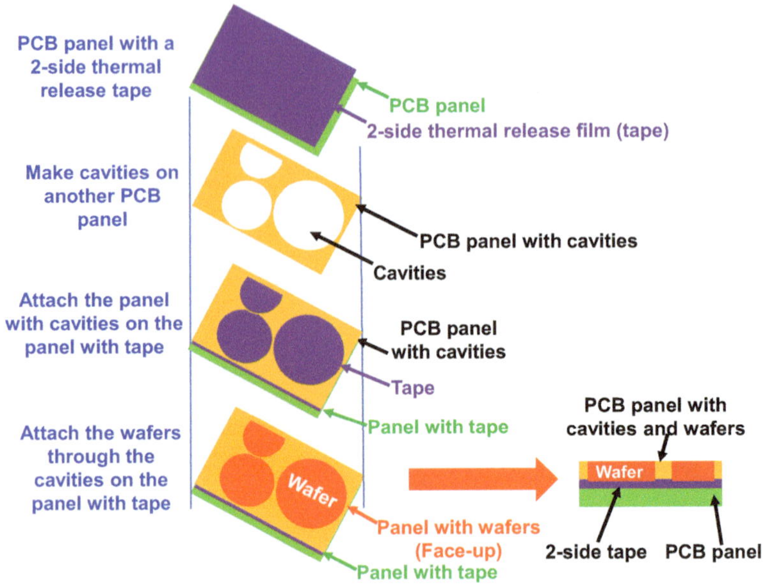

Fig. 3.38 Panels setup for the PLCSP made from the diced wafers

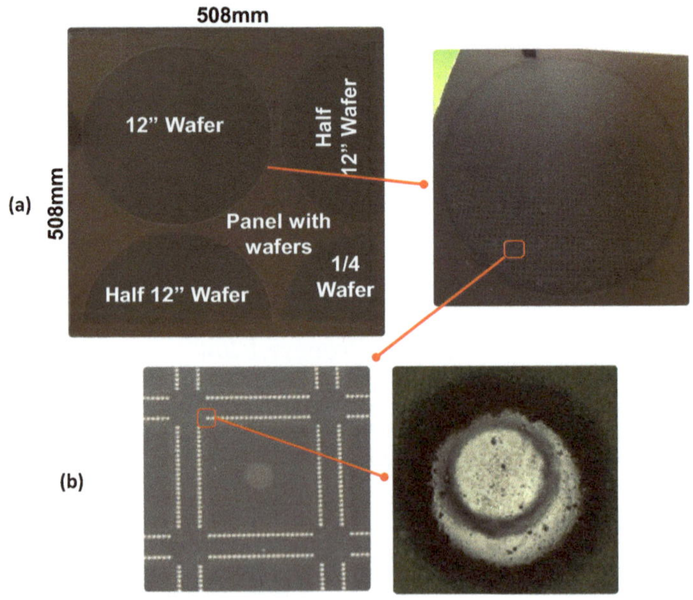

Fig. 3.39 PLCSP made from a 508 × 508-mm panel with one 12" wafer, two-half 12" wafer, and 1/4 12" wafer. **a** Cover with ABF. **b** Laser drill via

Figure 3.38 schematically shows the panel preparation. First, a PCB panel is laminated with a two-side thermal release tape at room temperature. Then, the cavities (300.15 mm in diameter) are made on another PCB panel and attached to the PCB with cavities to the PCB with the tape. It is followed by attaching the diced and undiced device wafers (face-up) through the cavities on the PCB panel with the tape. These are the first three steps of the process flow shown in Fig. 3.40a.

3.7.5 *Fabrication of Redistribution Layer*

Next, a 25-mm Ajinomoto build-up film (ABF) is laminated on the whole panel with wafers. There are two operating stages of the ABF: (1) at the first stage, the temperature was 120 °C for 30 s at vacuum condition and then pressed (0.68 MPa) for 30 s with the temperature and vacuum on, and (2) at the second stage, the temperature is 100 °C and pressed (0.58 MPa) for 60 s. Figure 3.39a shows the panel with the diced and undiced wafers covered with the ABF. Then, the vias are laser drilled on the ABF, and the results are shown in Fig. 3.39b. The typical diameter of the top via is 63 μm, the bottom via is 43 μm, and the via thickness is 18 μm. The target values are via diameter = 50 μm and via thickness = 20 μm.

It is followed by desmearing, electroless Cu plating for seed layers, photoresist dry film lamination, laser direct imaging (LDI) and development, and PCB Cu plating. Then, the dry film is stripped off and the seed layer etched off to obtain the RDL as shown in Fig. 3.41a. The typical metal line width of the RDL is between 16 and 25 μm. The target value is 20 μm. Thus, there is room for improvements.

3.7.6 *Solder Resist Opening and Solder Ball Mounting*

Next, a 25-μm ABF is laminated on the whole panel (for the solder mask), laser drilled for solder resist opening (SRO), desmeared, and surface finished with an electroless nickel electroless palladium immersion gold as shown in Fig. 3.41b. The diameter of the SRO is targeted for 0.22 μm and is operated by a carbon dioxide (CO_2) laser. The speed in making the vias is 1,500 vias/s, and it takes 5.75 min to complete all the (517,671) vias in one panel. The results (top via diameter 5 202.6 μm, bottom via diameter 5 165.2 μm, and the via thickness 5 17 μm) are acceptable. Then (230 °C for 10 min), the wafers thermal debonded from the panels, and the solder balls are mounted on the wafers with the equipment shown in Fig. 3.41b. Figure 3.41c shows an x-ray image of the ordinary PLCSP. The average solder ball height is equal to 150 μm.

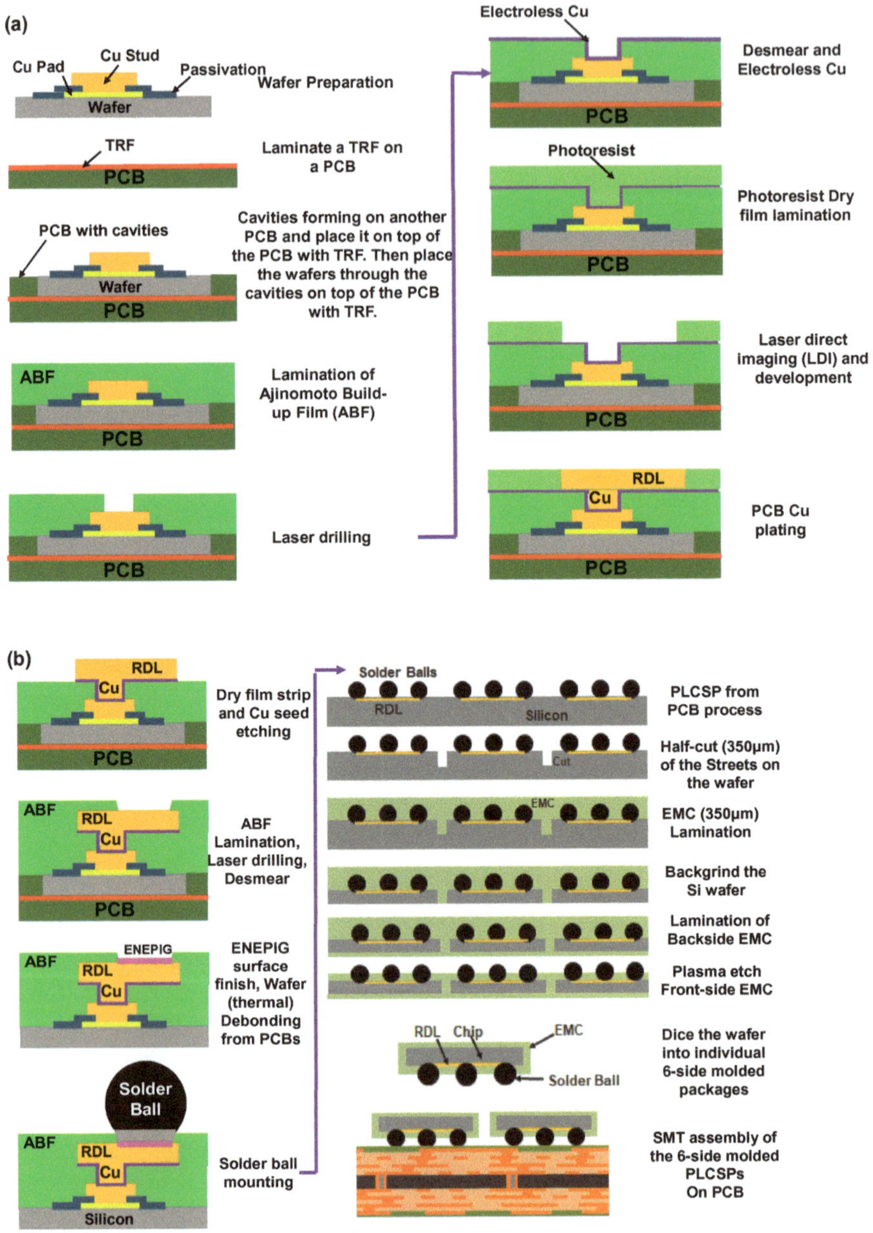

Fig. 3.40 **a** Key process steps for the PLCSP made from the diced wafers. **b** Key process steps for the six-side molded PLCSP made from the diced wafers

(a)

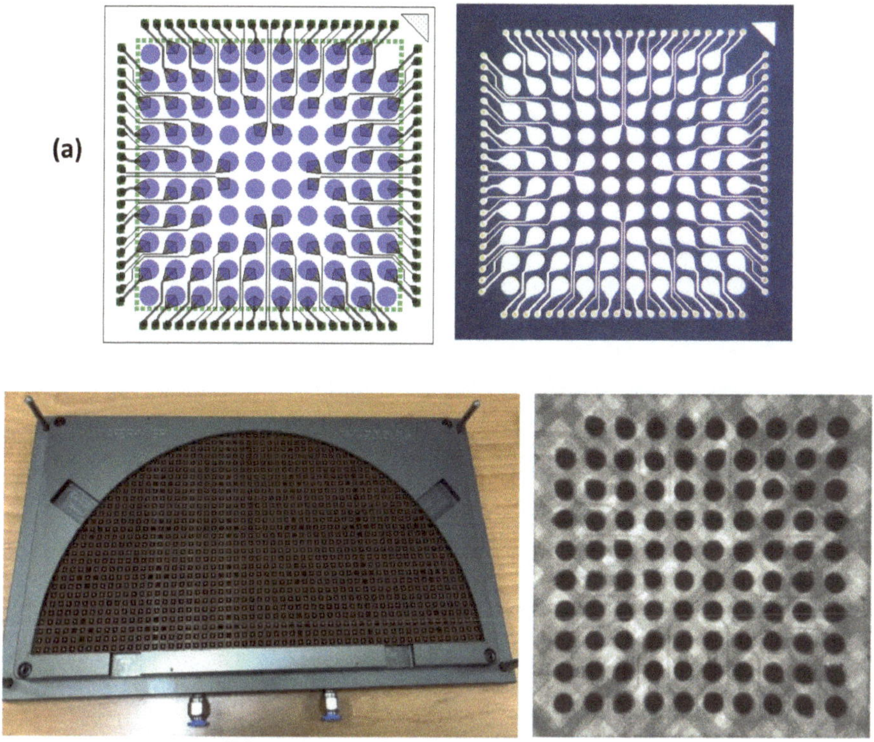

(b) (c)

Fig. 3.41 **a** Fabricated PLCSP from a 508 × 508-mm panel. **b** Solder ball mounting equipment. **c** x-ray image of PLCSP with solder balls

3.7.7 Cutting and Epoxy Molding Compound Molding of Wafer from the Front Side

Next, the wafer is half-cut with RDL and solder balls from its front side with a mechanical dicing saw (Fig. 3.40b. Laser grooving [63–65] is better, but optional. The average trench width is equal to 170 μm, and the average depth is equal to 394 μm, as shown in Fig. 3.42. Then, it is followed by laminating an EMC (ABF) with 200 μm thickness on the front side of the wafer, Fig. 3.42b. Because of the lamination pressure, the EMC is filled in the trench, as shown in Fig. 3.42b. It can be seen that the thickness of the EMC is about 150 μm (not 200 μm because some of the EMC materials go to fill the trenches in the saw streets), which is equal to the total height of EMC (202.4 μm) minus the (solder mask + dielectric = 50 μm) thickness.

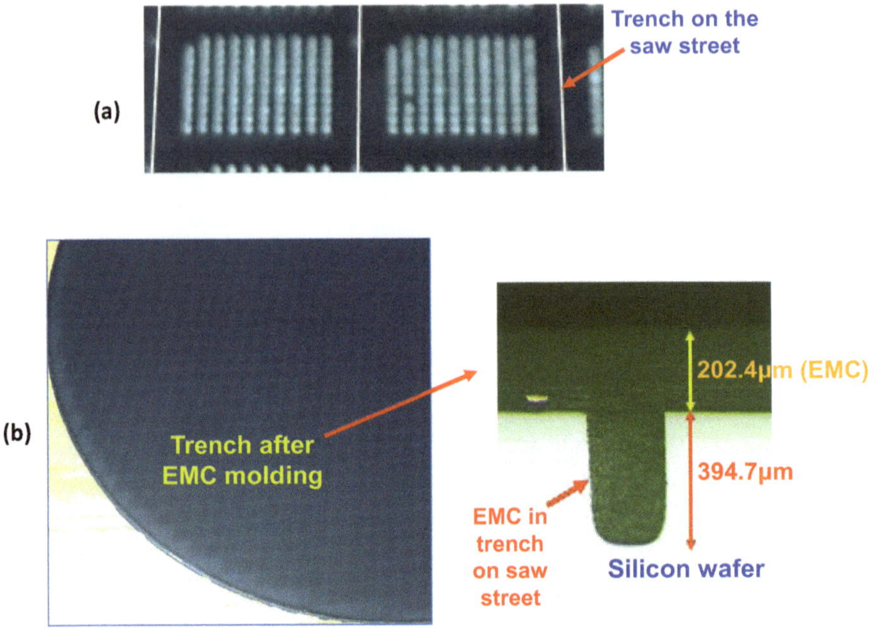

Fig. 3.42 **a** Dicing the front side of the wafer. **b** After EMC molding of the front side of the wafer

3.7.8 Backgrinding and Wafer Backside Molding

Now, it is time to perform the backgrinding of the wafer from its backside. This continues until the EMC in the trench is seen (about 390 μm), as shown in Fig. 3.43. Next, a 25- μm EMC (ABF) is laminated at the backside of the wafer.

3.7.9 Plasma Etching and Dicing

Next, it is time for plasma etching to remove some of the EMC from the solder balls, as shown in Fig. 3.40b. The plasma conditions are temperature = 132 °C, power = 2,700 W, pressure = 0.33 torr, and O2/CF4 = 2,000 sccm. The etch rate is about 1.76 μm/min. Figure 3.44a shows the ordinary PLCSP or WLCSP. Figure 3.44b shows the six-side molded PLCSP with front side EMC molding. Figure 3.44c shows the solder joint of the six-side molded PLCSP after plasma etching. The average solder ball height of an ordinary PLCSP or WLCSP is 148 μm (Fig. 3.44d), and the average solder ball height of a five-side or six-side molded PLCSP is 103 mm (Fig. 3.44e).

Then, the wafer is diced into individual six-side molded PLCSPs. The 3-D view and the cross-sectional view of the six-side molded PLCSP are shown, respectively, in Fig. 3.45a and b. It can be seen that the average sidewall molding is 78 μm, the average front-side molding is 53 μm, and the average standoff height of the solder ball is 100 μm.

Fig. 3.43 Backgrinding the backside of the wafer until seeing the EMC on the trench of the saw street

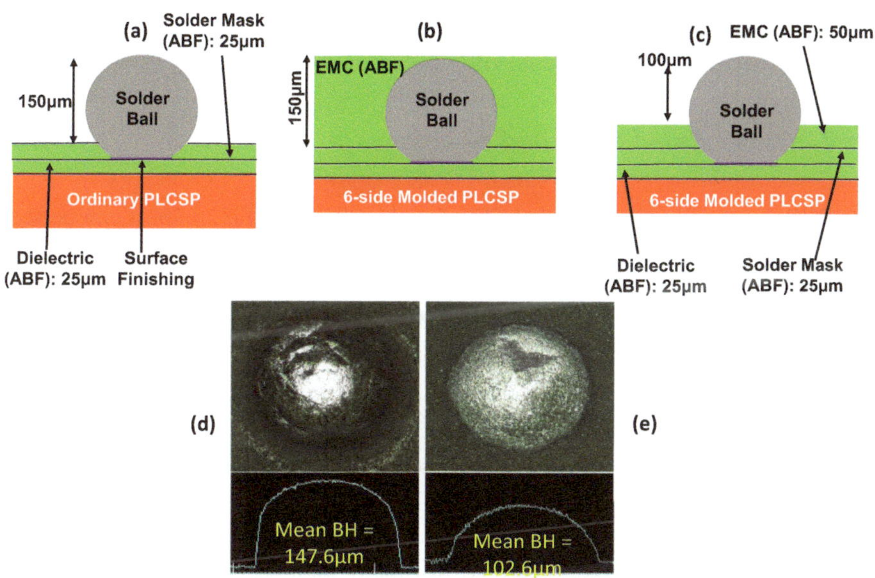

Fig. 3.44 Solder ball height. **a** Ordinary PLCSP. **b** Right after front-side molding. **c** Right after plasma etching. **d** Ordinary PLCSP solder ball image. **e** Five-Side or six-side molded PLCSP solder ball image

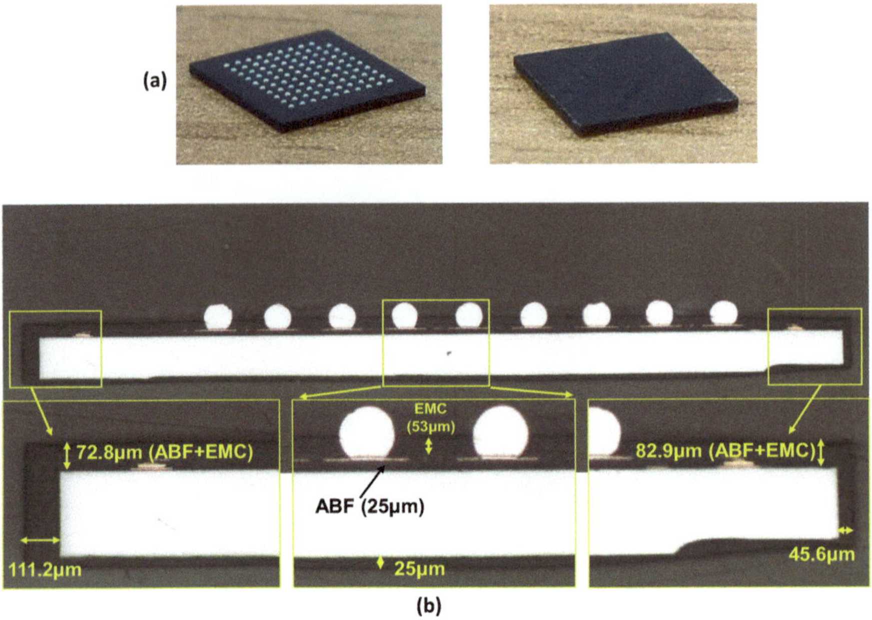

Fig. 3.45 **a** 3-D view. **b** Cross section view of the six-side molded PLCSP

3.8 Reliability (Drop Test) of Six-Sides Molded PLCSP

3.8.1 Test Printed Circuited Board

The PCB for the PLCSP is made of FR-4 and is shown in Fig. 3.46. It can be seen that there are 42 PLCSP sites on the board. The dimensions of the PCB are 132 × 77 × 0.65 mm, and there are six layers. There are 99 pads (with a pitch = 0.4 mm) for each PLCSP. The pad with a diameter equal to 0.3 mm is non-solder mask defined, and its surface finish is an organic solderability preservative.

3.8.2 Surface Mount Technology Assembly of the PLCSP on PCB

A lead-free SMT process is used to assemble the PLCSPs on the test PCB. Figure 3.47a shows the reflow temperature profile. Figure 3.47b shows the PCB assembly of the ordinary PLCSP, and Fig. 3.47c shows the PCB assembly of the six-side molded PLCSP.

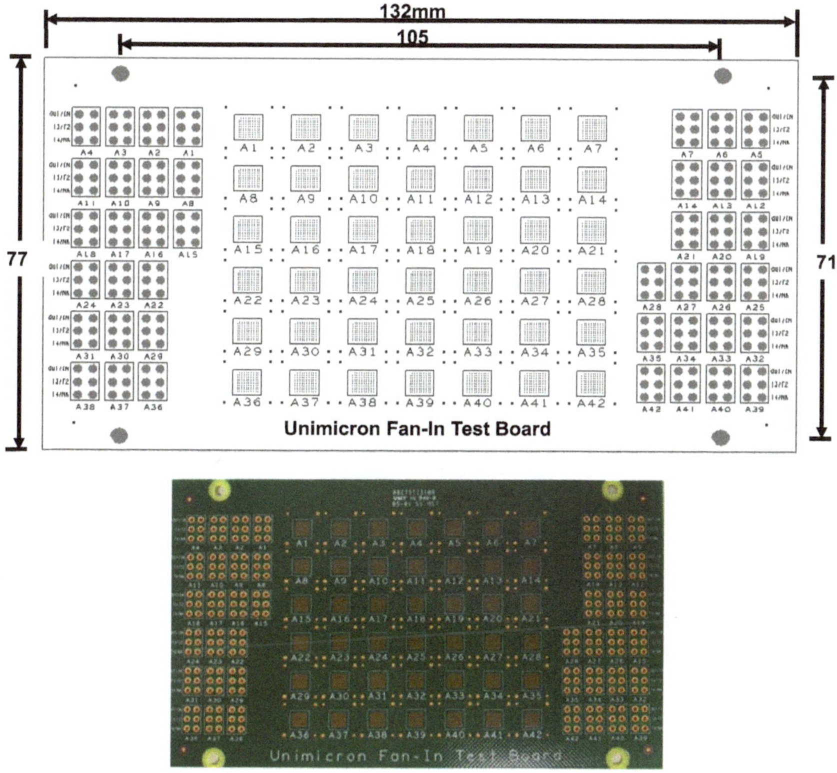

Fig. 3.46 Reliability test PCB

3.8.3 Drop Test Setup and Spectrum

The reliability assessment of the ordinary PLCSP on PCB assembly without underfill is by a shock (drop) test. The test setup is according to the JEDEC Standard JESD22-B111, as shown in Fig. 3.48. It consists of a drop tower, a drop table for the PCB with samples. There are 25 channels plus one common, and a data acquisition system (DAS). The height of the drop table is 69.85 mm, which yields the drop spectrum with 1,500 G/ms (1,500 Gs, 0.5 ms half-sine pulse), as shown in Fig. 3.48. The drop condition is 30 drops. The failure criterion is when the measured resistance during the drop test reaches 1,000 ohms, as shown in Fig. 3.49b. Less than that, it is considered no failure, as shown in Fig. 3.49a.

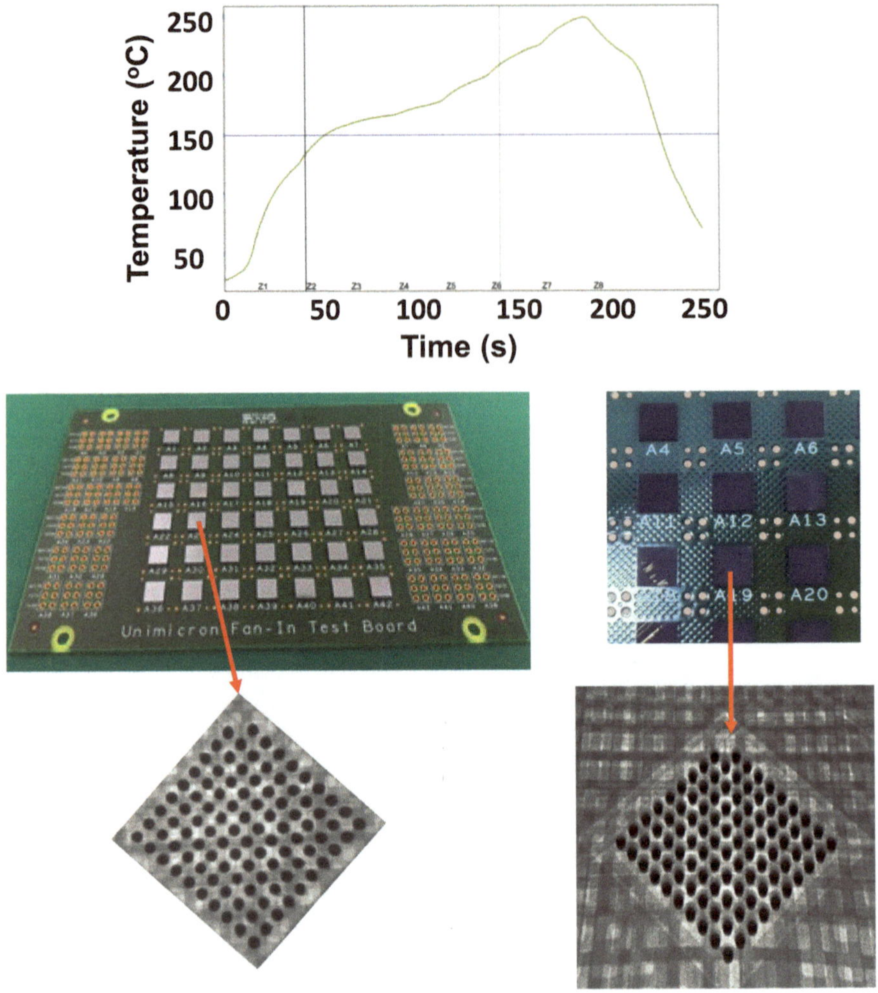

Fig. 3.47 SMT assembly of the six-side molded PLCSP

3.8.4 Drop Test Results and Failure Analysis

After 30 drops, there is not any failure, i.e., the measurement resistance data (<1,000ohms are like those shown in Fig. 3.49a. Thus, the PLCSP assembly passes the drop test. However, the cross section of some of the unfailed samples shows that actually there are some small cracks, as shown in Fig. 3.50. The test continues to 50 drops, and there is one failure at 31 drops, i.e., the measurement resistance date (>1,000 ohms) is like that shown in Fig. 3.49b. The cross section of the failed sample shows that the solder joint is totally cracked (opened), as shown in Fig. 3.51.

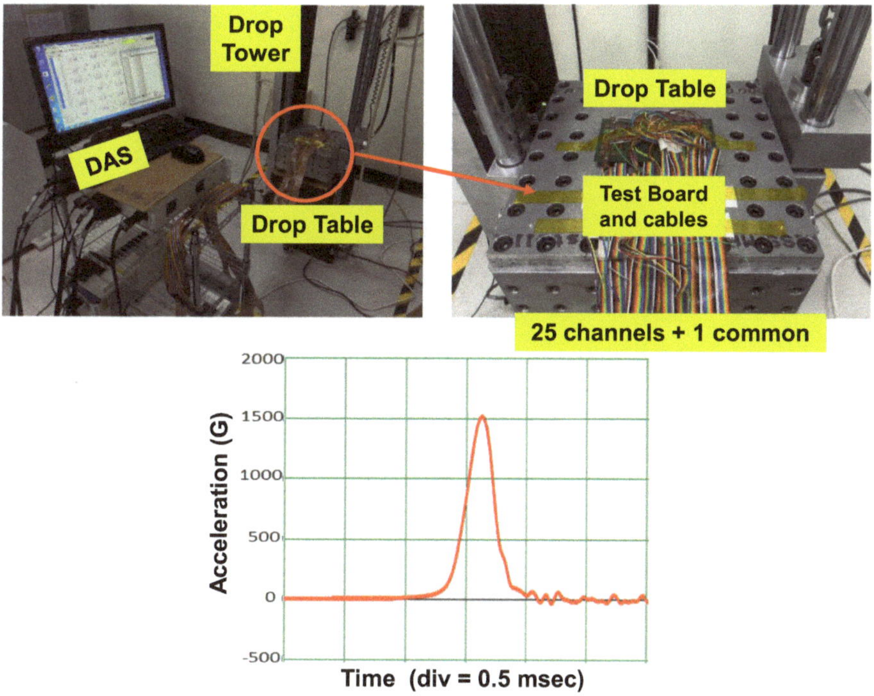

Fig. 3.48 Drop test setup and the drop spectrum

3.9 Reliability (Thermal Cycling Test and Simulation) of Six-Sides Molded PLCSP

3.9.1 Thermal Cycling Test Failure Criteria

The most important factor in thermal cycling reliability tests is the failure criteria [66]. During reliability tests, we continuously perform the resistance measurements. The failure criterion is defined as the resistance increases to certain (e.g., 1 to ∞) percentages of the original resistance. Usually, daisy chains, which connect the solder interconnects of the chip/package and substrate/ PCB, allow the measurements. Because most solder joints under thermal fatigue loadings will go through crack initiation, crack propagation, and crack rupture eventually, the daisy chains' resistance will increase accordingly, i.e., from small to large, and become ∞ when the solder joint is totally cracked (opened). (Most people unintentionally pick the last one.) Since the exact "resistance versus crack length" relations of various lead-free solder joints do not exist today, people just randomly pick a number, which is from 1 to ∞ percent of the initial resistance. That is one of the reasons why there are so many different Weibull/lognormal plots in the lectures, even with the same package, PCB, solder paste, sample size, test condition, and test period. Thus, the failure criteria should be

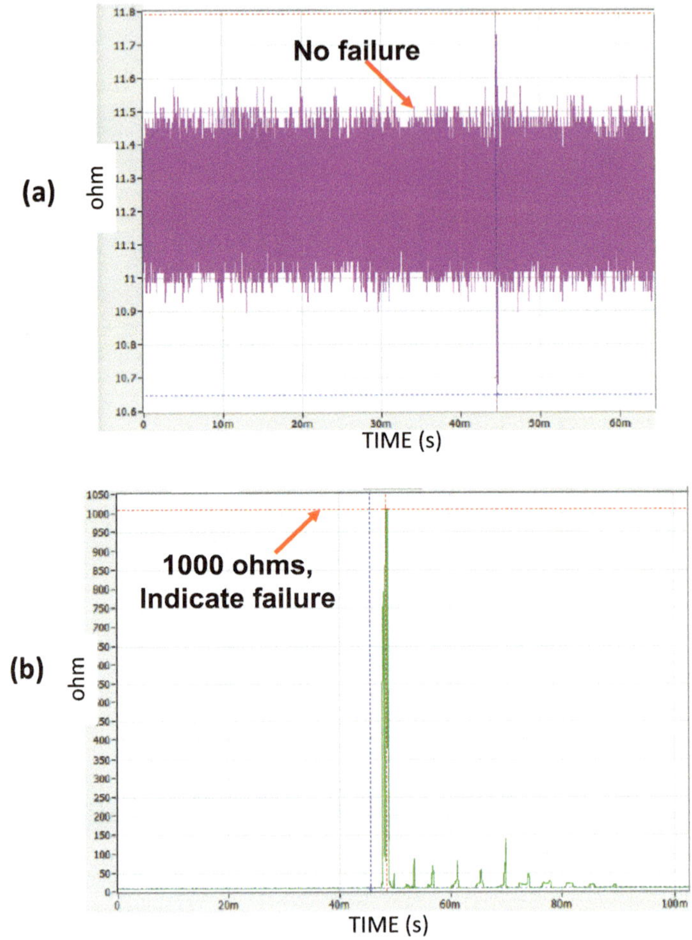

Fig. 3.49 Failure criteria. **a** No failure. **b** Sample failed

clearly stated. In this study, the failure criterion is when the resistance of the daisy chain of the PLCSP PCB assembly increases by 50%. The cycle at which the first solder joint of a PLCSP failed is considered as the cycle to failure of the PLCSP.

3.9.2 Test Results of the Six-Sides Molded PLCSP

The PCB assembly of the ordinary PLCSP and six-side molded PLCSP are put into a thermal cycling chamber as shown in Fig. 3.52. The temperature in the chamber during test is measured through a thermal couple (in air) and is shown in Fig. 3.53. It can be seen that the temperature profile is basically −55 °C ↔ 125 °C at 50-min

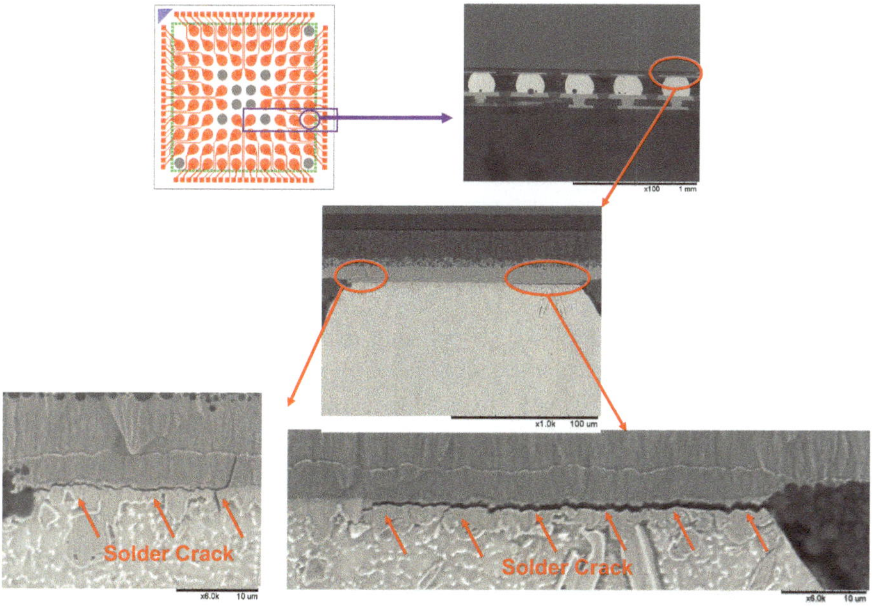

Fig. 3.50 SEM shows there is no failure, but actually there are small cracks of the solder joint

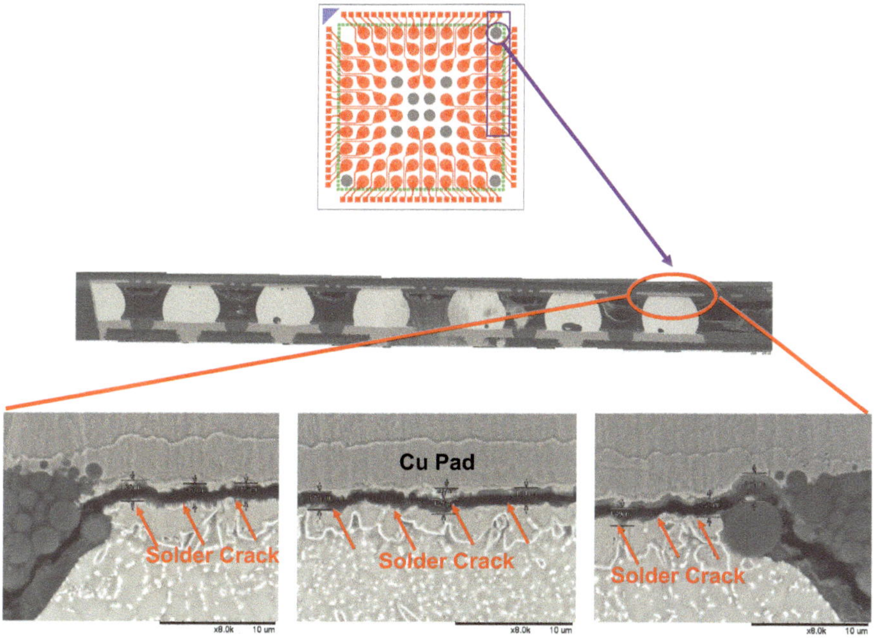

Fig. 3.51 SEM shows there is a failure and actually the solder joint is totally opened (cracked)

cycle with 26 °C/min (ramp-up) and 29 °C/min (ramp-down). The sample size of the ordinary PLCSP is 48 and that of the six-side molded PLCSP is 24.

The thermal cycling test results of the six-side molded PLCSP are tabulated in Table 3.5. It can be seen that there are 15 six-side molded PLCSP failures. (Again, in this study, the cycle at which the first solder joint of a six-side molded PLCSP failed is considered as the cycle-to-failure of the six-side molded PLCSP.) The median (50%) rank is determined from F(x) = (j – 0.3)/(n + 0.4), where j = failure order number and n = sample size = 24. These data are plotted into a Weibull distribution and is shown in Fig. 3.54. It can be seen that the characteristic life (θ) (63.2% failed) = 1,037 cycles and the Weibull slope (β) is = 5.53.

For other required ranking (G), the percent rank (z) can be determined by (z is the percent rank of the jth value in n samples) [66]

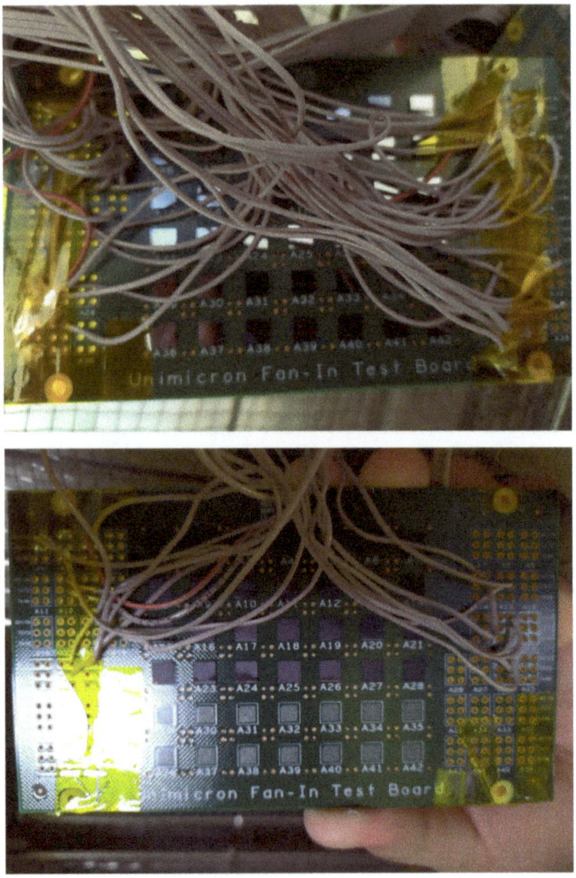

Fig. 3.52 Test boards with six-side molded PLCSPs and ordinary PLCSPs inside the thermal cycling chamber

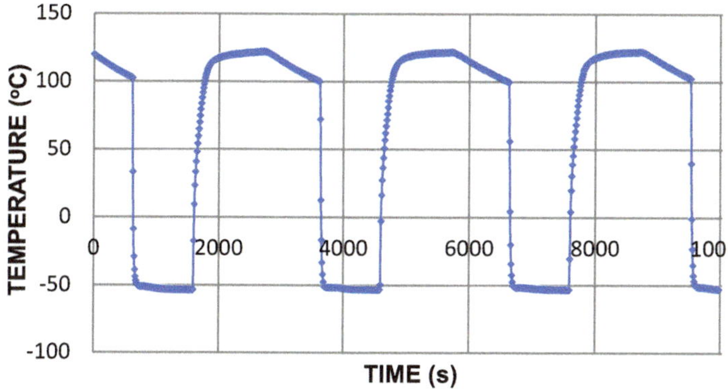

Fig. 3.53 Thermal cycling temperature profile

Table 3.5 Test results of the six-side molded PLCSP at 50, 5, and 95% ranks

Failure order	Cycles-to-failure	$F(x)$		
		Median (50%) rank	90% confidence	
			5% rank	95% rank
1	638	2.88	0.21	11.73
2	638	6.98	1.50	18.29
3	712	11.08	3.50	23.98
4	712	15.17	5.90	29.23
5	788	19.27	8.59	34.18
6	872	23.37	11.49	38.91
7	872	27.46	14.57	43.47
8	872	31.56	17.80	47.87
9	872	35.66	21.26	52.14
10	872	39.75	24.64	56.29
11	872	43.85	28.24	60.32
12	872	47.95	31.94	64.24
13	872	52.04	35.76	68.06
14	1096	56.14	39.68	71.76
15	1096	60.24	43.71	75.36

Sample Size = 24 Failed = 15

$$1-(1-z)^n - nz(1-z)^{n-1} - \frac{n(n-1)}{2!}z^2(1-z)^{n-2} - \cdots$$
$$-\frac{n(n-1)\cdots(n-j+1)}{(j-1)!}z^{j-1}(1-z)^{n-j+1} = G$$

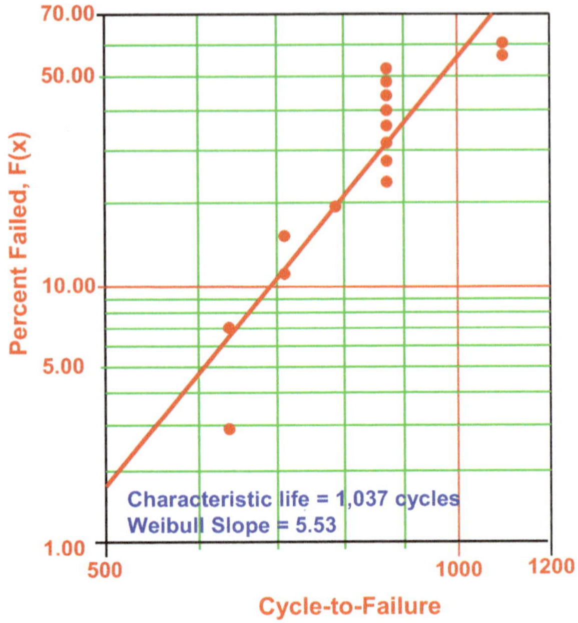

Fig. 3.54 Weibull plot of the six-side molded PLCSP solder joint at median rank

For 90% confidence level, we need the 5% rank ($G = 5\%$) and 95% rank ($G = 95\%$), and they are also listed in Table 3.5. For example, with $N = 24$ and $G = 5\%$, the percent failure rank (z) is 0.21 for $j = 1$ and is 1.5 for $j = 2$. Another example, with $N = 24$ and $G = 95\%$, the percent failure rank (z) is 11.73 for $j = 1$ and is 18.29 for $j = 2$.

Confidence is defined as the probability that a given interval determined from the test data will contain the population parameters, such as (using Weibull as an example) the characteristic life and Weibull slope. Confidence applies to the test itself, and reliability applies to the product!

Figure 3.55 shows the Weibull plot of the six-side molded PLCSP at 90% confidence. It can be seen that the true characteristic life (θ_t) at 90% confidence is $958 \leq \theta_t \leq 1{,}069$ cycles.

To predict the true Weibull slope (β_t) of the population, it is necessary to estimate the Weibull slope error in relation to the sample size and the required confidence level. The error (E) in the Weibull slope depends on the number of failures (N) and the required confidence level (C), which can be estimated from Fig. 3.56 [30]. In the case of the six-side molded PLCSP, $C = 90\%$ and $N = 15$, then $E = 0.29$. Thus, the true Weibull slope is $3.93 \leq \beta_t \leq 7.13$.

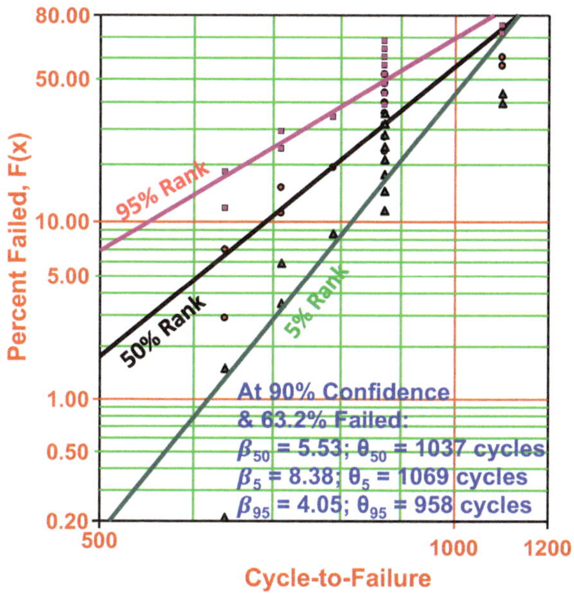

Fig. 3.55 Weibull plot of the six-side molded PLCSP solder joint at 90% confidence

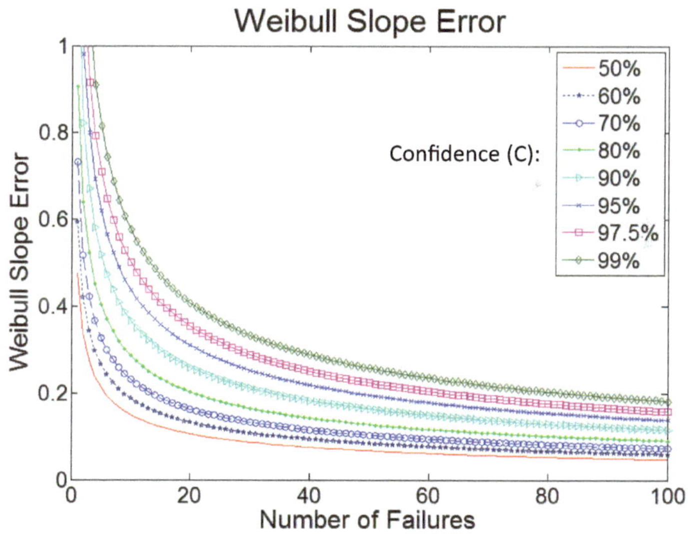

Fig. 3.56 Weibull slope error

3.9.3 Failure Location and Failure Mode of the Six-Sides Molded PLCSP PCB Assembly

The typical failure location occurs at the outer lows (near the corners) of the solder joints of the six-side molded PLCSP, as shown in Fig. 3.57. The failure mode of most of the solder joints is the cracking of the solder near the interface between the PCB and the bulk solder, as shown in Fig. 3.57.

3.9.4 Test Results of the Ordinary PLCSP

The thermal cycling test results of the ordinary PLCSP are shown in Table 3.6 and are Weibull plotted (median rank) in Fig. 3.58. It can be seen that the Weibull slope = 3.34 and the characteristic life = 386 cycles. Figure 3.59 shows the Weibull plot of the ordinary PLCSP solder joints at 90% confidence. It can be seen that the true characteristic life (θ_t) at 90% confidence is $326 \leq \theta_t \leq 403$ cycles. One of the reasons for the low reliability is because of the poor quality due to using a panel for the very first time as a temporary carrier with multiple diced and undiced wafers [29]. The other reason is that the silicon thickness of the ordinary PLCSP is 0.79 mm. The Weibull slope error of the ordinary PLCSP can be estimated from Fig. 3.56 with $C = 0.9$ and $N = 48$. Then, $E = 0.22$ and the true Weibull slope is $2.6 \leq \beta_t \leq 4.1$.

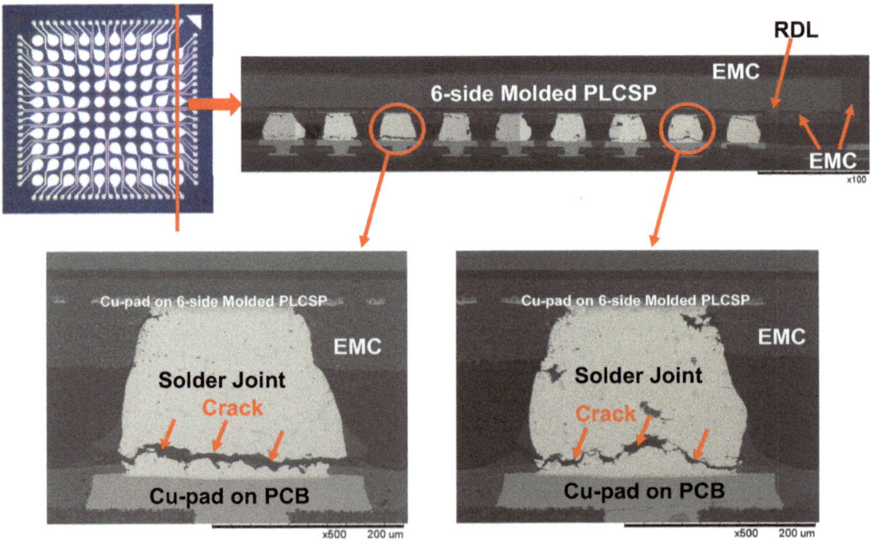

Fig. 3.57 Failure location and failure mode of the six-side molded PLCSP solder joints

3.9.5 Failure Location and Failure Mode of the Ordinary PLCSP PCB Assembly

The failure location and failure mode of the ordinary PLCSP solder joints are shown in Fig. 3.60. It can be seen that just like the case of six-side molded PLCSP, the failure location of ordinary PLCSP occurs along the outer rows near the corners of solder joints. However, the failure mode of all solder joints of the ordinary PLCSP is very different from that of six-side molded PLCSP.

The cracking of solder of the ordinary PLCSP occurs near the interface between the chip and the bulk solder as shown in Fig. 3.60, and there is not any cracking near the interface between the bulk solder and PCB. More on this characteristic will be elaborated in the thermal cycling simulation section.

3.9.6 Comparison Between the Six-Sides Molded PLCSP and the Ordinary PLCSP

The Weibull plots of the six-side molded PLCSP and the ordinary PLCSP at median rank are shown in Fig. 3.61. From these plots, it is obvious that the characteristic life of the six-side molded (1,037 cycles) PLCSP is better than that (368 cycles) of the

Table 3.6 Test results of the ordinary PLCSP at 50, 5, and 95% ranks

Failure order	Cycles-to-failure	F(x)		
		Median (50%) rank	90% confidence	
			5% rank	95% rank
1	155	1.44	0.11	6.05
2	177	3.51	0.75	9.51
3	180	5.57	1.73	12.54
4	186	7.64	2.90	15.37
5	193	9.70	4.20	18.06
6	195	11.77	5.59	20.66
7	199	13.84	7.05	23.19
8	200	15.90	8.57	25.65
9	203	17.97	10.15	28.07
10	205	20.04	11.76	30.44
11	213	22.10	13.42	32.77
12	213	24.17	15.10	35.07
13	250	26.23	16.83	37.34
14	250	28.30	18.57	39.57

(continued)

Table 3.6 (continued)

Failure order	Cycles-to-failure	F(x)		
		Median (50%) rank	90% confidence	
			5% rank	95% rank
15	250	30.37	20.35	41.78
16	250	32.43	22.15	43.97
17	250	34.50	23.98	46.13
18	250	36.56	25.83	48.27
19	300	38.63	27.70	50.38
20	300	40.70	29.59	52.48
21	300	42.76	31.50	54.55
22	300	44.83	33.43	56.60
23	300	46.90	35.39	58.63
24	300	48.96	37.36	60.65
25	350	51.03	39.35	62.64
26	350	53.09	41.37	64.61
27	350	55.16	43.40	66.57
28	350	57.23	45.45	68.50
29	350	59.29	47.52	70.41
30	350	61.36	49.62	72.30
31	400	63.43	51.73	74.17
32	400	65.49	53.87	76.02
33	400	67.56	56.03	77.85
34	400	69.62	58.22	79.65
35	400	71.69	60.43	81.43
36	400	73.76	62.66	83.17
37	450	75.82	64.93	84.90
38	450	77.89	67.23	86.58
39	450	79.16	69.56	88.24
40	450	82.02	71.93	89.85
41	450	84.09	74.35	91.43
42	450	86.15	76.81	92.95
43	500	88.22	79.4	94.41
44	500	90.29	81.94	95.80
45	500	92.35	84.63	97.10
46	500	94.42	87.46	98.27
47	500	96.48	90.49	99.25
48	500	98.55	93.95	99.89

Sample Size = 48 Failed = 48

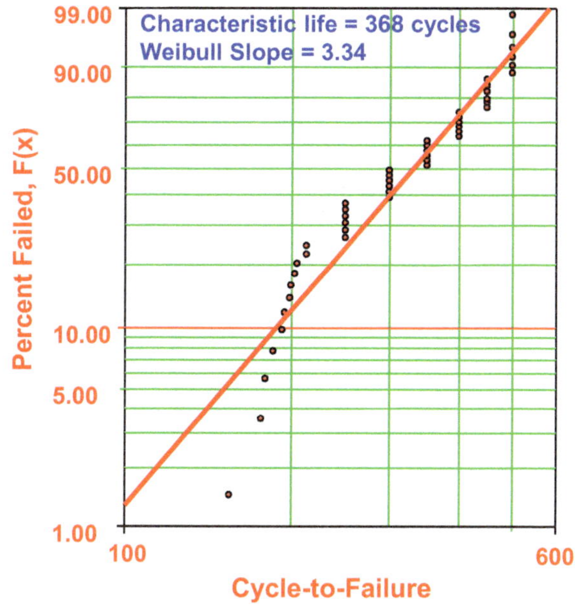

Fig. 3.58 Weibull plot of the ordinary PLCSP solder joint at median rank

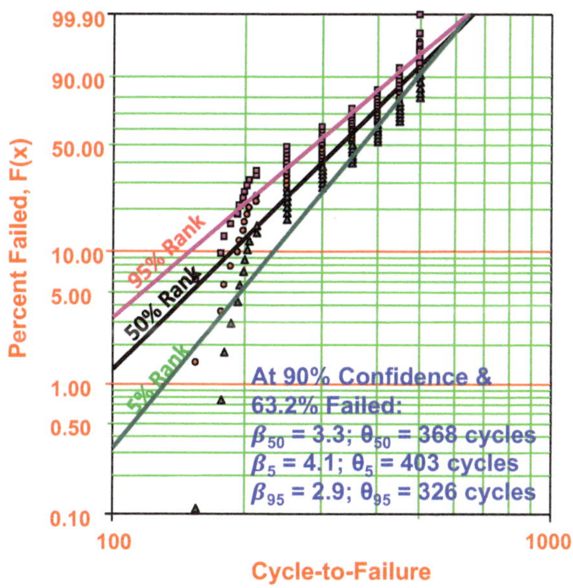

Fig. 3.59 Weibull plot of the ordinary PLCSP solder joint at 90% confidence

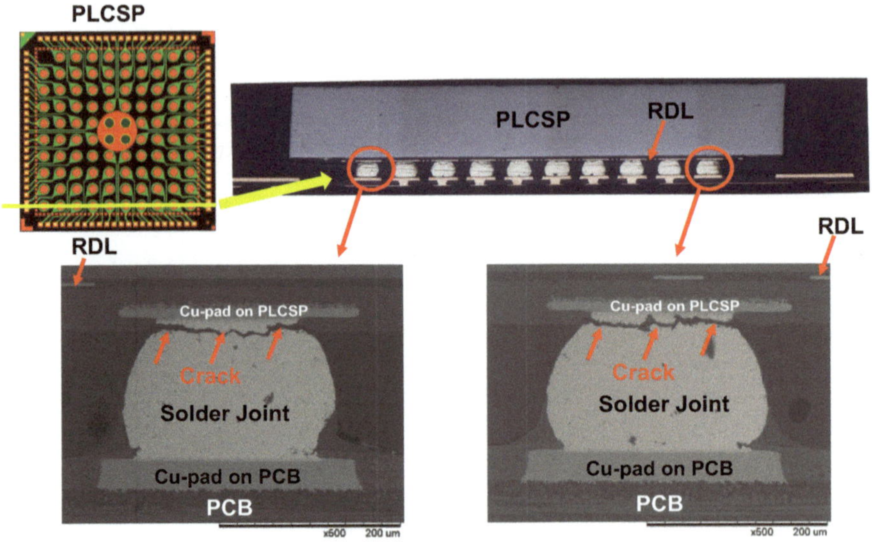

Fig. 3.60 Failure location and failure mode of the ordinary PLCSP solder joints

ordinary PLCSP. Also, the mean life of the six-side molded PLCSP (958 cycles) is better than that (330 cycles) of the ordinary PLCSP [mean life = $\theta\Gamma(1 + 1/\beta)$].

If one product is found to be superior to another from testing, how CONFIDENT (P) can it be that the same is true of their populations? Herein, a simple approach

Fig. 3.61 Comparison between the six-side Molded PLCSP and the ordinary PLCSP solder joints

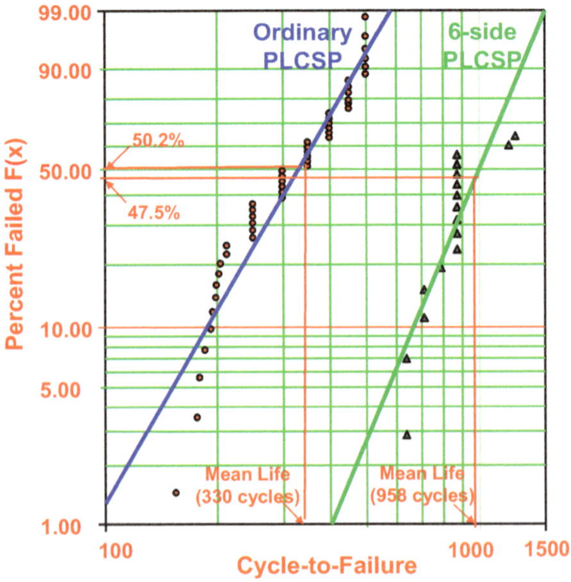

is used to determine whether the mean life of one product is better than that of the other, without inquiring what the actual differences are [66].

$$P = \frac{1}{1 + \frac{\log 1/q}{\log 1/(1-q)}}$$

where

$$q = 1 - \frac{1}{\left[1 + \left(\frac{t+4.05}{6.12}\right)^5\right]^{40/7}}$$

$$t = \frac{\sqrt{1 + \sqrt{T}(\rho - 1)}}{\rho\Omega_2 + \Omega_1}$$

$$\rho = \frac{M_2}{M_1}$$

$$T = (r_1 - 1)(r_2 - 1)$$

$$\Omega_1 = \sqrt{\frac{\Gamma(1 + 2/\beta_1)}{\Gamma^2(1 + 1/\beta_1)} - 1}$$

$$\Omega_2 = \sqrt{\frac{\Gamma(1 + 2/\beta_2)}{\Gamma^2(1 + 1/\beta_2)} - 1}$$

and M_1 is the mean life of Sample 1 (S1), M_2 is the mean life of Sample 2 (S2), r_1 is the number of failures in S1, r_2 is the number of failures in S2, β_1 is the Weibull slope of S1, and β_2 is the Weibull slope of S2. In our case, $M_1 = 958$, $M_2 = 330$, S1 = 24, S2 = 48, $r_1 = 15$, $r_2 = 48$, $\beta_1 = 5.53$, and $\beta_2 = 3.34$. A simple calculation yields $M_1/M_2 = 2.9$ and $P = 0.999$, i.e., in 999 of 1,000 cases, the mean life of the six-side molded PLCSP solder joint is superior (by 2.9 times) to that of the ordinary PLCSP.

3.9.7 Simulation Results of the Six-Sides Molded PLCSP Solder Joint

(A) Structure and Kinematic Boundary Condition

The PCB assembly of the six-side molded PLCSP is schematically shown in Fig. 3.62. Because of the symmetry about the x-axis, the y-axis, and the diagonal axes, only one-eighth of the structure is modeled. From the concept of distance to neutral point (DNP) [21], the corner solder joint does not carry any power/ground/signal but functions

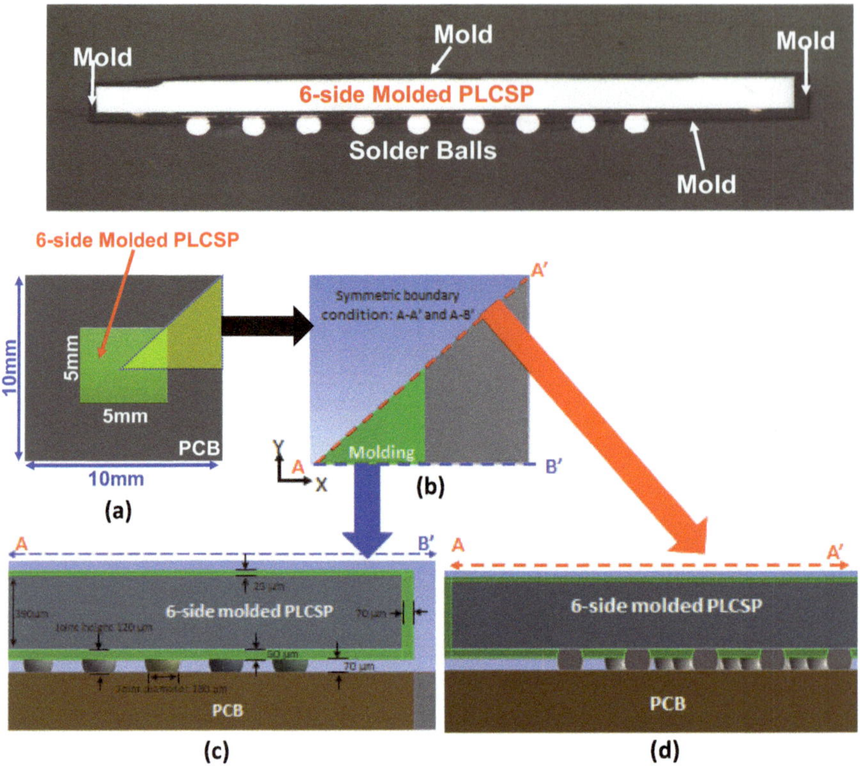

Fig. 3.62 Geometry of the six-side molded PLCSP

as a dummy (mechanical) solder joint used for absorbing the maximum stress and strain.

Despite the overall economy of elements in the one-eighth of model, selective mesh refinement is used to concentrate highly refined elements in the solder joints where failure is anticipated. In the present PCB assembly, failure would be predicted in the solder joints with the greatest DNP and near the chip corners, as shown in Fig. 3.63. Thus, highly refined meshes are applied to these solder joints. The other joints are coarsely meshed. The 3- D Solid186 element of ANSYS 2020 is used.

(B) Material Properties

The material properties of the six-side molded PLCSP PCB assembly are given in Table 3.2. All the material properties are assumed to be constant, except for those of the solder. The Sn3Ag.5Cu is assumed to obey the generalized Garofalo creep equation [66]:

$$\frac{d\varepsilon}{dt} = 500,000\sinh(0.1\sigma)^5\exp\left(-\frac{5,800}{\mathrm{T}}\right)$$

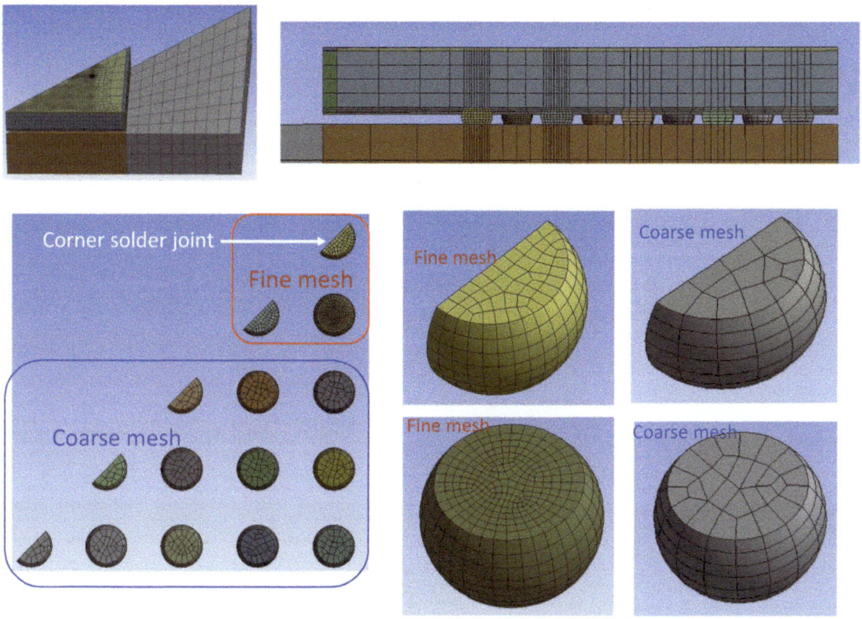

Fig. 3.63 3D-FEM of the six-side molded PLCSP

where ε is the strain, σ is the stress in MPa, and T is in Kelvin. The coefficient of thermal expansion (CTE) and Young's modulus of the solder are 21.3 + 0.017 T and 49 −0.07 T, respectively, and T is the temperature in Celsius.

Kinetic Boundary Condition—the temperature profile shown in Fig. 3.64 is to be imposed on the six-side molded PLCSP PCB assembly. Five temperature cycles are executed. It can be seen that the temperature condition is from −40 °C ↔ 85 °C. The cycle time is 60 min; the ramp-up and ramp-down are 15 min, and the dwell-at-hot and dwell-at cold are each 15 min.

(C) Thermal Cycling Simulation Results

Figure 3.65 shows the shear stress and creep shear strain hysteresis loops at the corner solder joint of the six-side molded PLCSP PCB assembly. It can be seen that after the second cycle, the creep responses converged (become stabilized). The deformed shape (color contours) and the undeformed shape (dark lines) of the six-side molded PLCSP PCB assembly are shown in Fig. 3.66a at 450 s (85 °C) and in Fig. 3.66b at 2,250 s (−40 °C). It can be seen that (1) at 85 °C, the PCB expanses more than the six-side molded silicon chip and curves outward; the solder joint (especially the corner one with the largest DNP) is subjected to mainly shear deformation, and the structure is deformed to a convex shape (smiling face), and (2) at −40 °C, the PCB shrinks more than the molded silicon chip and curves inward; the structure is deformed to a concave shape (crying face).

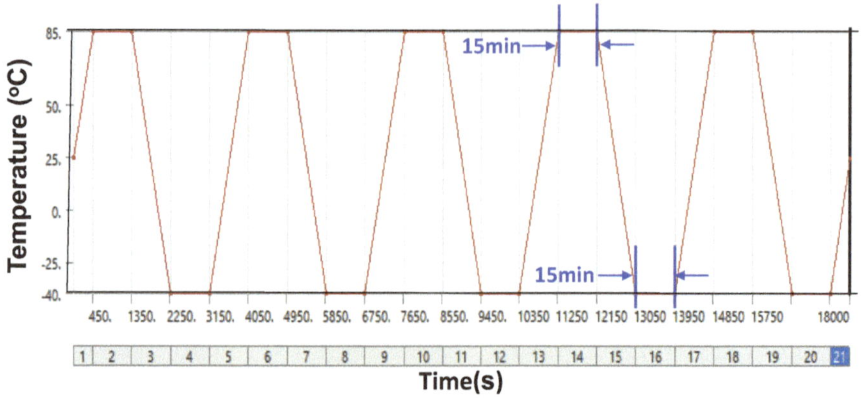

Fig. 3.64 Temperature boundary condition

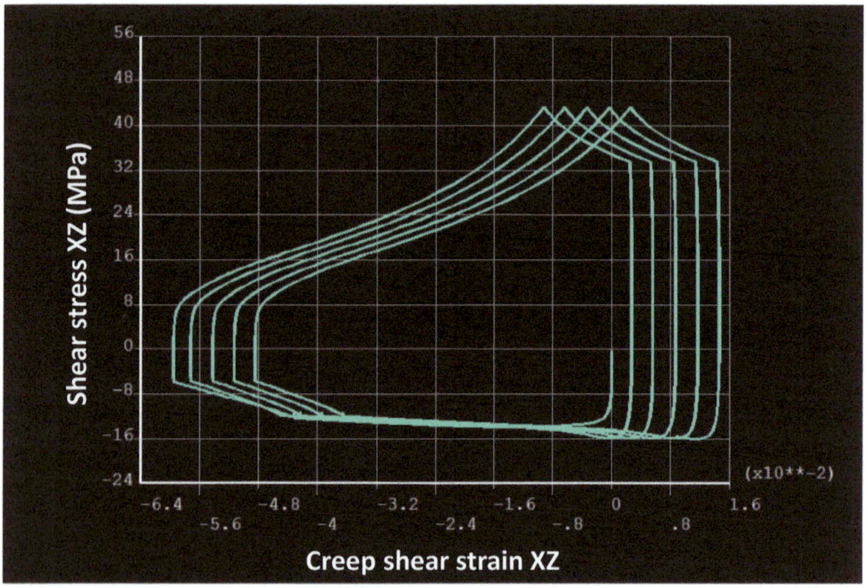

Fig. 3.65 Shear stress versus creep shear strain hysteresis loops at the six-side molded PLCSP corner solder joints

The largest accumulated creep strain occurs (the failure location) at the corner solder joint (Fig. 3.67) of the six-side molded PLCSP assembly. Figure 3.67a shows the distribution of accumulated creep strain contours at 450 s (85 °C) and Fig. 3.67b at 2,250 s (−40 °C) in the corner solder joint. It can be seen that the maximum accumulated creep strain occurs near the interface between the PCB and the bulk solder of the corner solder joint, and it is the initial driving force for cracking (the

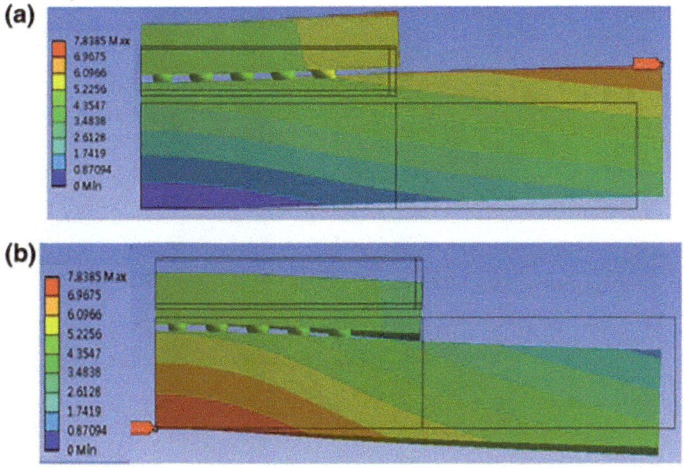

Fig. 3.66 Deformed shape (color contours) and undeformed (dark lines) shape of the six-side molded PLCSP PCB assembly. **a** At 85 °C. **b** At −40 °C

failure mode) of the solder joint. Thus, any failure should occur (initially) at this location. However, the maximum accumulated creep strain only occurs in a very small (red) volume of the whole corner solder joint. Actually, the accumulated creep strains in most of the (blue) volume of the corner solder joint (especially near the interface between the chip and the bulk solder) are very small. This is because the upper portion (about 50 mm) of the corner solder joint is embedded in the EMC.

3.9.8 Simulation Results of the Ordinary PLCSP Solder Joint

An ordinary PLCSP PCB assembly subjected to the same boundary conditions is also analyzed. The largest accumulated creep strain also occurs at the corner solder joint (the same failure location as in the six-side molded case). However, the distribution of the accumulated creep strain (failure mode) in the corner solder (Fig. 3.68) is very different from that of the six-side molded PLCSP (Fig. 3.67).

First of all, the site of the maximum accumulated creep strain in the corner solder joint has been switched from the interface between the chip and the bulk solder for the ordinary PLCSP (Fig. 3.68) to the interface between the PCB and the bulk solder for the six-side molded PLCSP (Fig. 3.67). This is because there is more local thermal expansion mismatch between the silicon chip and the bulk solder for the ordinary PLCSP. (The local thermal expansion mismatch between the six-side molded silicon chip and the bulk solder is smaller.) Also, the upper portion (about 50 μm) of the six-side molded solder joint is embedded in the EMC.

Second, the corner solder joint failure mode is different between the ordinary PLCSP and the six-side molded PLCSP. For the ordinary PLCSP, the crack should

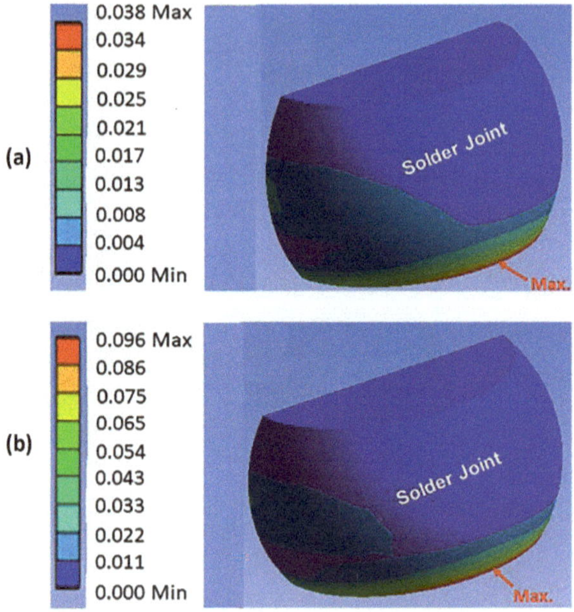

Fig. 3.67 Accumulated creep strain at the corner solder joint of the six-side molded PLCSP PCB assembly. **a** At 85 °C (450 s). **b** At −45 °C (2,250 s)

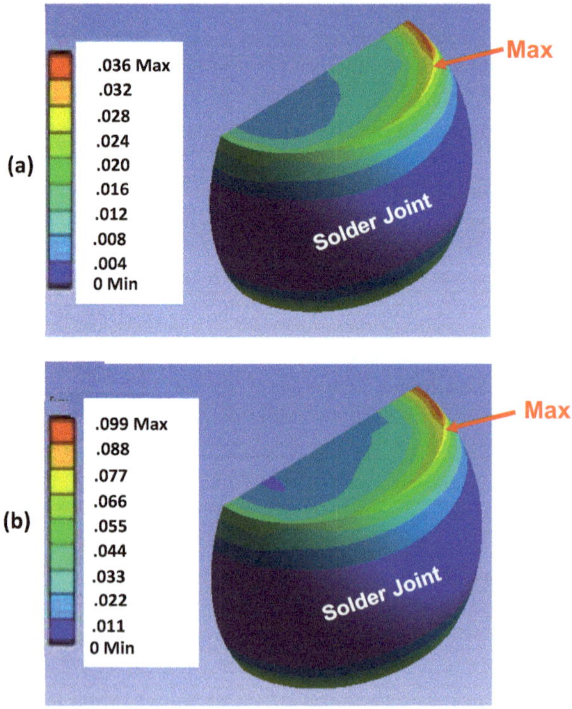

Fig. 3.68 Accumulated creep strain at the corner solder joint of the ordinary PLCSP PCB assembly. **a** At 85 °C (450 s). **b** At −45 °C (2,250 s)

be initially at the interface between the chip and the bulk solder. On the other hand, for the six-side molded PLCSP, the crack should be initially at the interface between the PCB and the bulk solder.

3.9.9 Summary

Some important results and recommendations are summarized as follows:

- The feasibility of the design, materials, process, fabrication, and reliability of a six-side molded PLCSP has been demonstrated.
- The RDL of the PLCSP has been fabricated on a 508 × 508-mm panel with diced and undiced wafers 2.25 times of 300-mm device wafers. It is a very high throughput process.
- The RDL of the PLCSP has been fabricated with an all-PCB processes and equipment on the large panel with diced and undiced device wafers. It is a very low-cost process.
- After RDL fabrication, the wafers have been debonded from the panel. Then, solder ball mounting, half-cutting the wafer from the front side to make the trenches on the saw streets, wafer front-side EMC laminating, backgrinding the backside of the wafer until the EMC on the trenches was seen, backside laminating of EMC, plasma etching some of the front-side EMC to expose a portion of the solder ball, then dicing the wafer into individual six-side molded PLCSPs. With laser dicing such as [17–19], the saw street of device wafers can be as little as 60 mm.
- Based on the JEDEC Standard JESD22-B111, a drop test of the ordinary PLCSP on PCB assembly has been performed. All the samples passed 30 drops, thus the PLCSP passed the drop test.
- Based on a nonlinear, time- and temperature-dependent finite element analyses of the six-side molded PLCSP and the ordinary PLCSP on PCB assembly, it has been found that (1) they have the same failure location, which is at the corner solder joint; (2) they have different failure mode; for the ordinary PLCSP, the maximum accumulated creep strain occurs near the interface between the chip and the bulk solder, but for the six-side molded PLCSP, it occurs near the interface between the PCB and the bulk solder; (3) because it is less compliant, the maximum accumulated creep strain in the six-side molded corner solder joint is larger than that in the ordinary corner solder joint; however, this happened in a very small volume of the corner solder joint; and (4) most volumes of the six-side molded PLCSP corner solder joint have much lower values of the accumulated creep strain than the ordinary PLCSP corner solder joint, and thus it is expected that the six-side molded PLCSP corner solder joint has a longer thermal fatigue life.
- Based on the thermal cycling testing results, the mean life of the six-side molded PLCSP solder joint is superior (by 2.9 times) to that of the ordinary PLCSP.

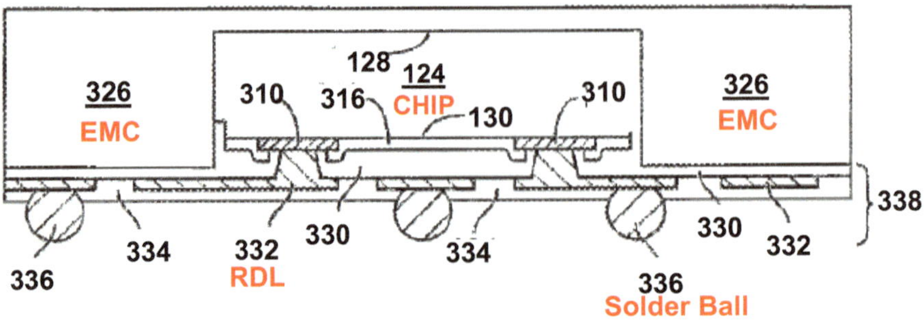

Fig. 3.69 Statschippad's patent on 5-side molded WLCSP

3.10 eWLCSP by Statschippac

The five-side molded WLCSP was first patented (e.g., US8456002) [47] in 2011 (Fig. 3.69), and published [48–51] by STATS ChipPAC. Figure 3.70 shows the images of Statschippac's eWLCSP, which can be seen that basically it is the fan-out chip-first with face-down method. They test and then dice the original device wafer into KGDs (known good dies), pick and place the individual KDGs on a temporary wafer carrier, and then compression mold an EMC (epoxy molding compound) on the re-constituted wafer as shown in Fig. 3.70. It is followed by debonding the temporary carrier, fabricating the redistribution-layer (RDL), solder ball mounting, and dicing the re-constituted wafer into individual eWLCSPs (Fig. 3.70). It can be seen that the backside and four sides of the chip are protected by EMC.

3.11 WLCSP by UTAC

UTAC's method [52] in making the 5-side molded WLCSP is very different from that of Statschippac. They use the original device wafer (without dicing and P&P of the KGDs on a temporary wafer) to make the RDL; solder ball mounting, dicing, molding and then grinding. It is followed by dicing the device wafer into individual WLCSPs (Fig. 3.71).

3.12 mWLCSP by SPIL

SPIL's method on 5-side protected (four side walls and the front side of the chip) is very similar to that of UTAC. They called it mWLCSP (molded WLCSP) [53] (Fig. 3.72). SPIL also proposed a 6-side protection by laminating an epoxy film at the backside of the chip.

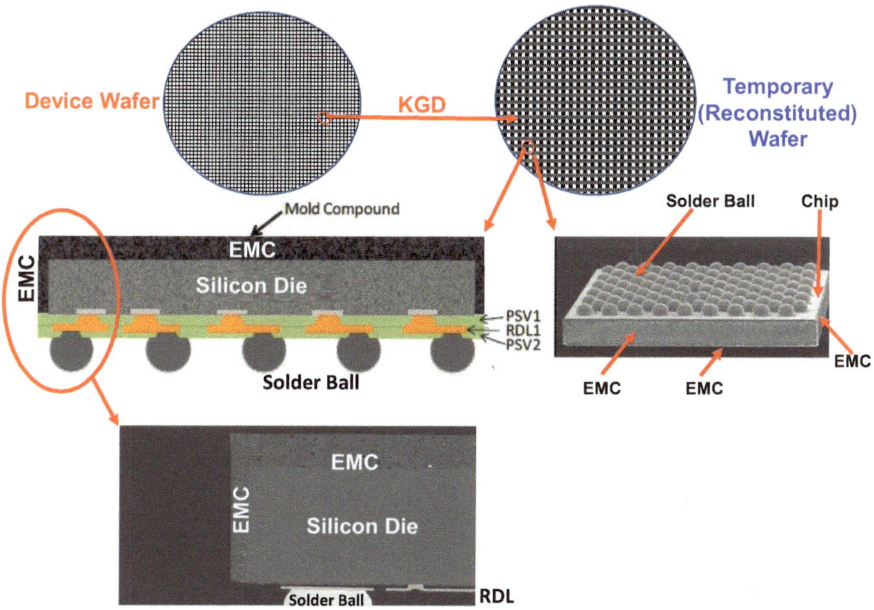

Fig. 3.70 Statschippad's eWLCSP

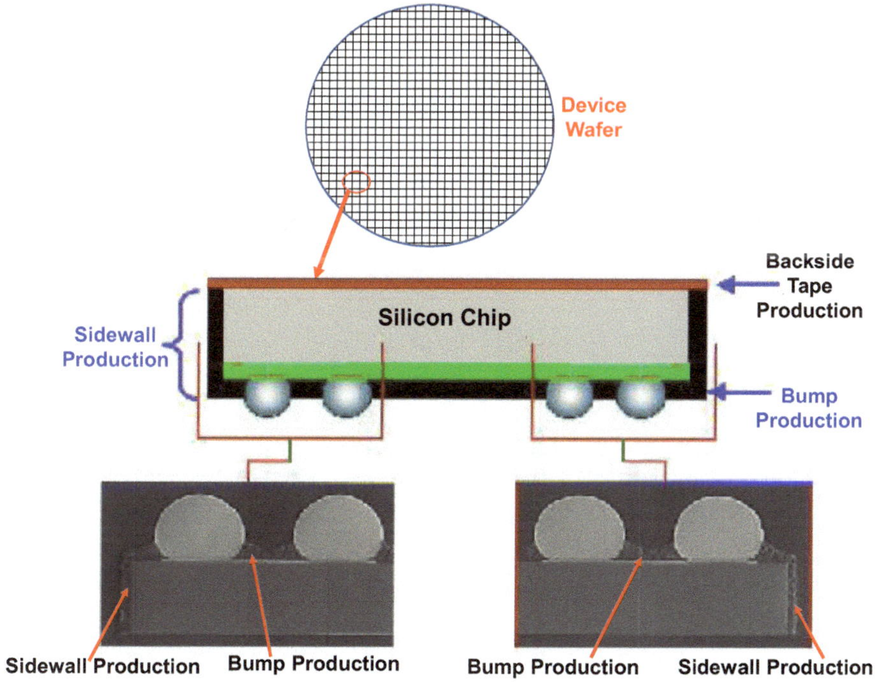

Fig. 3.71 UTAC's WLCSP

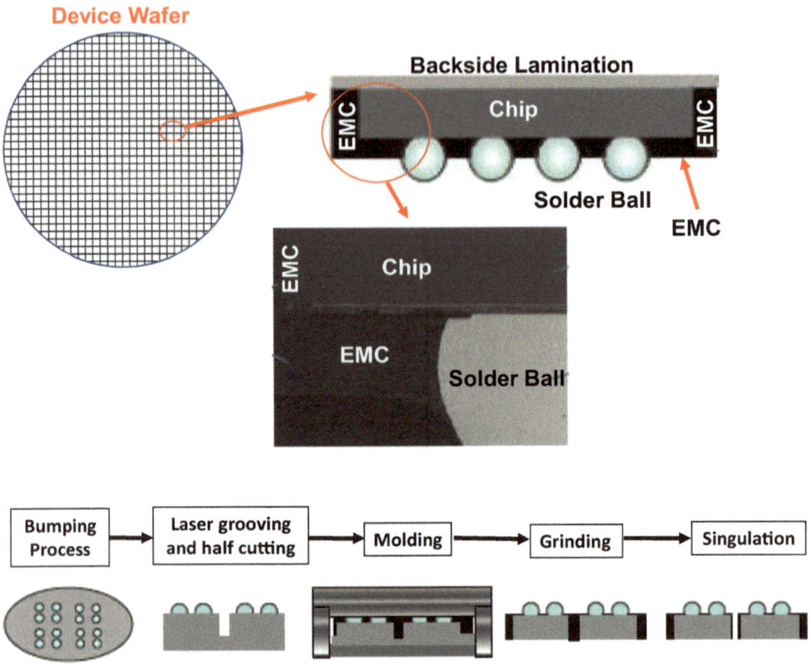

Fig. 3.72 SPIL's mWLCSP

3.13 WLCSP by Huatian

Huatian's method [54–56] in making the 6-side molded WLCSP (four side walls, front side, and back side of the chip) is very similar to that of UTAC [52] and SPIL [53]. Their structure is shown in Fig. 3.73 and process flow is shown in Fig. 3.74. With laser dicing such as those given by ASM [63–65], the saw street can be as little as 60 μm.

3.14 mWLCSP by SPIL and MediaTek

Figure 3.75 shows the 5-side molded mWLCSP and 6-side molded mWLCSP by SPIL and MediaTek [57]. The process for the 5-side molded WLCSP is shown in Fig. 3.76 and for the 6-side molded WLCSP is shown in Fig. 3.77. Similar to Statschippac [48–51], which use a temporary carrier, but their process is quite different [57].

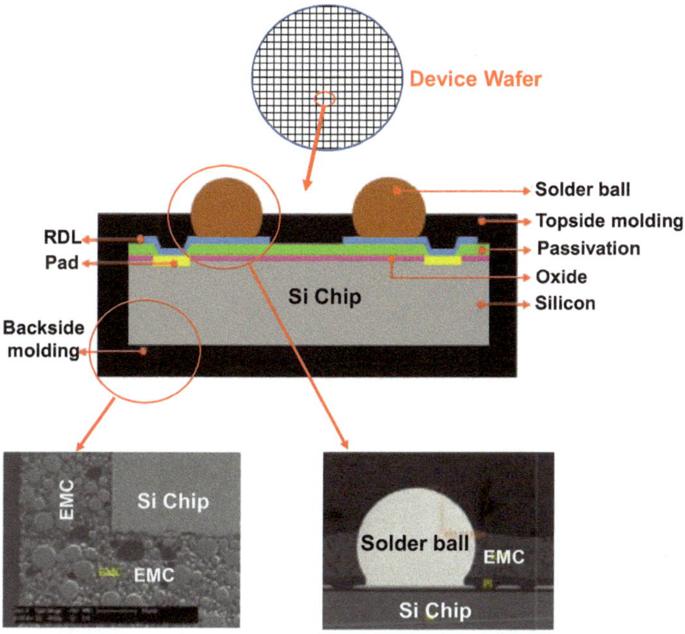

Fig. 3.73 Huatian's WLCSP

3.15 Summary

Some important results and recommendations are summarized as follows:

- The feasibility of the design, materials, process, fabrication, and reliability of a six-side molded PLCSP has been demonstrated.
- The RDL of the PLCSP has been fabricated on a 508 × 508-mm panel with diced and undiced wafers 2.25 times of 300-mm device wafers. It is a very high throughput process.
- The RDL of the PLCSP has been fabricated with an all PCB processes and equipment on the large panel with diced and undiced device wafers. It is a very low-cost process.
- After RDL fabrication, the wafers have been debonded from the panel. Then, solder ball mounting, half-cutting the wafer from the front side to make the trenches on the saw streets, wafer front-side EMC laminating, backgrinding the backside of the wafer until the EMC on the trenches was seen, backside laminating of EMC, plasma etching some of the front-side EMC to expose a portion of the solder ball, then dicing the wafer into individual six-side molded PLCSPs.
- Based on the JEDEC Standard JESD22-B111, a drop test of the ordinary PLCSP on PCB assembly has been performed. All the samples passed 30 drops, thus the PLCSP passed the drop test.

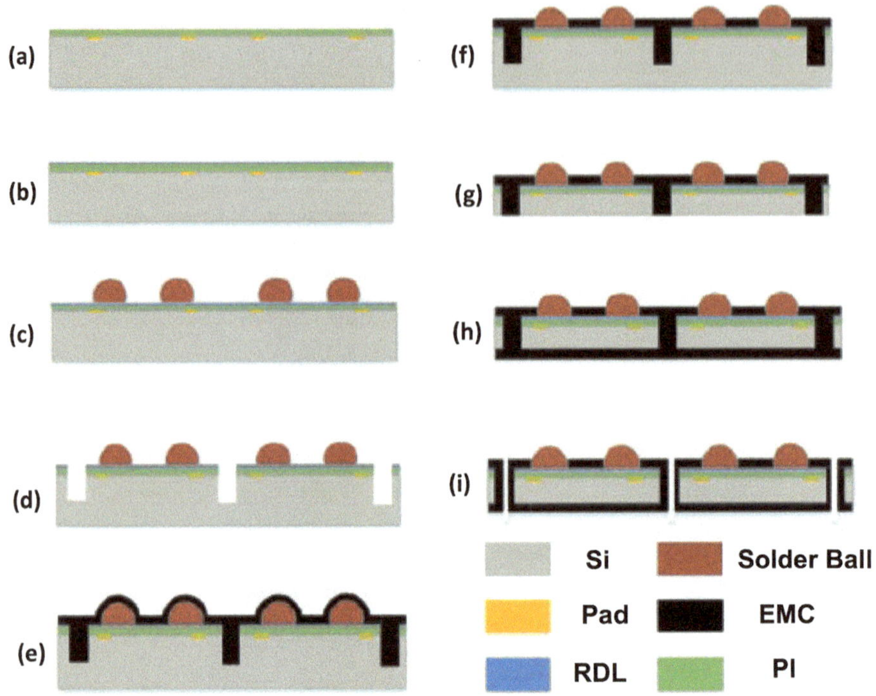

Fig. 3.74 Key process steps in Huatian's WLCSP

- Based on a nonlinear, time- and temperature-dependent finite element analyses of the six-side molded PLCSP and the ordinary PLCSP on PCB assembly, it has been found that (1) they have the same failure location, which is at the corner solder joint; (2) they have different failure mode; for the ordinary PLCSP, the maximum accumulated creep strain occurs near the interface between the chip and the bulk solder, but for the six-side molded PLCSP, it occurs near the interface between the PCB and the bulk solder; (3) because it is less compliant, the maximum accumulated creep strain in the six-side molded corner solder joint is larger than that in the ordinary corner solder joint; however, this happened in a very small volume of the corner solder joint; and (4) most volumes of the six-side molded PLCSP corner solder joint have much lower values of the accumulated creep strain than the ordinary PLCSP corner solder joint, and thus it is expected that the six-side molded PLCSP corner solder joint has a longer thermal fatigue life.
- PCB assemblies of the six-side molded PLCSP and the ordinary PLCSP have been subjected to thermal cycling test. The sample size of the six-side molded PLCSP is 24 and that of ordinary PLCSP is 48. The test stopped at 1,300 cycles. There were 15 failures of the six-side molded PLCSP samples and 48 failures of the ordinary PLCSP samples.

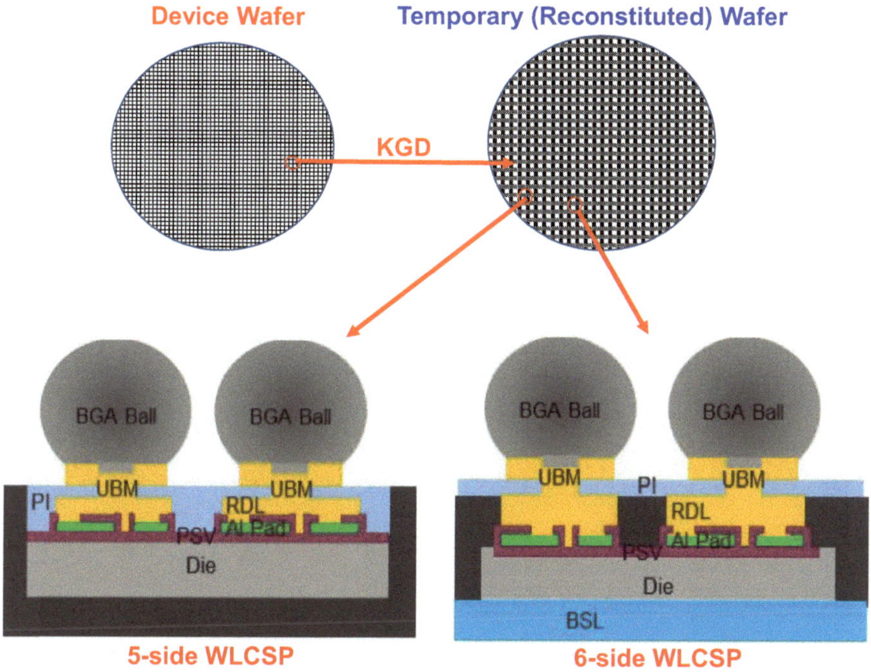

Fig. 3.75 SPIL/MediaTEK's 5-side and 6-side molded mWLCSP

- The failure criterion is when the resistance of the daisy chain of the PLCSP PCB assembly increases by 50%.
- For median rank, the Weibull slope and the characteristic life of the six-side molded PLCSP solder joints are, respectively 5.53 and 1,037 cycles. For 90% confidence, the true Weibull slope is $3.93 \leq \beta_t \leq 7.13$ and the true characteristic life is $958 \leq \theta_t \leq 1{,}069$ cycles.
- The failure location of the six-side molded PLCSP is along the outer row (near the corner) of solder joints. The failure mode is the cracking of solder near the interface between the PCB and the bulk solders.
- For median rank, the Weibull slope and the characteristic life of the ordinary PLCSP solder joints are, respectively 3.34 and 368 cycles. For 90% confidence, the true Weibull slope is $2.6 \leq \beta_t \leq 4.1$ and the true characteristic life is $326 \leq \theta_t \leq 403$ cycles.
- The failure location of the ordinary PLCSP is along the outer row (near the corner) of solder joints. The failure mode is the cracking of solder near the interface between the chip and the bulk solders.
- By comparing the mean life between the six-side molded PLCSP solder joint (958 cycles) with the ordinary PLCSP solder joints (330), it has been determined that in 999 of 1,000 cases, the mean life of the six-side molded PLCSP is 2.9 times of that of the ordinary PLCSP.

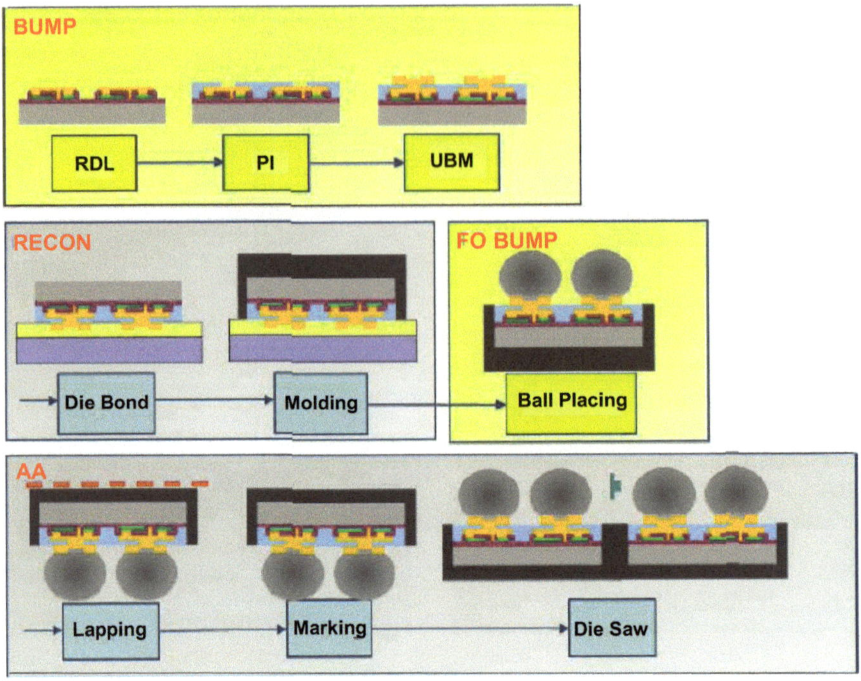

Fig. 3.76 SPIL/MediaTEK's 5-side molded mWLCSP key process steps

- A nonlinear, time- and temperature-dependent 3-D finite element simulation of the six-side molded PLCSP PCB assembly showed that the maximum accumulated creep strain occurs near the interface between the PCB and the bulk solders. Thus, any failure should occur (initiate) at this location. This confirms with the failure mode from the thermal cycling test results.
- The simulation of the ordinary PLCSP PCB assembly showed that the maximum accumulated creep strain occurs near the interface between the chip and the bulk solders. Thus, any failure should occur (initiate) at this location. This confirms with the failure mode from the thermal cycling test results.
- The reason for the switch of the failure mode from the interface between the PCB and the bulk solders (for the six-side molded PLCSP) to the interface between the chip and the bulk solders (ordinary) is because there is protection of the upper solder joint from the EMC molding.
- The maximum values of the accumulated creep strain (DW_i) in the corner solder joint for the six-side molded PLCSP and the ordinary PLCSP are about the same. However, these maximum values only occurred at a very small volume of the six-side molded PLCSP solder joint. The accumulated creep strain in most of the volumes of the six-side molded solder joint is smaller than those in the ordinary PLCSP solder joins. Thus, the thermal fatigue life of the six-side molded PLCSP should be longer than that of the ordinary PLCSP.

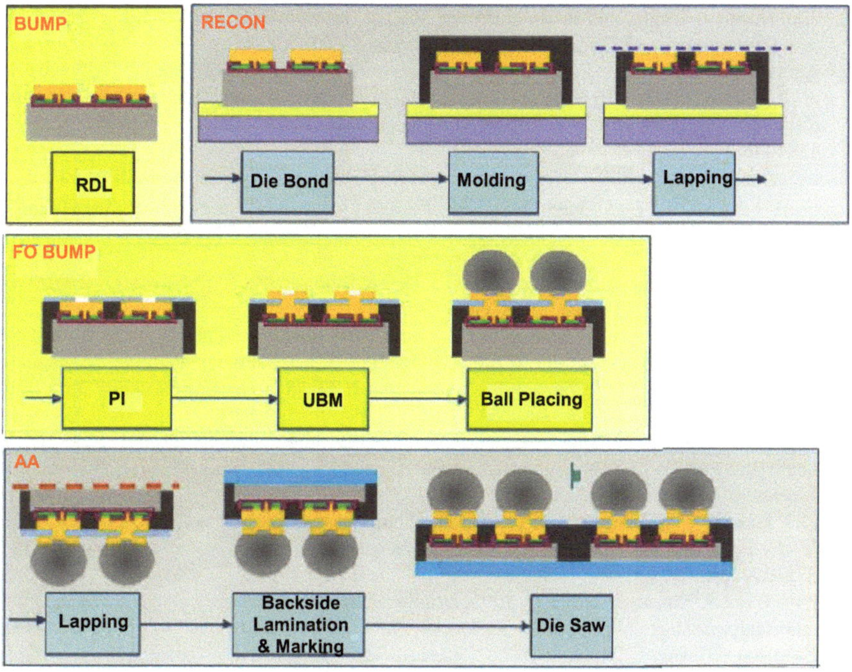

Fig. 3.77 SPIL/MediaTEK's 6-side molded mWLCSP key process steps

References

1. Elenius, P., and H. Yang. 1998. The UltraCSP wafer-scale package. *High-Density Interconnect* 1 (6): 36–40.
2. Lau, J.H., and S.W.R. Lee. 1999. *Chip Scale Package*. New York: McGraw-Hill Book Company.
3. Lau, J. H., T. Chung, R. Lee, C. Chang, and C. Chen. 1999. A novel and reliable Wafer-Level Chip Scale Package (WLCSP). In *Proceedings of the Chip Scale International Conference, SEMI*, September 1999, H1–8.
4. Wu, M., S. Chung, and D. Yu. 2016. UFI (UBM-free integration) fan-in WLCSP technology enables large die fine-pitch packages. In *IEEE/ECTC Proceedings*, May 2016, 1154–1159.
5. Yasunaga, M., S. Baba, M. Matsuo, H. Matsushima, S. Nakao, and T. Tachikawa. 1994. Chip scale package: A lightly dressed LSI chip. In *Proceedings of IEEE/CPMT International Electronics Manufacturing Technology Symposium*, La Jolla, CA, 1994, 169–176.
6. Marcoux, P. 1994. A Minimal Packaging Solution for Known Good Die And Direct Chip Attachment. In *Proceedings of Surface Mount International Conference*, San Jose, CA, August 1994, 19–26.
7. Chanchani, R., K. Treece, and P. Dressendorfer. 1995. A new mini ball grid array (m-BGA) multichip module technology. In *Proceedings of NEPCON West*, February 1995, 938–945.
8. Badihi, A., and E. Por. 1995. Shellcase—A true miniature integrated circuit package. In *Proceedings of International FC, BGA, Advanced Packaging Symposium*, San Jose, CA, February 1995, 244–252.
9. Baba, S., Y. Tomita, M. Matsuo, H. Matsushima, N. Ueda, and O. Nakagawa. 1996. Molded chip scale package for high pin count. In *Proceedings of IEEE/ECTC*, Orlando, FL, May 1996, 1251–1257.

10. Topper, M., J. Simon, and H. Reichl. 1996. Redistribution technology for chip scale package using photosensitive BCB. *Future Fab International*, 363–368.
11. Auersperg, J., D. Vogel, J. Simon, A. Schubert, and B. Michel. 1997. Reliability evaluation of chip scale packages by FEA and microDAC. In *Proceedings of Symposium on Design and Reliability of Solder and Solder Interconnections*, TMS Annual Meeting, Orlando, FL, 1997, 439–445.
12. DiStefano, T. 1997. Wafer-level fabrication of IC Packages. In *Chip Scale Review*, 20–27.
13. Kohl, J. E., C. W. Eichellerger, S. K. Phillips, and M. E. Rickley. 1997. Low cost chip scale packaging and interconnect technology. In *Proceedings of the CSP Symposium*, San Jose, CA, 1997, 37–43.
14. Elenius, P. 1998. Flip chip bumping for IC packaging contractors. In *Proceedings of NEPCON West,* Anaheim, CA, February 1998, 1403–1407.
15. Lau, J. H., S. W. R. Lee, C. Chang, and C. Chen. 1999. Solder joint reliability of Wafer Level Chip Scale Packages (WLCSP): A time-temperature-dependent creep analysis. ASME Paper No. 99-IMECE/EEP-5. In *International Mechanical Engineering Congress and Exposition*, November 1999.
16. Lau, J.H., S.W.R. Lee, and C. Chang. 2000. Solder joint reliability of Wafer Level Chip Scale Packages (WLCSP): A time-temperature-dependent creep analysis. *ASME Transactions Journal of Electronic Packaging* 122 (4): 311–316.
17. Lau, J. H., S. W. R. Lee, and C. Chang. 2001. Solder joint crack propagation analysis of wafer-level chip scale package on printed circuit board assemblies. *IEEE Transactions on CPMT* 285–292.
18. Lau, J. H., S. Pan, and C. Chang. 2001. Nonlinear fracture mechanics analysis of wafer level chip scale package solder joints with cracks. *The International Journal of Microcircuits and Electronic Packaging* 24(1): 1–10.
19. Lau, J.H., S. Pan, and C. Chang. 2002. A new thermal-fatigue life prediction model for Wafer Level Chip Scale Package (WLCSP) solder joints. *ASME Transaction Journal of Electronic Packaging* 124 (3): 212–220.
20. Lau, J.H. 2001. Elastic, plastic, and creep analysis of Wafer Level Chip Scale Package (WLCSP) solder joints on Microvia build-up PCB. *Circuit World* 27 (1): 20–31.
21. Sarto, M., L. Maggi, T. Chiarillo, E. Duqi, L. Baldo, A. Abbisogni, and F. Daniele. 2023. Wafer level chip scale package technology applied to MEMS pressure sensor. In *Proceedings of ECTC*, May 2023, 1273–1278.
22. Chen, C., Y. S. Wu, K. H. Chen, W. S. Tseng, P. H. Tsao, and S. T. Leu. 2022. Investigation of LowK WLCSP die strength impact induced by singulation process. In *Proceedings of ECTC*, May 2022, 1025–1029.
23. Lim, J., B. Kim, R. Gacho, and B. Dunlap. 2022. Component level reliability evaluation of low cost 6-sided die protection versus wafer level chip scale packaging with 350um ball pitch. In *Proceedings of ECTC*, May 2022, 1791–1797.
24. Chen, L., X. Fan, and Y. Liu. 2021. Testing and modeling of board level reliability of WLCSP under UHAST conditions. In *Proceedings of ECTC*, May 2021, 1329–1333.
25. Zhang, H., Z. Wu, J. Malinowski, M. Carino, K. Young-Fisher, J. Trewhella, and P. Justiso. 2020. 45RFSOI WLCSP Board Level Package Risk Assessment and Solder Joint Reliability Performance Improvement. In *Proceedings of ECTC*, May 2020, 2151–2156.
26. Chen, C., K. Chen, Y. Wu, P. Tsao, and S. Leu. 2020. WLCSP Solder Ball Interconnection Enhancement for High Temperature Stress Reliability. In *Proceedings of ECTC*, May 2020, 1212–1217.
27. Ma, S., Y. Liu, F. Zheng, F. Li, D. Yu, A. Xiao, and X. Yang. 2020. Development and Reliability Study of 3D WLCSP for Automotive CMOS Image Sensor using TSV Technology. In *Proceedings of ECTC*, May 2020, 461–466.
28. Machani, K., F. Kuechenmeister, D. Breuer, C. Klewer, J. Cho, and K. Young-Fisher. 2020. Chip Package Interaction (CPI) Risk Assessment of 22FDX® Wafer Level Chip Scale Package (WLCSP) Using 2D Finite Element Analysis Modeling. In *Proceedings of ECTC*, May 2020, 1100–1105.

29. Chiu, J., K.C. Chang, S. Hsu, P. Tsao, and M. J. Lii. 2019. WLCSP Package and PCB Design for Board Level Reliability. In *Proceedings of ECTC*, May 2019, 763–767.
30. Yu, D., Y. Zou, X. Xu, and A. Shi. 2019. Development of 3D WLCSP with Black Shielding for Optical Finger Print Sensor for the Application of Full Screen Smart Phone. In *Proceedings of ECTC*, May 2019, 884–889.
31. Zhou, Y., L. Chen, Y. Liu, and S. Sitaraman. 2019. Thermal Cycling Simulation and Sensitivity Analysis of Wafer Level Chip Scale Package with Integration of Metal-Insulator-Metal Capacitors. In *Proceedings of ECTC*, May 2019, 1521–1528.
32. Lai, C., G. Lin, T. Chao, Y. Chen, and F. Chien. 2019. Board Level Reliability Study of Fan-Out Single Die Package with 350um Bump Pitch. In *Proceedings of ECTC*, May 2019, 1170–1174.
33. Chou, P. H., H. Y. Hsiao, and K.N. Chiang. 2019. Failure Life Prediction of Wafer Level Packaging using DoS with AI Technology. In *Proceedings of ECTC*, May 2019, 1515–1520.
34. Chen, Z., B. Lau, Z. Ding, E. Wai, B. Han, L. Bu, H. Chang, and T. Chai. 2018. Development of WLCSP for Accelerometer Packaging with Vertical CuPd Wire as Through Mold Interconnection (TMI). In *Proceedings of ECTC*, May 2018, 1188–1193.
35. Tsao, P., T. H. Lu, T. M. Chen, K. C. Chang, C. M. Kuo, M. J. Lii, and L. H. Chu. 2018. Board Level Reliability Enhancement of WLCSP with Large Chip Size. In *Proceedings of ECTC*, May 2018, 1200–1205.
36. Ramachandran, V., K. C. Wu, C. C. Lee, and K. N. Chiang. 2018. Reliability Life Assessment of WLCSP Using Different Creep Models. In *Proceedings of ECTC*, May 2018, 1017–1022.
37. Sheikh, M., A. Hsiao, W. Xie, S. Perng, E. Ibe, K. Loh, and T. Lee. 2018. Multi-Axis Loading Impact on Thermo-Mechanical Stress-Induced Damage on WLCSP and Components With Via-in Pad Plated Over (VIPPO) Board Design Configuration. In *Proceedings of ECTC*, May 2018, 911–915.
38. Yang, T. L., S. B. Yang, W. L. Huang, C. C. Chen, C. C. Kuo, G. C. Huang, K. Y. Wu, T. C. Chang, C.C. Hsu, C. L. Chang, H. L. Huang, E. Chen, K. C. Liu, M. Liao, A. Kalnitsky, and H. Ku. 2018. Realization of High Electrical Performance On-chip Thick Copper Inductor Package by Via Interface Process Improvement for Metal Contact. In *Proceedings of ECTC*, May 2018, 1692–1699.
39. Lau, J. H., S. Pan, and C. Chang. 2002. Creep Analysis of Wafer Level Chip Scale Package (WLCSP) with 96.5Sn-3.5Ag and 100In Lead-Free Solder Joints and Microvia Build-Up Printed Circuit Board. *ASME Transactions, Journal of Electronic Packaging* 124: 69–76.
40. Darveaus, R., and K. Banerji. 1992. Constitutive Relations for Tin-Based Solder Joints. In *Proceedings of IEEE Electronic Components & Technology Conference*, May 1992, 538–551.
41. Lau, J. H. 1993. Creep of 96.5Sn-3.5Ag Solder Interconnects. *Soldering & Surface Mount Technology* 15: 45–49.
42. Fung, Y.C. 1965. *Foundations of Solid Mechanics*. New York: Prentice Hall.
43. ANSYS *User's Manual*, Release 5.6.1, 2000.
44. Lau, J. H., and S. W. R. Lee. 2001. Effects of Microvia Build-Up Layers on the Solder Joint Reliability of a Wafer Level Chip Scale Package (WLCSP). In *Proceedings of ECTC*, May 2001, 1207–1215.
45. Lau, J.H. 2000. Elastic, Elastic-Plastic, and Creep Analyses of Wafer Level Chip Scale Package Solder Joints on Microvia Build-Up Printed Circuit Boards. *Circuit World* 27 (1): 20–31.
46. Kazama, A., T. Satoh, Y. Yamaguchi, I. Anjoh, and A. Nishimura. 2001. Development of Low-cost and Highly Reliable Wafer Process Package. In *Proceedings of ECTC, May 2001*, 40–46.
47. Lin, Y., P. Marimuthu, K. Chen, H. Goh, Y. Gu, I. Shim, R. Huang, S. Chow, J. Fang, and X. Feng. 2011. Semiconductor device and method of forming insulating layer disposed over the semiconductor die for stress relief. *US Patent No. 8,456,002*, filling date: December 21, 2011.
48. Strothmann, T., S. Yoon, and Y. Lin. 2014. Encapsulated wafer level package technology (eWLCSP). In *Proceedings of IEEE/ECTC*, Orlando, FL, May 2014, 931–934,
49. Lin, Y., K. Chen, K. Heng, L. Chua, and S. Yoon. 2014. Encapsulated wafer level chip scale package (eWLCSP) for cost effective and robust solutions in FlexLine. In *Proceedings of IEEE/IMPACT*, Taipei, Taiwan, October 2014, 316–319.

50. Lin, Y., E. Chong, M. Chan, K. Lim, and S. Yoon. 2015. WLCSP1 and eWLCSP in FlexLine: innovative wafer level package manufacturing. In *Proceedings of IEEE/ECTC*, San Diego, CA, May 2015, 865–870.
51. Lin, Y., K. Chen, K. Heng, L. Chua, and S. Yoon. 2016. Challenges and improvement of reliability in advanced wafer level packaging technology. In *Proceedings of IEEE 23rd International Symposium on the Physical and Failure Analysis (IPFA)*, Singapore, July 2016, 47–50.
52. Smith, L., and J. Dimaano, Jr. 2015. Development approach & process optimization for sidewall WLCSP protection. In *Proceedings of IWLPC*, San Diego, CA, October 2015, 1–4.
53. Tang, T., A. Lan, J. Wu, J. Huang, J. Tsai, J. Li, A. Ho, J. Chang, and W. Lin. 2016. Challenges of ultra-thin 5 sides molded WLCSP. In *Proceedings of IEEE/ECTC*, Las Vegas, NV, May 2016, 1167–1771.
54. Ma, S., T. Wang, Z. Xiao, and D. Yu. 2018. Process development of five-and six-side molded WLCSP. In *Proceedings of China Semiconductor Technology International Conference* (CSTIC), Shanghai, China, March 2018, 1–3.
55. Zhao, S., F. Qin, M. Yang, M. Xiang, and D. Yu. 2019. Study on warpage evolution for six-side molded WLCSP based on finite element analysis. In *Proceeding of the International Conference on Electronic Packaging Technology (ICEPT)*, Changsha, China, August 2019, 1–4.
56. Qin, F., S. Zhao, Y. Dai, M. Yang, M. Xiang, and D. Yu. 2020. Study of warpage evolution and control for six-side molded WLCSP in different packaging processes. *IEEE Transactions on CPMT* 10 (4): 730–738.
57. Chi, Y., C. Lai, C. Kuo, J. Huang, C. Chung, Y. Jiang, H. Chang, N. Liu, and B. Lin. 2020. Board level reliability study of WLCSP with 5-Sided and 6-sided protection. In *Proceedings of IEEE/ECTC*, Orlando, FL, May 2020, 807–810.
58. Lau, J.H. 2018. *Fan-out Wafer-Level Packaging*. Germany: Springer.
59. Lau, J.H. 2019. Recent advances and trends in fan-out wafer/panel-level packaging. *ASME Transactions, Journal of Electronic Packaging* 141: 1–27.
60. Lau, J.H., C. Ko, T. Tseng, T. Peng, K. Yang, T. Xia, P. Lin, E. Lin, L. Chang, N. Liu, C. Lin, D. Cheng, and W. Lu. 2020. Six-side molded panel-level chip-scale package with multiple diced wafers. *IMAPS Transactions, Journal of Microelectronics and Electronic Packaging* 17: 111–120.
61. Lau, J.H., C. Ko, C. Peng, T. Tseng, K. Yang, T. Xia, P. Lin, E. Lin, L. Chang, H. Liu, C. Lin, Y. Fan, D. Cheng, and W. Lu. 2021. Thermal cycling test and simulation of six-side molded panel-level chip-scale packages (PLCSPs). *IMAPS Transactions, Journal of Microelectronics and Electronic Packaging* 18: 67–80.
62. Lau, J.H., C. Ko, T. Tseng, K. Yang, T. Peng, T. Xia, P. Lin, E. Lin, L. Chang, H. Liu, and D. Cheng. 2020. Panel-level chip-scale package with multiple diced wafers. *IEEE Transactions on CHMP* 10 (7): 1110–1124.
63. Borkulo, J., E. Tan, and R. Stam. 2017. Laser multi beam full cut dicing of dicing of wafer level chip-scale packages. In *Proceedings of IEEE/ECTC*, Orlando, FL, May 2017, 338–342.
64. Borkulo, J., and R. Stam. 2018. Laser-based full cut dicing evaluations for Thin Si wafers. In *Proceedings of IEEE/ECTC*, San Diego, CA, May 2018, 1945–1949.
65. Borkulo, J., R. Evertsen, and R. Stam. 2019. A more than moore enabling wafer dicing technology. In *Proceedings of IEEE/ECTC*, Las Vegas, NV, May 2019, 423–427.
66. Lau, J.H., and N.C. Lee. 2020. *Assembly and Reliability of Lead-Free Solder Joints*. New York, NY: Springer.

Chapter 4
Fan-Out Technology

4.1 Introduction

One of the major functions of semiconductor packaging is to fan-out the circuitries from the chip and talk to circuitries from another chip [1]. Kauffman [2] of The Jade Corporation proposed the use of a lead frame to fan-out the circuitry from a chip. Today, just about all electronic products use lead frames such as the gull-wing lead (e.g., plastic quad flat pack), J-lead (e.g., plastic leaded chip carrier), and/or dual in-line package to fan-out the circuitries from a chip to a printed circuit board (PCB).

Lin et al. [3] of Motorola proposed the use of an organic carrier or substrate with area array solder balls (instead of lead frames) to fan-out the circuitry from a chip to a PCB, which is called a plasticball grid array (PBGA) package. Amkor [1] led the outsourced semiconductor assembly and test providers (OSATs) to license this packaging technology from Motorola—and the BGA era began.

Elenius and Hollack [4] of Flip Chip Technologies proposed the use of redistribution layers (RDLs) to fan-in the circuitry from the original peripheral bond pads of a chip on a wafer and of solder balls to connect to a PCB without underfill. The packages made by the fan-in wafer-level packaging (WLP) are called wafer-level chip-scale package (WLCSP) or fan-in technology (Chap. 3 of this book), and one of the most famous is UltraCSP developed and patented by Flip chip Technologies [5]. In 2001, Amkor led the OSATs and foundries to license the UltraCSP, and the WLP era began.

Hedler et al. [6] of Infineon proposed the use of RDLs to fan-out the circuitry from the metal pad of the chip on a wafer and solder ball to the metal pads on a PCB without underfill. [At that time, they called it embedded wafer-level ball (eWLB) grid array [7, 8]. Today, we call it fan-out wafer-level packaging (FOWLP).] Some of the RDLs have a portion that extends beyond (fan-out) the edges of the chip. Even though some of the knowledge of this technology has been patented by General Electric [9, 10] and EPIC [11], however, Infineon's patent [6] specifically pointed out the use of RDLs to fan-out the circuitries from the metal pad of the chip on a

J. H. Lau, *Flip Chip, Hybrid Bonding, Fan-In, and Fan-Out Technology*,
https://doi.org/10.1007/978-981-97-2140-5_4

wafer and solder ball to the metal pads on a PCB. Infineon also specifically pointed out that some of the RDLs have a portion that extends beyond (fan-out) the edges of the chip. These are the major claimed in [6], which were not claimed by GE and EPIC.

The advantages of FOWLP over PBGA packages with solder bumped flip chip are [12]: (1) lower cost; (2) lower package profile; (3) eliminating the substrate; (4) eliminating the wafer bumping; (5) eliminating the flip-chip reflow; (6) eliminating the flux cleaning; (7) eliminating the underfill; (8) better electrical performance; (9) better thermal performance; and (10) easier to go for system-in-package (SiP) and 3D IC packaging [13, 14]. The advantages of FOWLP over fan-in WLP are [12]: (1) the use of known good die; (2) better wafer-level yield; (3) using the best of silicon; (4) single or multichip; (5) embedded integrated passive devices; (6) more layer of RDLs; (7) higher pin counts; (8) better thermal performance; (9) easier to go for SiP and 3-D IC packaging; and (10) higher PCB level reliability.

The biggest difference between flip chip and fan-in technology with the fan-out technology is that fan-out technology needs a temporary carrier, which can be round such as a wafer or rectangular such as a panel (Fig. 4.1). The reason is because the saw street (spacing) between chips on a device wafer is very small (~50 μm) and there is not enough space to fan out the RDLs of the chip. Thus, for fan-out packaging, the device wafer is first tested and then diced into individual known good dies (KGDs). It is followed by picking and placing the KGDs on a temporary carrier (either round or rectangular) with certain distance apart so the RDLs can fan-out from the chip edges.

Basically, there are two methods to form (build) the FOWLP [15]. One is the chip first, and the other is the chip last. For the chip-first method, there are two options: face-down (die-down) and face-up (die-up). On the other hand, there is only one option for the chip-last method (also called RDL-first): die face down. The chip-last

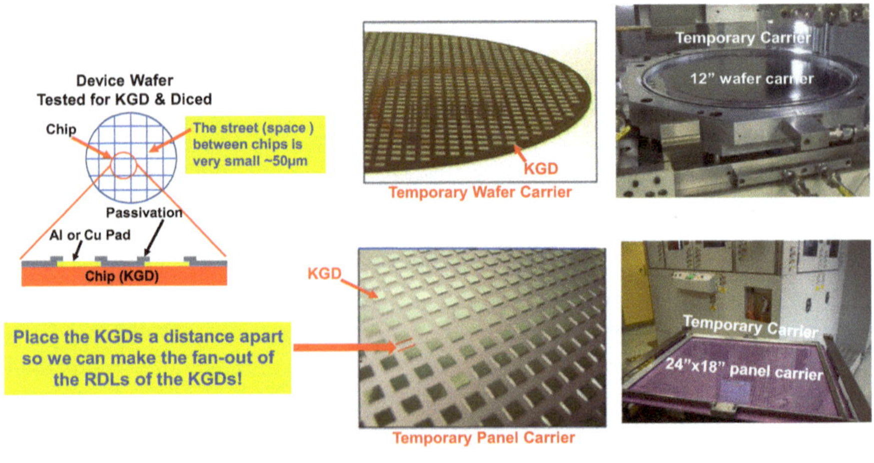

Fig. 4.1 Temporary carrier for fan-out technology

method [16–28] is the most expensive and complicated, which is suitable for very high-density, very high-performance, very large chip size, and very small (≤2 μm) RDL linewidth and spacing applications. The chip-first with die face-down is the eWLB [7, 8, 29–40], which is for low-end, small chip size, and large (10–15 μm) RDL linewidth and spacing applications. The chip-first with die face-up [41–47] is for medium chip size, medium (2–8 μm) RDL linewidth and spacing, and medium-to high-density/performance applications.

4.2 Topics in This Chapter

There are eight different topics in this chapter. (1) chip-first with die face-down for heterogeneous integration of 4 chips and 4 capacitors [36–39], (2) chip-first with die face-down for heterogeneous integration of mini-LED RGB display [40], (3) chip-first with die face-up for a large 10 mm × 10 mm chip [45–47], (4) panel-level fan-out RDL-first packaging for heterogeneous integration [18–20], (5) hybrid substrate with PID by RDL-first panel-level packaging [21, 22], (6) hybrid substrate with ABF by RDL-first panel-level packaging [23], (7) hybrid substrate with ultralarge organic interposer for heterogeneous integration [24], and (8) hybrid substrate with interconnect-layer for heterogeneous integration [25].

4.3 Chip-First with Die Face-Down for Heterogeneous Integration of 4 Chips and 4 Capacitors

In this section, the feasibility of a chip-first and die face-down FOWLP of a heterogeneous integration of four chips and four capacitors is demonstrated [36–39]. In order to have a very low-profile package and save the expensive epoxy molding compound (EMC) cost, a new process is developed to fabricate the RDLs. Reliability assessments, such as the thermal cycling and shock (drop) test, are performed.

4.3.1 Test Chip

Figures 4.2, 4.3 and 4.4 show the test chips under consideration. The layout of the large test chip and the fabricated large chip are shown in Fig. 4.3. It can be seen that the larger chip sizes are 5 mm × 5 mm × 150 μm, and there are 160 pads with a pitch = 100 μm (the inner rows). The SiO_2 passivation opening of the Al-pad is 50 μm × 50 μm, and the size of the Al-pad is 70 μm × 70 μm. The dimensions of the small chip are 3 mm × 3 mm × 150 μm, and the fabricated chip is shown in Fig. 4.4. It can be seen that there are 80 pads and are on 100 μm pitch (inner rows).

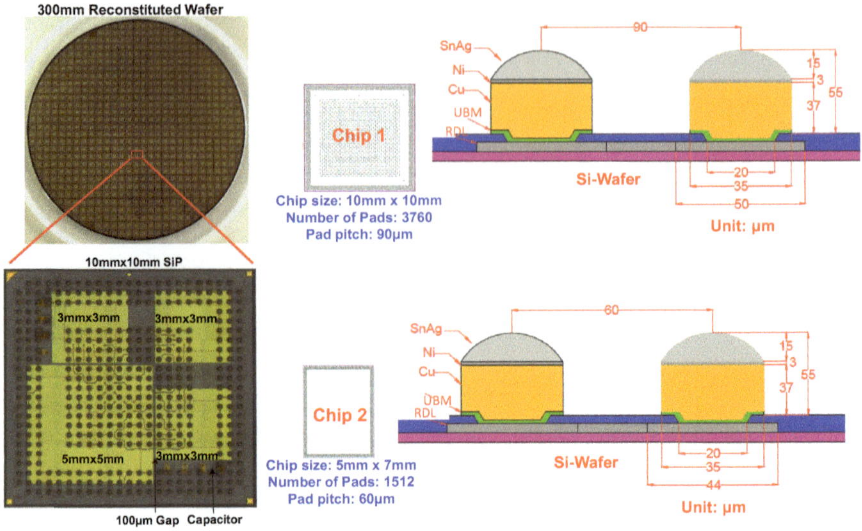

Fig. 4.2 Test chips and test package

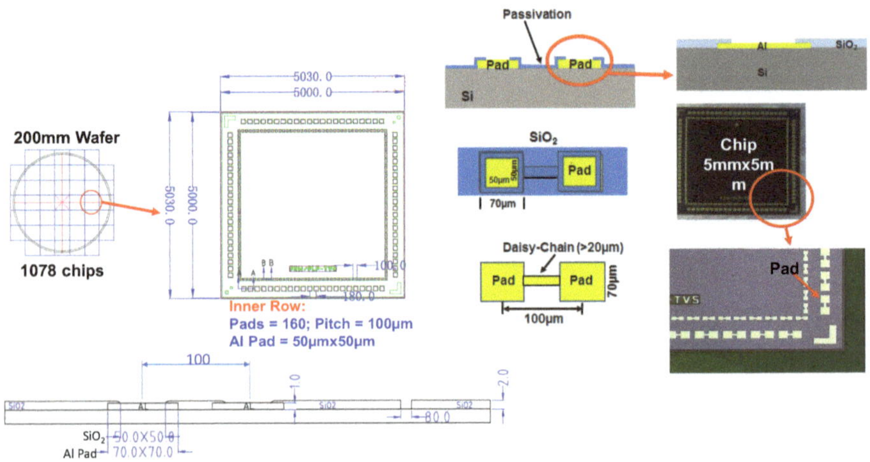

Fig. 4.3 Fabricated 5 mm × 5 mm test chip with 160 pads on a 100 μm pitch (inner rows)

The cross section and dimensions of the pads of the small chip are the same as those of the large chip.

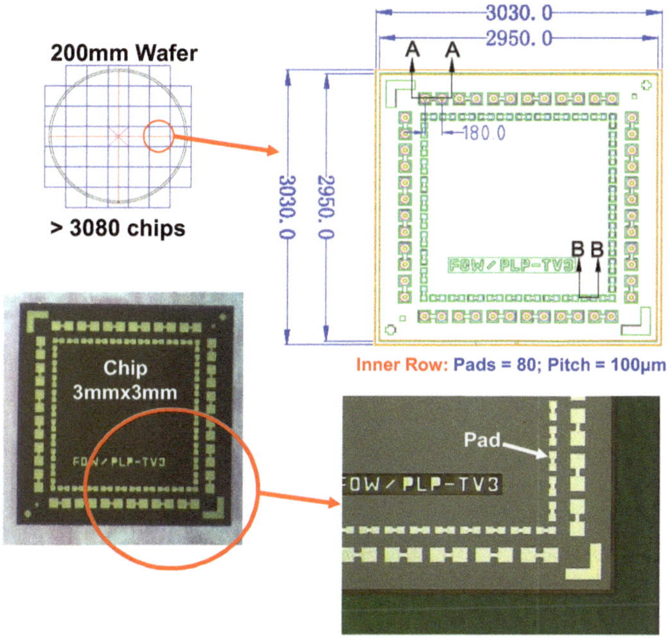

Fig. 4.4 Schematic and the fabricated 3 mm × 3 mm test chip with 80 pads on a 100 μm pitch (inner rows)

4.3.2 Test Package

Figure 4.5 schematically shows the test package under consideration. The dimensions of the test package are: 10 mm × 10 mm, and it consists of one large chip (5 mm × 5 mm), three small chips (3 mm × 3 mm), and four (0402) capacitors. The spacing (gap) between the large chip and the small chip is 100 μm. These packages are to be made from a 300 mm metal reconstituted wafer, as shown in Fig. 4.6. The pitch of the test package on the wafer is 10.04 mm.

Figure 4.7 schematically shows the cross-sectional view of the test package. It can be seen that there are two RDLs and the thickness of RDL1 is 3 μm and RDL2 is 7.5 μm. The linewidth and spacing of RDL1 are 10 μm and those of RDL2 are 15 μm. The dielectric layer thickness of DL1 and DL2 is 5 μm, and DL3 is 10 μm. The via (V_{C1}), through the first dielectric layer (DL1), connecting the Cu contact pad of the test chip to the first RDL (RDL1) is 20–30 μm in diameter. The pad diameter on the RDL1 is 55 μm, which is connected to RDL2 through the via (V_{12}) with a diameter of 30–40 μm. Similarly, the pad diameter on the RDL2 is 65 μm. Finally, 220 μm solder ball Cu pads are formed on RDL2. The opening of the passivation (DL3) is 180 μm. The solder ball size is 200 μm, and the ball pitch is 0.4 mm.

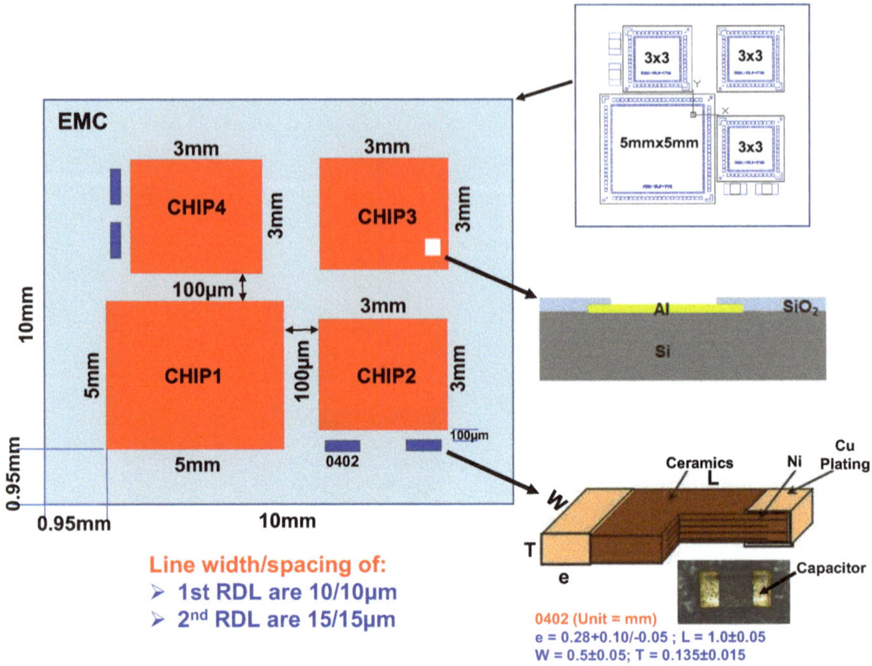

Fig. 4.5 Schematic of 10 mm × 10 mm test package with four chips and four capacitors

Figure 4.8 shows schematically the conventional FOWLP with the chip-first and die face-down assembly process. Basically, work must be done on the device (test chip) wafer [see Fig. 4.8 (left)] and the reconstituted wafer (see Fig. 4.8a–f).

4.3.3 *Conventional Chip-First (Die Face-Down) Wafer Process*

(A) On the Test Chip Wafer

First, test for known good die (KGD) of the device wafer. Then, dice the wafer into individual KGDs or chips.

(B) On the Reconstituted Wafer

Figure 4.8a–f show the conventional FOWLP with chip-first and die face-down process steps on the reconstituted wafer, and they will be briefly mentioned as follows.

(1) Metal Carrier Wafer: The temporary wafer carrier is made of the stainless-steel metal with the coefficient of thermal expansion = 10.6×10^{-6}/°C. The thickness of the metal carrier is about 1 mm. The advantages of 1 mm metal

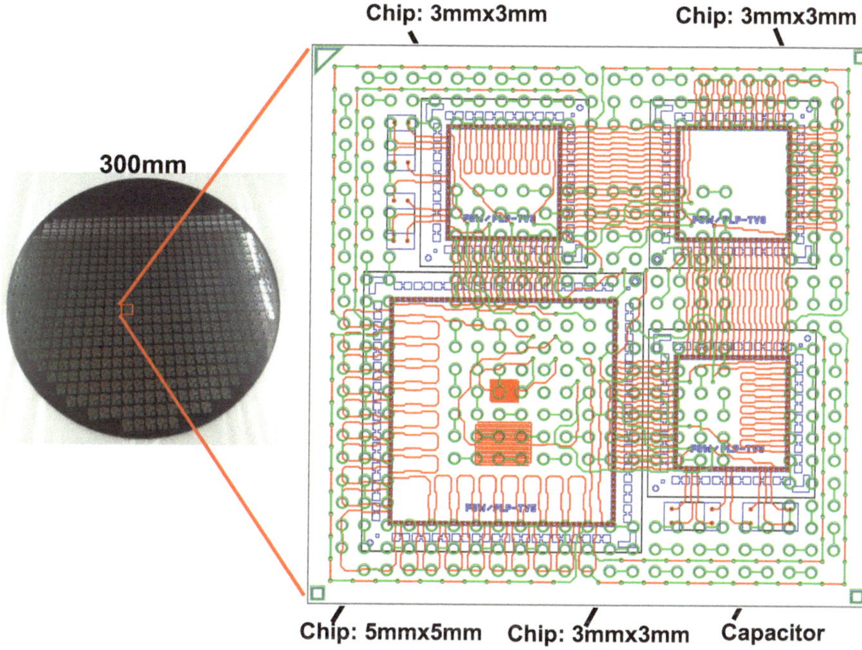

Fig. 4.6 300 mm wafer with test packages

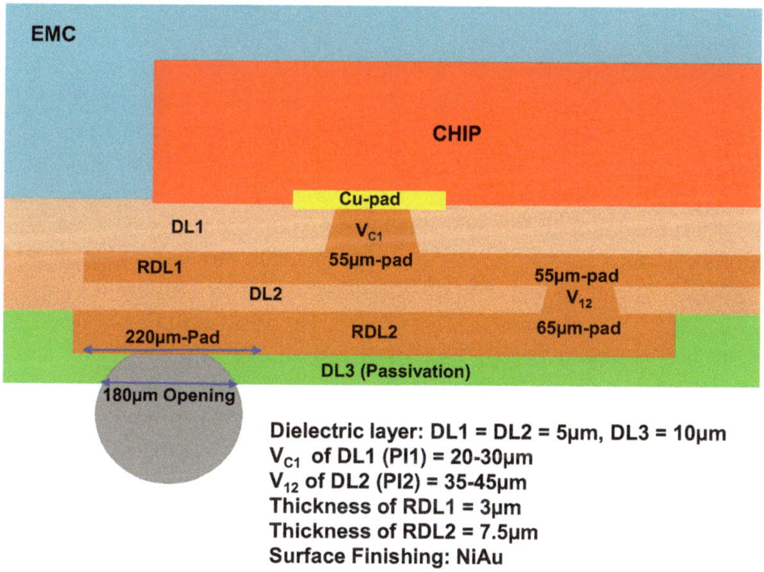

Fig. 4.7 Schematic cross section of the test package

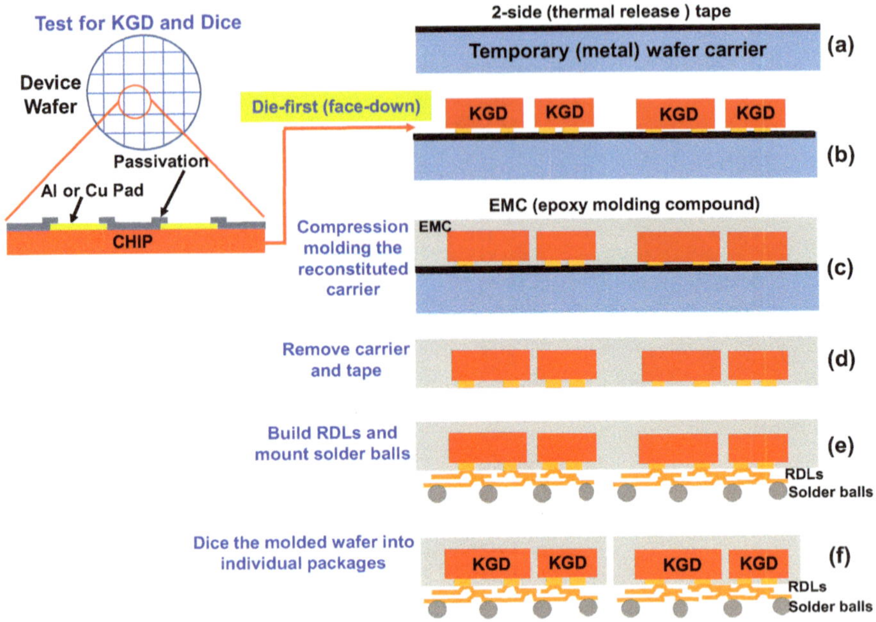

Fig. 4.8 Conventional chip-first die face-down process. **a** Temporary panel carrier with two-side tape. **b** Die face-down. **c** Compression molding the EMC. **d** Remove the panel carrier and tape. **e** Build RDLs and mount solder balls. **f** Dice the molded panel into individual packages

carrier are: (1) not easy to be broken when it is cleaned and reused, and (2) a thicker metal carrier can increase the stiffness and is more effective to resist to bending (warpage) of the reconstituted carrier during pick and place of the chips and compression molding and post-mold cure (PMC) of the EMC.

(2) Two-Side Thermal-Release Tape: A two-side thermal release tape as shown in Fig. 4.9 is needed for bonding and debonding the reconstituted carrier. The most commonly used tape is the REVALPHA provided by Nitto Denko, as shown in Fig. 4.9. It can be seen that this tape consists of five different layers: (1 and 2) the top (0.038 mm) and bottom (0.075 mm) release liners; (3) the pressure-sensitive adhesive layer (0.01 mm); (4) the polyester film (0.1 mm); and (5) the thermal-release adhesive layer (0.048 mm). This tape is bonded (by peeling off the release liner) to the metal carrier at room temperature, as shown in Fig. 4.8a. Then, peel off the other release liner and pick and place the KGDs (facedown) on the top of the thermal-release adhesive, as shown in Fig. 4.8b. The debonding temperature of the carrier is 170 °C.

(3) EMC, Compression Molding, and PMC: The EMC is a liquid-like material (Nagase R4507), and its material properties are shown in Table 4.1. It can be seen that the filler content is 85%. The average size of the filler is 8 μm, and the maximum size of the filler is 25 μm. The flexural modulus of the EMC is 19 GPa, and the viscosity is 250 Pa/s. The transition temperature (Tg) is 150 °C.

Fig. 4.9 Nitto Denko REVALPHA two-side tape

Table 4.1 Nagase liquid epoxy molding compound (EMC)

Item	R4507
Filler content (%)	85
Filler to cut (μm)	25
Filler average size (μm)	8
Specific gravity	1.96
Viscosity (Pa s)	250
Flexural modulus (GPa)	19
Tg (DMA) (°C)	150
CTE1 (ppm/K)	10
CTE2 (ppm/K)	41

- High flowability as suitable for large surface areas and thin film molds
- Liquid at room temperature and able to be dispensed, and dust-free as suitable for clean room environments
- Capable of low temperature form molding (125 °C)
- Delivers low reflection in large surface area molds with low stress design
- High reliability
- High purity
- Low α line

As shown in [37], the circular pattern of EMC dispensing (Fig. 4.10a) leads to a fewer amount of flow marks.

After EMC dispensing, it is followed by the compression molding, as shown in Fig. 4.8b, c. The keep-out-zone method is used, and the mold cap diameter is 300 mm (6 mm clearance from the 312 mm wafer metal carrier edge). The metal carrier with dies attached is fixed onto the top mold chase by vacuum, as shown in Fig. 4.10. The dispensed EMC on a releasing film is placed on the bottom plunger. After a few experiments, the optimal compression molding parameters are: temperature =

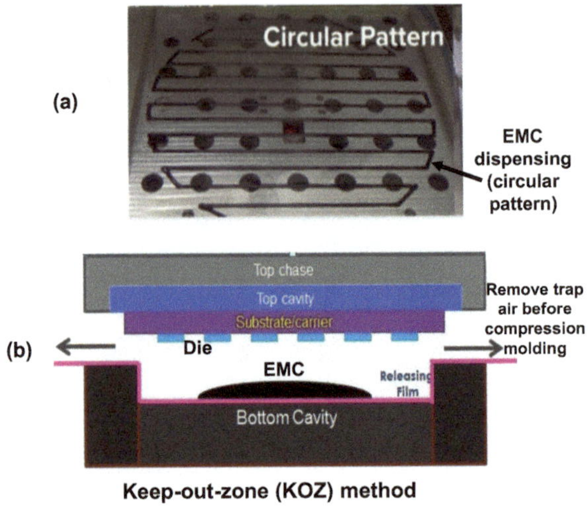

Fig. 4.10 Keep-out-zone compression molding. **a** EMC dispensing (circular pattern). **b** Compression molding

125 °C, pressure = 45 kg/cm^2, and time = 10 min and removing trap air before compression molding. It is followed by PMC with a temperature = 150 °C, time ≥ 60 min, and a dead weight = 15 kg for better warpage control. The thickness of the EMC above the backside of the chips is 150 μm.

Inspections for molding voids are carried out by C-mode scanning acoustic microscopy (C-SAM). In order to balance the resolution and signal penetration depth, a transducer of 75 MHz is selected for the void's observation. After a couple of parametric studies, there is not any void in the optimal wafers, as shown in Figs. 4.11 and 4.12.

4.3.4 RDLS and Solder Ball Mounting by a New Process

In this section, we will not use the conventional method such as Fig. 4.8e to build the RDLs. Instead, we will use the new process shown in Fig. 4.13.

(A) Debonding the Metal Carrier

In the conventional FOWLP, after debonding the carrier and peeling off the tape, Fig. 4.8d, the total thickness of the reconstituted wafer is usually ≥450 μm. It is followed by building the RDLs and mounting the solder balls (see Fig. 4.8e). However, in this section, in order to save the expensive EMC materials and have a very low-profile package, the total thickness of our reconstituted wafer without the carrier is only 300 μm (Figs. 4.11 and 4.12). Thus, the reconstituted wafer is too fragile to fabricate the RDLs and mounts the solder balls.

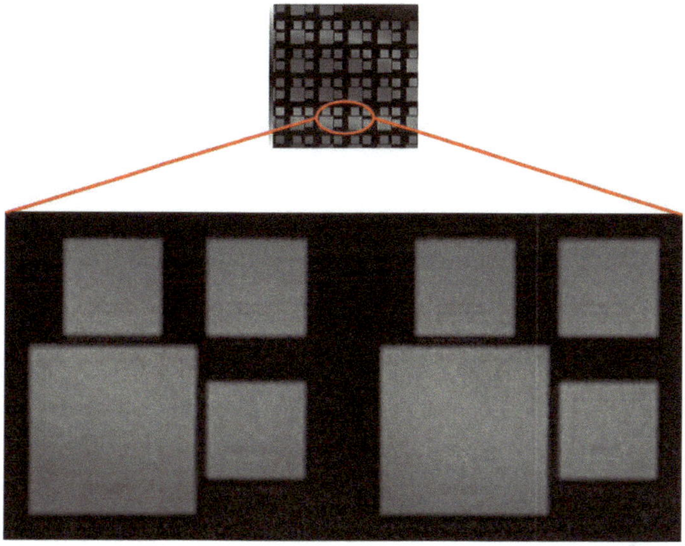

Fig. 4.11 C-SAM image of the molded wafer

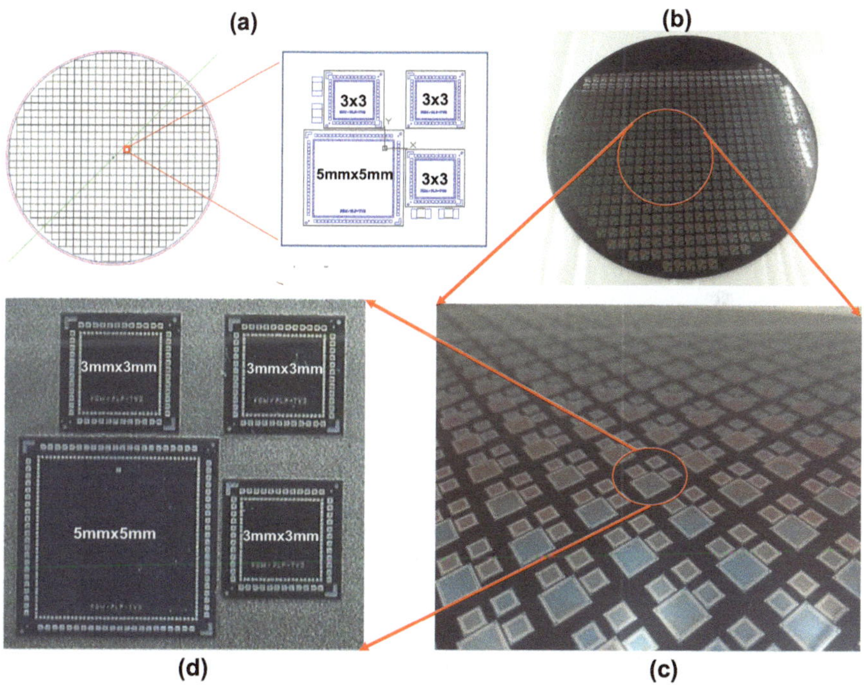

Fig. 4.12 a Schematic of the reconstituted wafer. b Molded reconstituted wafer without carrier. c Zone-in on the reconstituted wafer. d Individual package without RDLs and solder balls

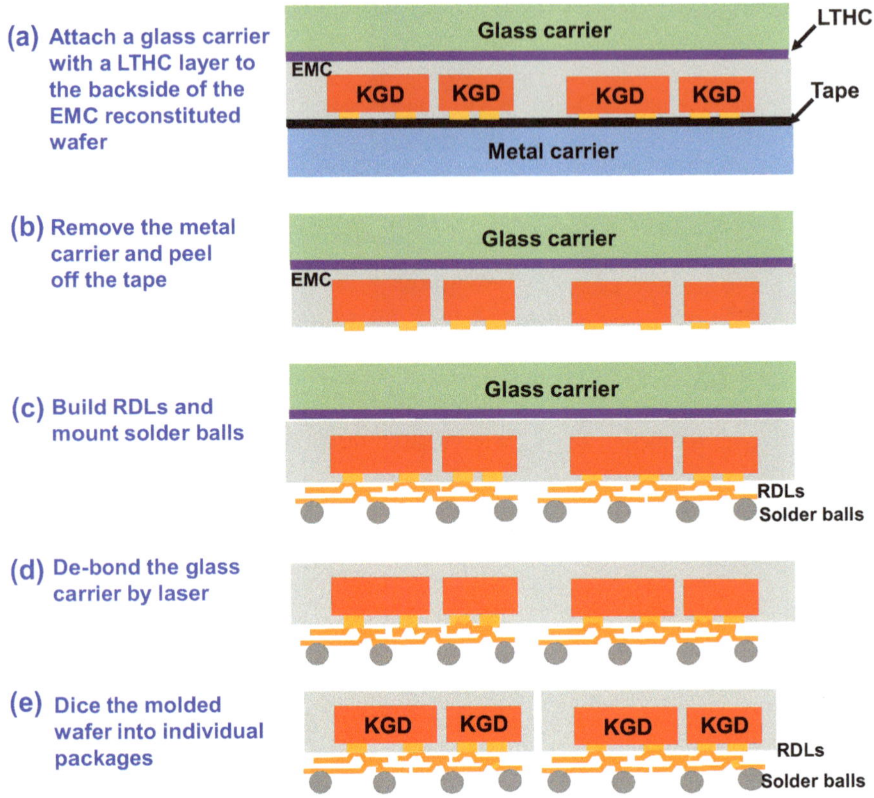

Fig. 4.13 New assembly process of the RDLs. **a** Attach a glass carrier with an LTHC layer to the backside of the EMC reconstituted wafer. **b** Remove the metal carrier and peel off the tape. **c** Build RDLs and mount solder balls. **d** Debond the glass carrier by a laser. **e** Dice the molded wafer into individual packages

(B) Temporary Bonding of Another Glass Carrier

One of the solutions is to attach the thin reconstituted wafer on another temporary glass wafer with a coated light-to-heat-conversion (LTHC) layer as shown in Fig. 4.13a and then remove the metal carrier and peel off the tape (see Fig. 4.13b). It is followed by making the RDLs as shown in Fig. 4.13c with the method shown as follows.

(C) RDLs

Figure 4.14 shows the key process steps in making the RDLs. First, spin-coat a photosensitive polyimide (PI) on the reconstituted wafer. Then, apply a stepper (every four test packages as a unit) and use photolithography techniques to align, expose, and develop the vias of the PI. Finally, cure the PI at 200 °C for 1 h. This will form a 4–5 μm-thick PI layer (PI development). It is followed by sputtering Ti (for adhesion

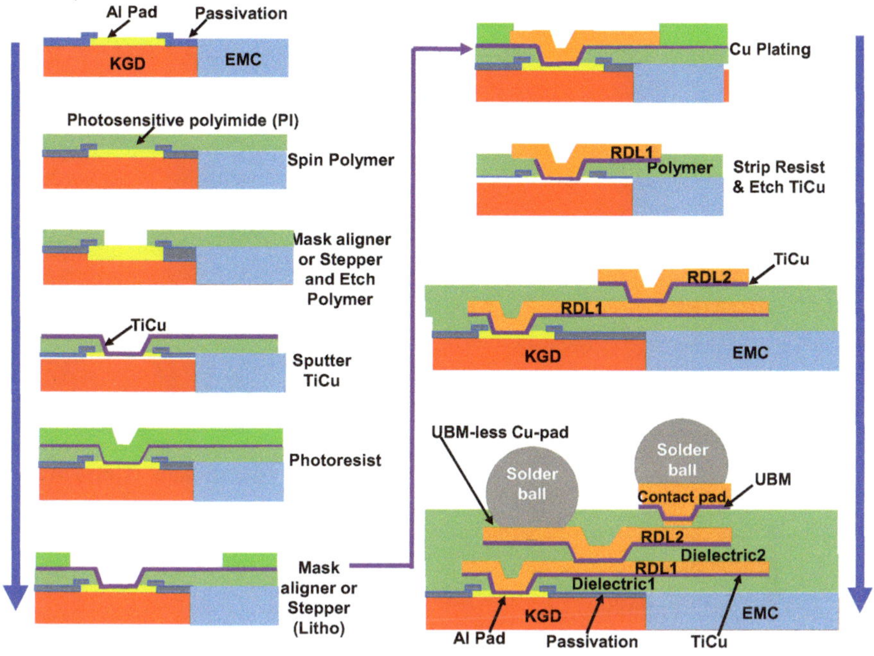

Fig. 4.14 RDLs process steps

and barrier layer) and Cu (for a seed layer) by physical vapor deposition (175–200 °C) over the entire wafer. Apply a photoresist and a stepper and then use photolithography techniques to open the redistribution trace locations. Then, electroplate Cu by electrochemical deposition (room temperature) on Ti/Cu in photoresist openings. Strip off the photoresist and etch off the Ti/Cu. (RDL1 is obtained.) Repeat the above-mentioned steps to obtain RDL2. Figure 4.15 shows the linewidth and spacing of the first RDL. It can be seen that the average linewidth and spacing are about 5% of the designed value. This could be due to the etching of the Ti.

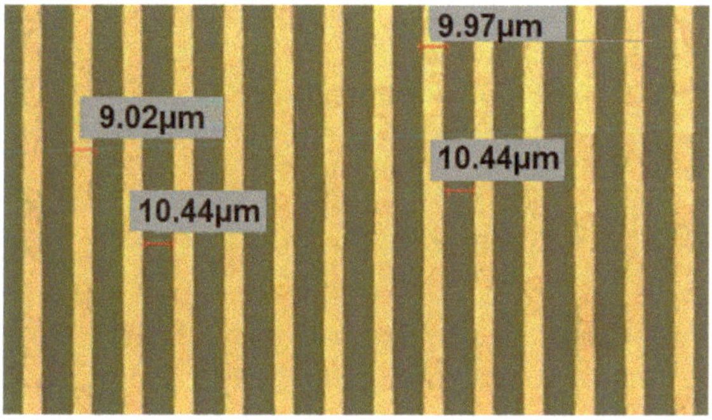

Fig. 4.15 Linewidth and spacing of RDL1

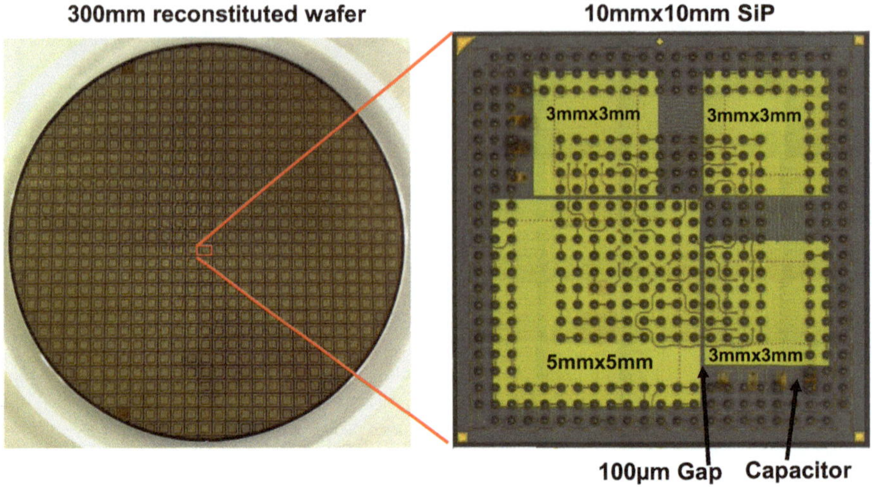

Fig. 4.16 Molded reconstituted wafer and close-up view

(D) Solder Ball Mounting

There are two different stencils for the solder ball mounting: one is for stencil printing the flux and the other is for stencil mounting the solder balls. The solder (Sn3wt%Ag0.5wt%Cu) balls (200 μm diameter) used are from Indium and are on 0.4 mm pitch. The peak temperature for solder reflows is 245 °C.

(E) Final Debonding

The debonding of the glass carrier as shown in Fig. 4.13d is by scanning a laser (355 nm DPSS Nd:YAG UV laser source is used) from the glass carrier side. The laser spot size is 240 μm, the scanning speed is 500 mm/s, and the scanning pitch is 100 μm. When the LTHC layer "sees" the laser light, it converts into powders and the glass carrier is easily removed. It is followed by chemical cleaning. Figure 4.16 shows the reconstituted wafer without any carrier and a close-up on one of the packages. It can be seen that there are four chips and four capacitors in a package, and they are properly fabricated. The reconstituted wafer is diced, as shown in Fig. 4.13e, into individual packages as shown in the X-ray image of Fig. 4.17, where it also shows the RDL1 and RDL2 of the package.

4.3.5 PCB Assembly

The design of the printed circuit board (PCB) and stencil will be briefly mentioned. The stencil solder paste printing, pick and place, and solder reflow will also be presented. The X-ray images and the cross-sectional images of the PCB assembly are provided.

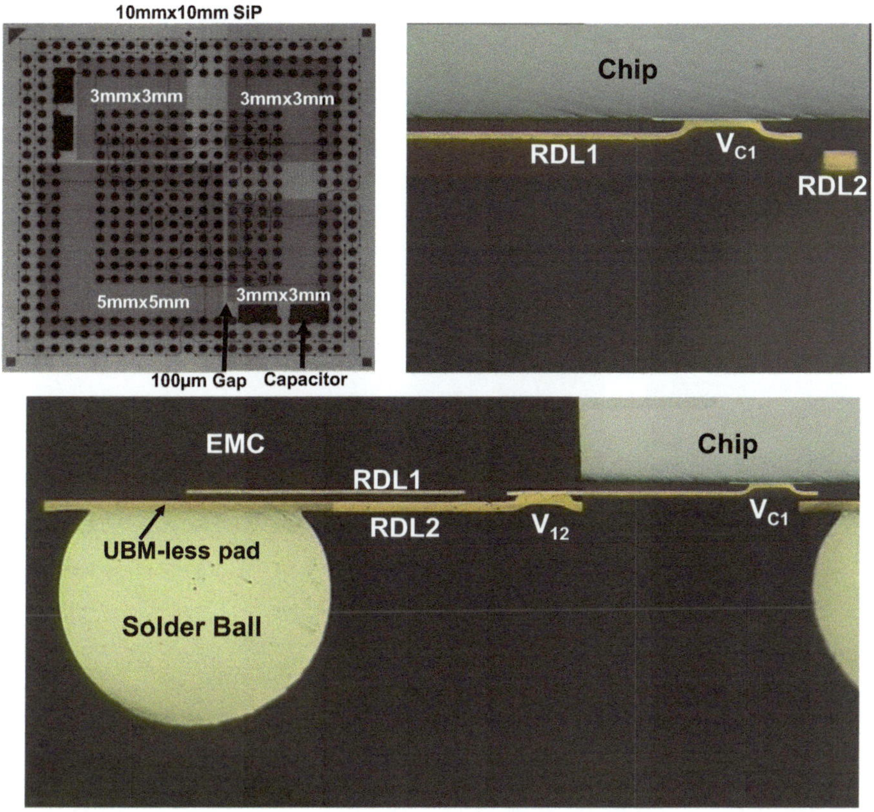

Fig. 4.17 X-ray image and the cross sections of the individual package

(A) PCB

The PCB for the fan-out panel-level package is made of FR-4 and is shown in Fig. 4.18. It can be seen that there are four package sites on the board. The dimensions of the PCB are 103 mm × 52 mm × 0.65 mm, and there are six layers. There are 405 pads (with a pitch = 0.4 mm) for each package. The pad with a diameter = 0.2 mm is non-solder mask defined, and its surface finish is an organic solderability preservative. The solder mask opening diameter is 0.28 mm.

(B) Stencil and Printing

The stencil is made of stainless steel with a grain size of 2 μm (which will benefit for the solder paste transfer) and is 0.08 mm thick. The opening is fabricated by laser and electrical/chemical polishing. The opening is increased from a 0.2286 mm-diameter circle to a 0.2413 mm^2 (a paste volume increased by 41.8%). The solder paste printing is by DEK Horizon 9.

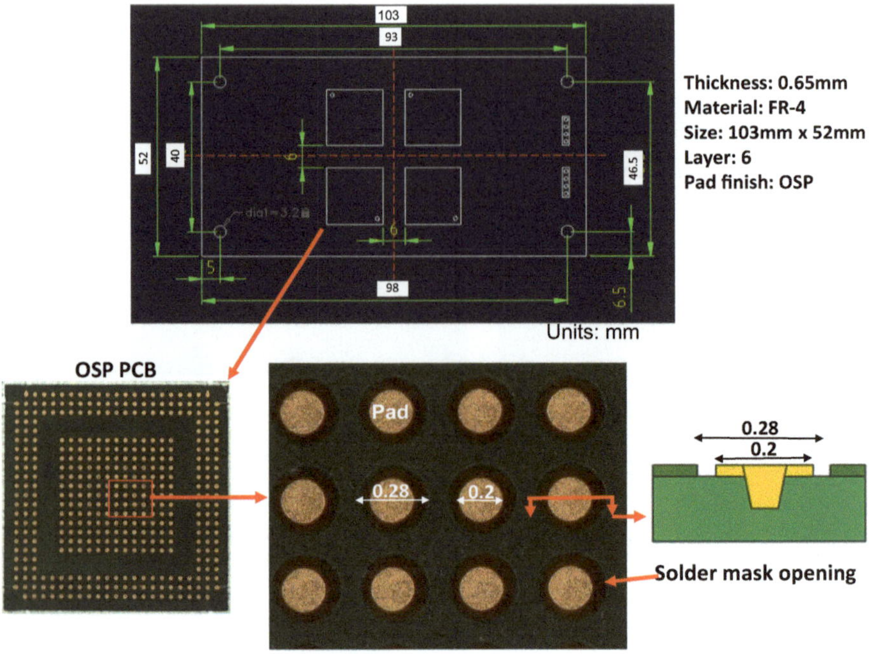

Fig. 4.18 PCB layout and fabricated PCB

(C) Pick and Place and Reflow

The pick and place are by the SiPlace x4s. The 10-temperature zones BYU Pyram nitrogen 150 N is used for the reflow. The temperature profile is shown in Fig. 4.19. It can be seen that the maximum temperature is 245 °C, and the time above 217 °C is 85 s.

4.3.6 Reliability Assessments—Thermal Cycling Test

Twelve boards (each with four packages) are used for the temperature cycling tests. The sample size is 48 packages. Thermal cycling is performed in a Votsch 7027-15 environmental chamber. This unit is capable of achieving a chamber temperature as high as 190 °C and as low as −70 °C. Heating and cooling are achieved by forced convection, and the maximum ramp rate is 15 °C/min. All of the boards are placed vertically in the chamber, as shown in Fig. 4.20. The temperature input to the chamber (measured in the air of the chamber) is shown in Fig. 4.20. It can be seen that the temperature cycling is from room temperature to 85 °C and stays there for 15 min, then ramps down to −40 °C and stays there for 15 min, then ramps up to 85 °C, and stays there for 15 min, and so on. The ramp-up and ramp-down times are 15 min each. The acquisition system is an Agilent 30970A data logger.

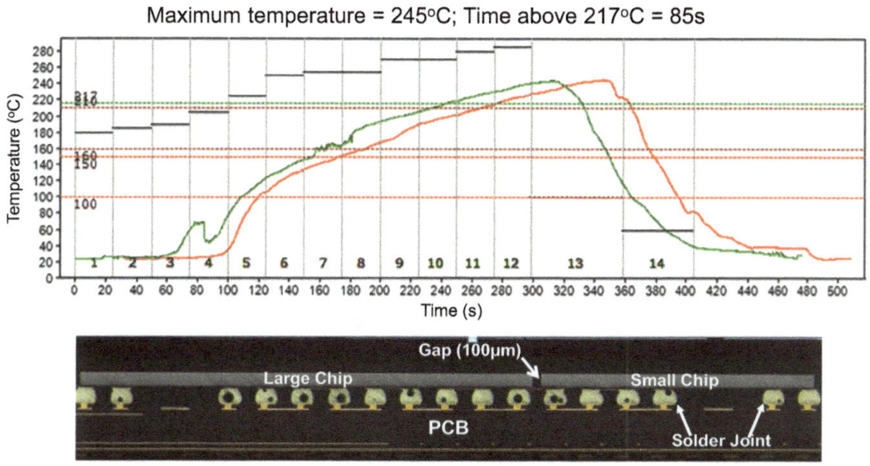

Fig. 4.19 **a** PCB assembly reflow temperature profile. **b** Cross section of the PCB assembly of a package

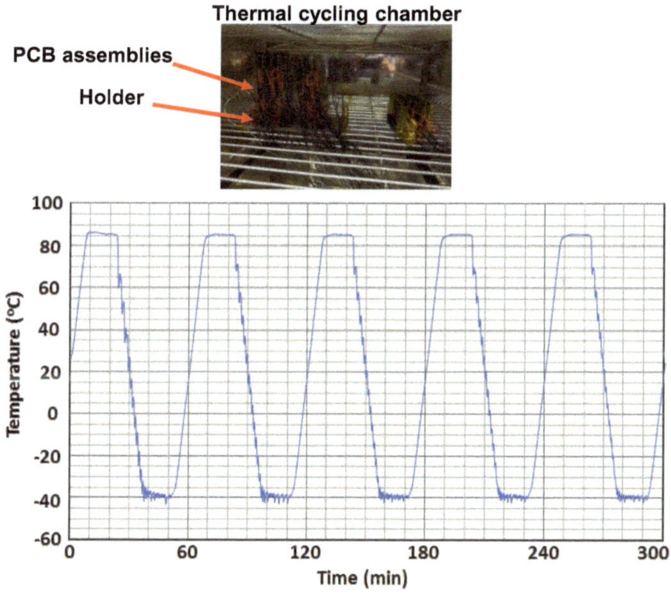

Fig. 4.20 Thermal cycling test setup and condition

The thermal cycling test results of the FOWLP solder joint (without underfill) reliability are shown in Fig. 4.21. It can be seen that the characteristic life (63.2% failed) of the Weibull plot is 1070 cycles, which is more than adequate for the

Fig. 4.21 Weibull plot of the solder joint under thermal cycling

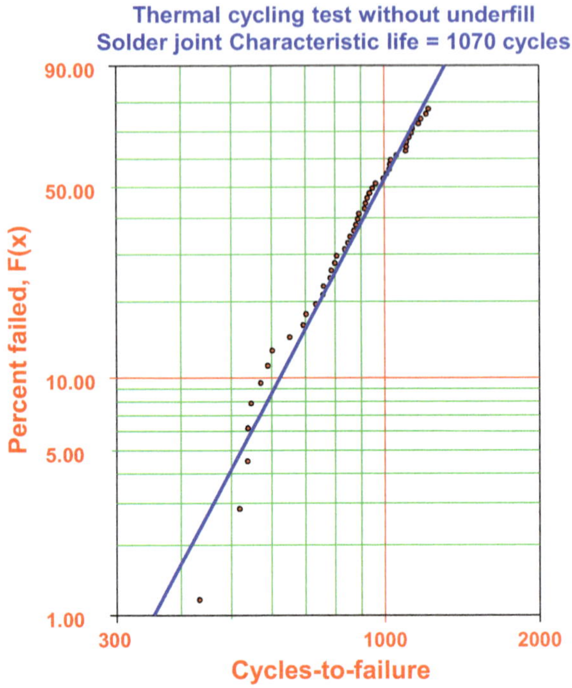

expecting life (usually is less than 3 years) of a mobile product, such as smartphones and tablets.

The thermal cycling test results of the FOWLP solder joint (without underfill) reliability are shown in Fig. 4.21. The thermal cycling test stops at 1,300 cycles. It can be seen that the characteristic life (63.2% failed) of the Weibull plot is 1,070 cycles which is more than adequate for the expecting life (usually is less than 3 years) of mobile products such as smartphones and tablets. The failure location and mode are shown in Figs. 4.22 and 4.23. It can be seen that the solder joint cracks near the interface between the bulk solder and the package contact pad and it occurs under the chips' corners near the package corners—the longest distance to neutral point (DNP) [48].

4.3.7 Reliability Assessments—Drop Test

The reliability assessments of the chip-first and die facedown FOWLP of the heterogeneous integration of four chips and four capacitors on the six-layer PCB assembly with underfill are by a shock (drop) test. The test setup is according to JEDEC Standard JESD22-B111, as shown in Fig. 4.24. After more than 20 tries, the right height

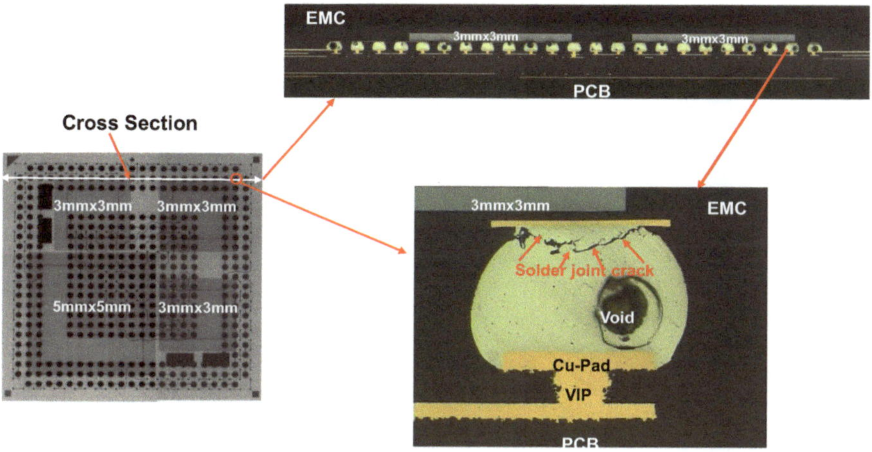

Fig. 4.22 Solder joint failure at the small chip corner

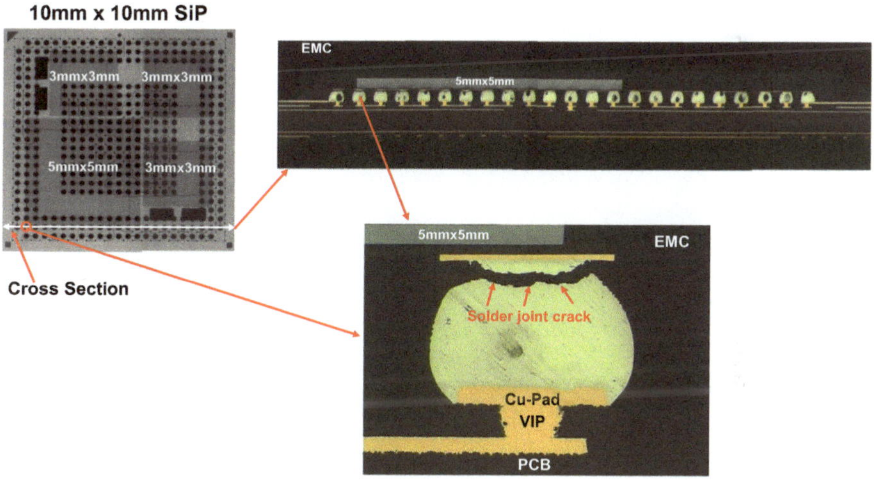

Fig. 4.23 Solder joint failure at the large chip corner

(1200 mm) of the drop table is obtained, which yields the drop spectrum with 1500 G/ms (1500 Gs, 0.5 ms half-sine pulse), as shown in Fig. 4.25.

There are 24 samples with underfill. The material properties of the underfill are: the filler content = 25%, the maximum filler size = 5 μm, the average filler size = 1–2 μm, and the curing time and curing temperature = 8 min at 135 °C or 5 min at 150 °C. The Young's modulus, Poisson's ratio, and coefficient of thermal expansion (CTE) are, respectively, 4–5 GPa, 0.35, and 50–52 × 10^{-6}/°C. The drop condition is 1,000 drops. After more than 3 weeks of drop tests, all these 24 samples passed 1,000

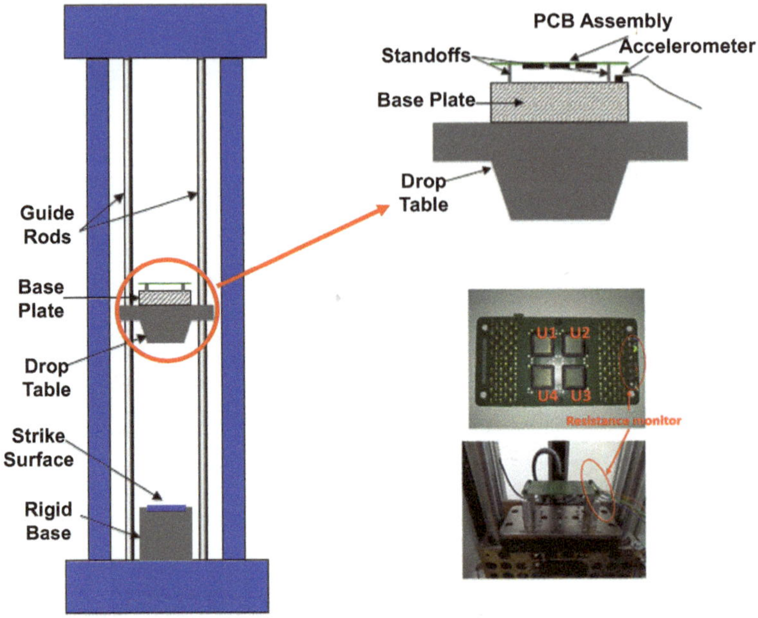

Fig. 4.24 Drop test setup

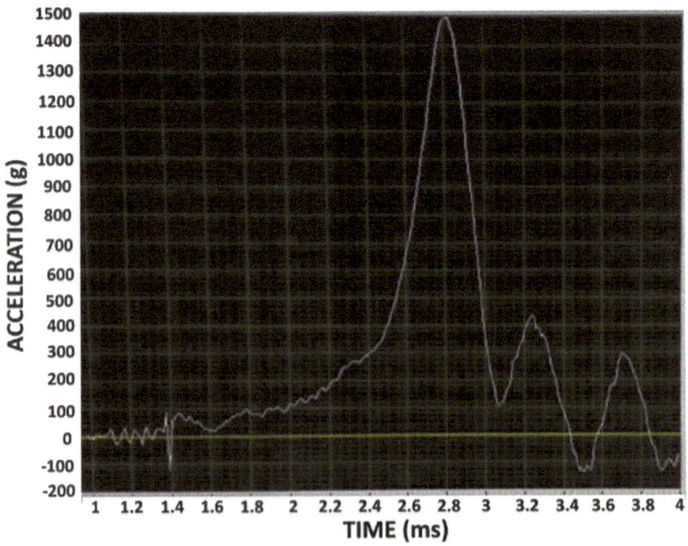

Fig. 4.25 Drop test spectrum—1500 G/ms

drops. Thus, the fabricated package samples in this section are considered reliable under drop conditions.

From a scientific point of view, unfortunately the test stops after 1,000 drops, thus there are no failure data to determine a life distribution, and consequently, the characteristic life, etc. On the other hand, because of resource availability, we cannot let the drop tests go on for longer.

4.3.8 Reliability Assessments—Thermal Simulations

(A) The Structure and Boundary Conditions

The PCB assembly of the fan-out SiP shown in Figs. 4.16, 4.17, 4.18 and 4.19 is modeled as a 3-D strip that captures the construction along a diagonal path from the assembly (Fig. 4.26). Because of the symmetry about the vertical midplane of a full strip, the model is actually a half-strip with the appropriate in-plane restraints placed on one symmetrical plane. Coupled in-plane translations are applied to the other symmetry plane to produce a state of generalized plane strain. Using exclusively hexahedral solid elements, the model can capture the precise shape of the packages' solder joint and potential DNP effects while retaining significant computational efficiency over full octant models.

Despite the overall economy of elements in the strip model, selective mesh refinement is used to concentrate highly refined elements in the solder joints where failure is anticipated. In the present PCB assembly, failure would be predicted in the solder joints with the greatest DNP (the package corner) and near the chip corners as shown in Fig. 4.26. Thus, highly refined meshes are applied to these solder joints. The other solder joints are coarsely meshed. ABAQUS 6.12 (C3D8R) is used for the model.

(B) Material Properties for Thermal Cycling

The material properties of the PCB assembly shown in Fig. 4.26 are given in Table 4.2. All the material properties are assumed to be constant except for those of the solder. The Sn3wt%Ag0.5wt%Cu is assumed to follow the generalized Garofalo creep equation [48]:

$$\frac{d\varepsilon}{dt} = 500,000 \text{ Sinh}^5\left(\frac{\sigma}{10^8}\right) \exp(-5,807/T(\text{K}))$$

where ε is the strain, σ is the stress in Pa, and T is the temperature in Kelvin. The CTE and Young's modulus of the solder are, respectively, $21.3 + 0.017\,T$ and $49 - 0.07\,T$, and T is the temperature in Celsius.

(C) Temperature Boundary Condition

The temperature profile shown in Fig. 4.20 is to be imposed on the PCB assembly. Five temperature cycles are executed.

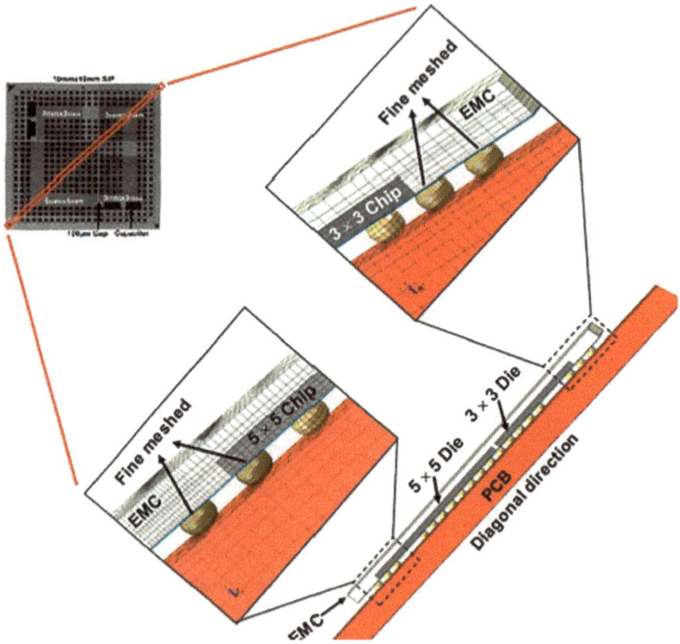

Fig. 4.26 Finite element model for thermal analysis

Table 4.2 Material properties for thermal cycling

	CTE (ppm/°C)	Young's modulus (GPa)	Poisson's ratio
Copper	163	121	0.34
PCB	$\alpha_1 = \alpha_2 = 18$, $\alpha_3 = 70$	$E_1 = E_2 = 22$, $E_3 = 10$	0.28
Silicon	2.8	131	0.278
Solder	21.3 + 0.017 T (°C)	49 − 0.07 T (°C)	0.3
PI	35	3.3	0.3
EMC	10 (<150 °C)	19	0.25

(D) Thermal Cycling Simulation Results

Figure 4.27 shows the shear stress and creep shear strain hysteresis loops at the solder joint under the 3 × 3 mm chip corner. It can be seen that after three cycles, the creep responses converged (become stabilized). Similar results are observed at the solder joint of the 5 × 5 mm chip corner. The largest Mises stress and accumulated creep strain occur at the solder joint under the 3 × 3 mm chip corner as shown in Figs. 4.28 and 4.29. The location is at the interface between the bottom of the package and the bulk solder. Thus, any failure should occur at this location. This correlates very well with the thermal cycling test failure location and mode as shown in Fig. 4.22.

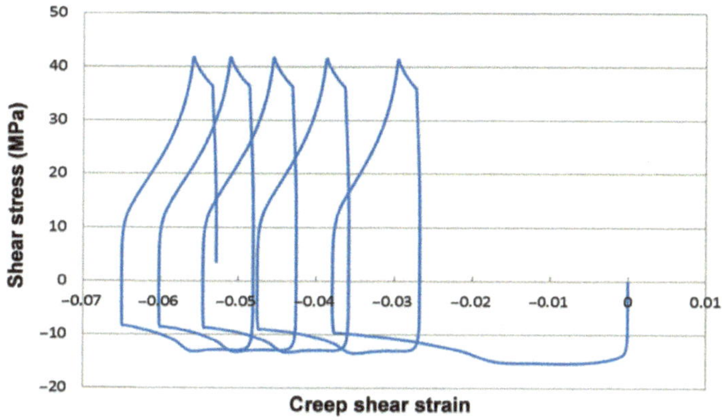

Fig. 4.27 Shear stress–shear strain hysteresis loops

The second largest Mises stress and accumulated creep strain occur at the solder joint under the 5 × 5 mm chip corner as shown in Figs. 4.28 and 4.29. Again, this correlates very well with the thermal cycling test failure location and mode as shown in Fig. 4.23. The maximum Mises stress (7.415×10^7 Pa, Fig. 4.28) occurs at −40 °C during the thermal cycling condition shown in Fig. 4.20. The maximum deformed shape (4.83×10^{-3} mm) of the strip model at −40 °C is shown in Fig. 4.30.

The accumulated creep strain versus time is shown in Fig. 4.31. It can be seen that the creep strain per cycle is equal to 0.018, which is too small to create solder joint reliability problem, especially for mobile products such as smartphones, for which the life is less than 3 years.

4.3.9 *Reliability Assessments—Drop Simulations*

(A) Structure and Boundary Conditions

The finite element model for drop test with underfill is shown in Fig. 4.32. It can be seen that, for simplicity, there is only one package at the center of the PCB and the package is subjected to the acceleration shown in Fig. 4.25. During impact, the most likely failure location is the RDLs near the package corners, and the most likely failure mode is the broken Cu wire in the RDLs. Thus, finer meshes are applied to the Cu wire of RDLs near the package corner as shown in Fig. 4.33.

(B) Material Properties for Drop

The material properties for drop analysis are given in Table 4.3. Because the focus of this analysis is not on the solder joint, all the materials are assumed to be constant. The ABAQUS explicit solver was applied for the simulation. Thus, the damping effect was included through the bulk viscosity method in ABAQUS explicit solver. The

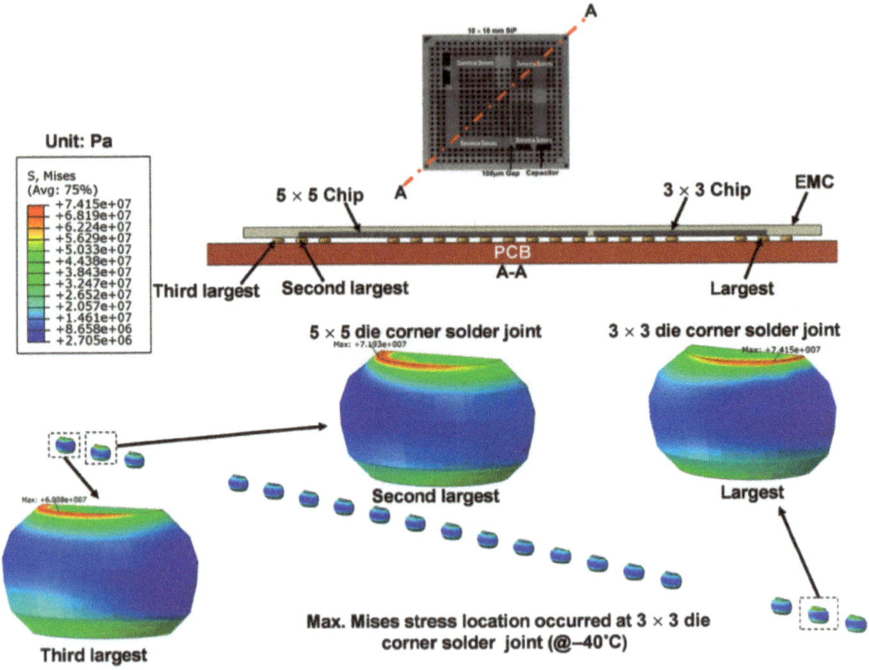

Fig. 4.28 Mises stress contours in critical solder joints

damping coefficients b_1 and b_2 applied in the analysis are assumed to be, respectively, 0.02 and 0.

(C) Drop Simulation Results

The applied acceleration is shown in Figs. 4.25 and 4.34. The time step for the calculation is 1×10^{-8} s. The time-histories of strain in the x-direction (ε_{11}) and y-direction (ε_{22}) and the maximum principal strain acting at the RDL of the package corner near the 5 × 5 mm chip are shown in Figs. 4.35. Basically, the ε_{11} and the principal strain are the same, i.e., the strain is dominated in the x-direction. The maximum deflection (2.783 mm) of the PCB assembly occurs at the 0.0022 s (Fig. 4.36) and is shown in Fig. 4.36. The maximum principal strain = 0.0023 also occurred at that time. The deflection and strains are decreasing with time because of the damping of the structure as shown in Fig. 4.35. The maximum Mises stress occurs (Fig. 4.37) at the same location (i.e., at the Cu wiring of RDL near the package corner of the 5 × 5 mm chip) as the maximum principal strain as shown in Fig. 4.35, and is equal to 3.546×10^8 Pa. Thus, any failure should occur at this location.

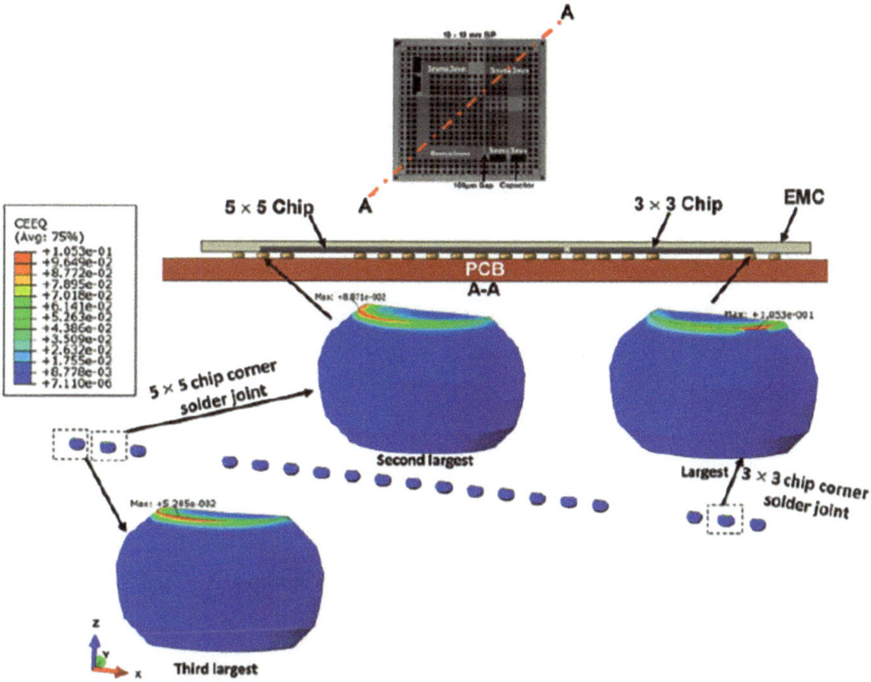

Fig. 4.29 Accumulated creep strain contours

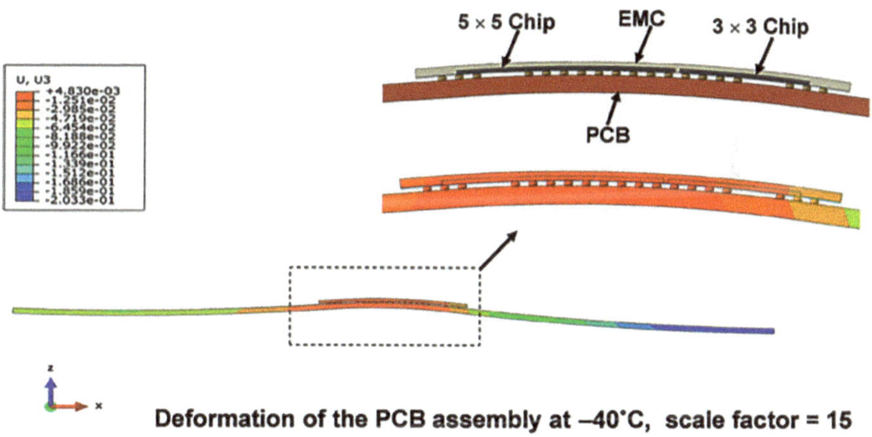

Fig. 4.30 The deflection (at 240 °C) of the PCB assembly

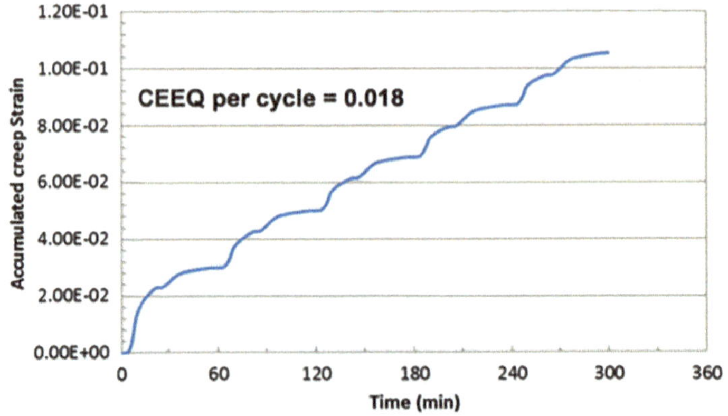

Fig. 4.31 Maximum accumulated creep strain versus time

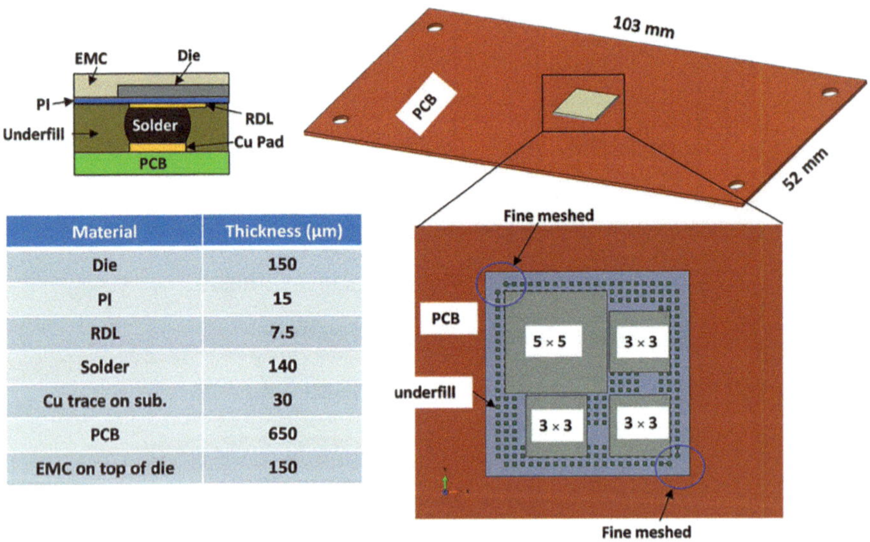

Material	Thickness (μm)
Die	150
PI	15
RDL	7.5
Solder	140
Cu trace on sub.	30
PCB	650
EMC on top of die	150

Fig. 4.32 3-D finite element modeling for drop analysis

4.3.10 Summary

Some important results and recommendations are summarized as follows:

- The feasibility of design, materials, process, fabrication, and reliability of a thin heterogeneous integration of four chips and four capacitors by a new FOWLP method have been demonstrated.
- Both X-ray and optical microscope images verified that the four chips and four capacitors have been assembled properly.

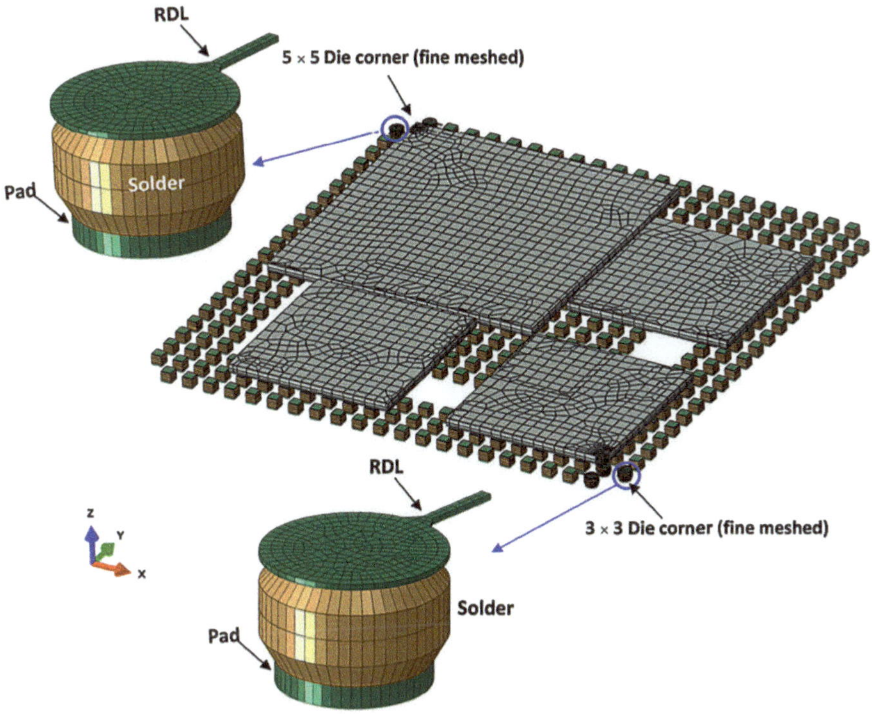

Fig. 4.33 Finite element meshes. Finer meshes for the corner solder joints and pads, and RDLs

Table 4.3 Material properties for shock (drop)

	Density (kg/m^3)	Young's Modulus (GPa)	Poisson's ratio
Copper	8,954	121	0.34
PCB	1,800	$E_1 = E_2 = 22$, $E_3 = 10$	0.28
Silicon	2,330	131	0.278
Solder	7,400	49	0.3
Underfill	1,560	4.5	0.35
PI	1,420	3.3	0.3
EMC	1,960	19	0.25

- Both X-ray and optical microscope images demonstrated that the RDLs have been properly fabricated.
- The gap between the large chip (5 mm × 5 mm) and the small chip (3 mm × 3 mm) is 100 μm. It has been found by the C-SAM method that there is not any void in the optimal EMC molding.

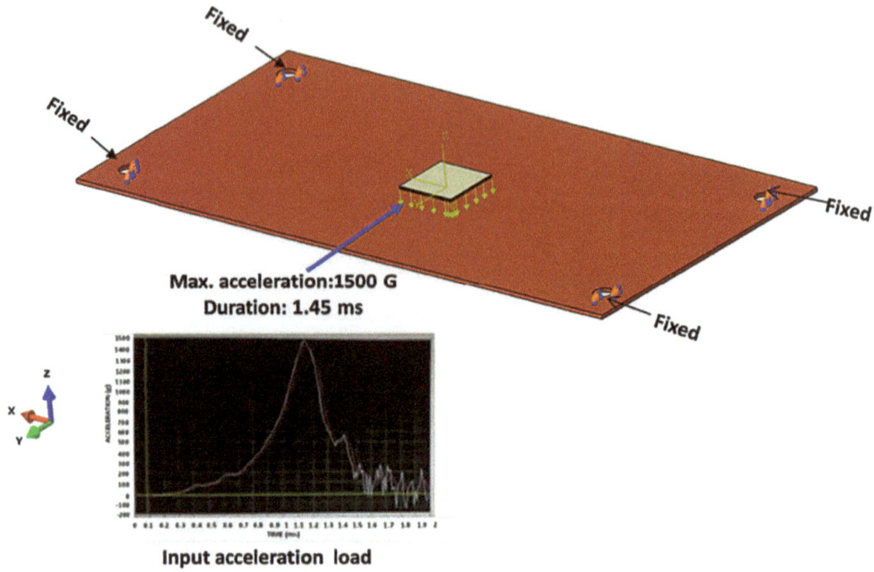

Fig. 4.34 Boundary value problem for drop analysis

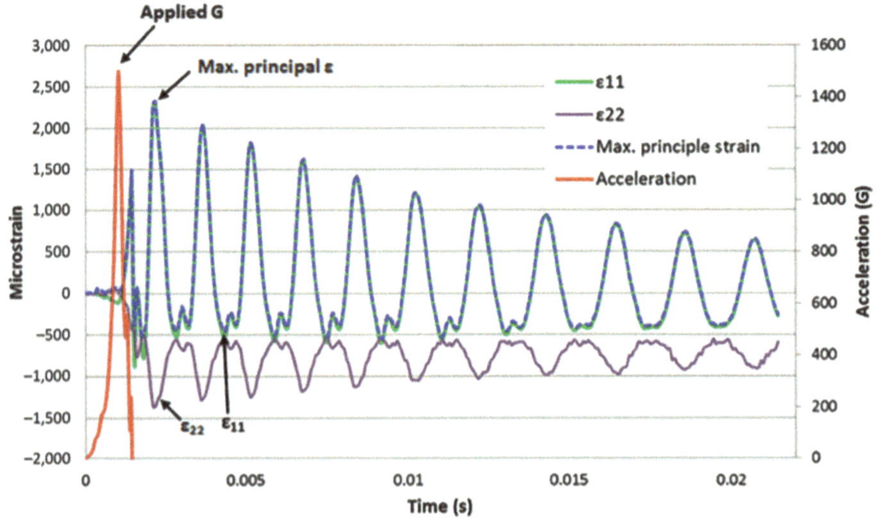

Fig. 4.35 Applied acceleration. Time-history of ε, ε11, ε22, at the RDL of the package corner near the 5 × 5 chip

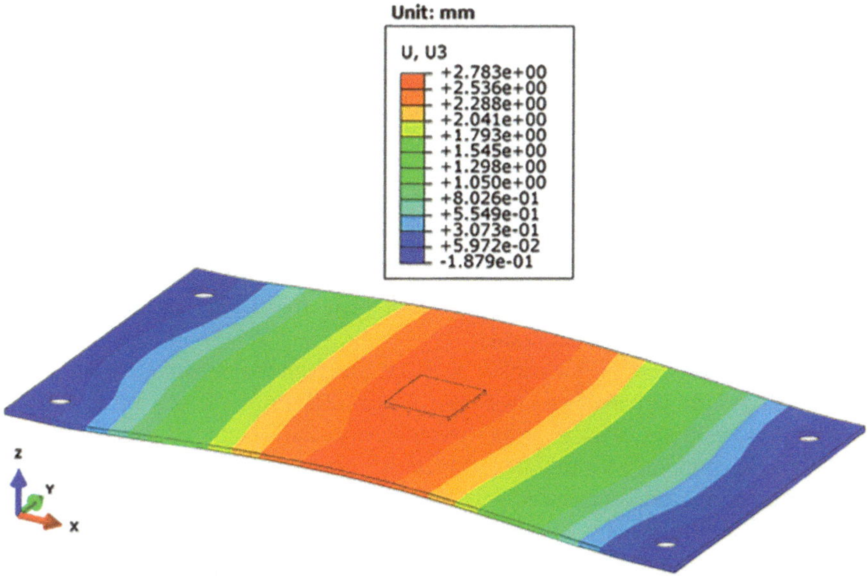

Fig. 4.36 Deflection of the PCB assembly (drop test)

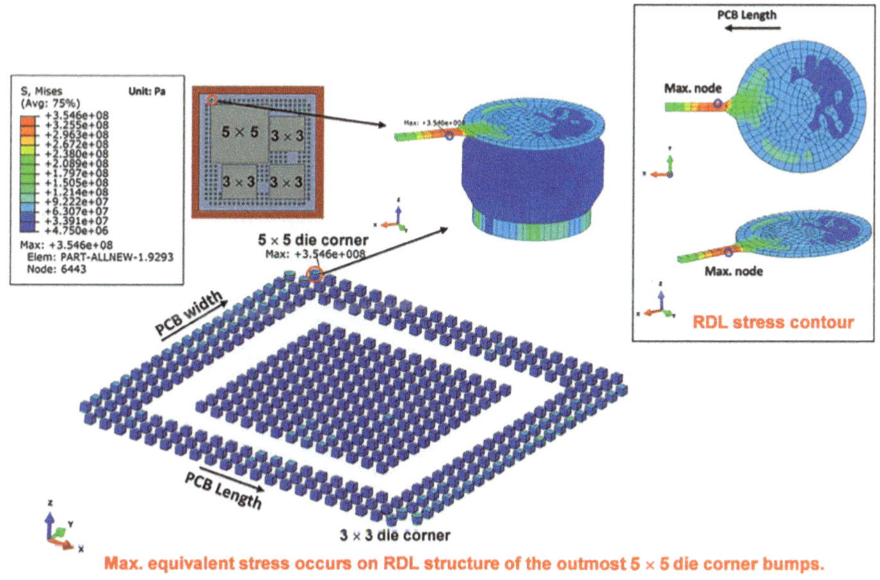

Fig. 4.37 Von Mises stress contours

- The heterogeneous integration package thickness is 300 μm (not including the solder balls). It is not only saved expensive EMC costs but also led to a very low-profile package.
- The proposed assembly process, i.e., by attaching the thin reconstituted molded wafer on another temporary glass wafer with a coated LTHC layer, provided a better warpage control, an opportunity for making supper low profile package, and a way to save the EMC material.
- The thin heterogeneous integration package has been assembled on a PCB without underfill and through a thermal cycling test. The characteristic life is more than 1000 cycles, which is more than adequate for mobile products such as smartphones, tablets, and wearables.
- The thin heterogeneous integration package has been assembled on a PCB with underfill and through a shock (drop) test. Up to 1000 drops, there is not any failure of all (24) samples. Thus, the fabricated test package samples are considered reliable (in dropping) for mobile products.
- Unfortunately, from a scientific point of view, all the tests stop after 1000 drops, and thus, there is not any failure data to estimate a life distribution, characteristic life, and failure rate of the package under drop test.
- The characteristic life of the solder joints without underfill under thermal cycling test has been estimated to be 1,070 cycles. This is more than adequate for most mobile products such as smartphones and tablets, whose life is usually less than 3 years.
- The failure locations of the solder joints have been found to be at: (1) the solder joint near the package corner and under the 3 × 3 mm chip corner along the diagonal direction of the large and small chips, and (2) the solder joint near the package corner and under the 5 × 5 mm chip corner.
- The failure mode of the solder joints without underfill has been found to be cracking between the Cu pad of the package and the bulk solder of the solder joint.
- The creep of solder joints under the thermal cycling condition has been determined by nonlinear and time dependent 3-D finite element simulation. Five temperature cycles have been executed and the hysteresis loops have shown that the creep behavior has been stabilized after the third temperature cycle. The maximum creep strain and Mises stress have been found to occur near the interface between the bulk solder and the Cu pad under the chip corners, near the package corners, and along the diagonal of the large and small chips. These failure locations and mode correlated very well with the experimental results (failure locations and mode).
- As expected, the DNP concept works for thermal cycling. The outermost two diagonal solder joints from all four package corners have been removed from the present section. It would be even better if the solder joint under the larger chip corner and the solder joint under the small chip corner (along the diagonal direction of the large and small chips) are removed.
- PCB assembly of the heterogeneous integration package with underfill has been through the drop test with an acceleration spectrum intensity = 1,500 G/ms

(1,500 Gs, 0.5 ms half-sine pulse). All 24 samples passed 1,000 drops. Thus, the fabricated RDLs of the package are reliable for mobile applications.

- The shock of the heterogeneous integration package RDLs under the given drop condition has been determined by time history analysis. It has been found that because of the structural damping, the dynamic responses are decreasing with time. The maximum stress (3.546×10^8 Pa) and strain (0.0023) occurred at 0.0022 s after the impact. The location of the maximum stress and strain has been found to be at the Cu conductor wiring of RDL2 near the package corner and the 5×5 mm chip corner.
- Unfortunately, the characteristic life of the RDLs cannot be estimated from the drop test, which stopped right after 1,000 drops for each sample, without failure.

4.4 Chip-First with Die Face-Down for Heterogeneous Integration of Mini-LED RGB Display

Light-emitting diode (LED) is a specialized semiconductor that glows when current passes through it. When the current is on, it emits light. While the current is off, it goes dark. The average size of a conventional LED is 1 mm [49], of a mini-LED is 75–300 μm [50], and of a micro-LED is ≤75 μm [50]. Since the shipment of micro-star international (MSI) Creator 17 gaming notebook with mini-LED backlighting liquid-crystal display (LCD) in April 2020 [51], the mini-LED has been getting lots of tractions. For examples, Apple [52] launched its first mini-LED backlit LCD iPad Pro in the first quarter of 2021 and mass produced mini-LED backlit LCD MacBook Pro in the second quarter of 2021. Samsung Electronics [53] shipped 2 million mini-LED backlit LCD televisions in 2021.

There are at least two different kinds of mini-LED display technology, namely, the mini-LED backlighting (backlit) LCD and the mini-LED RBG display. For LCD, each color filter defines a subpixel, and we control the color filter to generate an image. For mini-LED RBG display, each mini-LED is a subpixel, and we drive each mini-LED to generate an image. Mini-LED RGB display is the focus of this section.

Most of the current assembly methods of mini-LEDs' RGB display are picking up the individual mini-LED from a diced LED wafer and placing it on a PCB with solder paste and then surface mount technology (SMT) reflow [54]. Basically, this is the very old mini-LED (chip)-on-board technology [55]. One of the drawbacks of this method is the limitation (from PCB) of the pitch (or spacing) between the mini-LEDs and the other is the assembly yield (because of the flatness of the electroplated Cu pad on PCB).

Another possible assembly method is to use plastic ball grid array (PBGA) packages which attach some of the mini-LEDs on an organic package substrate with solder balls on PCB. In this case, the pitch between the mini-LEDs can be slightly smaller and the pressure on PCB is eliminated. However, the assembly yield loss due to the flatness of the electroplated Cu pad on the organic package substrate remains.

Also, there are additional costs on: (1) LED wafer bumping, or stencil printing solder paste on the package substrate or LED wafer and (2) the package substrate.

To have very high-density (e.g., ultra-fine pitch) and low-cost (e.g., minimum yield loss) mini-LED RGB display package, in this section, the fan-out (chip-first with die face-down) assembly process [12] is adopted. Because of the nature of fan-out technology, the spacing between individual mini-LEDs can go down to <50 μm. Also, because of the fan-out redistribution layers (RDLs), the mini-LED is not in contact with the electroplated Cu pad on PCB. Compared with the PBGA method, the costs and process steps of the package substrate and solder bumps between the LEDs and the package substrate are eliminated. Also, a lower package profile can be obtained from fan-out technology. Alternatively, the fan-out technology requires a temporary carrier and the fabrication of the RDLs.

There are at least two different kinds of temporary carrier for making the RDLs, namely, round (wafer) and rectangular (panel). To have high throughput, in this section, a rectangular panel is used as the temporary carrier [40]. A PCB is designed and fabricated for the drop test of the mini-LED package. Thermal cycling of the mini-LED RGB display SMD PCB assembly is also performed by a 3-D nonlinear temperature- and time-dependent finite-element simulation.

4.4.1 Test Mini-LEDs

The mini-LEDs under consideration are shown in Fig. 4.38 [40]. Their dimensions are red (125 × 250 × 100 μm), green (130 × 270 × 100 μm), and blue (130 × 270 × 100 μm).

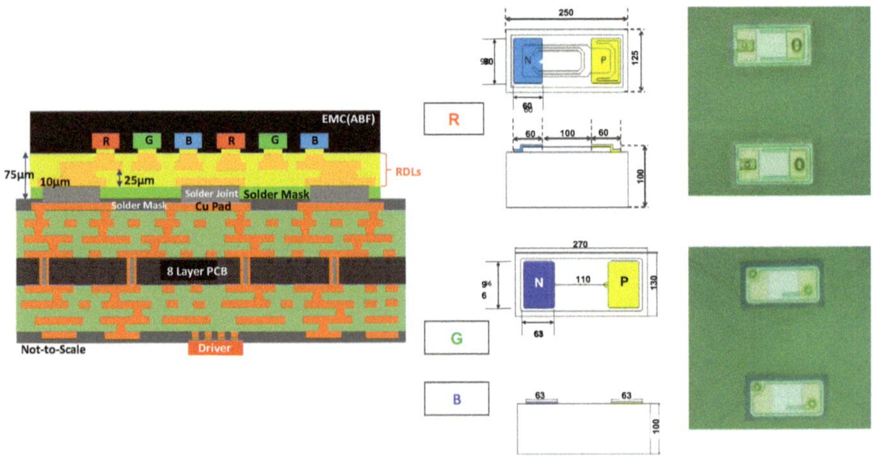

Fig. 4.38 Test LEDs: red (125 × 250 × 100 μm), green (130 × 270 × 100 μm), and blue (130 × 270 × 100 μm)

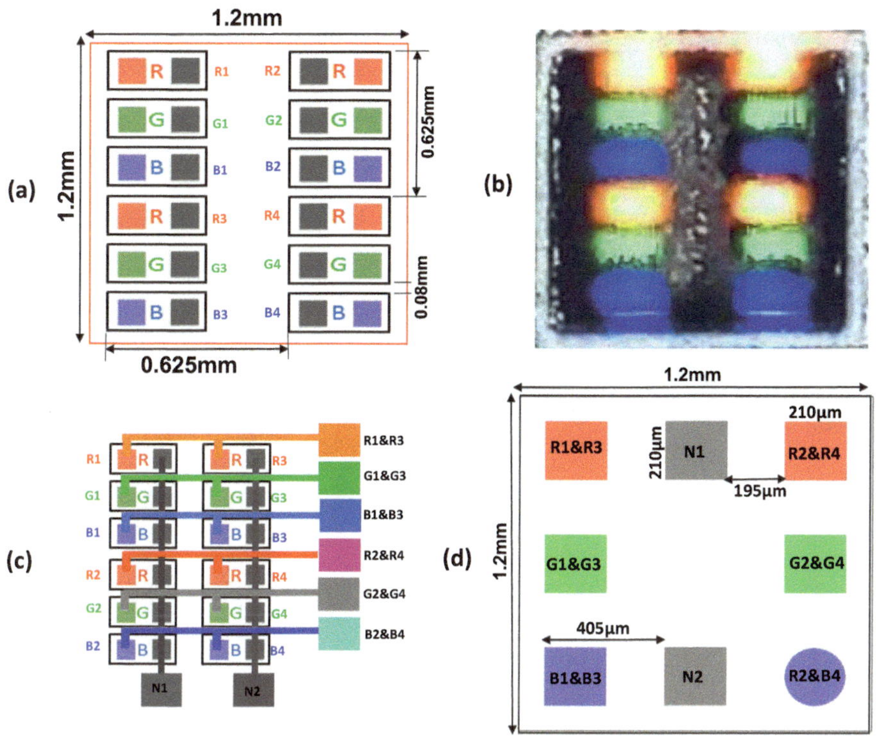

Fig. 4.39 SMD. **a** 4 (2 × 2 pixels)-in-one SMD, i.e., a total of 12 R, B, and G mini-LEDs. **b** Light-up SMD. **c** Daisy-chained layout of the SMD. **d** Bottom-side layout of the SMD

The pad surface finishing of the P-electrode and N-electrode of these mini-LEDs is Cr/Au. The pad size of the green and blue mini-LEDs is 96 μm × 63 μm and of the red mini-LED is 90 μm × 60 μm. To improve SMT assembly yield, 12 individual mini-LEDs are grouped into a 4-in-one surface mount device (SMD) as shown in Fig. 4.39a. It can be seen that the spacing among the RGB mini-LEDs is 80 μm and the pixel-to-pixel spacing is also ~80 μm. The pixel pitch is 625 μm. The mini-LEDs are daisy-chained as shown in Fig. 4.39c. Figure 4.39d shows the layout of the bottom side of the SMD.

4.4.2 Test Mini-LED Package

The cross section of the test mini-LED RGB display SMD package is schematically shown in Fig. 4.40. It can be seen that the mini-LEDs are packaged by a fan-out chip-first with die face-down process. There are two RDLs with metal linewidth (L)

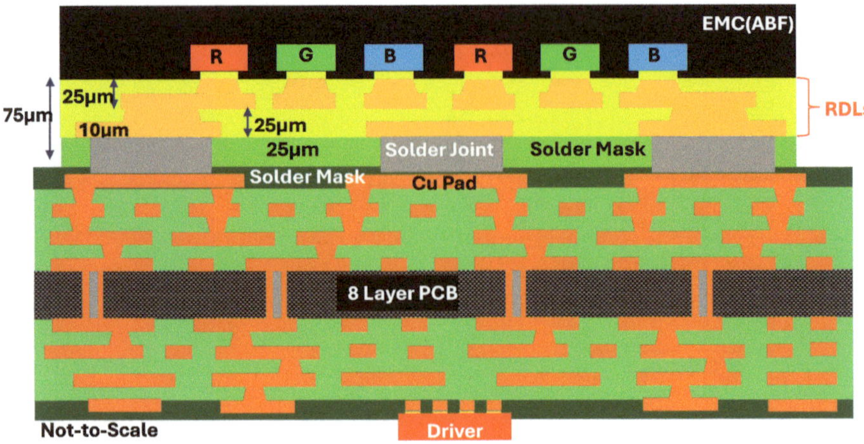

Fig. 4.40 Schematic cross section of the test mini-LED RGB SMD package PCB assembly

and spacing (S) equal to 35 μm/35 μm for RDL1 and 35 μm/50 μm for RDL2 as shown in Fig. 4.41.

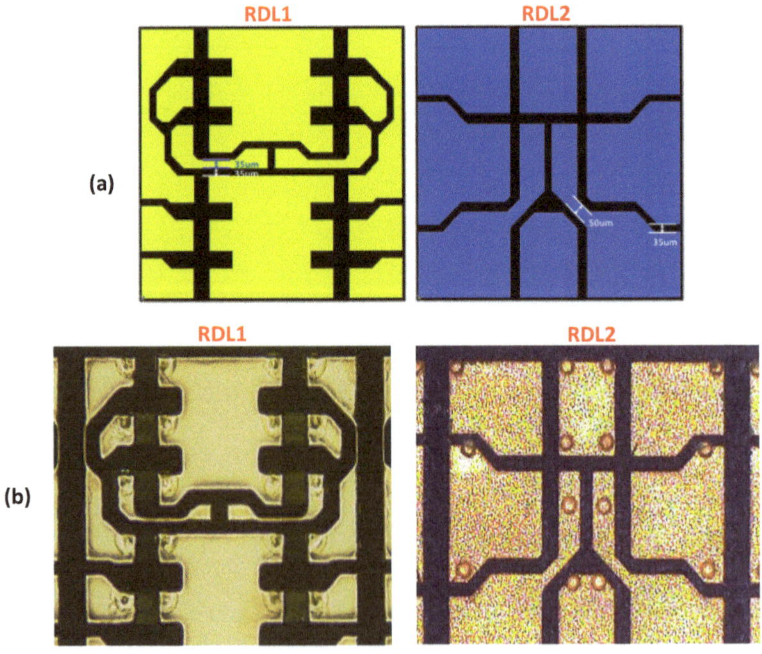

Fig. 4.41 RDLs. **a** Layout of RDL1 and RDL2. **b** Fabricated RDL1 and RDL2

4.4.3 RDL and Mini-LED RGB SMD Fabrication

Figure 4.41 shows the key process steps in fabricating the RDLs of the mini-LED RGB display package by the fan-out (chip-first and die face-down) process.

(A) Pick and Place of Mini-LEDs

After the sputtering of Ti by physical vapor deposition (PVD) on a 515 mm × 510 mm × 1.1 mm glass temporary carrier, liquid photoresist (LPR) lithography for the LED alignments, etching and striping, and Nitto Revalpha taping, pick and place (P&P) of the mini-LEDs is performed as shown in Fig. 4.42. There are 629,712 mini-LEDs per 515 mm × 510 mm panel. However, to save some of the mini-LEDs, only 74,064 mini-LEDs are P&P on the temporary carrier. It is performed by the ASM AD420 high-throughput chip shooter as shown in Fig. 4.43. Figure 4.44 shows the temporary panel with the P&P mini-LEDs.

(B) Molding With ABF

After P&P of the mini-LEDs, molding is done as shown in Fig. 4.42. In this section, the epoxy molding compound (EMC) is the Ajinomoto build-up film (ABF). The laminate condition of the ABF operates in two stages: (1) in the first stage, the temperature is 120 °C for 30 s at vacuum condition and then pressure (0.68 MPa) for 30 s with the temperature and vacuum on and (2) in the second stage, the temperature is 100 °C and pressure (0.58 MPa) for 60 s for flatness. Figure 4.45 shows the molded panel.

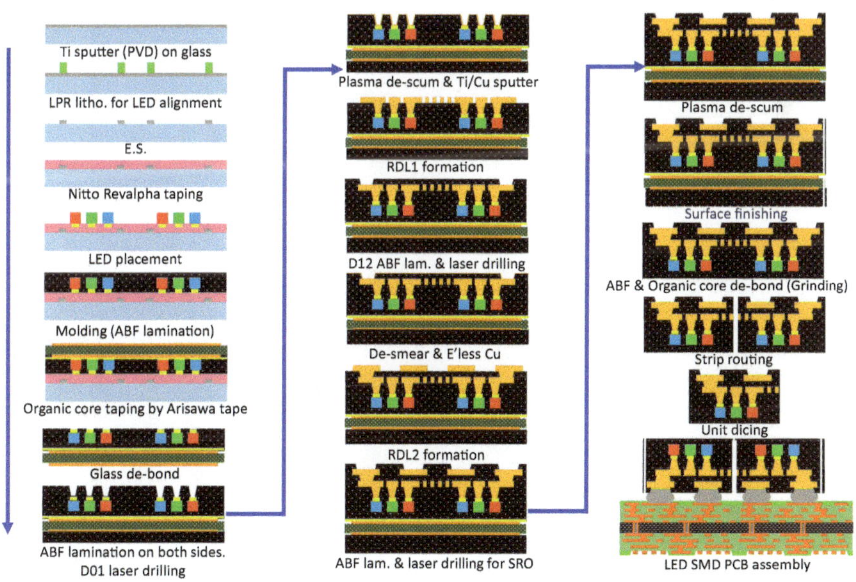

Fig. 4.42 Key process steps of fan-out panel-level packaging of mini-LEDs

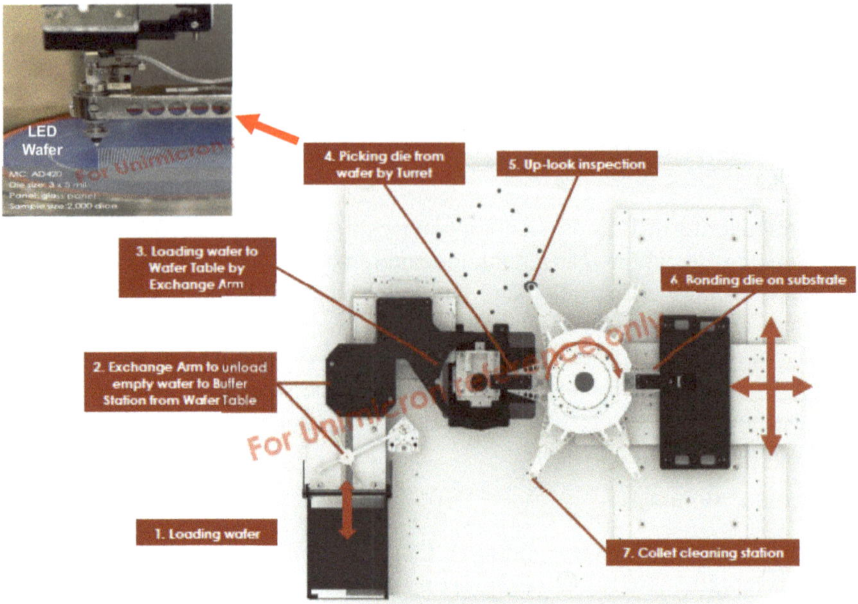

Fig. 4.43 Pick and place equipment for mini-LEDs by ASM

Fig. 4.44 Mini-LEDs on the temporary glass panel with the Revalpha tape

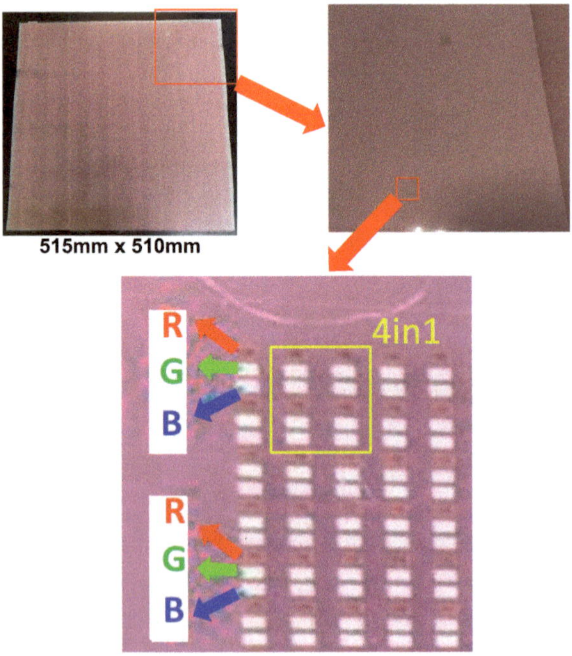

Fig. 4.45 EMC (ABF) molding

(C) Second Carrier Attachment

To save the ABF material and have a lower profile package, only 150 μm-thick ABF is used. Thus, before debonding the temporary glass carrier, an organic core is attached with an Arisawa tape to the backside of the ABF as shown in Fig. 4.42.

(D) Debond the First Temporary Glass Carrier

The debonding temperature of the temporary glass carrier is 170 °C. Under this temperature, it is easy to remove the glass carrier and peel off the tape. Then, the surface is cleaned with chemicals.

(E) RDLs Fabrication

First, the ABF is laminated on both sides of the reconstituted panel. It is followed by laser drilling, plasma de-scum, and Ti/Cu sputtering (because of Au surface finishing of the mini-LEDs) (Fig. 4.46). This is followed by dry film lamination, laser direct imaging (LDI) and develop, electroplating Cu, dry film stripping, and Ti/Cu etching to obtain the first RDL (RDL1), as shown in Fig. 4.41b. The dielectric layer is ABF, and the metal conductor layer is Cu. To fabricate RDL2, the ABF is laminated again, and laser drilling is done. This is followed by desmear and electroless Cu, dry film lamination, LDI and develop, electroplate Cu, dry film stripping, and seed layer etching to obtain the second RDL (RDL2) as shown in Fig. 4.41b.

(F) Solder Mask Opening and Surface Finishing

The ABF is laminated, and laser drilling is performed for solder mask opening (SMO) or solder resist opening (SRO and plasma de-scum. This is then followed by electroless palladium immersion gold (EPIG) as shown in Fig. 4.47.

(G) Backgrinding and Panel Routing

Finally, some of the ABF and the organic core are backgrinded away and the reconstituted panel is routed into individual mini-LED RGB display SMDs with two RDLs as shown in Fig. 4.48.

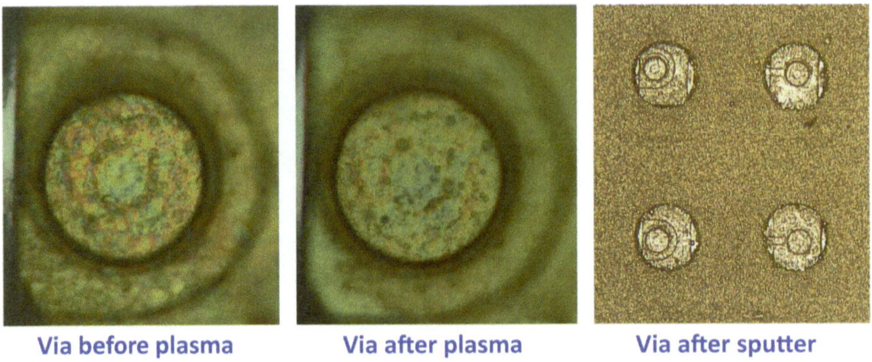

Fig. 4.46 Plasma de-scum and Ti/Cu sputtering of ABF vias

Fig. 4.47 SMO by laser drilling and EPIG surface finishing

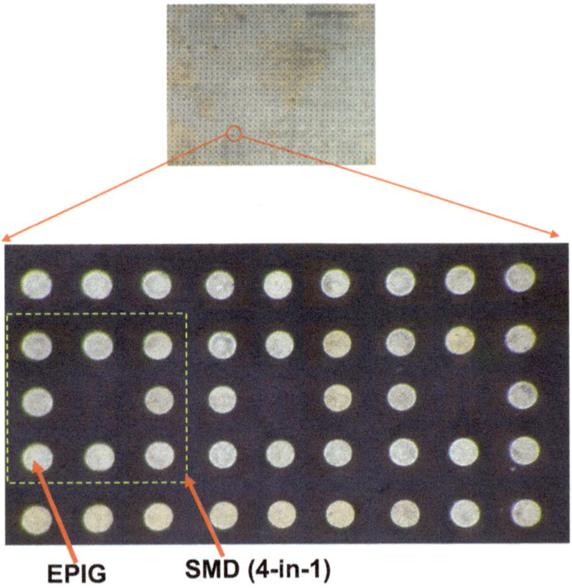

4.4.4 PCB Assembly

(A) PCB Layout

The layout of the test PCB (132 mm × 77 mm) to support the mini-LED RGB display SMDs for drop test and thermal cycling test is shown in Fig. 4.49. It is an eight-layer PCB with a few different mini-LED RGB display SMDs such as 4-in-1 and 16-in-1.

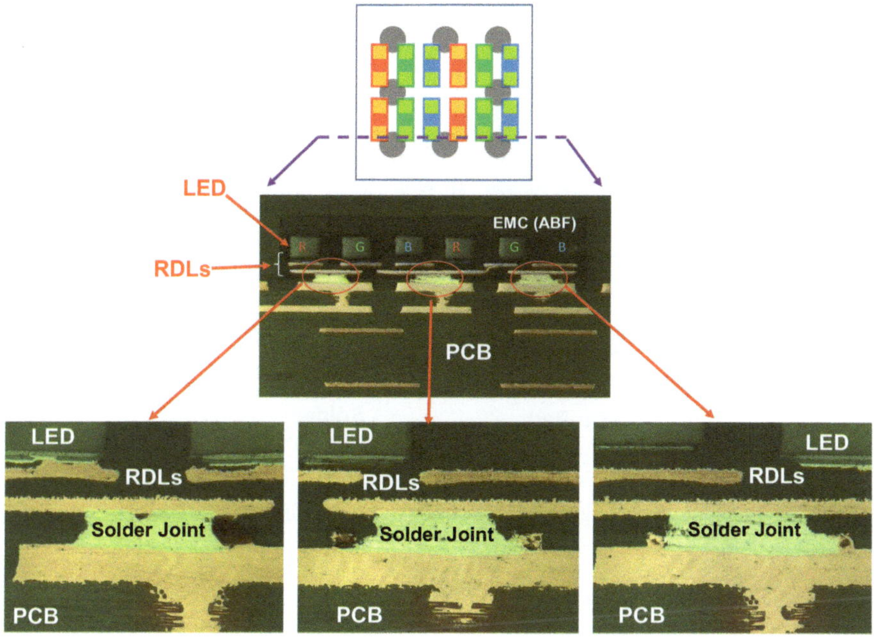

Fig. 4.48 Cross section image of the mini-LED SMD PCB assembly

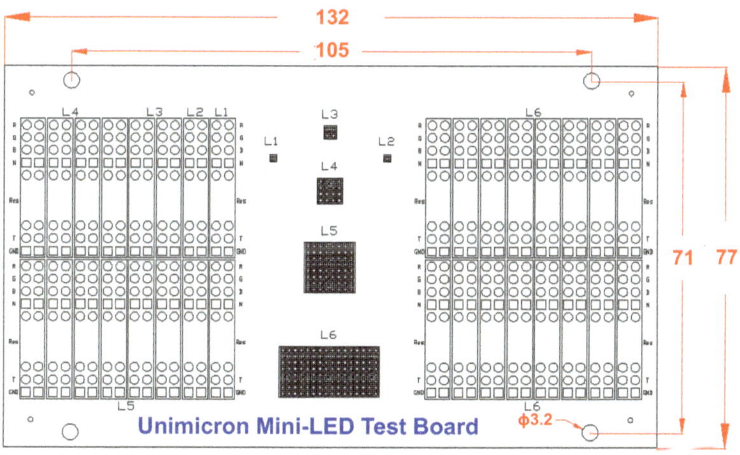

Fig. 4.49 PCB layout for the drop test and thermal cycling test of mini-LEDs' package

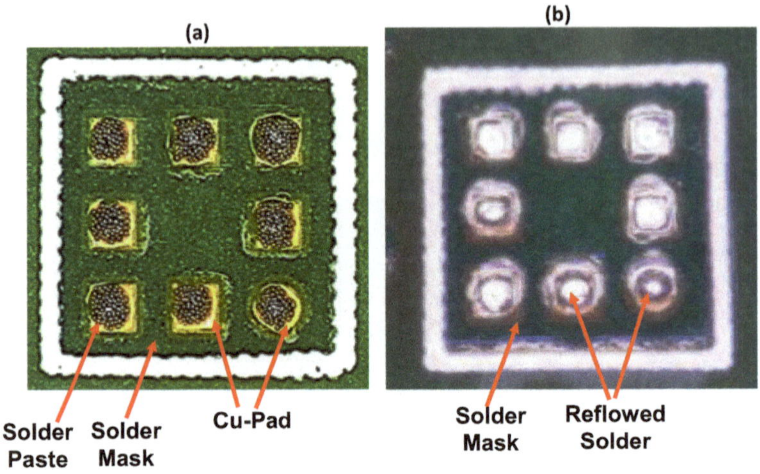

Fig. 4.50 **a** Solder paste printing on the PCB. **b** Reflow of the solder paste on the PCB

(B) Stencil Printing Solder Paste on PCB

A stainless-steel stencil with 170 μm-diameter opening is used to print the Sn3Ag0.5Cu solder paste on the PCB as shown in Fig. 4.50a. After reflowing, the solder balls are as shown in Figs. 4.50b and 4.51a.

(C) Final SMT Assembly

The mini-LED RGB display SMD is picked and placed on the PCB with solder balls plus flux. Then, the standard SMT reflow process is used. A typical cross section of the mini-LED RGB display SMD PCB assembly is shown in Fig. 4.48. To see the mini-LEDs, the solder joints are not cut along their center but near the edge. An X-ray image of the final assembly is shown in Fig. 4.51b.

4.4.5 Drop Test

(A) Test Setup

The mini-LED RGB display SMD PCB assembly is subjected to drop test. The test setup is according to JEDEC Standard JESD22-B111 as shown in Fig. 4.52. After more than ten tries, the right height (648 mm) of the drop table is obtained, which yields the drop spectrum with 1522 G/ms (1522 G and 0.5 ms half-sine pulse) (Fig. 4.53a). The sample size is 22.

(B) Test Results

After 100 drops, the SMDs light-up very well as shown in Fig. 4.53b and there is no failure. However, a closer look at the cross sections of the tested samples shows that

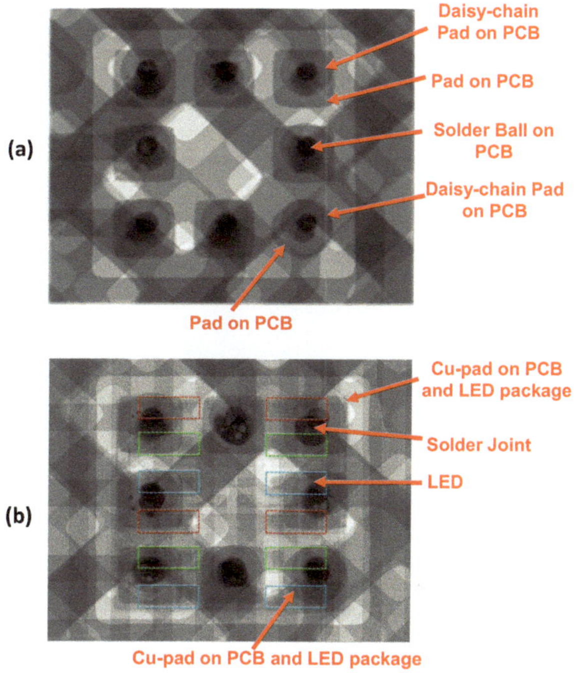

Fig. 4.51 **a** X-ray image of the PCB with reflowed solder. **b** X-ray image of the mini-LED SMD PCB assembly

one of the Cu metal lines of the SMDs is almost broken (just barely contacted) as shown in Fig. 4.54.

4.4.6 Thermal Cycling Simulation

(A) Structure and Kinematic Boundary Conditions

The PCB assembly of the mini-LED RGB display SMD is analyzed. Due to the symmetry of the x- and y-axes as shown in Fig. 4.55, only a quarter of the structure is modeled. Figure 4.56 shows the finite-element model in various views for simulation. The kinematic boundary conditions are: (1) there is not displacement (u_x) on the *yz* plane and rotation about the *y*-axis; (2) there is not displacement (u_y) on the *xz* plane and rotation about the *x*-axis; and (3) the center of the bottom of PCB is fixed in the *z*-direction.

(B) Materials Properties and Kinetic Boundary Conditions

The material properties of the PCB assembly of the mini-LED RGB display SMD are given in Table 4.4. All the material properties are assumed to be constant except for those of the solder (Sn3Ag0.5Cu) which is assumed to obey the generalized Garofalo creep constitutive equation and the coefficient of thermal expansion (CTE),

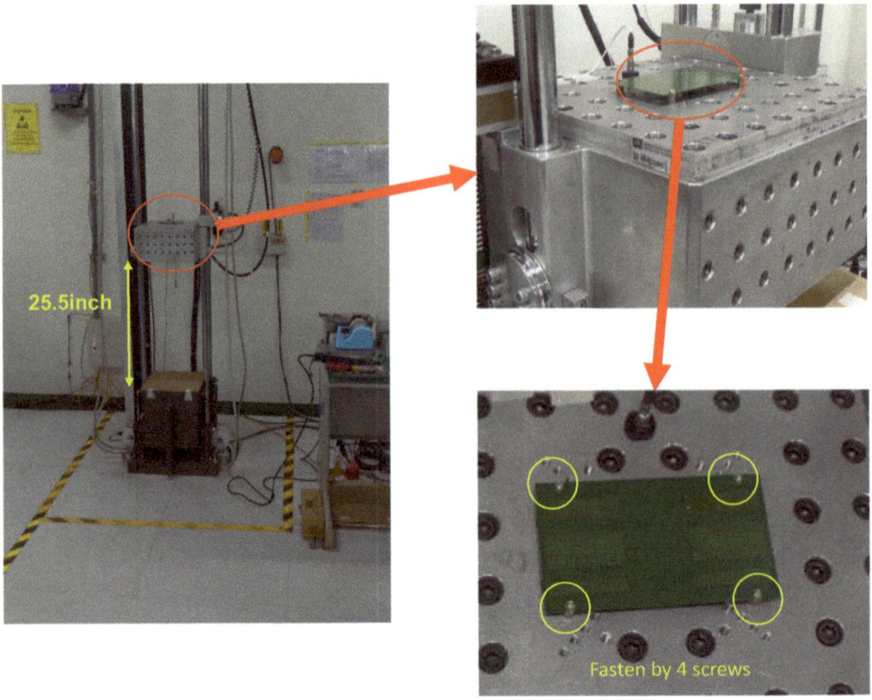

Fig. 4.52 Drop test of the mini-LED SMD PCB assembly

and Young's modulus of the solder are, respectively, 21 + 0.017 T and 49 − 0.07 T, [48].

The kinetic temperature boundary condition for the simulation is shown in Fig. 4.57. It can be seen that the temperature is −40 °C ↔ 85 °C. The cycle time is 60 min, and the ramp-up, ramp-down, dwell-at-hot, and dwell-at-cold are each 15 min.

(C) Simulation Results

The deformed shape (color contours) and the undeformed shape (dark lines) of the mini-LED RGB display SMD PCB assembly are shown in Fig. 4.58. It can be seen from Fig. 4.58b that at 85 °C, the PCB expands more than the SMD and the structure is deformed in a convex shape (smiling face). It can also be seen from Fig. 4.58c that at −40 °C, the PCB shrinks more than the SMD and the structure is deformed into a concave shape (crying face). These results show that the displacement, rotation, and temperature boundary conditions are prescribed correctly. Figure 4.59 shows the maximum accumulated creep strain contours' distribution, which occurs at the corner solder joint, at 450 s (85 °C), Fig. 4.59a, and at 2250 s (−40 °C), Fig. 4.59b. It can be seen that the maximum value occurs in a very small local area and the values of most of the other areas are very small. Figure 4.60 shows the maximum

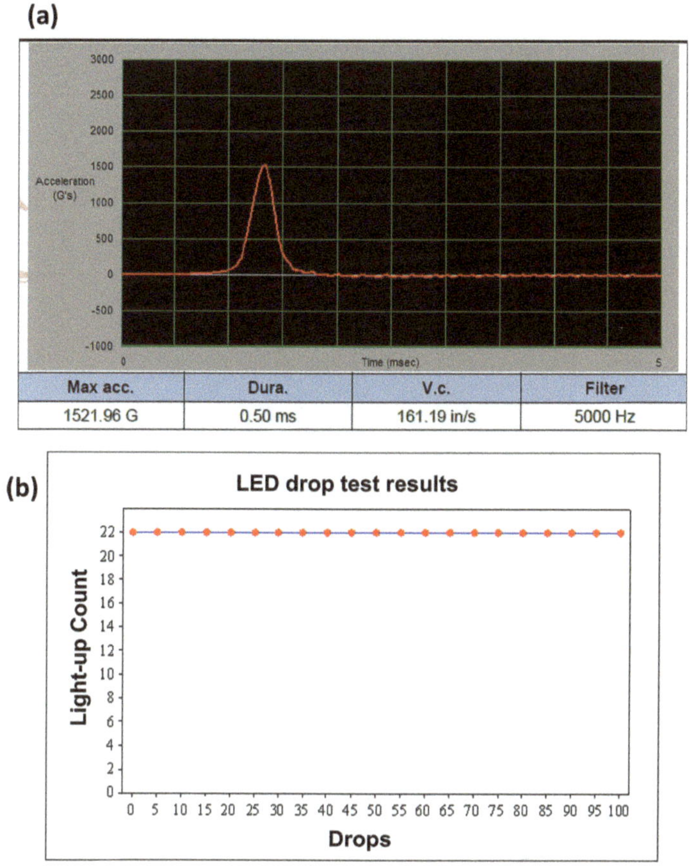

Fig. 4.53 **a** Drop test spectrum. **b** Drop test results

accumulated creep strain time history at the corner solder joint. Again, it can be seen that the maximum accumulated creep strain per cycle is very small (less than 1%).

Figure 4.61 shows the maximum creep strain energy density contour distribution, which occurs at the corner solder joint, at 450 s (85 °C), Fig. 4.61a, and at 2250 s (−40 °C), Fig. 4.61b. Again, it can be seen that the maximum value occurs in a very small local area and the values of most of the other areas are very small. Figure 4.62 shows the maximum at the corner solder joint. Again, it can be seen that the creep strain energy density per cycle is very small (less than 0.3 MPa). The maximum accumulated creep strain and creep strain energy density at the corner solder joints occur at the interfaces between the RDLs and the solder joint and between the solder joint and the PCB. Any potential failures subjected to thermal cycling test should occur at these locations.

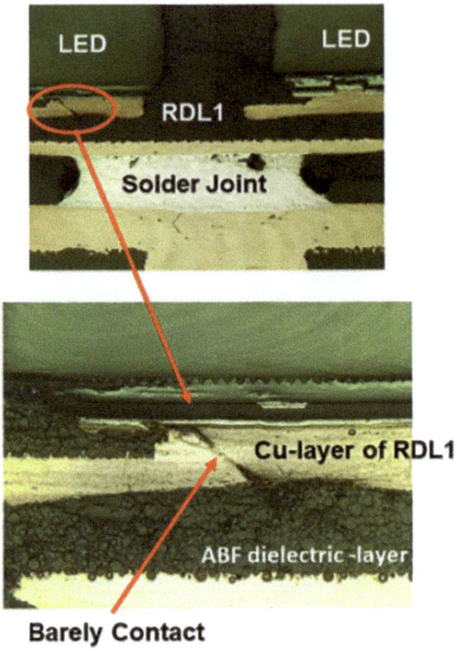

Fig. 4.54 Sample after 100 drops. The Cu layer of RDL1 is barely in contact

4.4.7 Summary

Some important results and recommendations are summarized as follows.

- The feasibility of the design, materials, process, fabrication, and reliability of a mini-LED RGB display SMD fabricated by a fan-out chip-first (face-down) panel-level packaging has been demonstrated.
- The present packaging method is not only very high-throughput and high assembly yield but also potentially a very low-cost method.
- Low-profile SMD has been obtained with the lamination of ABF, instead of the conventional compression of EMC.
- The reliability of the PCB assembly of the mini-LED RGB display SMD has been demonstrated by the drop test. After 100 drops, there is no failure.
- The reliability of the PCB assembly of the mini-LED RGB display SMD has been demonstrated by thermal cycling simulation. It is found that the maximum accumulated creep strain per cycle and the creep strain energy density per cycle are too small to create reliability concerns. Also, the maximum accumulated creep strain and creep strain energy density at the corner solder joints occur at the interfaces between the RDLs and the solder joint and between the solder joint and the PCB. Any potential failures subjected to thermal cycling test should occur at these locations.

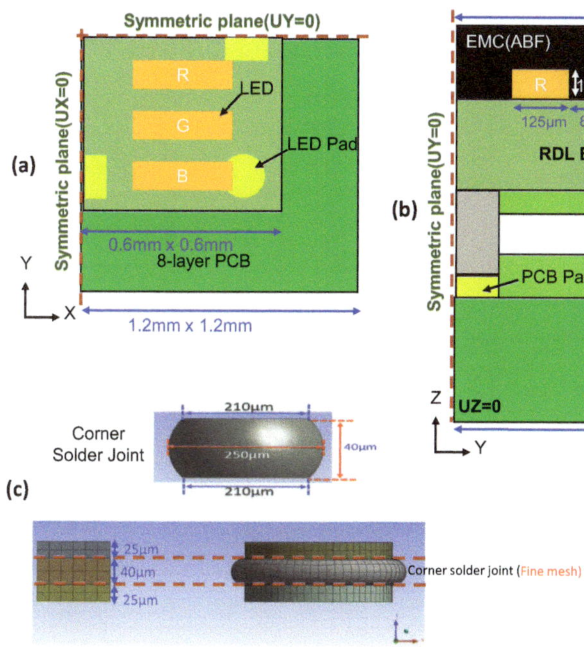

Fig. 4.55 **a** Quarter of structure. **b** Cross section of quarter assembly. **c** Fine meshes for corner solder joint

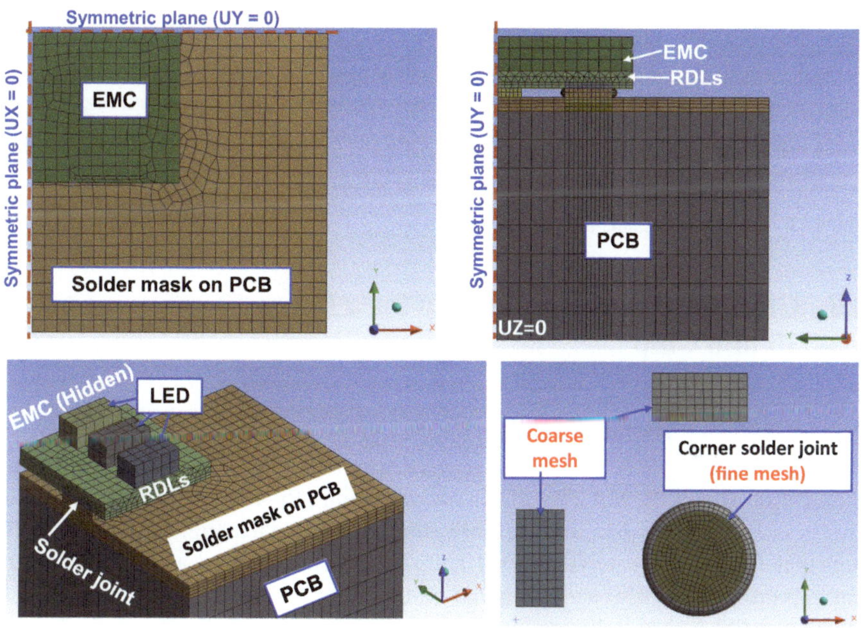

Fig. 4.56 Finite element modeling

Table 4.4 Material properties and constitutive equation for Sn3Ag0.5Cu solder alloy

Materials	CTE (10^{-6}/°C)	Young's modulus (GPa)	Poisson's ratio
Copper	16.3	121	0.34
PCB	$\alpha_x = \alpha_y = 18$ $\alpha_z = 70$	$E_x = E_y = 22$ $E_z = 10$	0.28
Solder mask (LED)	7.0	7.0	0.3
Solder	21 + 0.017 T (°C)	49 − 0.07 T (°C)	0.3
EMC (ABF)	7.0	7.0	0.3
AlGaInP	5.0	103	0.31
Solder mask (PCB)	39	4.1	0.3
RDL (equiv. block)	9.79	41.2	0.312

The constitutive eq. for solder is: $\frac{d\varepsilon}{dt} = 500,000[\sinh(0.01\sigma)]^5 \exp\left(-\frac{5,800}{T}\right)$ where T is in Kelvin and σ is in MPa

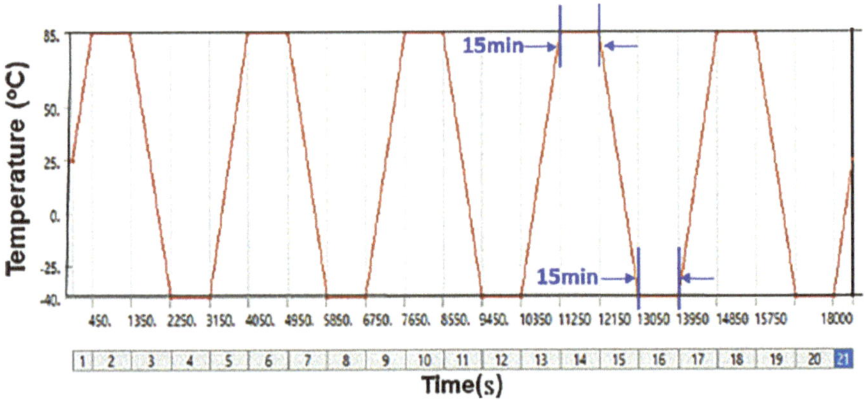

Fig. 4.57 Temperate boundary condition

4.5 Chip-First with Die Face-Up for a Large 10 mm × 10 mm Chip

4.5.1 Test Chip

Figure 4.63 shows the test chip under consideration [45–47]. The layout of the test chip is shown in Fig. 4.63a, and the fabricated chip is shown in Fig. 4.63b–d. It can be seen that the chip sizes are 10 mm × 10 mm × 150 μm, and there are 1988 pads with a minimum pitch = 150 μm staggered on the outer two rows. The polyimide (PI) opening of the Al-pad is 40 μm in diameter and is 5 μm thick. [Later on, because of the Cu smear under backgrinding of the epoxy molding compound (EMC) to expose

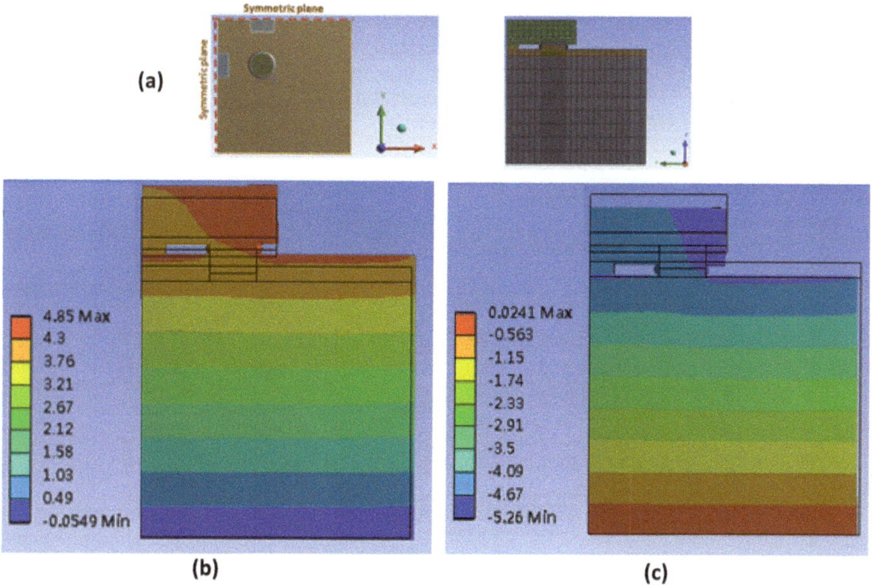

Fig. 4.58 **a** Top view and cross section view of the structure. **b** Deformed shape (color contours) at 450 s (85 °C). **c** Deformed shape (color contours) at 2250 s (−40 °C)

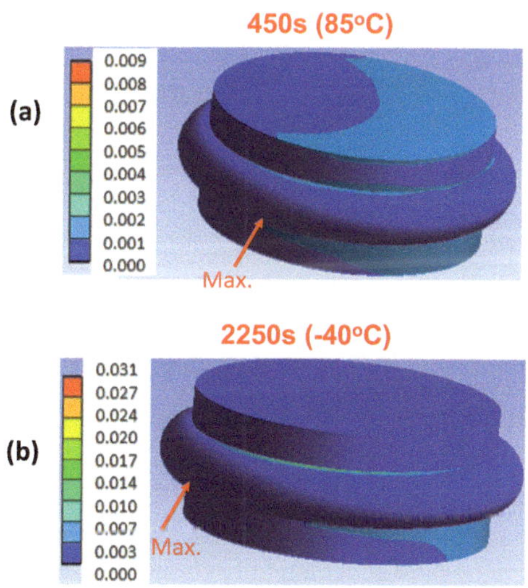

Fig. 4.59 Accumulated creep strain at the corner solder joint. **a** At 450 s (85 °C). **b** At 2250 s (−40 °C)

the Cu contact pad, the thickness of the PI has been increased to slightly taller than the Cu contact pad.] The SiO_2 passivation opening of the Al-pad is 50 μm × 50 μm, and the size of the Al-pad is 70 μm × 70 μm. The Cu contact pad is 60 μm in diameter and is 25 μm tall from the Al-pad.

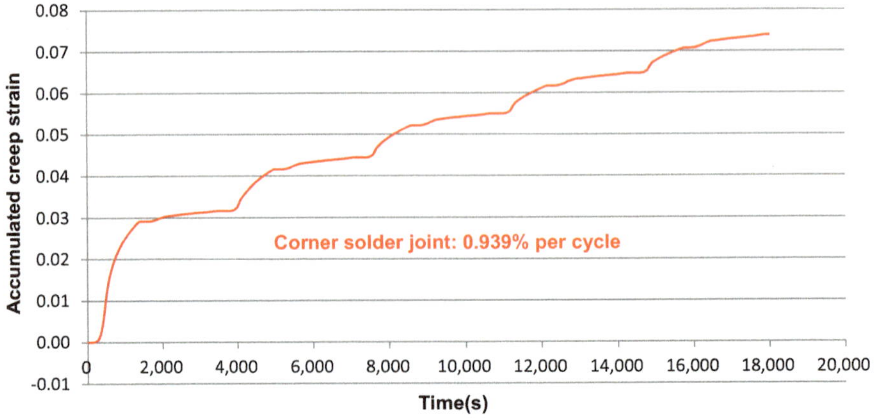

Fig. 4.60 Accumulated creep strain time history at the corner solder joint

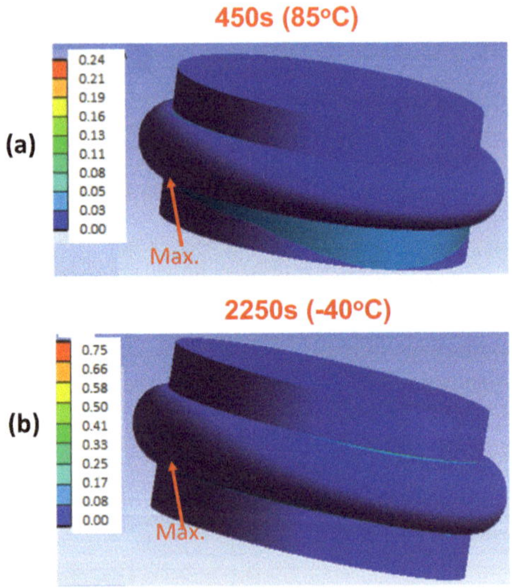

Fig. 4.61 Creep strain energy density at the corner solder joint. **a** At 450 s (85 °C). **b** At 2250 s (−40 °C)

4.5.2 Test Package

Figure 4.64 schematically shows the test package under consideration. It can be seen that the reconstituted (temporary) wafer is a 300 mm glass wafer, Fig. 4.64a, and there are 325 test packages with a pitch = 13.46 mm. The layout of the test package is shown in Fig. 4.64b. The dimensions of the test package are: 13.42 mm × 13.42 mm as shown in Fig. 4.64c. Thus, a package/chip area ratio = 1.8, and the test package is 1.71 mm all around larger than the test chip. The dimensions of a typical pad on the test package are shown in Fig. 4.64d.

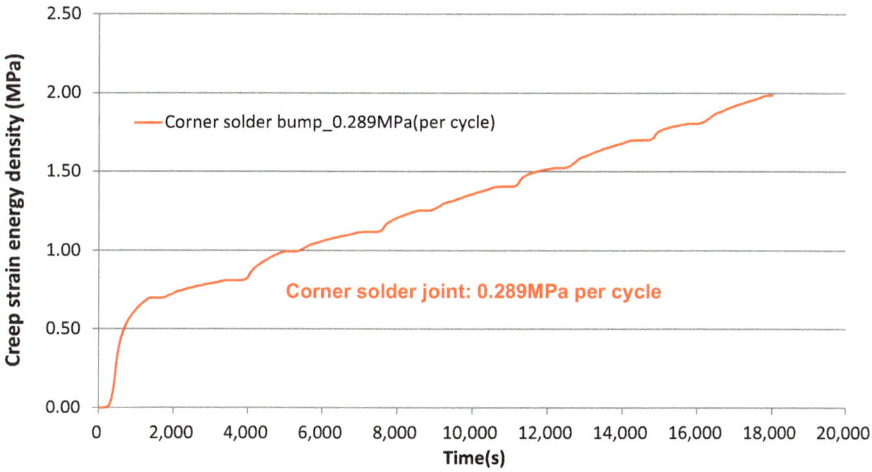

Fig. 4.62 Creep strain energy density time history at the corner solder joint

Figure 4.65 schematically shows the cross-sectional view of the test package. It can be seen that there are three RDLs and the thickness of RDL1 and RDL2 is 3 μm and of RDL3 is 7.5 μm. The linewidth and spacing of RDL1, RDL2, and RDL3 are, respectively, 5 μm/5 μm, 10 μm/10 μm, and 15 μm/15 μm. The dielectric layer thicknesses of DL1, DL2, and DL3 are 5 μm each and of DL4 is 10 μm. The via (V_{C1}), through the first dielectric layer (DL1), connecting the Cu contact pad of the test chip to the first RDL (RDL1) is 20 μm in diameter. The pad diameter on the RDL1 is 55 μm, which is connected to RDL2 through the via (V_{12}) with a diameter of 35 μm. Similarly, the pad diameter on the RDL2 is 65 μm, which is connected to RDL3 through V_{23} with a diameter of 45 μm. Finally, 220 μm Cu pads are formed on RDL3 for the solder ball mounting. The opening of the passivation (DL4) is 180 μm. The solder ball size is 200 μm, and the ball pitch is 0.4 mm.

4.5.3 Chip-First (Die Face-Up) FOWLP Assembly Process

Figure 4.66 shows schematically the FOWLP with the chip-first and die face-up assembly process. Basically, work must be done on the device (test chip) wafer (see Fig. 4.66a, b) and the reconstituted (temporary carrier) wafer (see Fig. 4.66c–h).

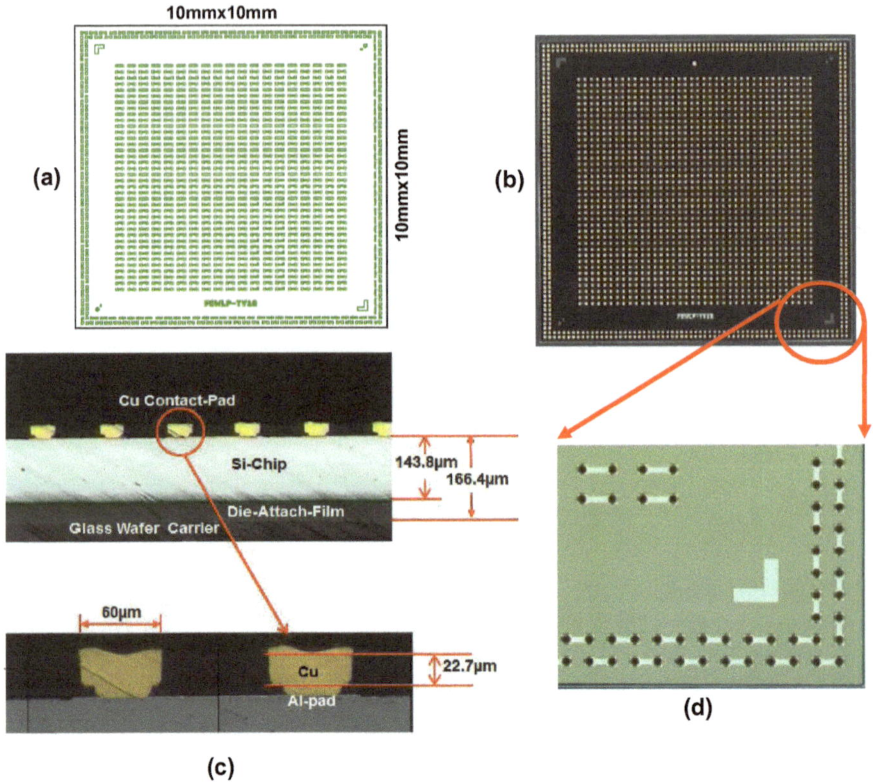

Fig. 4.63 Test chip. **a** 10 mm × 10 mm test chip layout. **b** Fabricated test chip. **c** Close-up look of the test chip. **d** Cross-sectional image of the test chip

4.5.4 On the Test Chip Wafer

First, the test chip wafer must be modified by sputtering a Ti/Cu as a bottom layer of under bump metallurgy (UBM) with a physical vapor deposition (PVD) on the Al (or Cu) pad, and a Cu contact pad (for building the RDLs later) is electroplated on the UBM, as shown in Fig. 4.66a. This step is followed by spin coating a polymer on the top of the test chip wafer and laminating (at ~70 °C) with a (~20 μm) die-attach film (DAF) provided by Hitachi at the bottom of the test chip wafer as shown in Fig. 4.66b. The test chip wafer is then diced into individual chips.

4.5.5 On the Reconstituted Wafer

Figure 4.66c–h show the FOWLP with chip-first and die face-up process steps on the reconstituted wafer, and they will be briefly mentioned as follows.

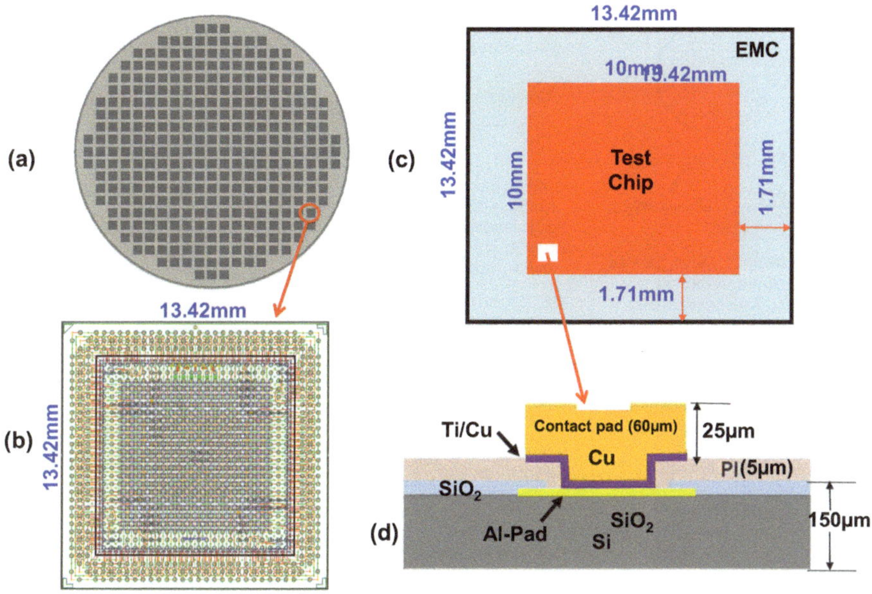

Fig. 4.64 Test package. **a** Schematic of the test packages on a 300 mm reconstituted wafer. **b** Layout of the test package. **c** Schematic of the test package (13.42 mm × 13.42 mm) for the test chip. **d** Schematic of the cross section of a pad on the test chip

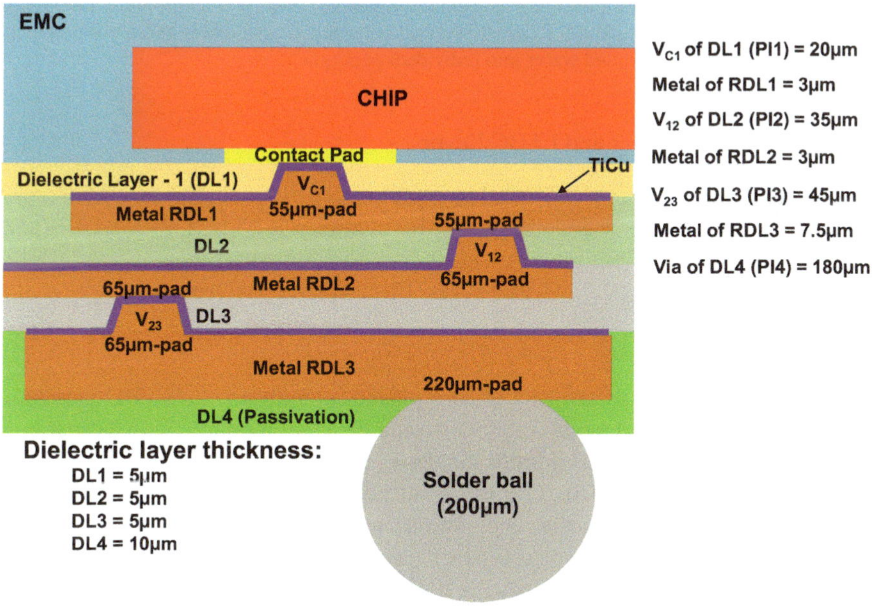

Fig. 4.65 Schematic of the cross section of the test package with three RDLs

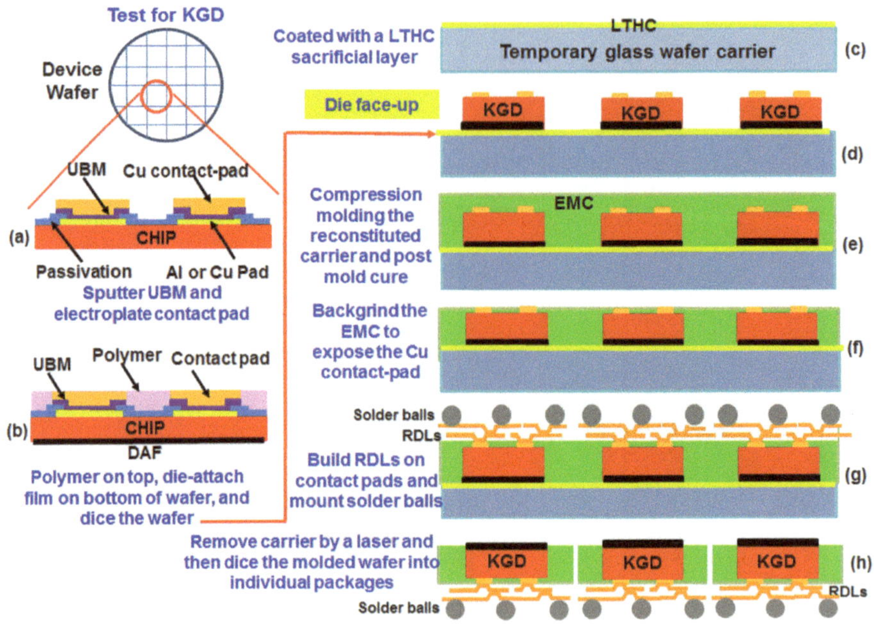

Fig. 4.66 FOWLP with chip-first and die face-up process. **a** Sputter UBM and ECD of Cu contact pad. **b** Polymer on top and DAF on bottom of wafer, and then dice the wafer. **c** Spin-coat an LTHC layer on the top of the temporary glass wafer carrier. **d** Pick and place the die face-up on the LTHC layer carrier. **e** Compression molds the reconstituted wafer and PMC. **f** Backgrinding the EMC to expose the Cu contact pad. **g** Build RDLs on contact pads and mount solder balls. **h** Remove the carrier and dice the reconstituted wafer into individual packages

(A) Glass Carrier Wafer and LTHC Layer

A light-to-heat conversion (LTHC) layer (about 1 μm) provided by 3M is spin-coated onto the temporary glass carrier wafer as shown in Figs. 4.66c and 4.67a. In this section, three different glasses with thermal coefficient of expansion (TCE) = 3.25×10^{-6}/°C, 7.6×10^{-6}/°C, and 8.1×10^{-6}/°C are considered.

(B) Pick and Place

The individual test chips are picked and placed face-up on the LTHC carrier as shown in Figs. 4.66d and 4.67a. In order to cure the DAF, a bonder with temperature and pressure such as the multipurpose high-accuracy NUCLEUS is used. The DAF process is carried out at 120 °C (both bond head and bond stage) with the bond force of 2 kg for 2 s. Thus, the reconstituted wafer will expand during chips pick and place. The X and Y pitch offsets (the measured pitch subtract from the designed pitch) of chips are plotting against the column and row as shown in Fig. 4.68. It can be seen that the chip placement offsets can be as large as ±100 μm. However, during patterning/photolithography of the RDLs, it is operated at room temperature. Thus, pitch compensation due to the DAF heating is needed and can be calculated

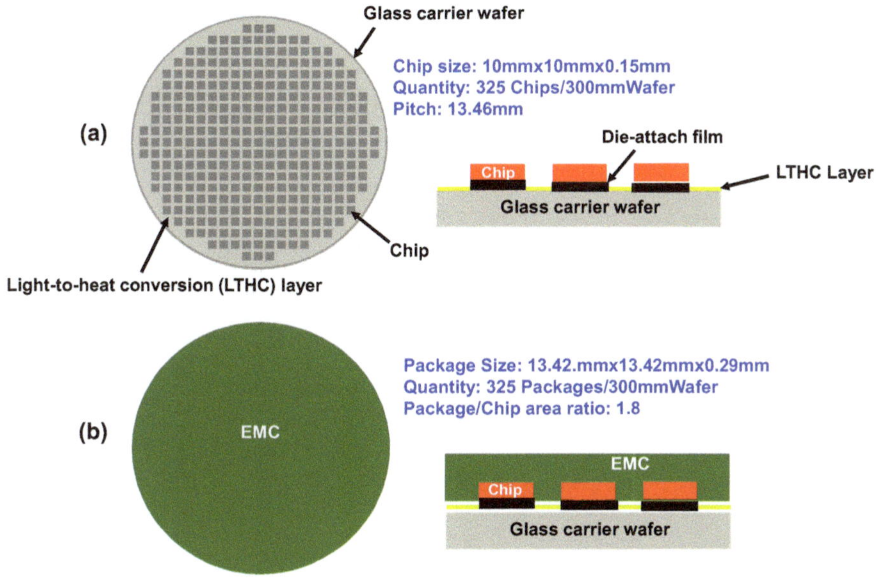

Fig. 4.67 **a** Spin-coat an LTHC layer on the carrier, and pick and place the chip face-up on the carrier. **b** EMC molding

by the following equation: $P_A = P_R + \alpha L \Delta T$ where P_A is the die-attach pitch, P_R is the designed RDL pitch, αL is the linear TCE of the glass carrier, ΔT is the temperature difference between the DAF curing temperature and RDL patterning/ photolithography temperature. After the calculation, the information is reentered to the bonder. The X and Y pitch offsets of chips after pitch compensation are shown in Fig. 4.69. It can be seen that the placement offset is reduced from ±100 μm (without pitch compensation) to ±4.0 μm (with pitch compensation), and this is a good performance that can meet the process specifications.

(C) EMC Dispensing

The EMC is a liquid-like material (Nagase R4507), and its material properties are shown in Table 4.5. It can be seen that the filler content is 85%. The average size of the filler is 8 μm, and the maximum size of the filler is 25 μm. The flexural modulus of the EMC is 19 GPa, and the viscosity is 250 Pa s. The transition temperature (Tg) is 150 °C. There are many ways to dispense the EMC such as the glob pattern, line pattern, circular pattern, and star pattern (Fig. 4.70). Experimental results show that there are many flow marks from the glob pattern. (One of the key reasons to have flow marks is the epoxy resin and filler separation and the silica fillers concentrate along those "streamline" shown in Fig. 4.71.) Also, experimental results show that the flow mark length of the glob pattern is very much larger than that from the line (rectangular) pattern as shown in Fig. 4.71. This is because the flow path of liquid EMC for the line pattern is shorter than that of the glob pattern. It should be pointed

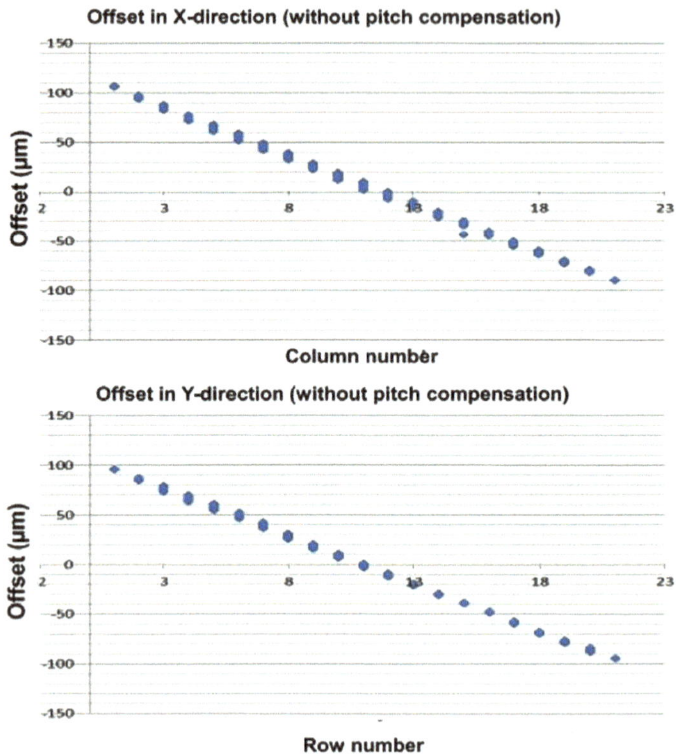

Fig. 4.68 Position offset data: x- and y-directions (without pitch compensation)

out that even though the experiments are carried out in square carrier, the results are also applied to circular carrier.

(D) Compression Molding and PMC

After EMC dispensing, it is followed by the compression molding as shown in Figs. 4.66e and 4.72. The keep-out zone method is used, and the mold cap diameter is 295 mm (2.5 mm clearance from the 300 mm glass carrier wafer edge). The glass carrier with dies attached is fixed onto the top mold chase by vacuum as shown in Fig. 4.72. The dispensed EMC on a releasing film is placed on the bottom plunger. After a few experiments, the optimal compression molding parameters are: temperature = 125 °C, pressure = 45 kg/cm^2, time = 10 min, and removing trapped air before compression molding. It is followed by post-mold cure (PMC) with a temperature = 150 °C, time $\geq$ 60 min, and a dead weight = 15 kg for better warpage control. The compression molding is performed by the high-coplanarity accuracy machine over-molding of resin epoxy compression auto-panel encapsulation system.

Inspections for molding voids are carried out by C-mode scanning acoustic microscopy. In order to balance the resolution and signal penetration depth, a transducer of 75 MHz is selected for the void's observation. Before compression molding

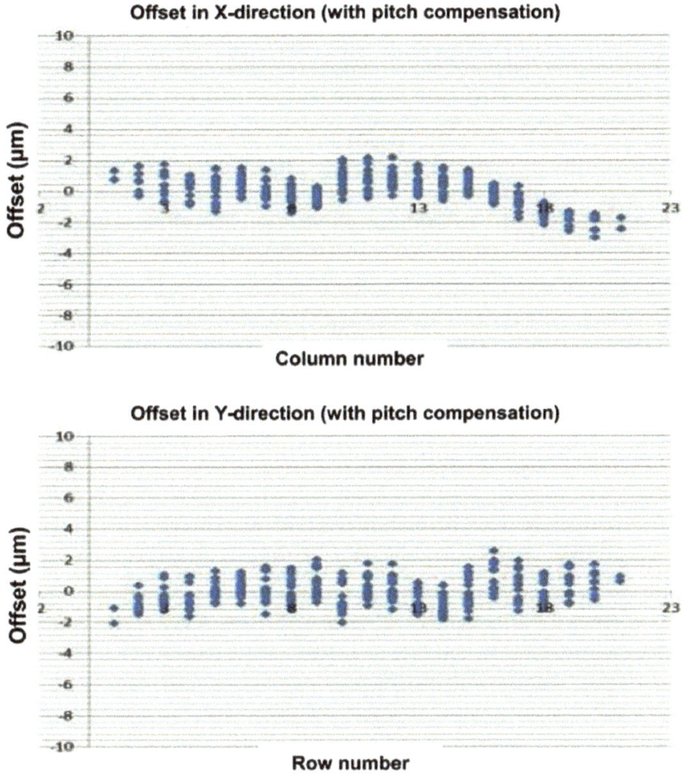

Fig. 4.69 Position offset data: x- and y-directions (with pitch compensation)

Table 4.5 Material properties of the EMC

Items	Units	R4507
Filler content	%	85
Filler to cut	μm	25
Filler average size	μm	8
Specific gravity		1.96
Viscosity	Pa.s	250
Flexural modulus	GPa	19
Tg (DMA)	°C	150
CTE1	ppm/K	10
CTE2	ppm/K	41
Application		Fan-out
Special feature		Fine filler

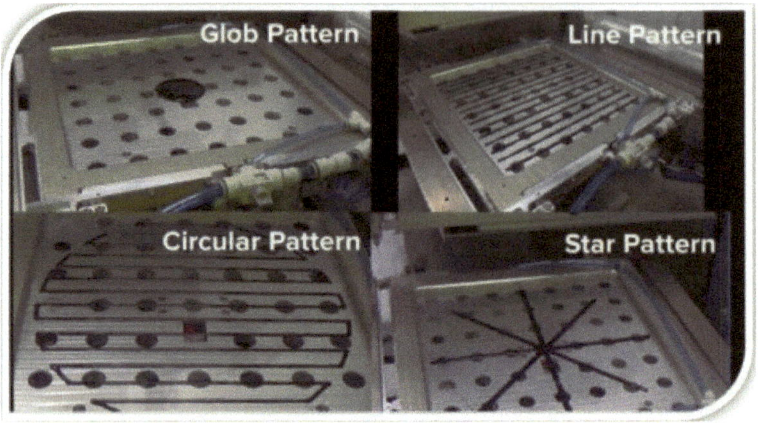

Fig. 4.70 Different liquid dispensing patterns

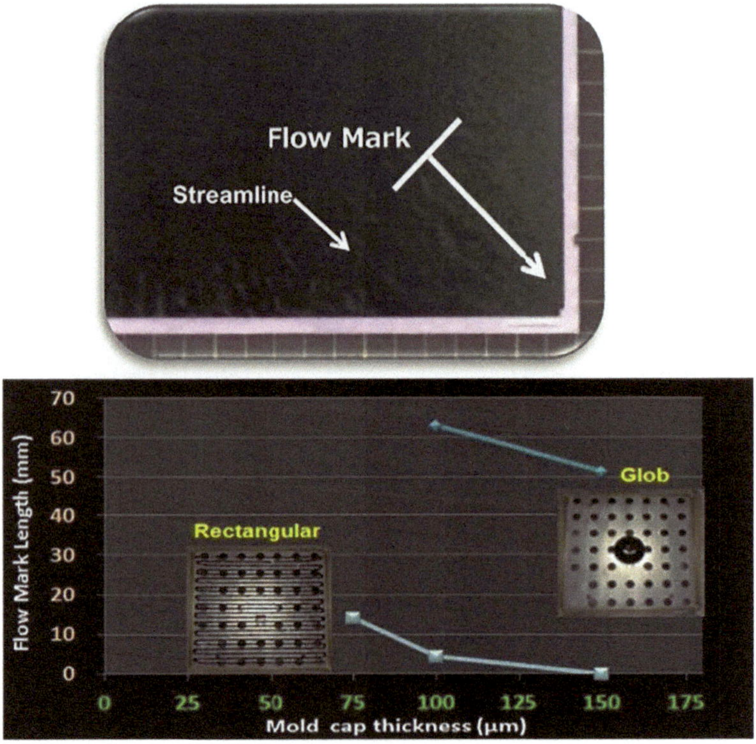

Fig. 4.71 Glob versus rectangular dispensing pattern. Flow mark point of measurement

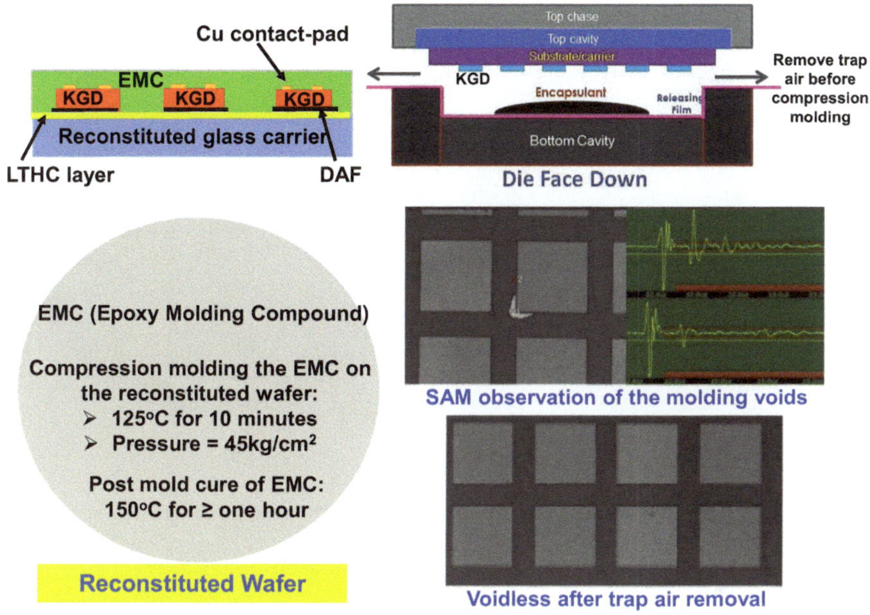

Fig. 4.72 Compression molding and PMC of EMC

parameter optimization, around 10 voids are detected (especially located at the die edge) in each of the molded wafers as shown in Fig. 4.72. After optimization, the voids are eliminated.

Figure 4.73a shows the top side of the molded test packages on the glass reconstituted wafer with the LTHC layer. In order to see the test chips, some of the reconstituted carrier wafers are without the LTHC layer, such as the one shown in Fig. 4.73b. In this case, we can determine the die shift due to compression molding by measuring the position of each chip before and after molding. Figure 4.74 shows the statistical plots of the x-position die shift and y-position die shift caused by the compression molding. It can be seen that because of the DAF (which solidly holds the chip to the carrier), the die shift (can be controlled within ±4 μm) is too small to be an issue on making the RDLs.

(E) Warpages

The total thickness variations (TTVs) of a few typical molded reconstituted wafers after PMC are given in Table 4.6. This is a measurement of eight points around the peripheral and one point at the center of the reconstituted wafer. It can be seen from Table 4.6 that the TTV is about 10 μm. The warpages of several reconstituted wafers after PMC are measured by the shadow Moiré method. The results are shown in the table of Fig. 4.75. It can be seen that the maximum average warpage for the glass coefficient of thermal expansion (CTE) = 3.25×10^{-6}/°C is 4.003 mm (0.7 mm thick), CTE = 7.6×10^{-6}/°C is 0.38 mm (1 mm thick), and CTE = 8.1×10^{-6}/°C is

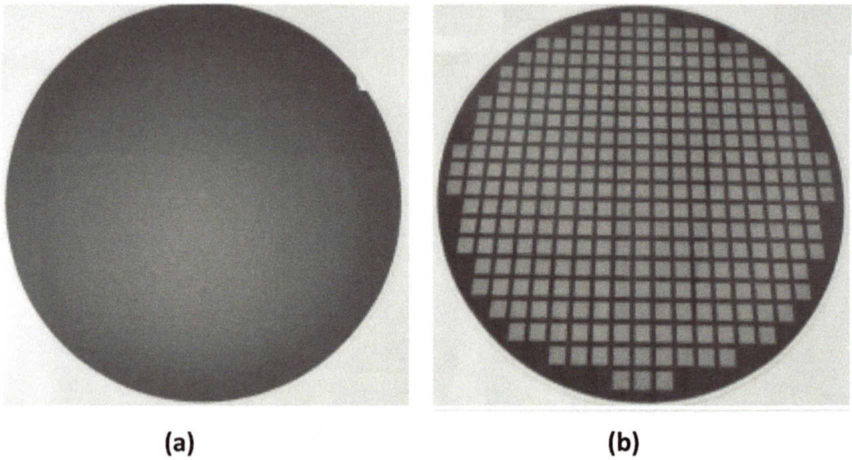

Fig. 4.73 **a** Top view of a molded reconstituted wafer. **b** Bottom view of a molded reconstituted wafer without the LTHC layer so we can see the test chips

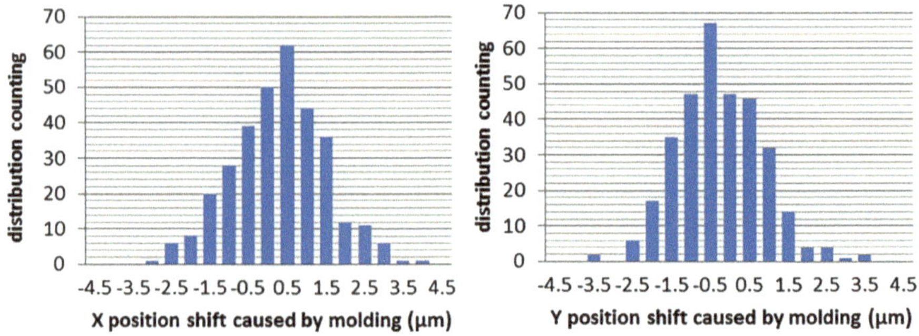

Fig. 4.74 Compression molding induced die-shift distribution data. x-Direction (left). y-Direction (right)

0.25 mm (1 mm thick). It is noted that the CTE of the EMC is 10×10^{-6}/°C. Thus, as expected, the smaller the CTE mismatch between the glass carrier and the EMC, the smaller the warpage.

Table 4.6 TTV of typical molded reconstituted wafers

Position	Wafer-1	Wafer-2	Wafer-3
1	1301	1301	1299
2	1307	1307	1300
3	1308	1308	1303
4	1305	1310	1305
5	1310	1312	1309
6	1307	1312	1307
7	1305	1312	1310
8	1306	1302	1307
9	1309	1310	1306
Range (μm)	9	11	11

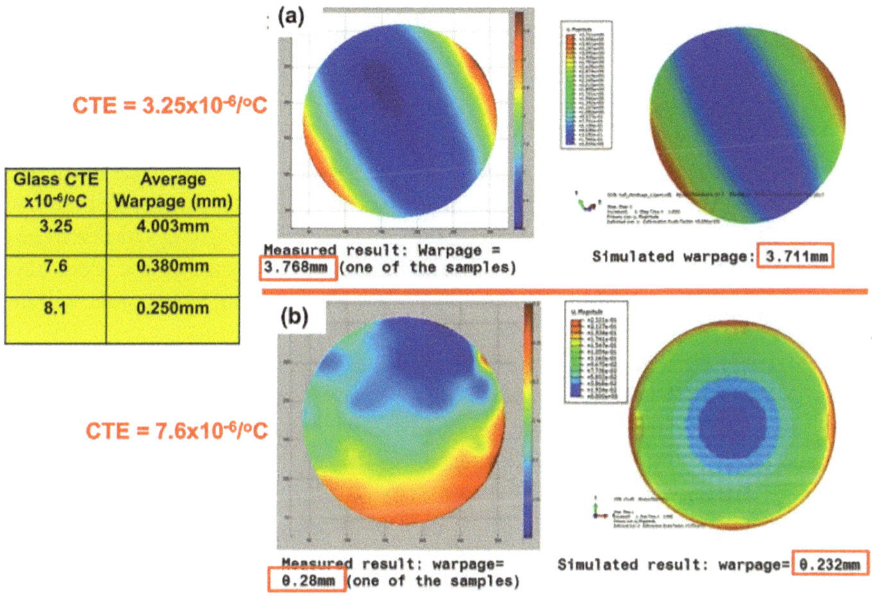

Fig. 4.75 Measured warpage and simulation contours of the molded wafer (after PMC) with **a** carrier CTE = 3.25 × 10^{-6}/°C and **b** carrier CTE = 7.6 × 10^{-6}/°C

Finite-element simulation results for the glass with CTE = 3.25 × 10^{-6}/°C (0.7 mm thick) and the glass with CTE = 7.6 × 10^{-6}/°C (1 mm thick) are reported, respectively, in Fig. 4.75a, b. The measurement results for these two cases are also reported in Fig. 4.75a, b. It can be seen that the measurement and simulation results compare very well.

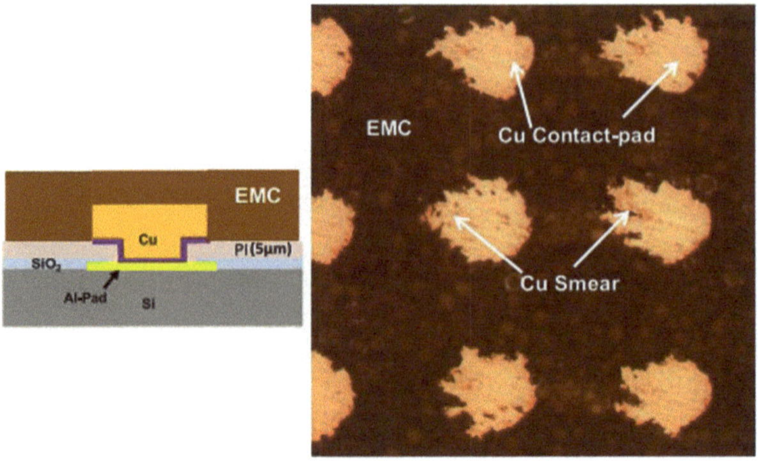

Fig. 4.76 Cu smear during backgrinding of the EMC to expose the Cu contact pad

(F) Cu Revealing

For FOWLP with chip-first and die face-up process, in order to make the RDLs and then mount the solder balls, the molded EMC above the Cu contact pad must be removed (Cu revealing) as shown in Fig. 4.66f. In this section, DISCO's backgrinding machine is used to remove the EMC. During backgrinding, some of the areas have Cu smearing as shown in Fig. 4.76. Because of the Cu smearing, the stepper will not work and the RDLs cannot be fabricated.

Another test chip is fabricated. This time the polymer is much taller (30 μm versus 5 μm, the first time shown in Fig. 4.63) and covers the Cu contact pad as shown in Fig. 4.77. After pick and place, EMC dispensing, compression molding, PMC, and backgrinding, the resulting surface of the reconstituted structure is shown in Fig. 4.77. It can be seen that because of the polymer, there is no Cu smearing, and the Cu contact pad is ready for the fabrication of the RDLs.

(G) RDLs

The key process steps in fabricating the RDLs of the FOWLP are shown in Fig. 4.78. It can be seen that for the PI development, the whole reconstituted wafer is spin-coated with a photosensitive PI. It is followed by applying a stepper (every four test packages as a unit), and then uses photolithography techniques to align, expose, and develop the vias of the PI. Finally, cure the PI at 200 °C for 1 h. This will form a 5 μm-thick PI layer (DL1). It is followed by sputtering Ti and Cu by PVD at 175 °C over the entire reconstituted wafer. Then, apply a photoresist and a stepper and then use photolithography techniques to open the redistribution trace's locations.

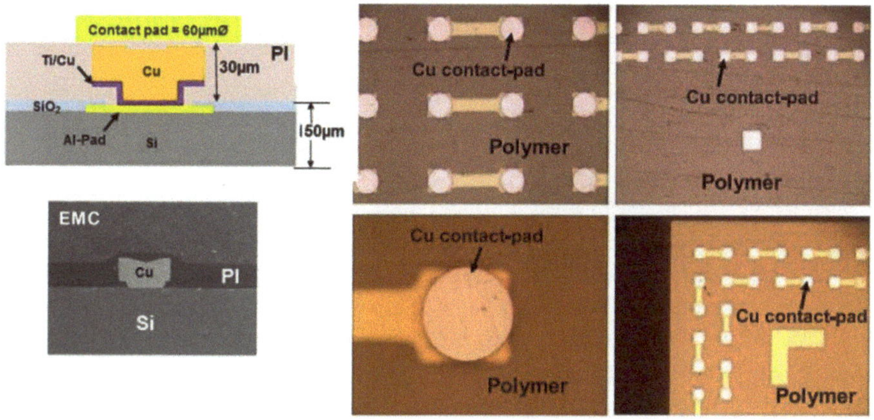

Fig. 4.77 (Left) Test chip with tall polymer (~30 µm) to cover the Cu contact pad and chip surface. (right) After backgrinding the EMC to expose the Cu contact pads (Cu revealing)

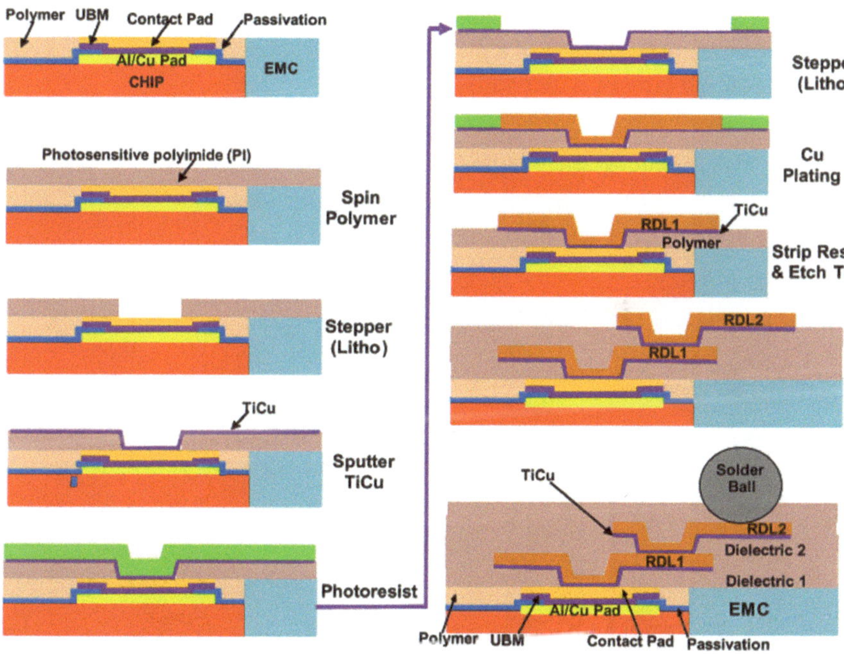

Fig. 4.78 Key process steps in fabricating the RDLs of FOWLP with UBM-less Cu pad for solder balls

Electroplate Cu by electrochemical deposition at room temperature on the Ti/Cu in photoresist openings. It is followed by stripping off the photoresist and etching off the Ti/Cu, then RDL1 is obtained. Repeat all the above steps to obtain DL2, RDL2 and DL3, RDL3. Finally, after the fourth PI development (DL4), then it is ready for solder ball mounting.

The thickness of DL1, DL2, and DL3 is 5 μm, and that of DL4 is 10 μm. The thickness of RDL1 and RDL2 is 3 μm, and that of RDL3 is 7.5 μm. (The thicker RDL3 is for the UBM-less Cu pads to "resist" the Cu consumption from the solder ball reflow and during operation.)

The RDLs are shown in Figs. 4.79 and 4.80. Overall, it can be seen that all the RDLs are properly done. However, a closer look at the RDL1, based on the top view as well as cross-sectional measurements, shows the actual fabricated linewidth/spacing is (on average) 4.68 μm/5.30 μm. This means that the linewidth is smaller than it should be while the line spacing is larger than the design values (5 μm/5 μm). These errors could be due to: (a) the small feature size of linewidth/spacing; (b) the photoresist opening; and (c) the seed layer (Ti) etching. For RDL2 and RDL3, they are very close to the designed values.

Solder Ball Mounting: There are two different stencils for the solder ball mounting: One is for stencil printing the flux, and the other is for stencil mounting the solder balls. The solder (Sn3wt%Ag0.5wt%Cu) balls (200 μm diameter) used are from Indium. The peak temperature for solder reflows is 245 °C. Figure 4.81a shows the reflowed reconstituted wafer, Fig. 4.81b shows the individual package, and Fig. 4.81c shows the close-up view of solder balls on the package.

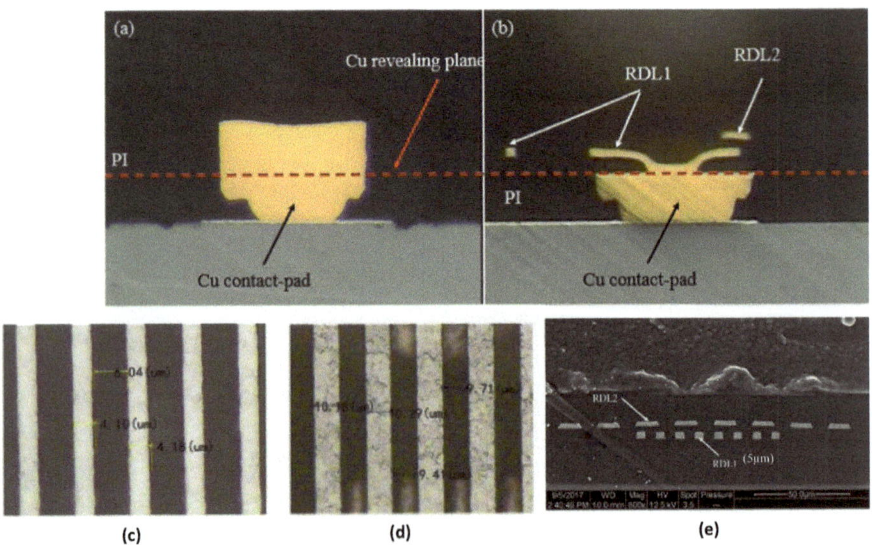

Fig. 4.79 **a** Before Cu revealing. **b** After Cu revealing and two RDLs. **c** RDL1 (5 μm/5 μm linewidth/spacing). **d** RDL2 (10 μm/10 μm linewidth/spacing). **e** Cross-sectional SEM images of RDL1 and RDL2

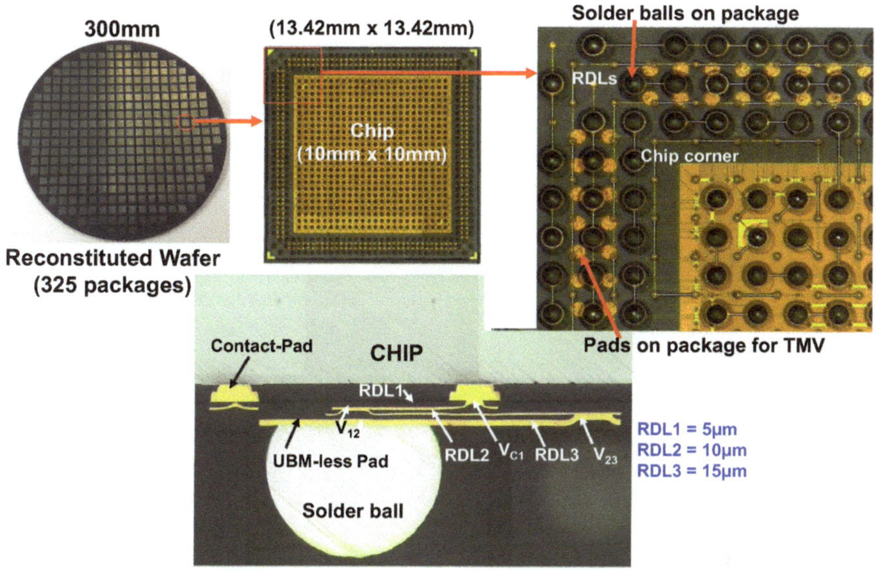

Fig. 4.80 Images of the assembled FOWLP with three RDLs

(H) Debonding

The debonding of the glass carrier is by scanning a laser (355 nm DPSS Nd:YAG UV laser source is used) from the glass carrier side. The laser spot size is 240 μm, the scanning speed is 500 mm/s, and the scanning pitch is 100 μm. When the LTHC layer "sees" the laser light, it converts into powders and the glass carrier is easily removed. It is followed by chemical cleaning. The DAF remains on the backside of the chip as shown in Fig. 4.82.

4.5.6 PCB Assembly

PCB Assembly The design of the PCB and stencil will be briefly mentioned. The stencil solder paste printing, pick and place, and solder reflow will also be presented. The X-ray images and the cross-section images of the PCB assembly are provided.

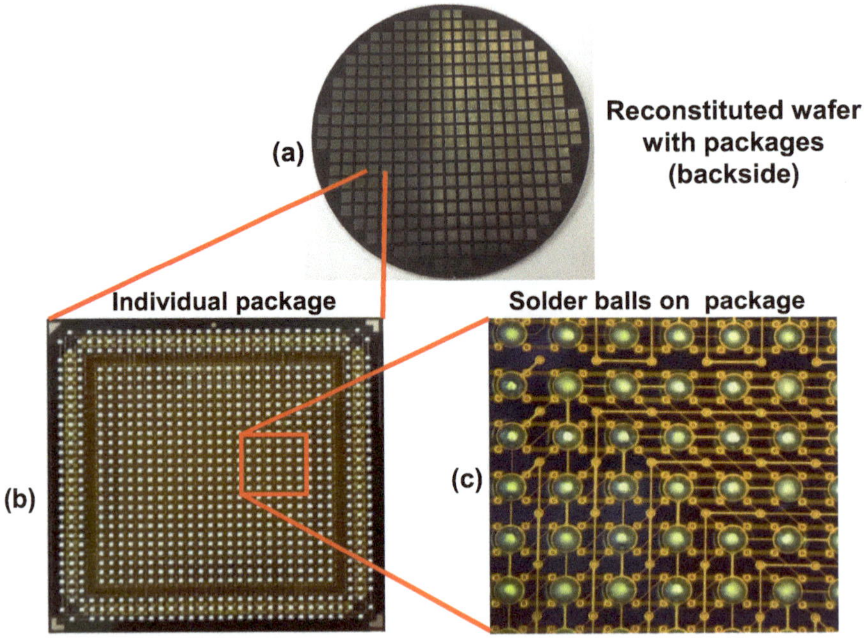

Fig. 4.81 **a** Image of the fabricated reconstituted wafer. **b** Individual FOWLP package. **c** Close-up view of the solder balls

Fig. 4.82 Reconstituted wafer after laser debonding

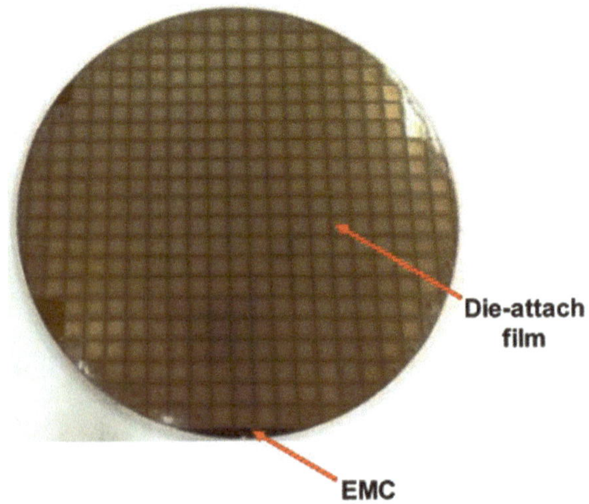

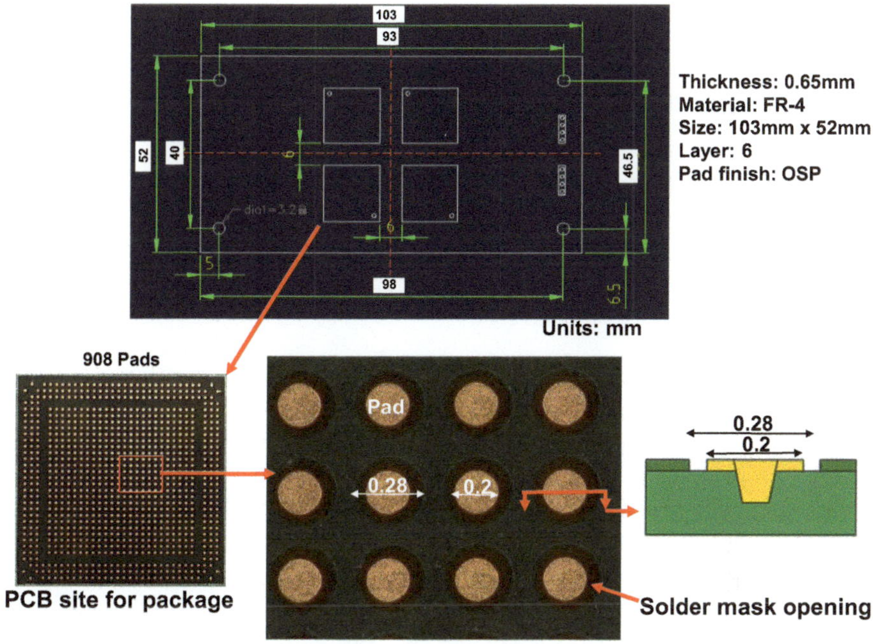

Fig. 4.83 PCB for package assembly and reliability tests

(A) PCB

The PCB for the fan-out panel-level package is made of FR-4 and is shown in Fig. 4.83. It can be seen that there are 4 package sites on the board. The dimensions of the PCB are 103 mm × 52 mm × 0.65 mm and there are 6 layers. There are 908 pads (with a pitch = 0.4 mm) for each package. The pad with a diameter = 0.2 mm is non-solder mask defined. The solder mask opening diameter is 0.28 mm.

(B) Stencil and Printing

The stencil is made of stainless steel with a grain size of 2 μm (which will benefit for the solder paste transfer) and is 0.08 mm-thick. The opening is fabricated by laser and electrical/chemical polishing. The opening is 0.2413 mm square with an aspect ratio = 0.754. The solder paste printing is by the DEK Horizon9.

(C) Pick and Place and Reflow

The pick and place are by the SiPlace x4s. The 10 temperature zones BYU Pyram nitrogen 150 N is used for the reflow. The temperature profile is shown in Fig. 4.84. It can be seen that the maximum temperature is 245 °C and the time above 217 °C is 85 s (it meets the JSTD-020 standard). Figure 4.85a shows the package PCB assembly without underfill. There are 4 packages on a PCB. Figure 4.85b shows an X-ray image of the PCB assembly. Figure 4.85c, d show the cross section of one of

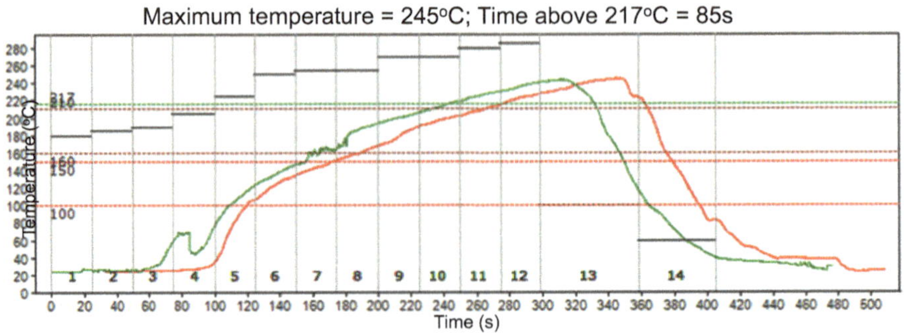

Fig. 4.84 Reflow temperature profile

the packages and the solder joint height measurement. It can be seen that the solder joints are properly made (no bridging and head-in-pillow) and the variation of solder joints height in three different cross sections is from 120 to 145 μm.

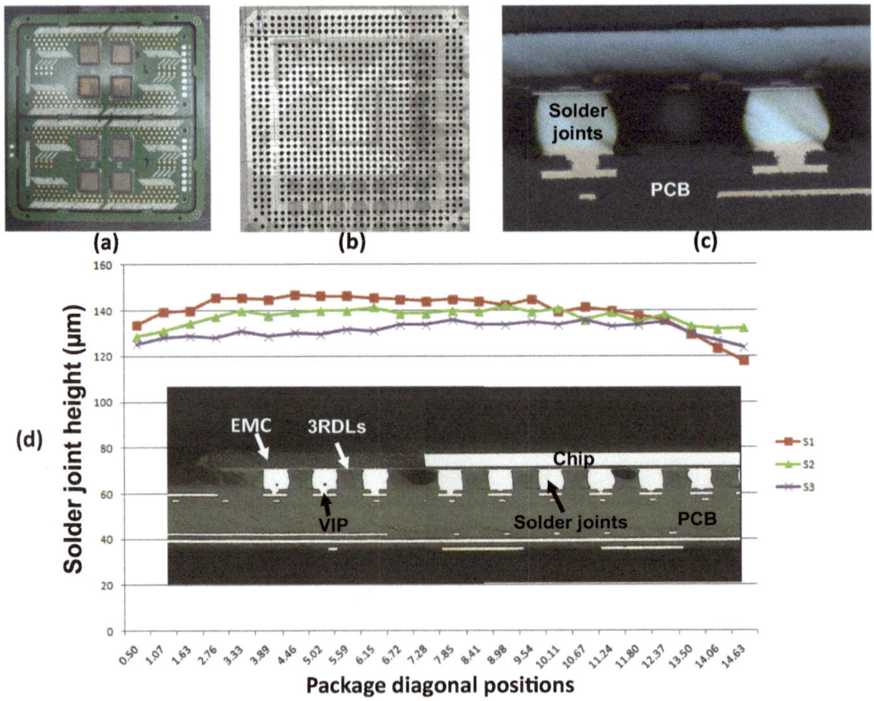

Fig. 4.85 Cross-section of PCB assembly and Solder joint height measurements

4.5.7 *Thermal Cycling Test and Results*

(A) Test Setup

Fifteen boards (each with 4 packages) are used for the temperature cycling tests. The sample size is 60 packages. Thermal cycling is performed in a Votsch 7027-15 environmental chamber. This unit is capable of achieving chamber temperatures as high as 190 °C and as low as −70 °C. Heating and cooling are achieved by forced convection, and the maximum ramp rate is 15 °C per minute. All of the boards are placed vertically in the chamber as shown in Fig. 4.86. The temperature input to the chamber (measured in the air of the chamber) is shown in Fig. 4.86. It can be seen that the temperature cycling is from room temperature to 85 °C and stay there for 15 min, then ramp down to −40 °C and stay there for 15 min, then ramp up to 85 °C and stay there for 15 min, and so for. The ramp up and ramp down times are 15 min each. The acquisition system is an Agilent 30970A data logger.

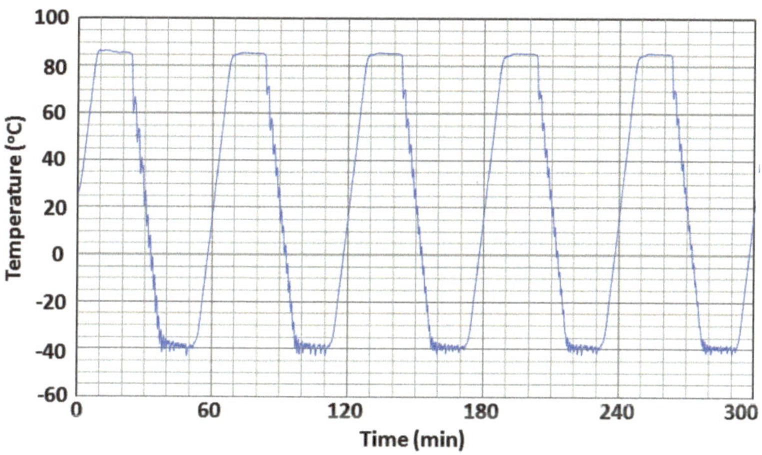

Fig. 4.86 Thermal cycling setup and profile

(B) Test Result

The thermal cycling test results of the FOWLP solder joint (without underfill) reliability are shown in Fig. 4.87. The thermal cycling test stops at 1,100 cycles and there are 14 failures (including one early failed at 58 cycles). It can be seen that the characteristic life (63.2% failed) of the Weibull plot is 2,382 cycles which is more than adequate for the expected life (usually is less than 3 years) of mobile products such as the smartphones and tablets. The failure mode is shown in Fig. 4.88. It can be seen that the solder joint cracks at the interface between the bulk solder and the RDL3.

4.5.8 *Drop Test and Results*

(A) Test Setup

The test setup is according to JEDEC Standard JESD22-B111 as shown in Fig. 4.89. After more than 20 tries, the right height of the drop table is obtained which yields the drop spectrum with 1500 G/ms as shown in Fig. 4.90.

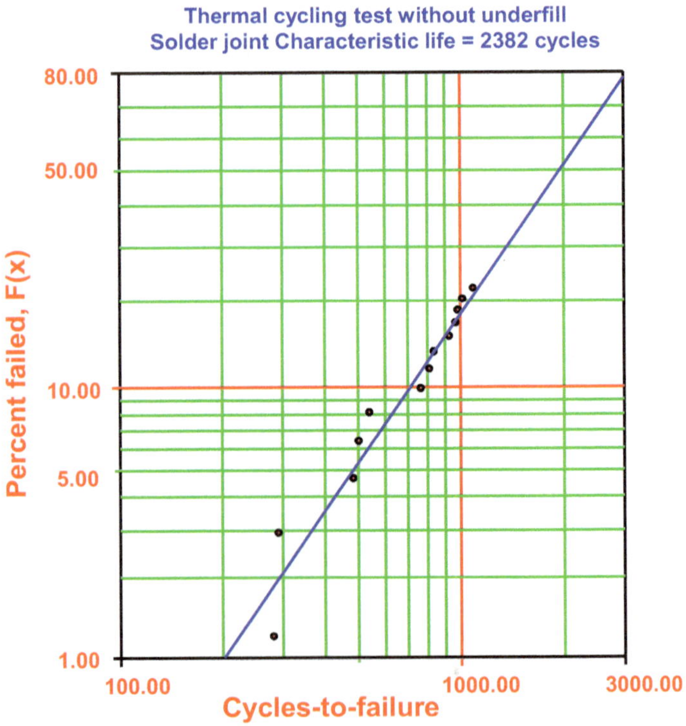

Fig. 4.87 Weibull plot of the solder joints under thermal cycling test

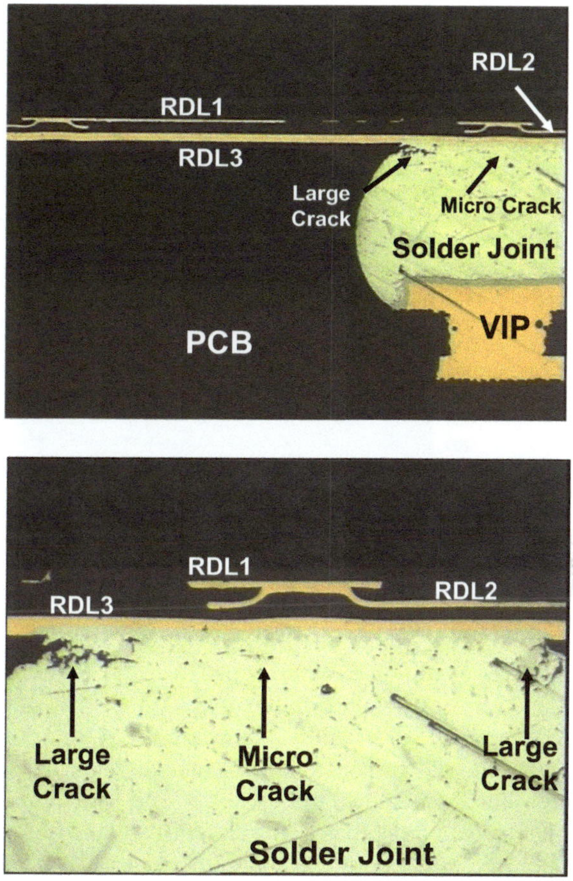

Fig. 4.88 Failure mode of solder joint under thermal cycling test (close-up in the bottom)

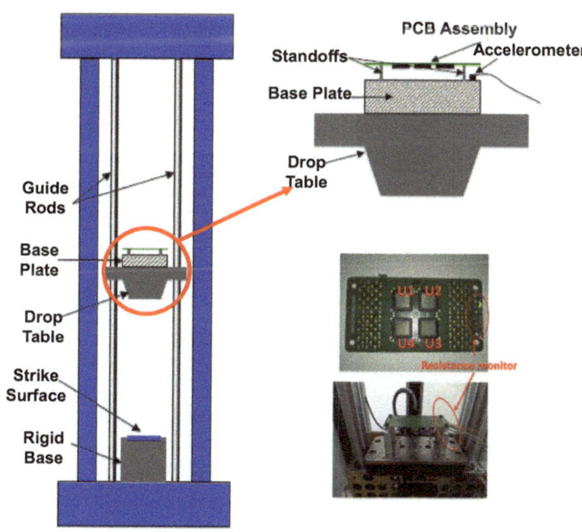

Fig. 4.89 Drop test setup

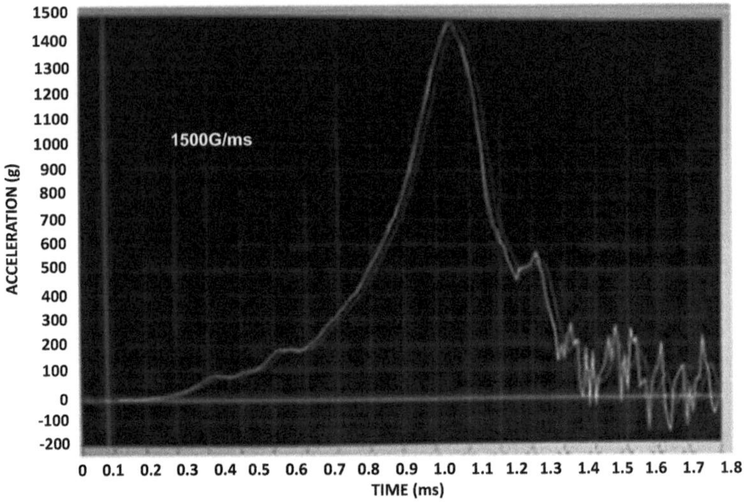

Fig. 4.90 Drop test acceleration profile

(B) Test Results

The drop condition is 1,000 drops. There are 24 samples. The ones without underfill failed very early and the failure mode is the break of the Cu conductor wiring of the RDL near the solder joint as shown in Fig. 4.91.

Another 24 PCB assemblies (samples) are made and underfilled and a typical cross section is shown in Fig. 4.92a. The X-ray images are shown in Fig. 4.92b, c. The material properties of the underfill are: the filler content = 25%, the maximum filler size = 5 μm, the average filler size = 1–2 μm, the curing time and curing temperature = 8 min @ 135 °C or 5 min @ 150 °C. The Young's modulus, Poisson's ratio, and CTE are respectively, 4–5 GPa, 0.35, and 50–52 × 10^{-6}/°C. The drop condition is also 1,000 drops. The results are shown in Fig. 4.93. It can be seen that the characteristic life of the RDLs under drop test is 1,187 drops and all 24 samples passed 500 drops without failure. Thus, the RDLs of the package should be reliable under drop for mobile applications. The first failure occurs after 550 drop and the failure modes are shown in Fig. 4.94. It can be seen that the RDLs are broken. In this case, the EMC also has cracks. However, the solder joint is not failed.

4.5.9 Thermal Cycling Simulation

(A) The Structure and Boundary Conditions

The PCB assembly of the fan-out wafer-level package shown in Fig. 4.85 is modeled as a three-dimensional strip that captures the construction along a diagonal path from the assembly's geometry center to a corner (Fig. 4.95). Due to the symmetry about the

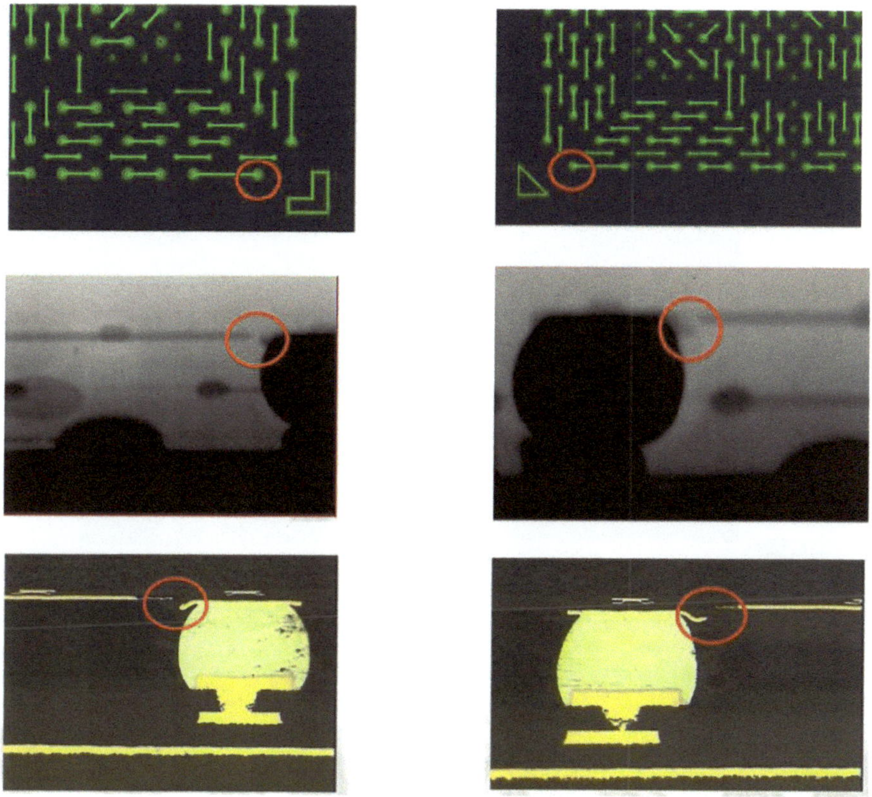

Fig. 4.91 Failure modes of assembly without underfill

vertical midplane of a full strip, the model is actually a half-strip with the appropriate in-place restraints placed on one symmetry plane. Coupled in-plane translations are applied to the other symmetry plane to produce a state of generalized plane strain. Using exclusively hexahedral solid elements, the model can capture the precise shape of the packages' solder joint and potential DNP (distance to neutral point) [48] effects while retaining significant computational efficiency over full octant models.

Despite the overall economy of elements in the strip model, selective mesh refinement is used to concentrate highly refined elements in the solder joints where failure is anticipated. In the present PCB assembly, failure would be predicted in the solder joints with the greatest DNP (the package corner) and near the chip corners as shown in Fig. 4.95. Thus, highly refined meshes are applied to these two solder joints. The other solder joints are coarsely meshed. The C3D8R element of ABAQUS is used for the model.

(B) Material Properties for Thermal Cycling

The material properties of the PCB assembly shown in Fig. 4.92 are shown in Table 4.7. All the material properties are assumed to be constant except for those

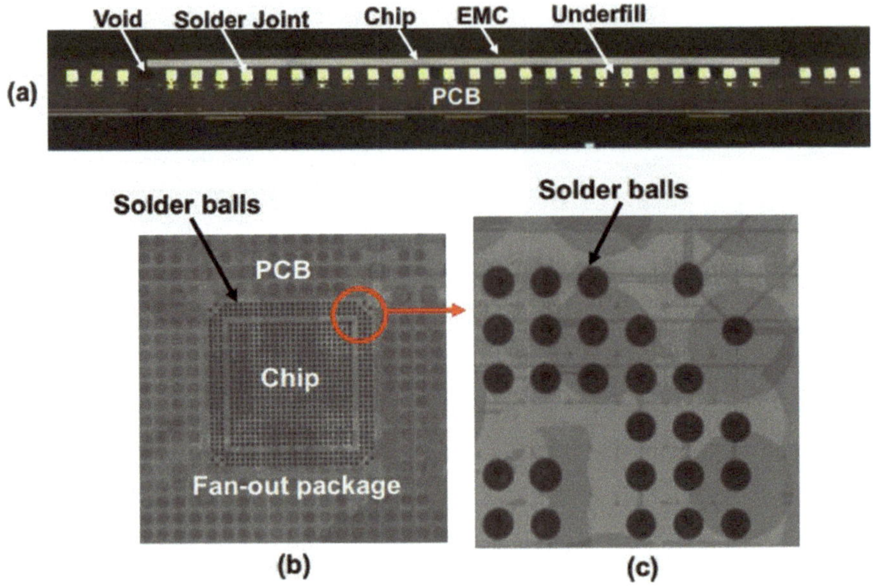

Fig. 4.92 **a** Cross section of the assembled PCB. **b** X-ray of a package assembly. **c** Close-up of a package assembly

Fig. 4.93 Weibull plot of PCB assembly with underfill under drop test

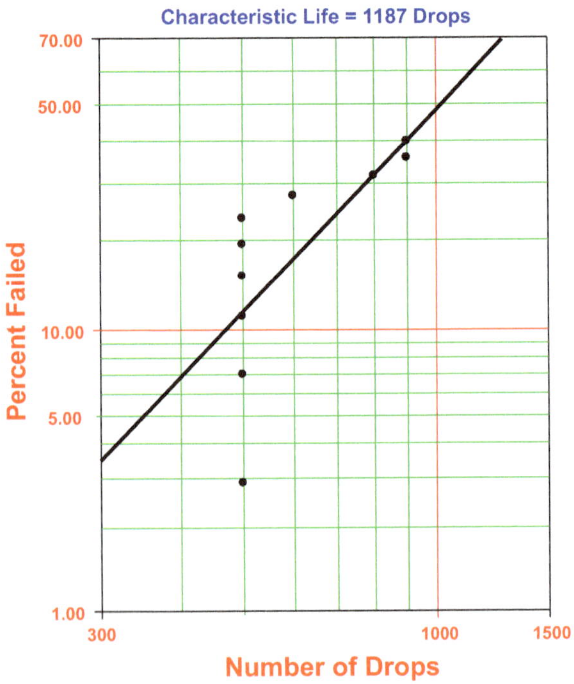

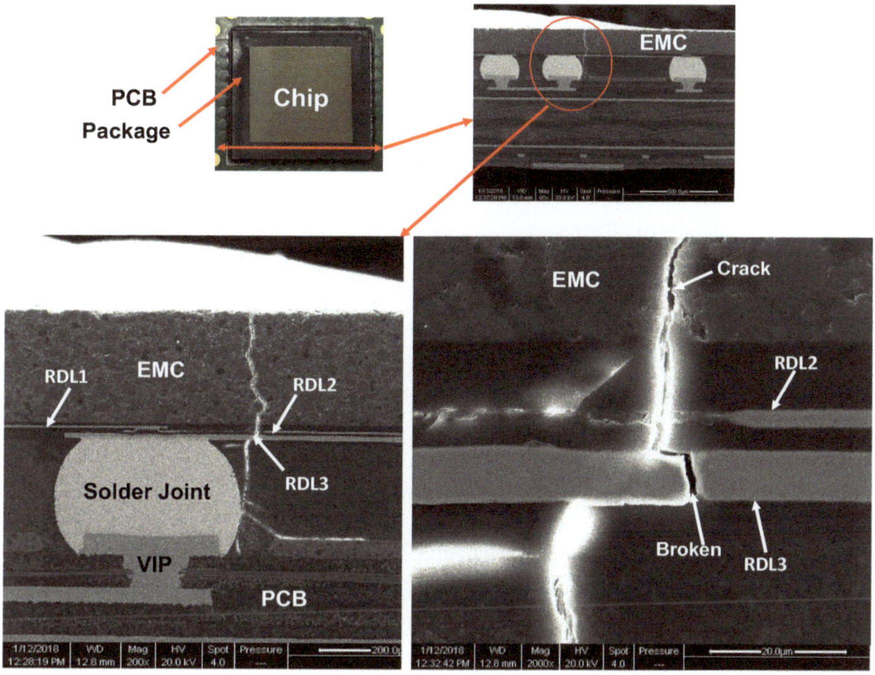

Fig. 4.94 PCB assembly for failure analysis (after 550 drops). Failure modes showing RDLs broken and EMC cracking

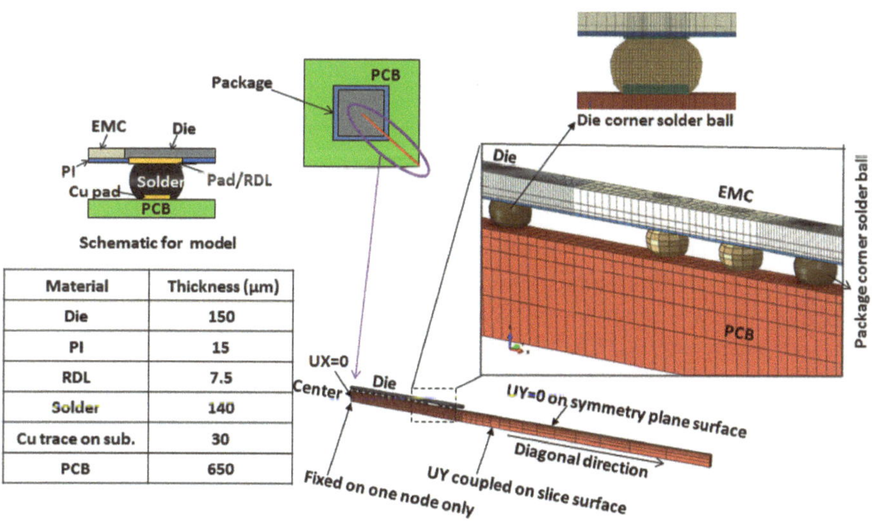

Material	Thickness (μm)
Die	150
PI	15
RDL	7.5
Solder	140
Cu trace on sub.	30
PCB	650

Fig. 4.95 The 3D strip model for thermal analysis

Table 4.7 Material properties for thermal cycling modeling

	CTE (ppm/°C)	Young's modulus (GPa)	Poisson's ratio
Copper	16.3	121	0.34
PCB	$\alpha_1 = \alpha_2 = 18, \alpha_3 = 70$	$E_1 = E_2 = 22, E_3 = 10$	0.28
Silicon	2.8	131	0.278
Solder	21 + 0.017 T (°C)	49 – 0.07 T (°C)	0.3
Polyimide	35	3.3	0.3
EMC	10 (<150 °C)	19	0.25

of solder. The Sn3wt%Ag0.5wt%Cu is assumed to follow the generalized Garofalo creep equation [4]:

$$d\varepsilon/dt = 500,000\,\mathrm{Sinh}^5\left(\sigma/1\times10^8\right)\,\exp(-5807/T(K))$$

where ε is the strain, σ is the stress in Pa, and T is the temperature in Kevin. The CTE and Young's modulus of the solder are respectively, 21.3 + 0.017 T and 49 – 0.07 T, and T is the temperature in Celsius.

(C) Temperature Boundary Condition

The temperature profile shown in Fig. 4.8 is to be imposed on the PCB assembly. Five temperature cycles are executed.

(D) Thermal Cycling Simulation Results

Figure 4.96 shows the shear stress and creep shear strain hysteresis loops at the solder joint under the chip corner. It can be seen that after three cycles, the creep responses converged (become stabilized).

The maximum Mises stress and accumulated creep strain occur at the solder joints near the chip corner and the package corner as shown in Figs. 4.97 and 4.98. The location is at the interface between the bottom of the package and the bulk solder. Thus, any failure should occur at this location. This correlates very well with thermal cycling test failure mode as shown in Fig. 4.88. The maximum Mises stress (Fig. 4.98) occurs at −40 °C during the thermal cycling condition shown in Fig. 4.86. The maximum deformed shape of the strip model at −40 °C is shown in the bottom of Fig. 4.98.

The accumulated creep strain vs. time is shown in Fig. 4.99. It can be seen that the creep strain per cycle is equal to 0.0084, which is too small to create solder joint reliability problem, especially for mobile products such as smartphones, which life is less than three years.

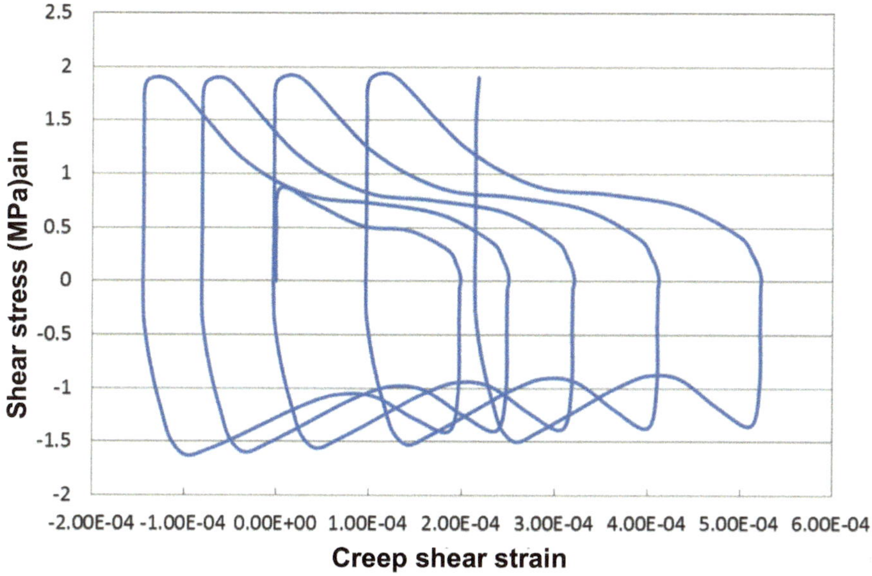

Fig. 4.96 Shear stress/creep shear strain hysteresis loops

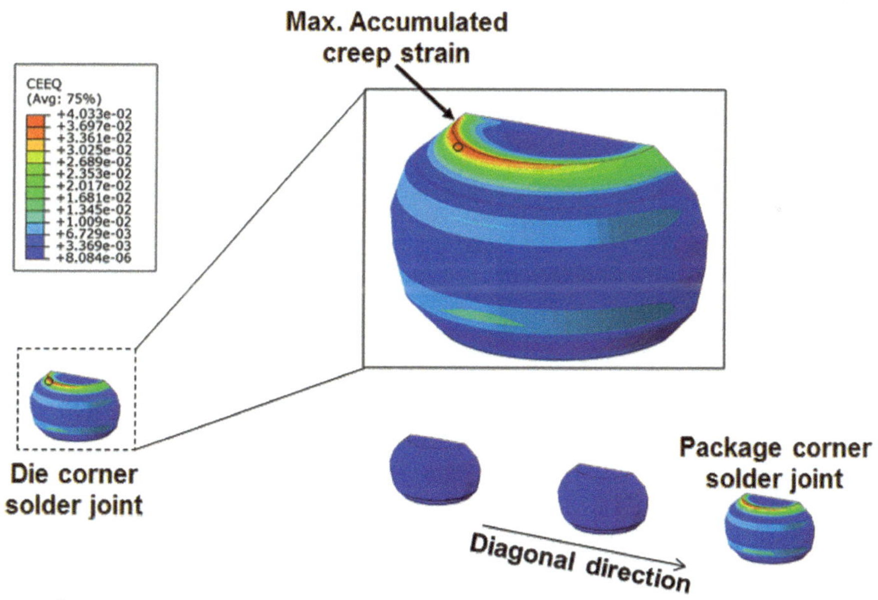

Fig. 4.97 Maximum accumulated creep strain

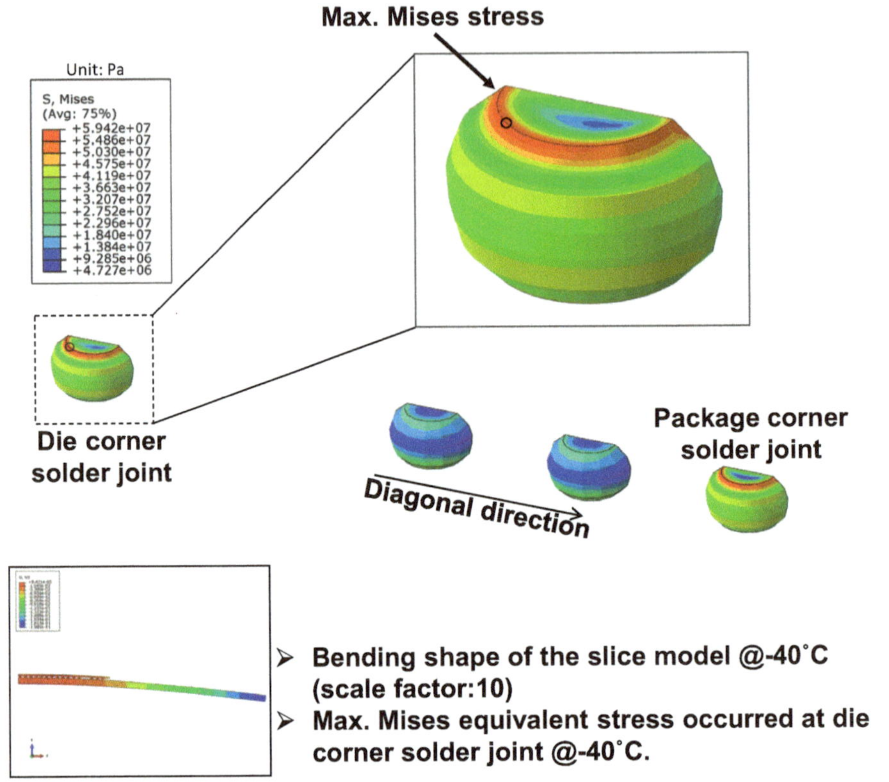

Fig. 4.98 (Top) Maximum mises stress at chip corner and package corner solder joints at −40 °C. (Bottom) The deformed shape of the quarter structure at −40 °C

4.5.10 Shock (Drop) Simulation

(A) Structure and Boundary Conditions

The finite element model for drop test with underfill is shown in Fig. 4.100. It can be seen that one-fourth of the PCB assembly is modeled. For simplicity, there is only one package at the center of the PCB and the package is subjected to the acceleration shown in Fig. 4.90.

During impact, the most likely failure location is the RDLs near the package corner and the most likely failure mode is the broken Cu wire in the RDLs and broken Cu trace in the PCB. Thus, finer meshes are applied to the Cu wire of RDLs and Cu trace of PCB near the package corner as shown in Fig. 4.101.

(B) Material Properties for Drop

The material properties for drop analysis are shown in Table 4.8. Since the focus of this analysis is not on the solder joint, all the materials are assumed to be constant.

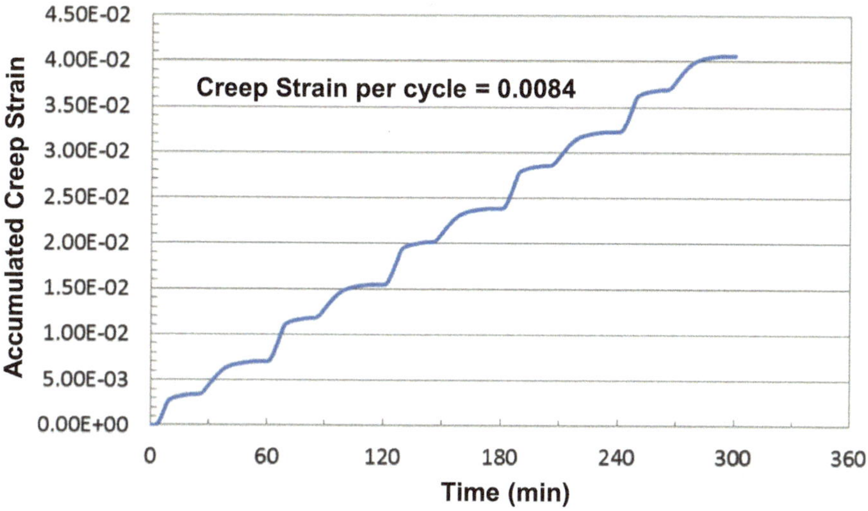

Fig. 4.99 Accumulated creep strain versus time

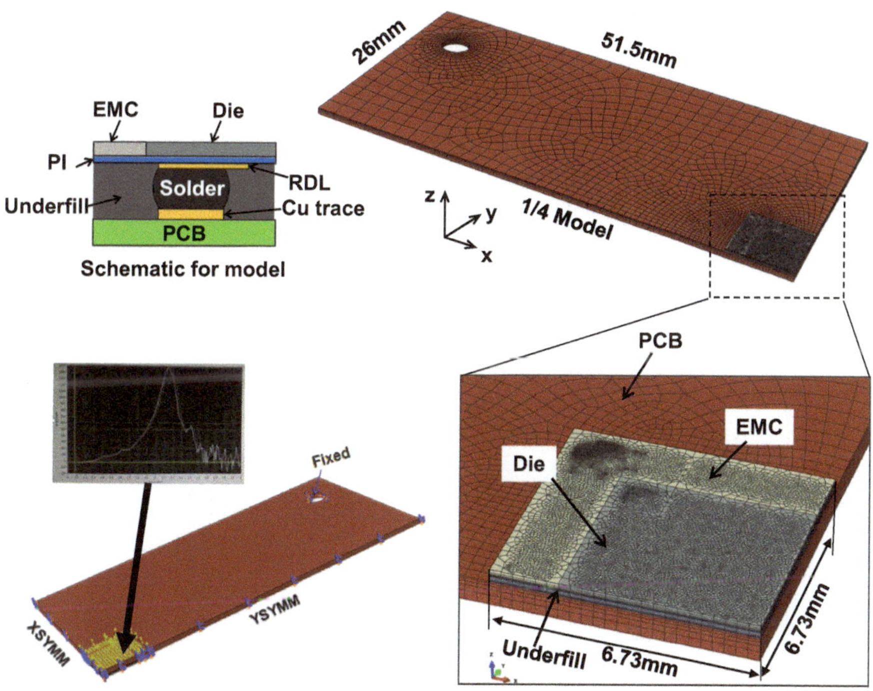

Fig. 4.100 Quarter model for drop analysis

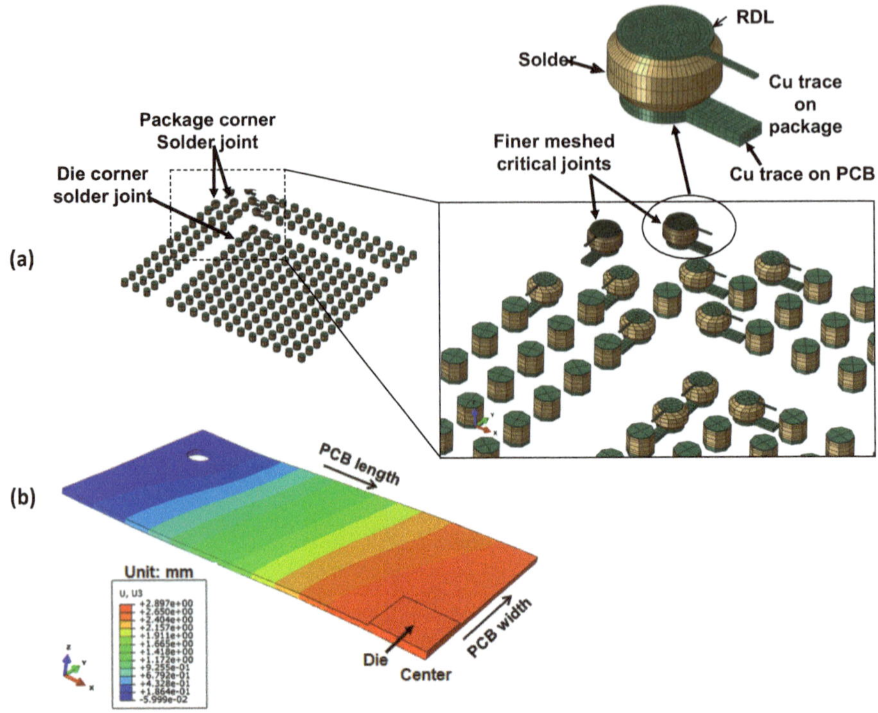

Fig. 4.101 **a** Mesh at corner solder joint, RDL on package, and Cu trace on PCB. **b** Deformed shape of the quarter of PCB assembly

Table 4.8 Material properties for drop modeling

Items	Density (kg/m^3)	Young's modulus (GPa)	Poisson's ratio
Copper	8954	121	0.34
PCB	1800	$E_1 = E_2 = 22$, $E_3 = 10$	0.28
Silicon	2330	131	0.278
Solder	7400	49	0.3
Underfill	1560	4.5	0.35
Polyimide	1420	3.3	0.3
EMC	1960	19	0.25

Since the ABAQUS explicit solver is applied for the simulation. Thus, the damping effect is included through the bulk viscosity method in ABAQUS explicit solver. The bulk viscosity introduces damping associated with volumetric straining. The damping coefficients b_1 and b_2 applied in the analysis are respectively, 0.02 and 0.

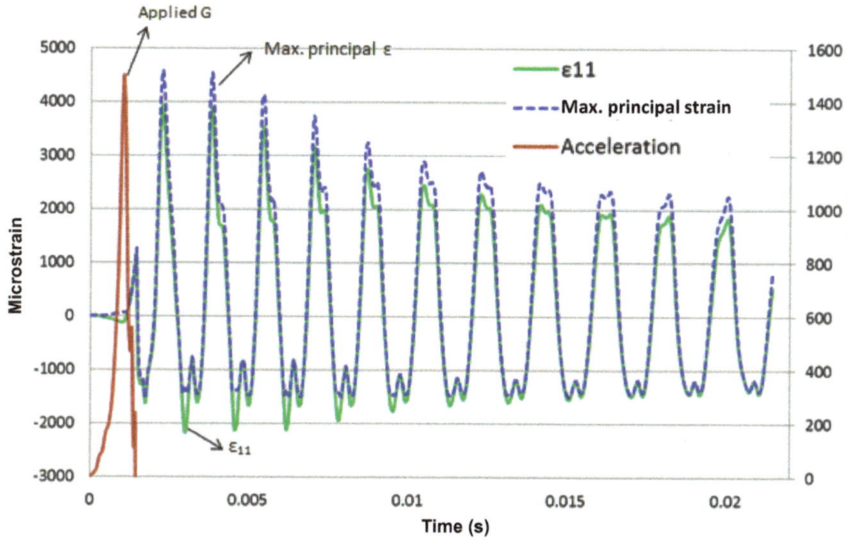

Fig. 4.102 Applied acceleration. Time-history maximum principal strain and ε_{11} at the RDL of the package

(C) Drop Simulation Results

The applied acceleration is shown in Figs. 4.90 and 4.102. The time step for the calculation is 1×10^{-8} s. The time-histories of strain in the x-direction, ε_{11} (Fig. 4.100) and maximum principal strain acting at the RDL near the package corner are shown in Fig. 4.102. Basically, they are the same, i.e., the strain is dominated in the x-direction. The maximum deflection of the PCB assembly occurs near the 0.0023 s (Fig. 4.102) and is shown in Fig. 4.101b. The maximum principal strain = 0.0045 also occurred at that time. The deflection and strains are decreasing with time due to the damping of the structure as shown in Fig. 4.102.

The maximum Mises stress occurs at the same location (at the Cu wiring of RDL near the package corner) as the maximum principal strain as shown in Fig. 4.103 and is equals to 7.724×10^8 Pa. Thus, any failure should occur at this location. This demonstrates the failure mode that the RDL is broken after the 550 drops test.

4.5.11 Summary

Some important results and recommendations are summarized in the following.

- The design, materials, process, fabrication, and reliability of FOWLP with chip-first and die face-up method have been investigated. The lessons learned from the problems and solutions during the process development have also been provided.

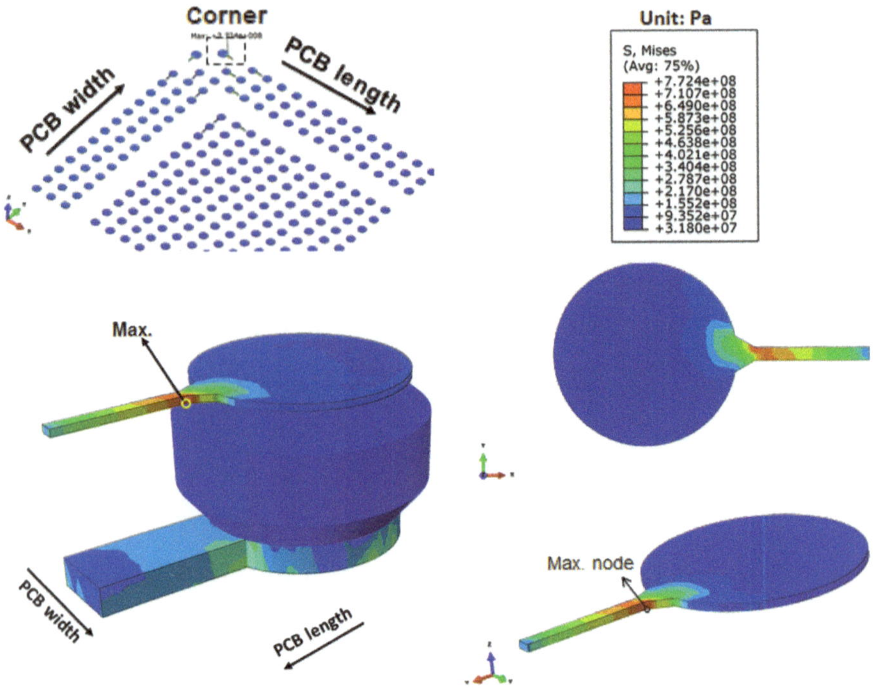

Fig. 4.103 Maximum Mises stress occurs at the RDL of the package

- A test package, 13.42 mm × 13.42 mm (with a package/chip ratio = 1.8 and a chip-top EMC cap = 100 μm), has been designed and fabricated. This test package has three RDLs: The linewidth/spacing of the first RDL is 5 μm/5 μm, the second RDL is 10 μm/10 μm, and the third RDL is 15 μm/15 μm.
- DAF helps! There is not any die shift issue for chip-first with die face-up FOWLP.
- The price we must pay is to do the P&P pitch compensation due to the thermal expansion mismatch between the silicon chip and the glass carrier wafer during the heating of the DAF (both bond-head and the bond-stage temperatures = 120 °C with the bond force of 2 kg for 2 s).
- For liquid EMC dispensing, in order to reduce the area and length of flow marks, the line pattern is recommended.
- Removal of trapped air before compression molding is a must, and the optimal conditions are: temperature = 125 °C, time = 2 min, and pressure = 45 kg/cm^2. The PMC is at 150 °C for ≥1 h, and a dead weight = 15 kg for better warpage control.
- For this study in order to have the minimum warpage (≤0.3 mm), the glass thickness is 1 mm and the glass thermal expansion coefficient is 7.6–8.1 × 10^{-6}/°C.

- During Cu revealing (backgrinding of the EMC to expose the Cu contact pads), there are Cu smears for the short (5 μm) PI, and there is no Cu smear for the tall (~30 μm) PI.
- The characteristic life of the FOWLP solder joints without underfill under thermal cycling test has been estimated to be 2,382 cycles. This is more than adequate for most mobile products such as smartphones and tablets, whose life is usually less than three years. The thermal-cycling condition is −40 °C ↔ 85 °C, one cycle per hours (the ramp up time and ramp down time are respectively, 15 min, and the dwell time at −40 °C and dwell time at 85 °C are respectively, 15 min). The sample size is 60.
- The failure location of the FOWLP solder joints without underfill under thermal cycling test has been found to be near the package corner and/or the chip corner, and the failure mode has been found to be cracked between the bulk solder and the RDL3.
- Creep of FOWLP solder joints under thermal cycling condition has been determined by non-linear and time dependent 3D finite element simulation. Five temperature cycles have been executed and the hysteresis loops have been shown that the creep behavior has been stabilized after the third temperature cycle. The maximum creep strain and Mises stress have been found to be occurred near the interface between the bulk solder and the RDL3 near the package corner solder joint and/or chip corner solder joint. This correlated very well with the experimental failure mode.
- The life of the FOWLP RDLs without underfill under drop test has been found to be very little. The Cu conductor wiring of RDLs have been broken under just a few drops. The drop setup and condition are according to the JEDEC Standard JESD22-B111 and the drop spectrum is with 1500 G/ms.
- Another 24 FOWLP PCB assemblies have been made with underfill and gone through the same drop test. All 24 samples passed 500 drops and the first RDL failure has been occurred after 550 drops (and all the solder joints are not failed). The characteristic life of the FOWLP RDLs with underfill under drop test has been estimated to be 1,271 drops. Again, this is more than adequate for most mobile products such as smartphones and tablets. The failure location of the FOWLP RDLs with underfill under drop test has been found to be in the Cu conductor wiring of RDL3 near the package corner, and the failure mode has been found to be broken of the Cu wiring in RDL3.
- Shock of FOWLP RDLs under the given drop condition has been determined by 3D dynamic time-history analysis. It has been found that due to the structural damping, the dynamic responses (stress and strain) are decreasing with time. The maximum stress and strain occurred at 0.0023 s after the impact. The location of the maximum stress and strain has been found to be at the Cu conductor wiring of RDL3 near the package corner. This correlated very well with the experimental failure mode.
- As expected, the DNP concept works for thermal cycling. The most outer two diagonal solder joints from all 4 package corners have been removed from the

present study. It would be even better if the solder joint near the 4 chip corners is removed.

4.6 Apple/TSMC's Application Processor with Chip-First (Die Face-Up) Process

Package-on-package (PoP) has been used for housing the application processor (AP) chipset in mobile phones for more than one decade as shown in Fig. 4.104. Usually, the top package is used to house the mobile memory and the bottom package is used to house the AP. In 2010, a joint industrial initiative between STATSChipPAC, ST Ericsson, and Infineon proposed a PoP for the AP chipset with the FOWLP technology and published in IEEE/ECTC2010 and 2012 proceedings [56, 57], Fig. 4.105.

During ECTC2016, TSMC presented two papers on FOWLP: one is their integrated fan-out (InFO) wafer-level packaging [58] for housing the most advanced AP for mobile applications [9], and the other is to compare the thermal and electrical performance between their InFO technology and the conventional flip chip on buildup package substrate technology [59]. In September 2016, TSMC put the PoP of AP chipset with FOWLP technology into high-volume manufacturing, Fig. 4.106.

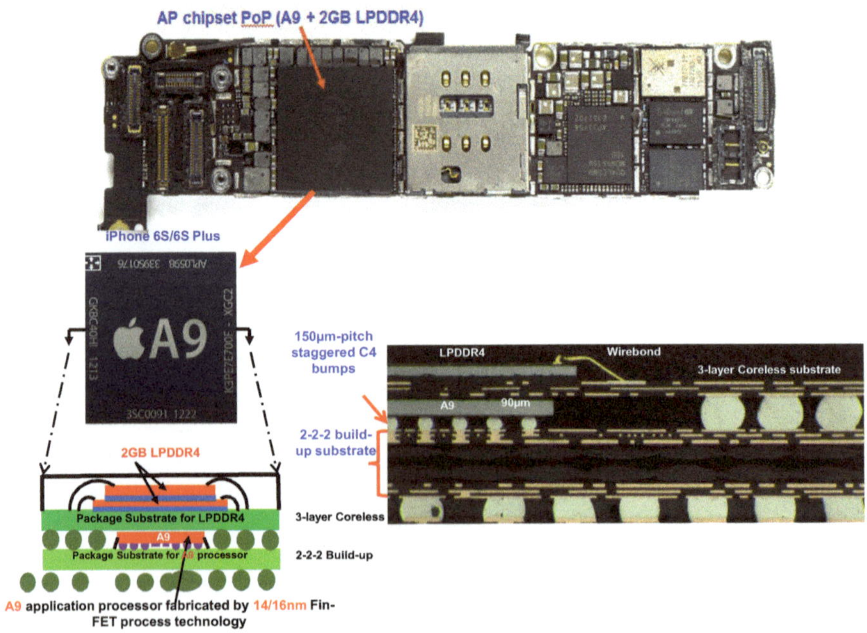

Fig. 4.104 AP in PoP with flip chip technology (Apple)

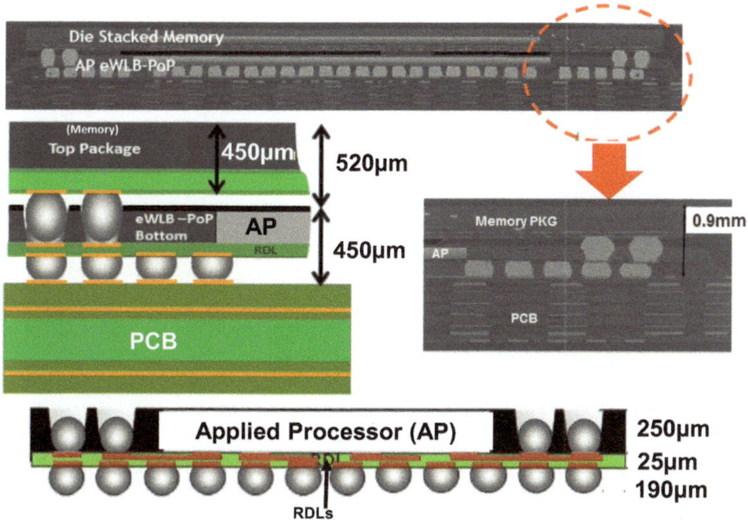

Fig. 4.105 AP in PoP with fan-out technology (Statschippac)

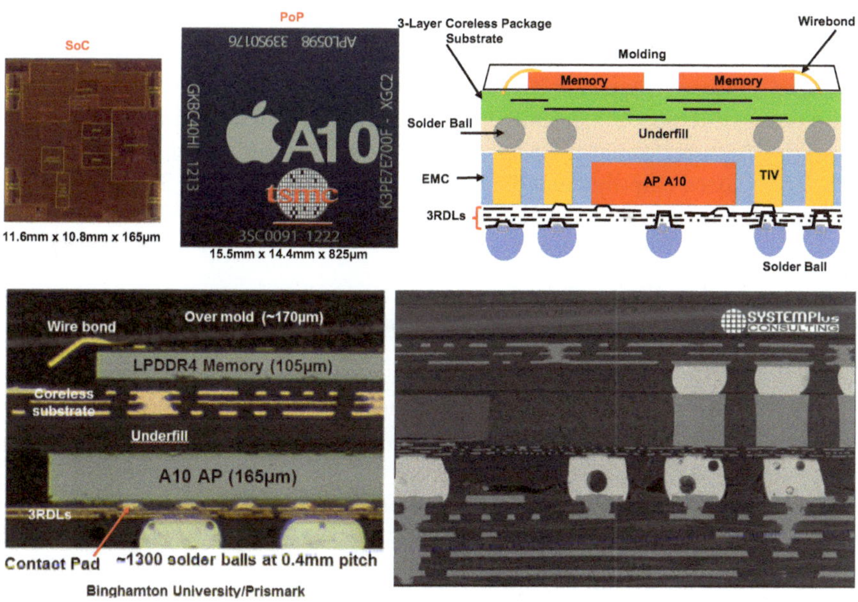

Fig. 4.106 A10 in PoP with fan-out technology (Apple/TSMC)

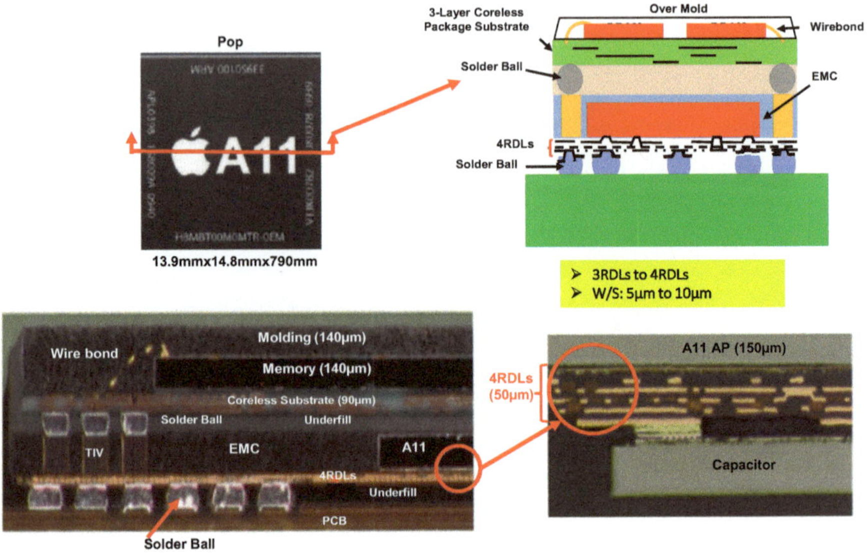

Fig. 4.107 A11 in PoP with fan-out technology (Apple/TSMC)

This is very significant since this means that FOWLP is not just only for packaging baseband, power management IC, radio frequency (RF) switch/transceiver, RF radar, audio codec, microcontroller unit, connectivity ICs, etc., but it can also be used for packaging high-performance and large (>120 mm^2) system-on-chip (SoC) such as APs. TSMC used chip-first and die face-up FOWLP. Since then, TSMC have been helping Apple to ship their iPhones such as those shown in Figs. 4.107 and 4.108.

4.7 Chip-Last or RDL-First Fan-Out Technology

Since 2006, NEC Electronics Corporation (now Renesas Electronics Corporation), Tokyo, Japan, has been developing a novel SMArt chip connection with feedthrough interposer (SMAFTI) packaging technology for inter-chip wideband data transfer [60, 61], 3D stacked memory integrated on logic devices [62–66], system-in wafer-level package (SiWLP) [16], and redistribution-layer (RDL)-first (or chip-last) fan-out wafer-level packaging [17]. The feedthrough interposer (FTI) used in SMAFTI is a film with ultrafine linewidth and spacing RDLs. The dielectric of the FTI is usually SiO_2 or a polymer, and the conductor wiring of the RDL is Cu. The FTI not only supports the RDL underneath the chip, but it also provides support beyond the edges of the chip. Area array solder balls are mounted at the bottom side of the FTI, which are to be connected to the package substrate or printed circuit board (PCB). Epoxy molding compound (EMC) is used to embed the chip and support the RDL and solder balls.

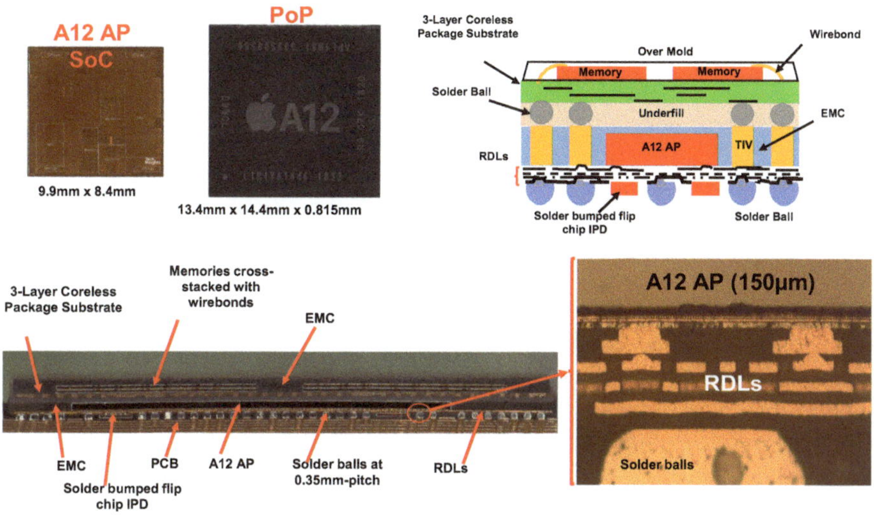

Fig. 4.108 A12 in PoP with fan-out technology (Apple/TSMC)

One of the key reasons for the RDL-first (or chip-last) [16, 17] is trying to prevent the loss of the embedded known good die (KGD) during the RDL fabrication of the chip-first process. This is true only if the RDL-first substrate for chip-last can be fully functionally tested at speed before other process steps, such as chip-to-substrate bonding, underfilling, and EMC molding; otherwise, the KGD still have to be thrown away for the case of a not good (NG) RDL-first substrate after a system test. Compared with the chip-first process, the chip-last process is more complex and higher cost. Besides making the RDL, chip-last also must perform the wafer bumping, chip-to-wafer or panel substrate bonding, and underfilling (or molded underfill). Thus, in general, RDL-first is meant for high-density fan-out wafer or panel-level packaging (HD FOWLP) or (HD FOPLP).

Since 2015, Amkor has been promoting a very similar RDL-first technology called silicon wafer integrated fan-out technology (SWIFT) [67]. Since 2015, the Institute of Microelectronics (IME) has been publishing a series of research and development papers on chip-last or RDL-first FOWLP and FOPLP [68–77]. In 2016, SPIL demonstrated the first hybrid integration of the fine line plasma-enhanced chemical vapor deposition (PECVD) RDL and the RDL with polymeric dielectrics [78]. In 2017, Amkor published a very similar hybrid RDL for its silicon-less integrated module (SLIM) [79, 80]. Also, in 2017, Unimicron published a paper on RDL-first with the minimum linewidth of 8 μm on a panel size of 370 mm × 470 mm [81]. In 2018, Samsung published two papers on using RDL-first to make Si-less RDL for high-performance computing (HPC) applications [82], and system-in-package (SiP) side-by-side (SbS) for advanced mobile applications [83]. Since 2018, ITRI has been publishing research and development papers in RDL-first FOWLP and FOPLP [84–86]. In 2019, ASE used RDL-first for its fan-out chip on substrate (FOCoS)

[87]. Also, in 2019, TSMC used RDL-first to make the RDL-interposer for its high-density heterogeneous large package [88]. For more recent developments on chip-last or RDL-first, please read the rest of this chapter.

4.8 Panel-Level Fan-Out RDL-First Packaging for Heterogeneous Integration

In this section, a fan-out RDL-first substrate fabricated on a panel size of 510 mm × 515 mm is investigated. There are three RDL and the minimum metal linewidth and spacing are 2 μm. The heterogeneous integration of three chips on the fan-out RDL-first substrate is investigated (Fig. 4.109). The chip sizes are one 10 mm × 10 mm and two 7 mm × 5 mm. One practical application of heterogeneous integration is for the application processor chipset, i.e., the large chip could be a processor and the small chips could be memories [18–20].

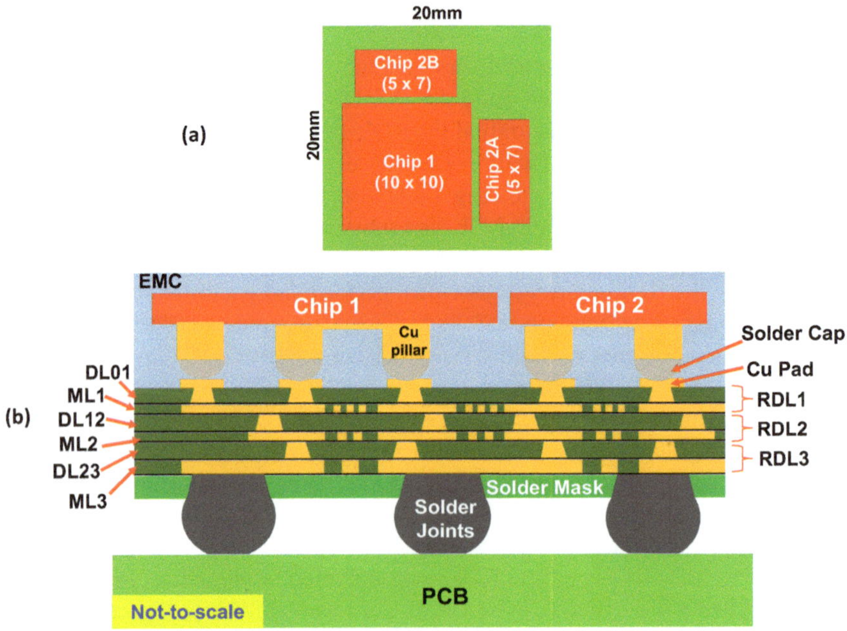

Fig. 4.109 Test package (20 mm × 20 mm). **a** Top view. **b** Cross-sectional view

4.8.1 Test Chips

Figure 4.110 shows the test chips under consideration. It can be seen that the size of the large chip is 10 mm × 10 mm. There are 3760 area array pads on the 90 μm pitch and are daisy chained. The small chip is 7 mm × 5 mm and with 1512 area array pads on the 60 μm pitch. The pads are also daisy chained.

The cross section of the chips is schematically shown in Fig. 4.110. It can be seen that for the large chip, the Cu pad size is 50 μm × 50 μm, the Ti/Cu (0.1/0.2 μm) under bump metallurgy (UBM) pad size is 35 μm diameter, the passivation (PI2) opening is 20 μm diameter, the Cu-pillar is 35 μm diameter and 37 μm tall, and the SnAg solder cap is 15 μm with a barrier (Ni = 3 μm), and for the small chip, the Cu pad size is 44 μm × 44 μm, the Ti/Cu (0.1/0.2 μm) UBM pad size is 35 μm diameter, the passivation (PI2) opening is 20 μm diameter, the Cu-pillar is 35 μm diameter and 37 μm tall, the Ni barrier is 3 μm, and the SnAg solder cap is 15 μm.

4.8.2 Test Package

Figure 4.109 schematically shows the test package. It can be seen from Fig. 4.109a that there are three chips in the heterogeneous integration, namely the large chip (Chip 1) and two small chips (Chips 2A and 2B). The spacing (gap) between the large chip and small chips is 200 μm. The RDL-substrate has three RDL, and each RDL consists of one dielectric layer (DL) and one metal layer (ML), Fig. 4.109b. The linewidth and spacing of ML1–ML3 of RDL1–RDL3, respectively, are 2/2, 5/5, and 10/10 μm, as shown in Fig. 4.111a. The thicknesses of ML1–ML3 are, respectively,

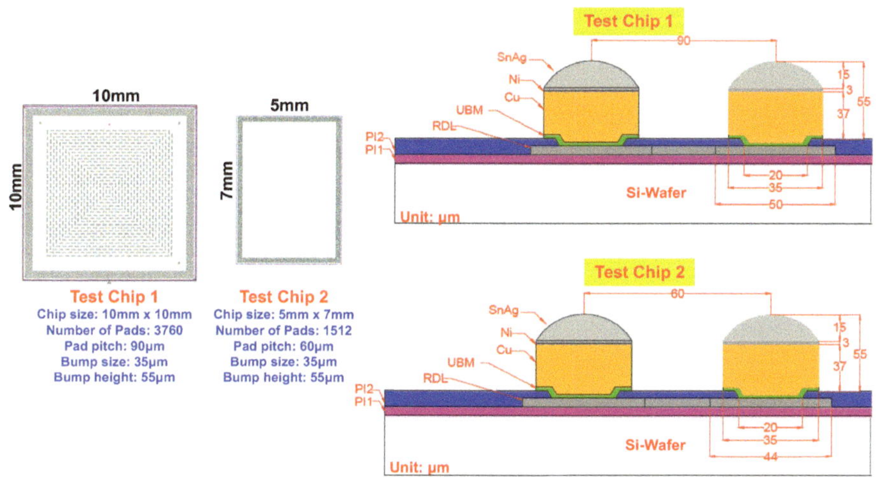

Fig. 4.110 Top view and cross section of test chips

2.5, 2.5, and 8 μm. The thicknesses of DL01 (DL between the Cu pad and ML1), DL12 (DL between ML1 and ML2), and DL23 (DL between ML2 and ML3) are, respectively 5, 6.5, and 6.5 μm as shown in Table 4.9 and Fig. 4.111a. The pad diameter and thickness are 35 and 8 μm, respectively. V_{01} (via connecting between the pad and ML1) is 22 μm (top diameter) and 17 μm (bottom diameter), V_{12} (via connecting between ML1 and ML2) is 30 μm (top diameter) and 25 μm (bottom diameter), and V_{23} (via connecting between ML2 and ML3) is 30 μm (top diameter) and 25 μm (bottom diameter). The solder mask (resist) opening, and thickness are, respectively, 245–250 and 5 μm.

The panel for fabricating the RDL-first substrate for the heterogeneous integration of three-chip packages is shown in Fig. 4.111b. It can be seen that the panel size is 515 mm × 510 mm and is made of glass. The panel is divided into 12 strips and each

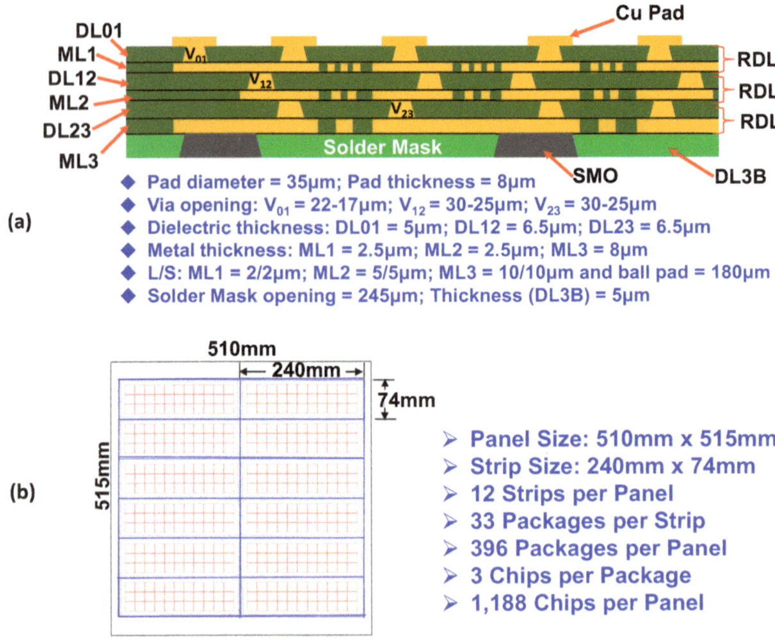

Fig. 4.111 **a** Three-layer RDL-first substrate. **b** Panel (515 mm × 510 mm) for making the RDL-first substrate

Table 4.9 RDL: metal layer (ML) and dielectric layer (DL)

RDL	ML (line width/spacing) (μm)	ML (thickness) (μm)	DL (thickness) (μm)
RDL1	ML1 = 2/2	2.5	DL01 = 5
RDL2	ML2 = 5/5	2.5	DL12 = 6.5
RDL3	ML3 = 10/10	8	DL23 = 6.5

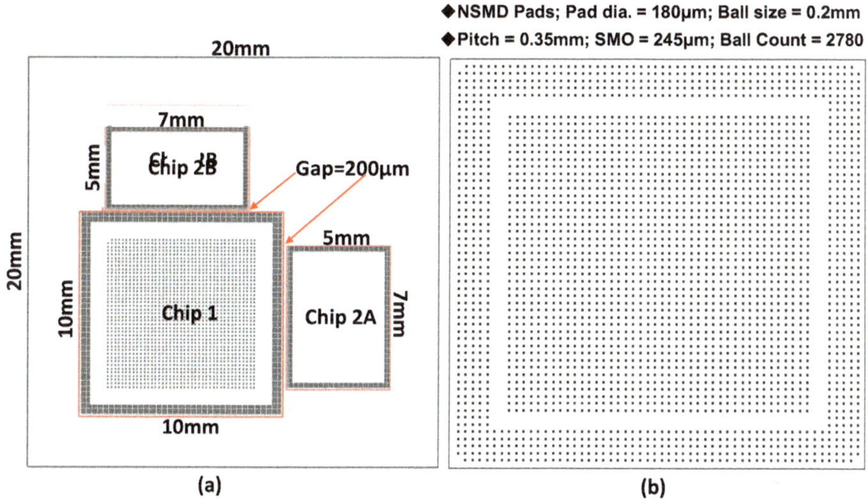

Fig. 4.112 RDL-first substrate of the test package. **a** Top view. **b** Bottom view

strip (24 mm × 74 mm) has 33 packages. Thus, in one shot, it can make the RDL-first substrate for 1188 chips in 396 heterogeneous integration packages. The top view and bottom view of the 20 mm × 20 mm RDL-first substrate of the heterogeneous integration of the three-chip test package are shown in Fig. 4.112. It can be seen that the following holds.

(a) There are 2 × 1512 + 3760 = 6784 pads on top of the RDL-first substrate of the package. These pads are for chip-to-substrate bonding.
(b) There are 2780 solder ball pads (180 μm diameter) on 0.35 mm pitch at the bottom of the RDL-first substrate of the package. The pads are non-solder mask defined (NSMD) and the solder mask opening is 245 μm. These pads are for solder ball mounting and the Sn3Ag0.5Cu lead-free solder ball diameter is 0.2 mm.

4.8.3 RDL-First Panel-Level Packaging for Heterogeneous Integration

The focus of this section is on the design, materials, process, and reliability of the RDL-first substrate for high-density fan-out panel-level heterogeneous integration applications. The process flow for fabricating the RDL-first substrate, chip-to-strip bonding, underfilling, EMC molding, solder ball mounting, and dicing is shown in Fig. 4.113a, b.

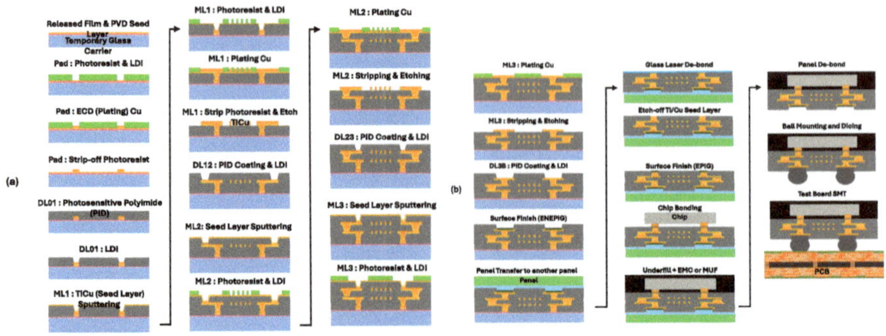

Fig. 4.113 **a** Process flow of the fan-out RDL-first panel-level package. **b** Process flow of the fan-out RDL-first panel-level package (continue)

(A) RDL-First Substrate Fabrication

The heart of this section is on the fabrication of the RDL-first substrate on a 515 mm × 510 mm temporary glass panel. First, a sacrificial layer (released film) is slit coated on the temporary glass carrier, and then, a Ti/Cu seed layer is formed by physical vapor deposition (PVD). It is followed by photoresist, laser direct imaging (LDI), and development. Then, electrochemical deposition (ECD) Cu and strip off the photoresist to form the contact pad (or called land).

It is followed by slit coating a PID (photosensitive polyimide) and LDI to get the first DL (DL01) of RDL1. Then, sputter the Ti/Cu seed layer, photoresist, LDI, and develop, and ECD the Cu. It is followed by stripping off the photoresist and etching off the TiCu seed layer to get the first ML (ML1) of RDL1. Figure 4.114 shows the photograph of the RDL1. It can be seen that the linewidth and spacing of ML1 are properly made. Repeat the same process steps to get DL12 and ML2 of RDL2, and DL23 and ML3 of RDL3.

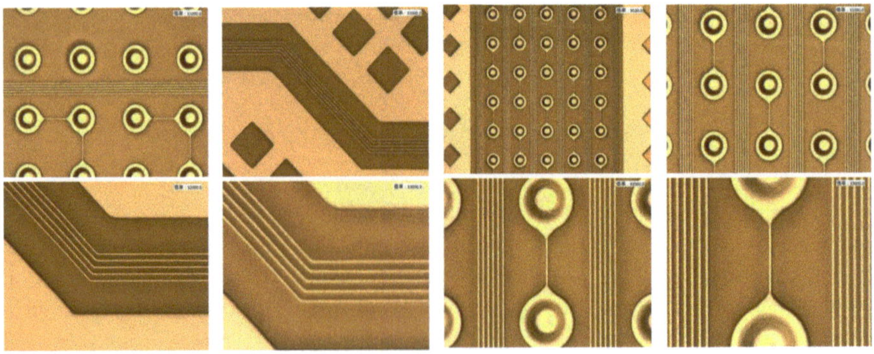

Fig. 4.114 Top view of the RDL1 with ML1 = 2/2 μm (L/S)

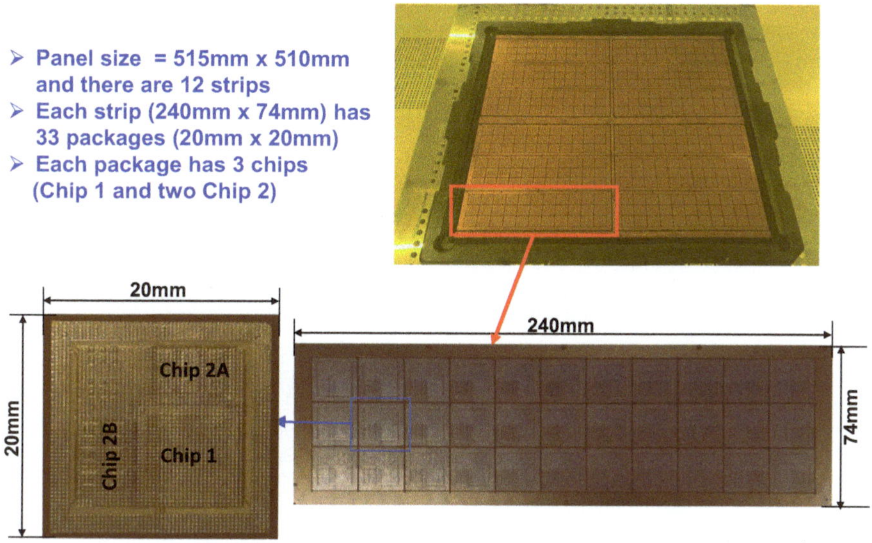

Fig. 4.115 Fabricated three-layer RDL-first substrate on a 515 mm × 510 mm panel with glass carrier

After making ML3, it is followed by slit coating a PID and LDI to get the passivation (or solder mask) and then surface finishing with electroless palladium immersion gold (EPIG). Figure 4.115 shows the 515 mm × 510 mm panel with the three-layer RDL-first substrate.

It is followed by cutting the panel with glass into 12 strips and transferring the strip to another temporary carrier and then debond the glass carrier from the strip carrier with a laser. The reason for this strip transfer is because it is necessary to make the RDL1 (with 2 μm metal linewidth and spacing) to be chip-to-strip bonding first. It is followed by etching the Ti/Cu seed layer and surface finishing with EPIG. The fabrication of the RDL-first substrate is completed.

Figure 4.116 shows the scanning electron microscope (SEM) images of a typical cross section of the RDL-first substrate. This cross section is showing (including) all three ML1–ML3 and the contact pad. A closer look at the left SEM image, it can be seen that for RDL1, the linewidth does not meet the 2 μm target but equals 2.43, 2.53 μm, …, and the line spacing also does not meet the 2 μm target but equals 1.69, 1.69 μm, …. Thus, there are rooms for improvements, such as a better estimation of the compensation of photoresist, LDI, ECD Cu, Cu etching, etc. The linewidth and spacing of ML2 are much better, and they are, respectively, 5.31 and 5.16 μm, and the target is 5 μm. This is understandable because the smaller the linewidth and spacing, the larger the error. The thicknesses of ML1–ML3 are, respectively, 2.13, 2.53, and 8.53 μm, which are close to the target values of 2.5, 2.5, and 8 μm.

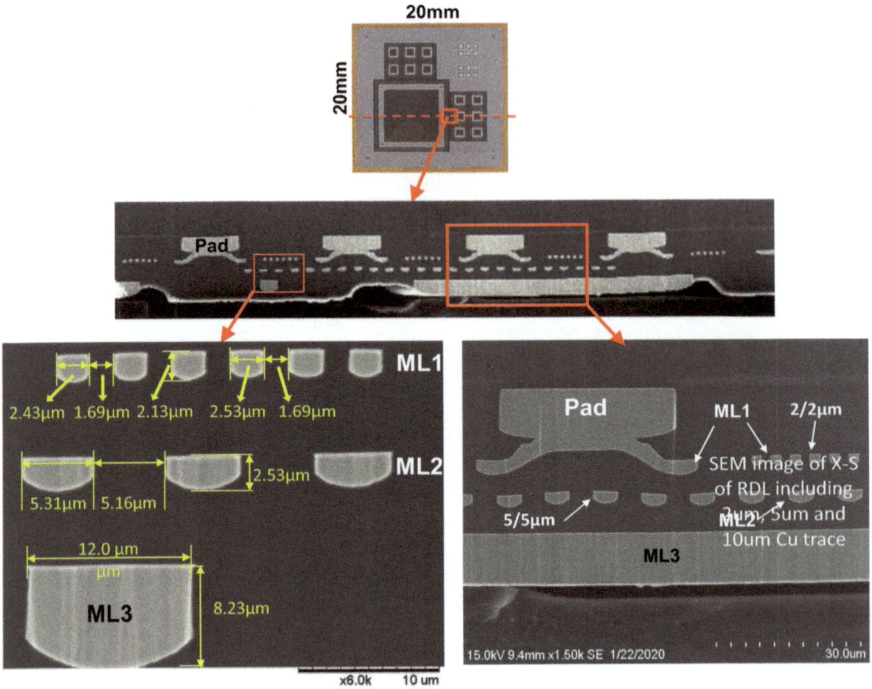

Fig. 4.116 Cross-sectional view of the RDL-first substrate showing ML1–ML3

(B) Wafer Bumping

In parallel, the wafer bumping of the large and small chips with the standard PVD and ECD Cu and solder process [10] is performed. Then, dice the wafers into individual chips. For all the chips, the bump consists of the Cu pillar, Ni barrier, and SnAg cap.

(C) Chip-to-RDL-First Substrate (Strip) Bonding

Before chip-to-RDL-first substrate bonding, as mentioned earlier, the RDL-first substrate in strip form is transferred to another temporary carrier and debond the glass temporary carrier (by laser), so the chips can be bonded on the pads (near to RDL1) of the substrate (strip). There are many ways to perform the chip-to-RDL-first substrate bonding. In this section, mass reflow and thermocompression bonding (TCB) are attempted.

First, mass reflow of chips on substrate is tried. With this method, all the chips are picked and placed on the substrate at room temperature, and then, the solder is reflowed in a 12-zone reflow oven, typically called mass reflow. Due to the warpage of the substrate and the thermal expansion mismatch between the packages and the substrate, there is not a good assembly yield.

Next, TCB is attempted. First, a flux is applied to the RDL-first substrate. Then, TCB of the large (10 mm × 10 mm) chip and then, the small chips (Chips 2A and 2B). A common TCB bonding profile is the substrate temperature of 120 °C and the bond-head temperature of 250 °C and then heating up until the solder melt and wet is applied. Figure 4.117a shows the chip-to-substrate bonding by TCB. The chip-to-substrate bonding assemblies pass the electrical continuity test. Figure 4.118 shows a typical cross section of the assembly. The chip, Cu-pillar, solder joint, contact pad, ML1, ML2, and ML3 of the RDL-first substrate are clearly shown.

(D) Underfilling and Curing

After the chip-to-substrate bonding of all the chips on the RDL-first substrate, it is followed by underfill dispensing. Then, the underfill is cured at 165 °C for 2 h. Figures 4.117b and 4.118 show the underfilled chips.

(E) EMC Lamination

There are at least three methods to perform the EMC molding, namely, transfer molding, compression molding, and lamination molding. Because of the size of the strip, transfer molding cannot do the job. Usually, for fan-out packaging, it requires compression molding. However, the compression molding machine is very expensive. Thus, in this section, the lamination method of EMC is adopted. After underfilling the assembly, the whole strip is laminated with an EMC. The temperature and pressure for the lamination are as follows: (1) first stage: 120 °C/vacuum for 30 s and press (0.68 MPa) for 30 s with the temperature and vacuum on and (2) second stage: 100 °C and press (0.58 MPs) for 60 s. Figure 4.117c shows the package after EMC lamination.

(F) Debonding

After EMC lamination of the packages on the strip, the second temporary strip carrier is debonded. Then, it is time for solder ball mounting.

Fig. 4.117 **a** Chip-to-substrate bonding. **b** With underfill. **c** With underfill and EMC

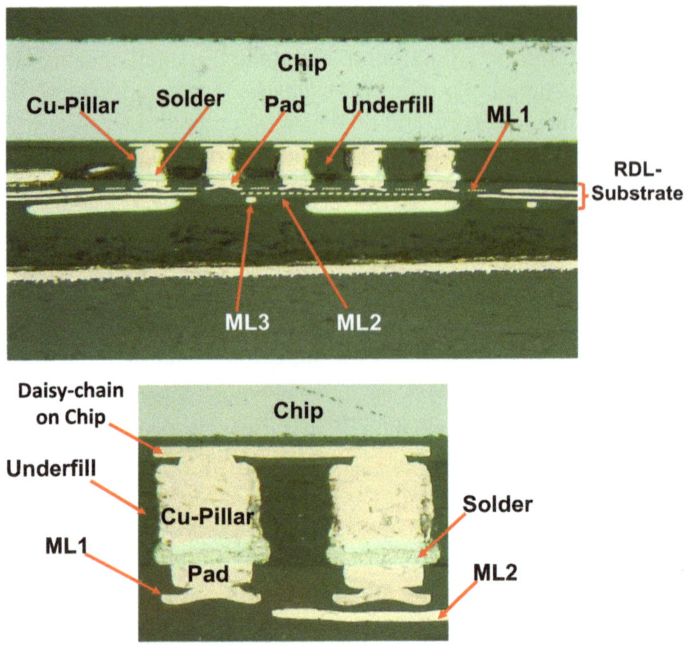

Fig. 4.118 Cross section of the chip-to-substrate bonding

(G) Solder Ball Mounting and Dicing

The solder ball mounting equipment for the strip is shown in Fig. 4.119. It consists of two stencils: one is for fluxing and the other is for solder ball mounting. The solder (Sn3wt%Ag0.5wt%Cu) balls are with a 200 μm diameter and at 0.35 mm pitch. The peak temperature for the solder reflows is 245 °C (see Fig. 4.120). After solder ball mounting, the strip is diced into individual packages.

4.8.4 *Thermal Cycling Simulation*

(A) Structure and Kinematic Boundary Conditions

The focus of the present thermal cycling simulation of the PCB assembly of the heterogeneous integration of the three-chip package is on the board-level solder-joint reliability. Thus, the Cu-pillar micro solder joints are not modeled.

Due to symmetry about the diagonal axes of the structure as shown in Figs. 4.121 and 4.122, only half of the structure is modeled. Proper kinematic boundary conditions, such as there is no displacement perpendicular to the symmetry-plane and there is no rotation about the diagonal axes (AA′), are specified.

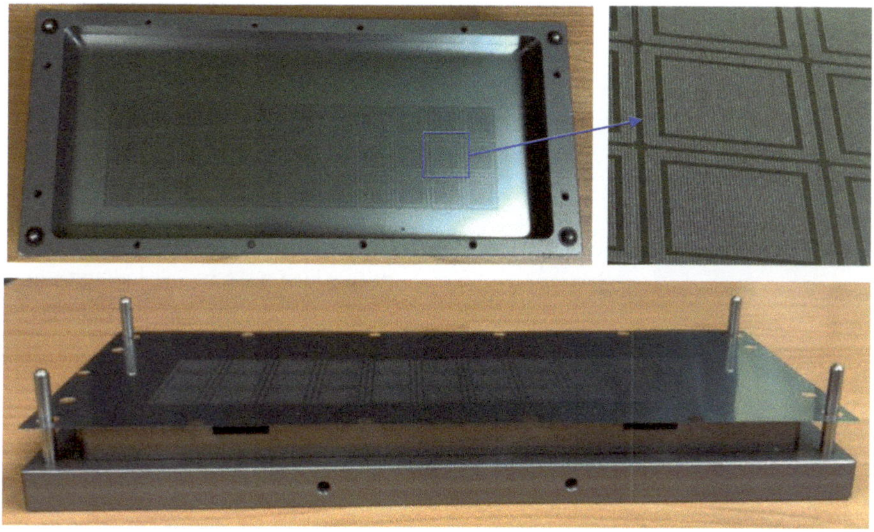

Fig. 4.119 Solder ball mounting fixture for a strip (240 mm × 74 mm)

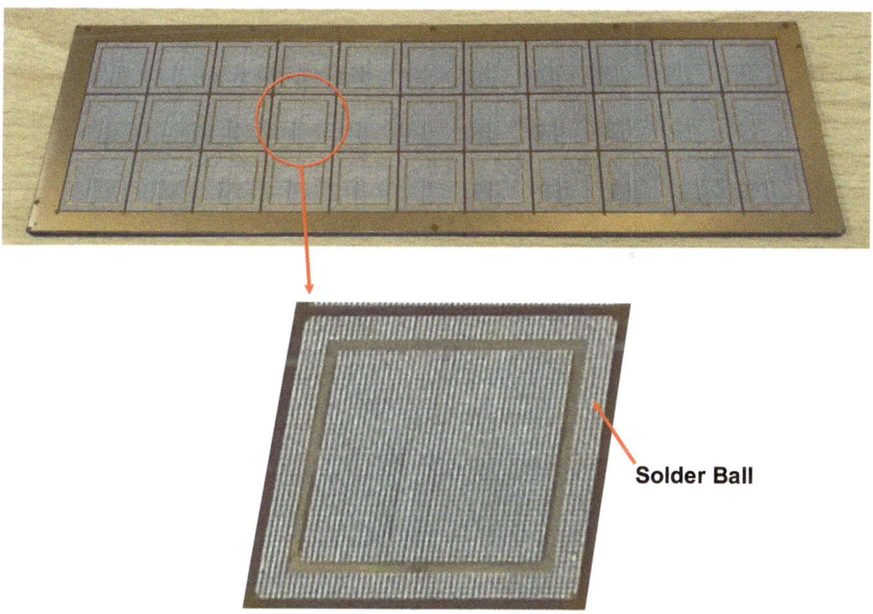

Fig. 4.120 Solder balls mounted on a strip with 33 packages

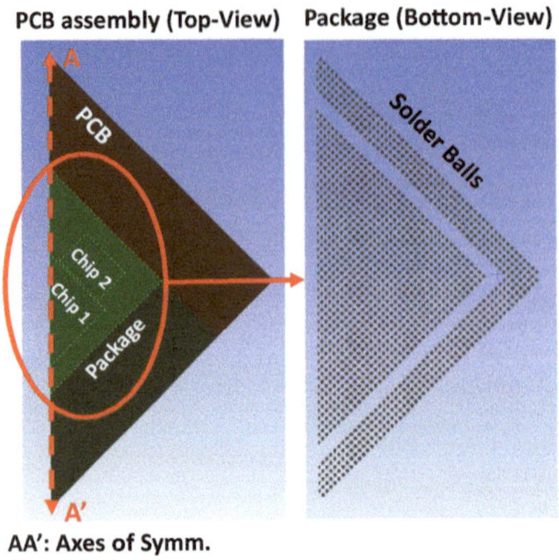

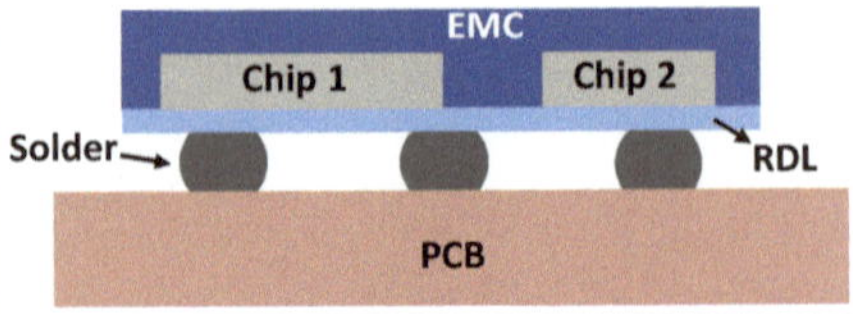

	Length & Width	Thickness (μm)
Chip 1	10 x 10 mm	275
Chip 2	4.8 x 6.9 mm	275
EMC	20 x 20 mm	400
RDL	20 x 20 mm	39
PCB	40 x 40 mm	1000
Solder	Φ = 180 μm	120

Fig. 4.121 Finite-element modeling of the PCB assembly of the heterogeneous integration of three-chip package

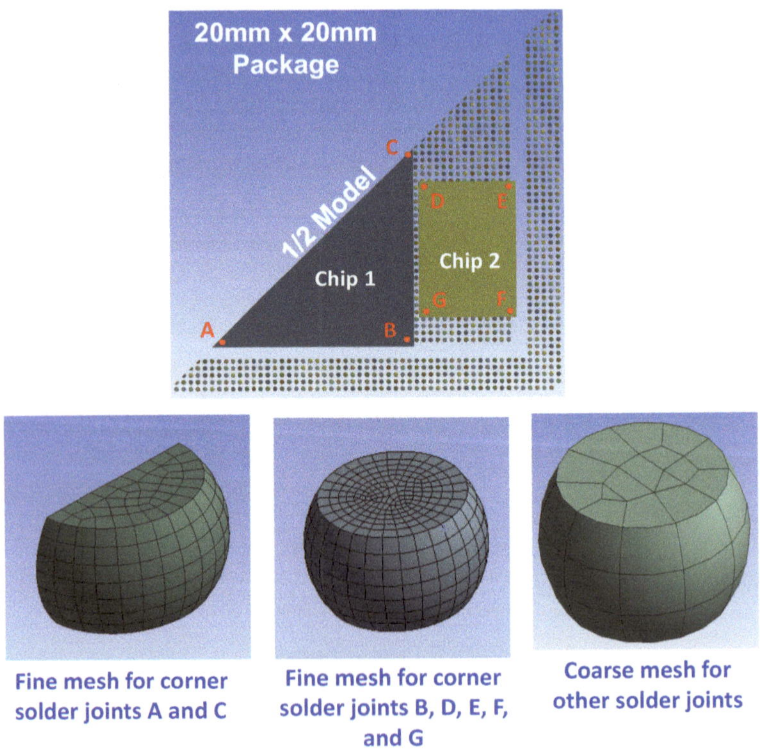

Fig. 4.122 Due to symmetry, only one-half of the PCB assembly is modeled. Different mesh sizes for different solder joints

Despite the overall economy of elements in the 1/2 model, selective mesh refinement is used to concentrate highly refined elements in the solder joints where failure is anticipated. In the present PCB assembly, failure would be predicted in the solder joints with the greatest DNP [36] and near the chip corners, as shown in Figs. 4.121 and 4.122. Thus, highly refined meshes are applied to these solder joints (A–G). The other joints are coarsely meshed. The 3D Solid186 element of ANSYS 2019 R3 code is used.

(B) Material Properties

The material properties of the PCB assembly are given in Table 4.10. All the material properties are assumed to be constant except for those of the solder. Sn3Ag0.5Cu is assumed to obey the generalized Garofalo creep equation [48]

$$d\varepsilon/dt = 500{,}000\,\mathrm{Sinh}^5\left(\sigma/1\times10^8\right)\,\exp[-5{,}807/T(K)]$$

where ε is the strain, σ is the stress in Pa, and T is the temperature in kelvin. The CTE and Young's modulus of the solder are, respectively, 21.3 + 0.017 T and 49 – 0.07 T, and T is the temperature in Celsius.

Again, since the focus of the present simulation is on the board-level solder joint reliability, in other to simplify the modeling and save the computing time, those three RDLs are modeled as an equivalent block as shown in Fig. 4.121 with the following effective material properties (see Table 4.10):

$$\begin{aligned} E_{eff} =& \left[(3+3+8)\times 121\times10^9 + (5+6.5+6.5+5)\times 3.3\times10^9\right] \\ & \div (3+3+8+5+6.5+6.5+5) = 47.8\times10^9\ \mathrm{Pa} \end{aligned}$$

Table 4.10 Material properties for simulation

Materials	CTE (10^{-6}/°C)	Young's modulus (GPa)	Poisson's ratio
Copper	16.3	121	0.34
PCB	$\alpha_x = \alpha_y = 18$ $\alpha_z = 70$	$E_x = E_y = 22$ $E_z = 10$	0.28
Solder mask (RDL)	7.0	7.0	0.3
Solder	21 + 0.017 T (°C)	49 – 0.07 T (°C)	0.3
EMC (ABF)	7.0	7.0	0.3
Silicon	2.8	131	0.278
Solder mask (PCB)	39	4.1	0.3
RDL equiv. block	14.08	70.53	0.32
Underfill	50	4.5	0.35

The constitutive equation of solder: $\frac{d\varepsilon}{dt} = 500,000[\sinh(0.01\sigma)]^5 \exp\left(-\frac{5,800}{T}\right)$
where T is in Kelvin and σ is in MPa

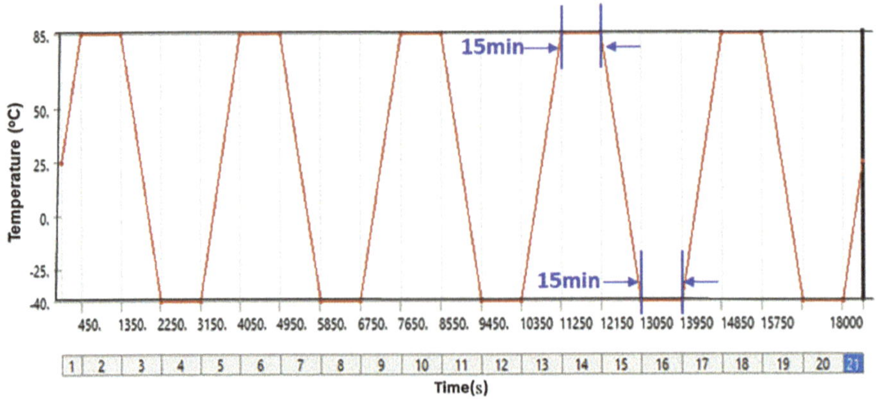

Fig. 4.123 Temperature boundary condition

$$\begin{aligned}\mathrm{CTE}_{\mathrm{eff}} &= \left[(3+3+8)\times 16.3\times 10^{-6} + (5+6.5+6.5+5)\times 35\times 10^{-6}\right]\\ &\quad \div (3+3+8+5+6.5+6.5+5) = 27.9\times 10^{-6}/^{\circ}\mathrm{C}\end{aligned}$$

where E_{eff} and CTE_{eff} are, respectively, the effective Young's modulus and effective coefficient of thermal expansion of the RDL-first substrate equivalent block. Its Poisson is assumed to be 0.3.

(C) Temperature Boundary Condition

The PCB assembly of the heterogeneous integration of the three-chip package is subjected to the kinetic temperature profile shown in Fig. 4.123. Five temperature cycles are executed. It can be seen that the temperature is −40 °C ↔ 85 °C. The cycle time is 60 min and the ramp-up, ramp-down, dwell-at-hot, and dwell-at-cold are each 15 min.

(D) Simulation Results

The deformed shape (color contours) and the undeformed shape (dark lines) of the heterogeneous integration package PCB assembly are shown in Fig. 4.124 (left) at 450 s (85 °C) and in Fig. 4.124 (right) at 2250 s (−40 °C). It can be seen that at 85 °C, the PCB expands more than the package and the structure is deformed in a convex shape (smiling face), and at −40 °C, the PCB shrinks more than the package and the structure is deformed into a concave shape (crying face). These results show that the displacement, rotation, and temperature boundary conditions are prescribed correctly.

Figure 4.125 shows the shear stress and creep shear strain hysteresis loops at solder joint A of Chip 1 and solder joint F of Chip 2. It can be seen that after three cycles, for both solder joints, the creep responses converged (become stabilized). The area within the loop (creep strain energy density per cycle) at solder joint A is larger than that at solder joint F.

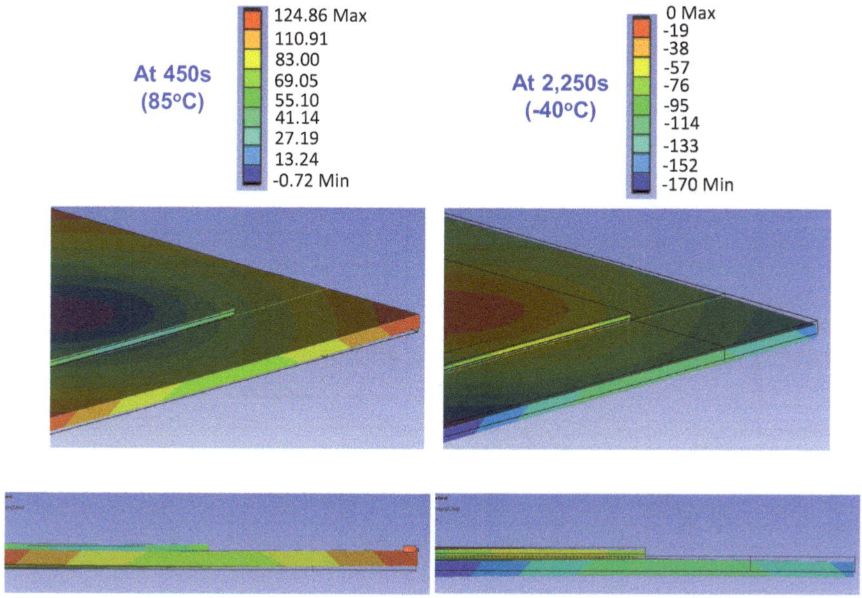

Fig. 4.124 Deformed and undeformed shapes of the PCB assembly of the heterogeneous integration package

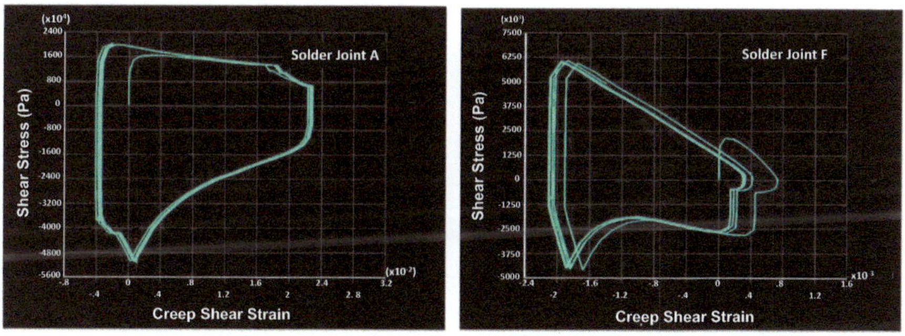

Fig. 4.125 Shear stress and creep shear strain hysteresis loops at solder joints A and F

The accumulated creep strain time history at various critical solder joints in Chips 1 and 2 is shown in Fig. 4.126. It can be seen that the maximum accumulated creep strain occurs at the corner joint A of the large chip (Chip 1) with the largest DNP (the failure location).

Figure 4.127 shows the maximum accumulated creep strain contours at solder joint A at 450 s (85 °C) and 2250 s (−40 °C). It can be seen that at both times and temperatures, the maximum accumulated creep strain occurs near the interface

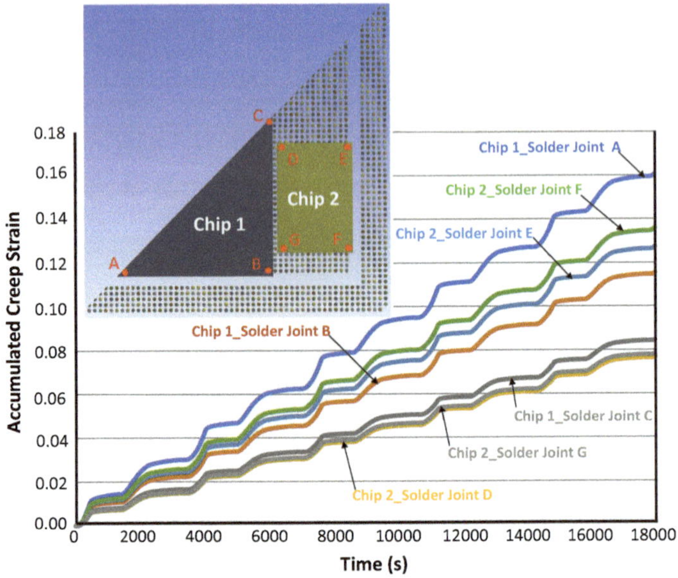

Fig. 4.126 Maximum accumulated creep strain time history at various critical solder joints

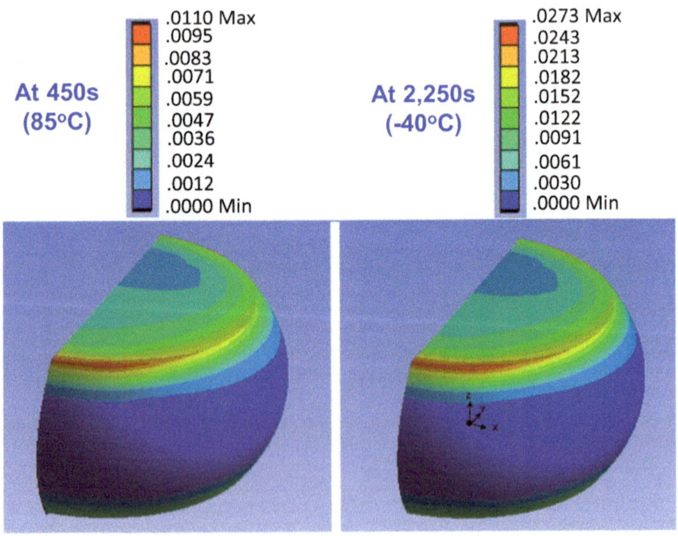

Fig. 4.127 Accumulated creep strain at solder joint A at 450 s (85 °C) and 2250 s (−40 °C)

between the RDL-substrate of the package and the bulk solder of solder joint A in Chip 1, and this is the driving force for cracking (the failure mode) of the solder joint. Thus, any failure should occur at this location. It can also be seen from Fig. 4.127 that the maximum accumulated creep strain per cycle at solder joint A of Chip 1 equals 0.0164 (0.0273–0.011), which is too small to create solder joint reliability problem, especially for mobile products such as smartwatches, smartphones, and tablets, for which the life is less than three years.

The second maximum accumulated creep strain of the heterogeneous integration PCB assembly occurs in solder joint F and solder joint E in Chip 2, as shown in Fig. 4.126. It can be seen that their magnitudes are very close since they have almost the same thermal expansion mismatch between the package and the PCB. The accumulated creep strain in solder joints C, B, D, and G is smaller than those in solder joints A, E, and F. One of the reasons could be due to the contraction and expansion between Chips 1 and 2.

Figure 4.128 shows the maximum accumulated creep strain contours at solder joint F in Chip 2 at 450 s (85 °C) and 2250 s (−40 °C). It can be seen that at both times and temperatures, the maximum accumulated creep strain occurs near the interface between the RDL-substrate of the package and the bulk solder of solder joint F. Thus, any failure should occur at this location. The creep strain energy density time history at solder joint A of Chip 1 and at solder joint F of Chip 2 is shown in Fig. 4.129. It can be seen that the creep strain energy density at solder joint A of Chip 1 is larger than that at solder joint F of Chip 2.

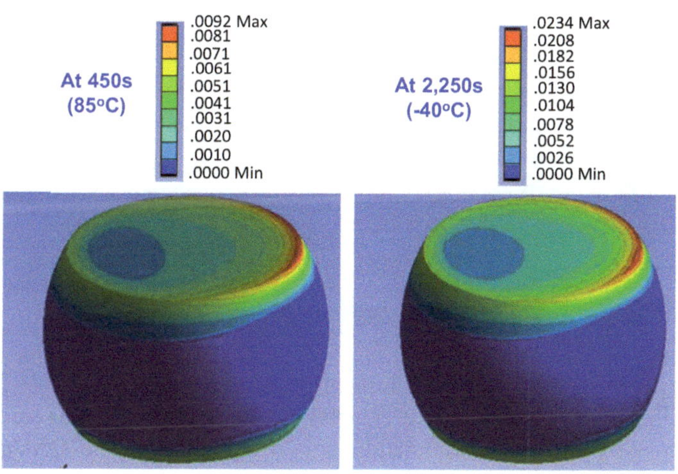

Fig. 4.128 Accumulated creep strain at solder joint F at 450 s (85 °C) and 2250 s (−40 °C)

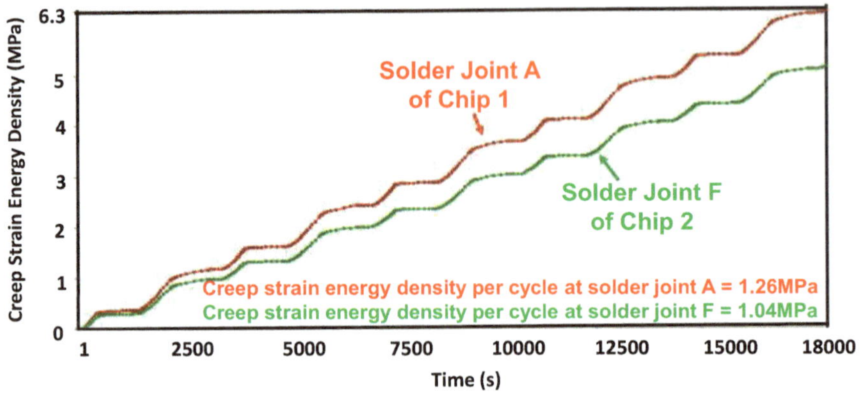

Fig. 4.129 Creep strain energy density time history at solder joints A and F

4.8.5 *PCB Assemble*

(A) PCB

The PCB is made of FR-4 with eight layers, and there are eight package sites on the board, Fig. 4.130. The pad with a diameter equal to 0.2 mm is non-solder mask defined, and its surface finish is an organic solderability preservative. The solder mask opening diameter is 0.245 mm. The package ball pad diameter is 0.18 mm.

(B) SMT Assembly

A lead-free surface-mount technology (SMT) process is used to assemble the packages on the test PCB. Figure 4.131a shows the reflow temperature profile. Figure 4.131b shows the PCB assembly of the packages, and Figure 4.131c shows a cross section of the PCB assembly of the heterogeneous integration package.

4.8.6 *Drop Test Results and Failure Analysis*

The shock (drop) test setup is according to JEDEC Standard ESD22-B111, as shown in Fig. 4.132. It consists of a drop tower, a drop table for the PCB with samples. There are 22 channels plus one common and a data acquisition system (DAS). The drop spectrum with 1,500 G/ms (1,500 Gs, 0.5 ms half-sine pulse) is shown in Fig. 4.133. The drop condition is 30 drops. The failure criterion is when the measured resistance during the drop test reaches 1,000 Ω as shown in Fig. 4.134. Less than that, it is considered no failure, as shown in Fig. 4.135.

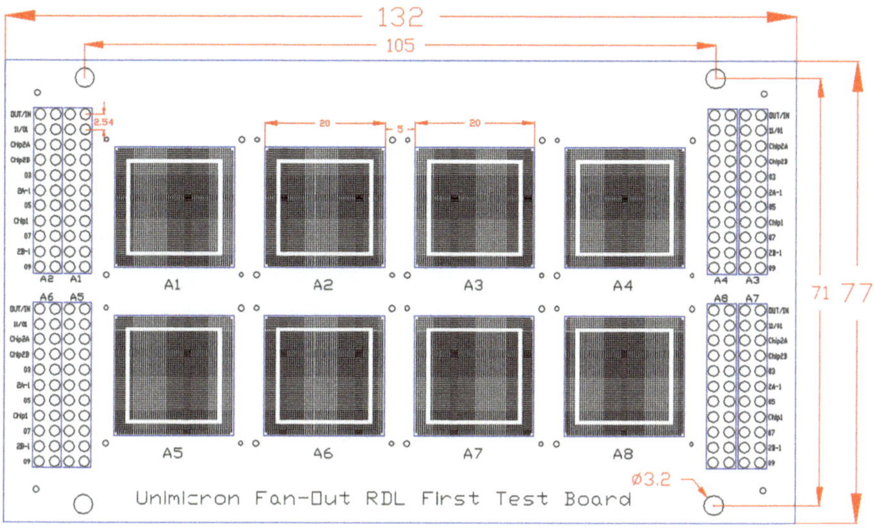

Fig. 4.130 Reliability test PCB with 8 20 mm × 20 mm heterogeneous integration packages

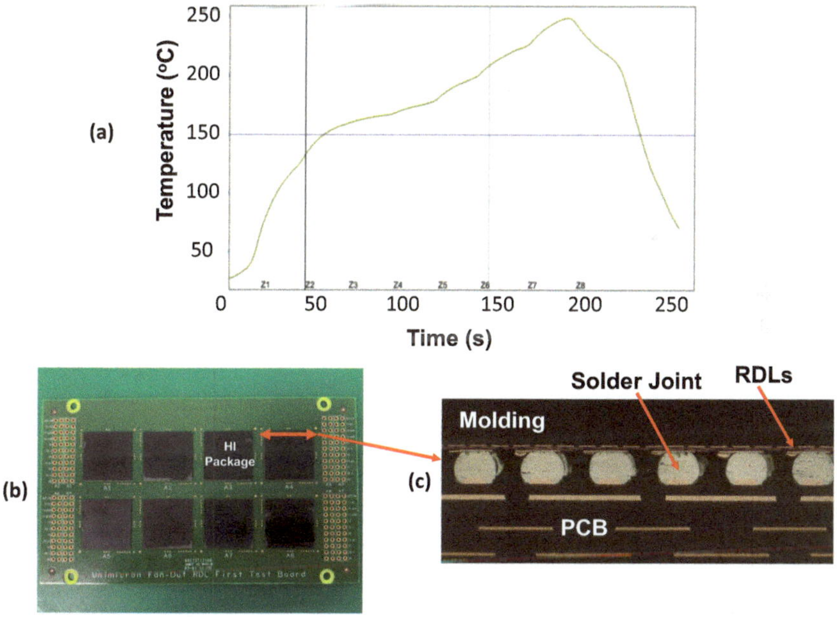

Fig. 4.131 **a** SMT reflow temperature profile. **b** The assembled test board. **c** Typical cross section of the PCB assembly

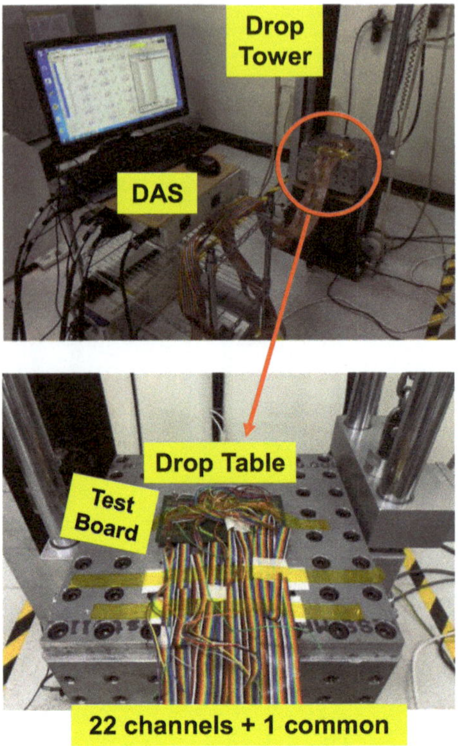

Fig. 4.132 Drop test set up, DAS, and samples

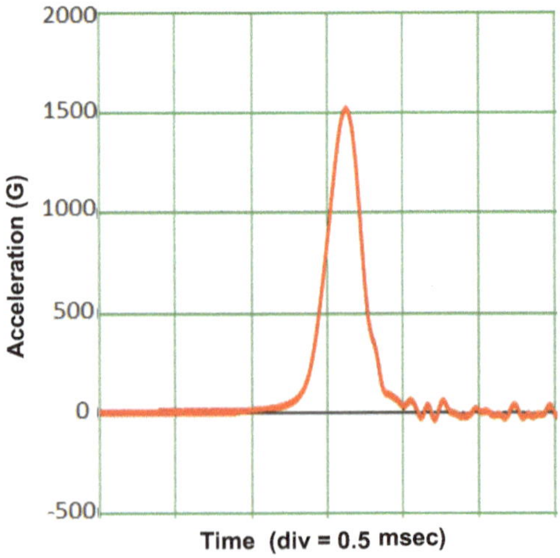

Fig. 4.133 The drop spectrum

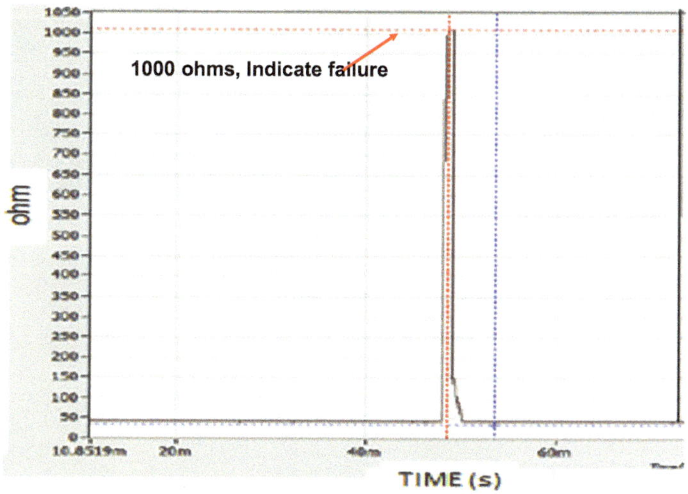

Fig. 4.134 Failure criteria. When the resistance ≧1000 Ω, it indicates failure

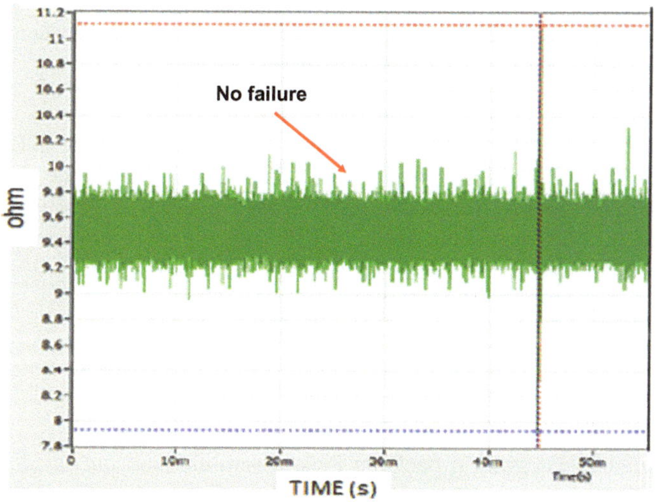

Fig. 4.135 Failure criteria. No failure

After 30 drops, there is only one failure which occurs after 23 drops. The DAS shows that the second package has failed (Fig. 4.136). Failure analysis by X-ray and cross sections indicates that the failure location of this package occurs near the corner solder joints and the edge near the middle of the PCB as shown in Fig. 4.137. However, this particular solder joint is bad to begin with. It is a head-in-pillow [48] solder joint. Before the drop test, the DAS cannot detect it and think it is a good solder joint. After 23 drops, the head is separated from the pillow, i.e., the solder joint is

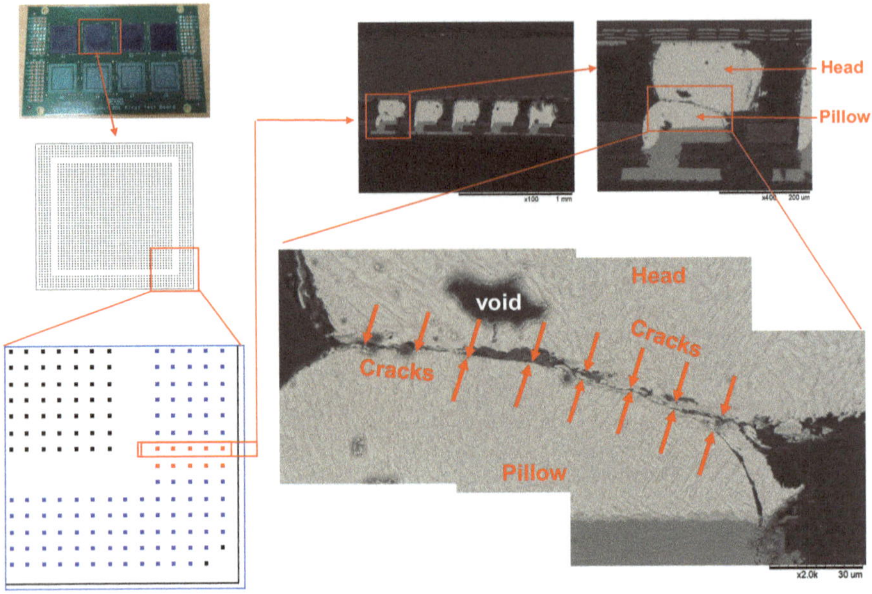

Fig. 4.136 The failure location and failure mode of the only failure sample under drop test. It turns out the failure sample is bad (head-in-pillow) to begin with

totally cracked. Thus, this sample should not be counted, and the heterogeneous integration package PCB assembly should be considered passing the drop test. On the other hand, failure analysis on some non-failure samples (the measured resistance is far less than 1,000 Ω as shown in Fig. 4.135) indicates that there are some small cracks in the solder joint as shown in Fig. 4.137.

4.8.7 Thermal Cycling Test and Results

The heterogeneous integration package PCB assemblies are put inside a thermal cycling chamber, as shown in Figs. 4.138 and 4.139. The temperature in the chamber during the test is measured through a thermal couple (in air) and is shown in Fig. 4.140. It can be seen that the temperature profile is basically −55 °C ↔ 125 °C at 50 min cycle. There are 49 channels, and each covers certain outer rows of solder joints.

(A) Failure Criterion

In this section, the failure criterion is when the resistance of the daisy chain of the heterogeneous integration PCB assembly increases by 50%. The cycle at which the first solder joint of a package failed is considered as the cycle-to-failure of the heterogeneous integration package.

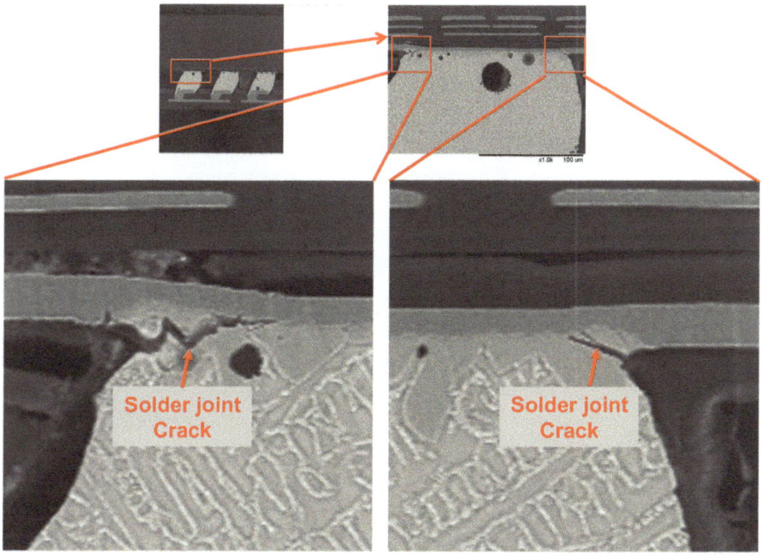

Fig. 4.137 No-failure sample showing small cracks after drop test

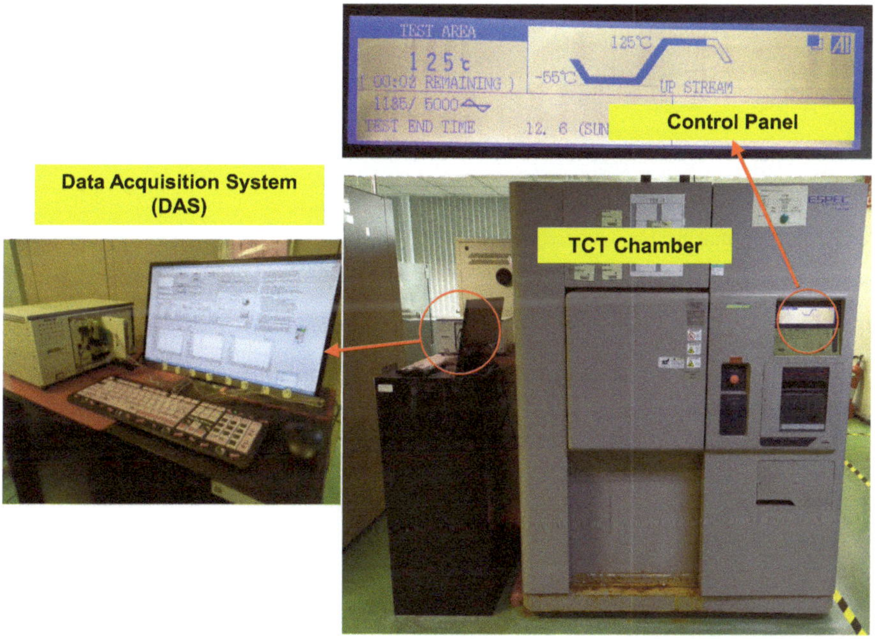

Fig. 4.138 Thermal cycling chamber and data acquisition system

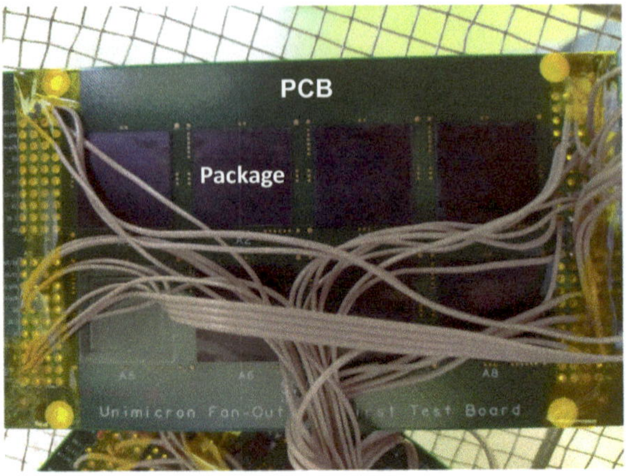

Fig. 4.139 Test boards with the packages inside the thermal cycling chamber

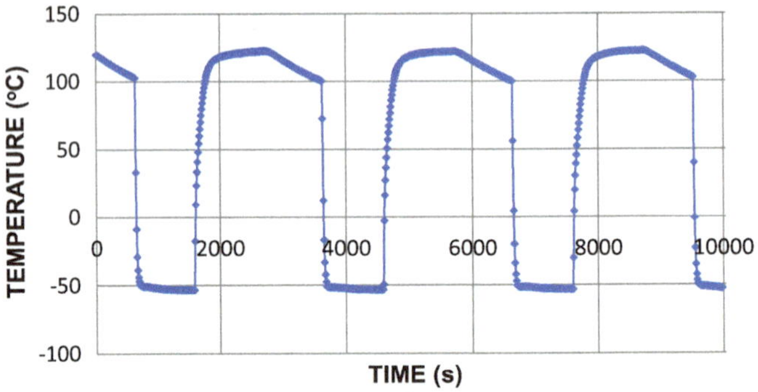

Fig. 4.140 Thermal cycling temperature profile

(B) Test Results of the Package at Median Rank

The thermal cycling test results of the heterogeneous integration package PCB assembly are tabulated in Table 4.11. It can be seen that there are 12 failures. The median (50%) rank is determined from F(x) = (j − 0.3)/(n + 0.4) [10], where j = failure order number and n = sample size 5 49. These data are plotted into a Weibull distribution and is shown in Fig. 4.141 for the median rank. It can be seen that the characteristic life (θ) (63.2% failed) = 1,702 cycles, and the Weibull slope (β) is = 2.28.

(C) Test Results of the Package at other Ranks

For other required ranking (G), the percent rank (z) can be determined by (z is the percent rank of the jth value in n samples) [48]:

Table 4.11 Test results of the heterogeneous integration package PCB assembly at 50, 5, and 95% ranks (sample size = 49)

Failure order	Cycles-to-failure	$F(x)$		
		Median (50%) rank	90% Confidence	
			5% Rank	95% Rank
1	300	1.41	0.10	5.93
2	400	3.43	0.73	9.32
3	552	5.46	1.69	12.30
4	552	7.48	2.84	15.07
5	552	9.51	4.11	17.71
6	552	11.53	5.47	20.27
7	624	13.56	6.90	22.74
8	788	15.58	8.39	25.16
9	872	17.60	9.93	27.54
10	872	19.63	11.51	29.86
11	1003	21.65	13.13	32.15
12	1003	23.68	14.78	34.41

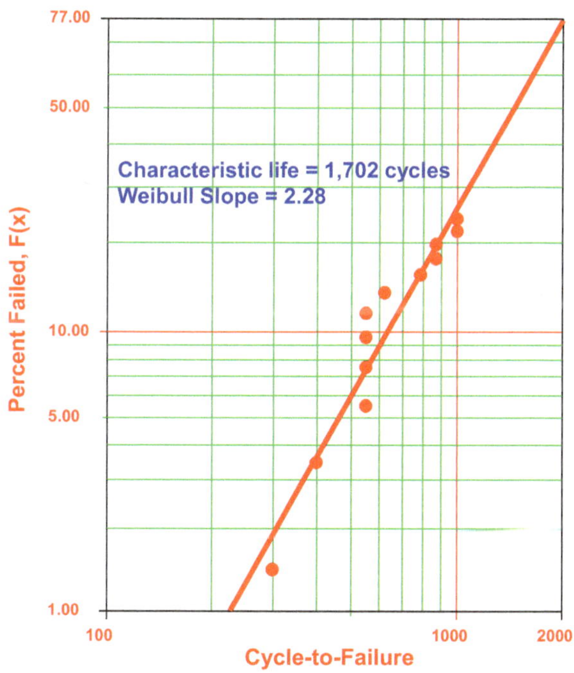

Fig. 4.141 Weibull plot of the heterogeneous integration solder joints at median rank

$$1 - (1-z)^n - nz(1-z)^{n-1} - \frac{n(n-1)}{2!}z^2(1-z)^{n-2} - \cdots$$
$$-\frac{n(n-1)\cdots(n-j+1)}{(j-1)!}z^{j-1}(1-z)^{n-j+1} = G$$

For 90% confidence level, we need the 5% rank (G = 5%) and 95% rank (G = 95%), and they are also listed in Table 4.11. For example, with N = 49 and G = 5%, the percent failure rank (z) is 0.1 for j = 1 and is 0.73 for j = 2. Another example, with N = 49 and G 595%, the percent failure rank (z) is 5.93 for j = 1 and is 9.32 for j = 2. Figure 4.142 shows the Weibull plot of the solder joints for the heterogeneous integration package PCB assembly at 90% confidence. It can be seen that at 90% confidence (i.e., nine in 10 cases it will happen), the true 10% failed life is 387 $\leq x_t$ $\leq$ 817 cycles.

(D) Failure Location and Failure Mode of the Heterogeneous Integration Package PCB Assembly

The typical failure location occurs near the solder joints underneath the Chip 1 and Chip 2A corners, as shown in Figs. 4.143 and 4.144. The failure mode is the cracking of the solder near the interface between the RDL-substrate and the bulk solders, as shown in Figs. 4.143 and 4.144.

(E) True Weibull Slope

Confidence is defined as the probability that a given interval determined from the test data will contain the population parameters, such as (using Weibull as an example)

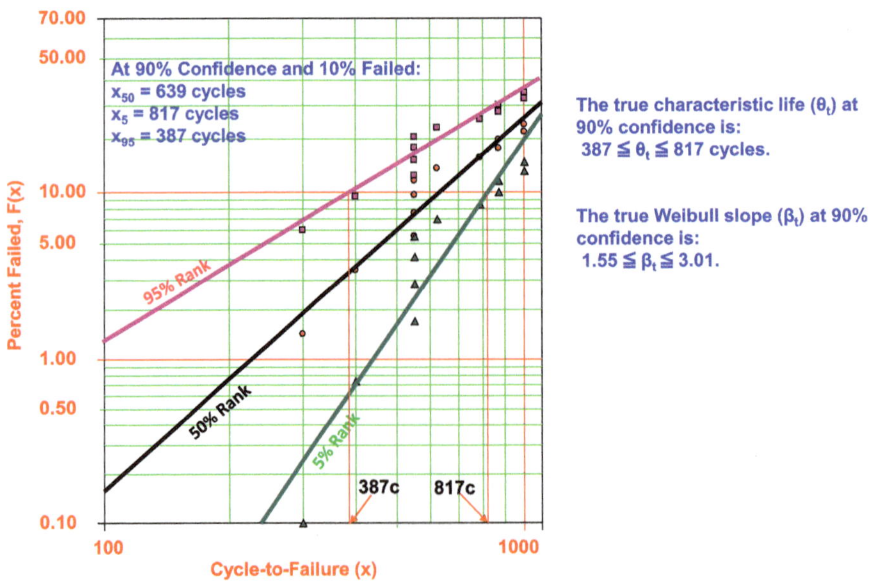

Fig. 4.142 Weibull plot of the heterogeneous integration solder joints at 90% confidence

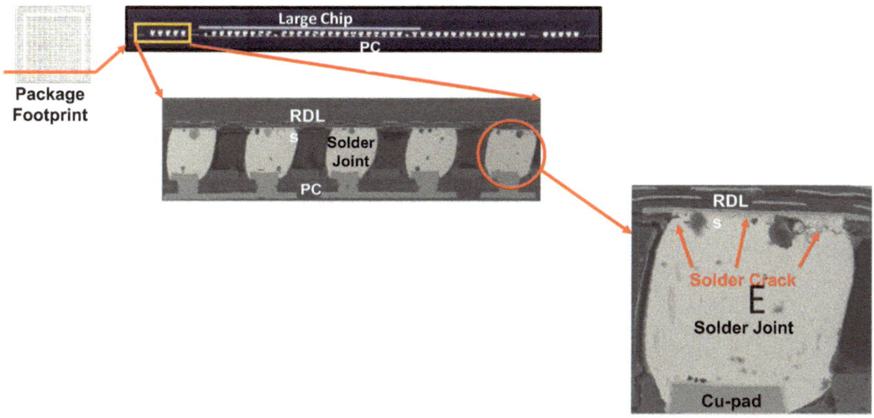

Fig. 4.143 Failure location and failure mode

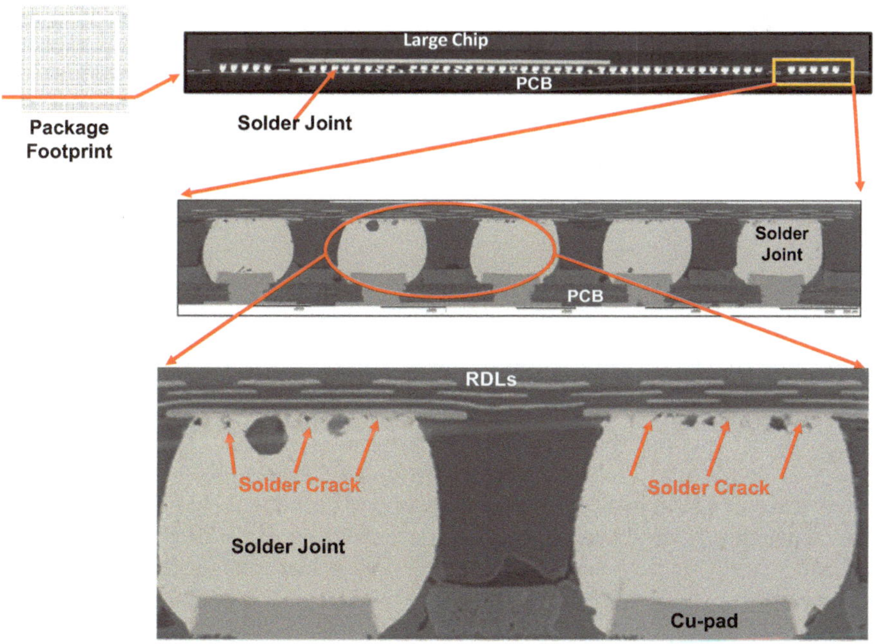

Fig. 4.144 Failure location and failure mode

the mean/characteristic life and Weibull slope. Confidence applies to the test itself and reliability applies to the product!

To predict the true Weibull slope (β_t) of the population, it is necessary to estimate the Weibull slope error in relation to the sample size and the required confidence level. The error (E) in the Weibull slope depends on the number of failures (N) and the required confidence level (C), which can be searched from equation [48]:

$$\frac{1}{\sqrt{\pi}} \int_{-\infty}^{E\sqrt{2N}} e^{-\frac{t^2}{2}} dt = \frac{1+C}{2}$$

This integral cannot be carried out in close form and is approximated by the following [48]

$$\frac{1}{\sqrt{2\pi}} \int_{-\infty}^{E\sqrt{2N}} e^{-\frac{t^2}{2}} dt = 1 - Z(x)\left(b_1 t + b_2 t^2 + b_3 t^3 + b_4 t^4 + b_5 t^5\right) + \varepsilon(x)$$

where

$$Z(x) = \frac{1}{\sqrt{2\pi}} e^{-\frac{x^2}{2}}$$

$$t = \frac{1}{1 + px}$$

$$x = E\sqrt{2N}$$

$p = 0.2{,}316{,}419$,
$b_1 = 0.31{,}938{,}153$,
$b_2 = 2.356{,}563{,}782$,
$b_3 = 1.781{,}477{,}937$,
$b_4 = 1.821{,}255{,}978$,
$b_5 = 1.330{,}274{,}429$,
$|\varepsilon(x)| < 7.5 \times 10^{-8}$.

Thus, for a given C (the confidence level) and N (the number of failures), the E (error) can be searched from the aforementioned equations. In our case, for 90% confidence, i.e., $C = 0.9$ and $N = 12$, we can search that $E = 0.32$. Thus, the true Weibull slope β_t of the heterogeneous integration solder joints will be happened in the intervals of $(2.28 - 0.32 \times 2.28) \le \beta_t \le (2.28 + 0.32 \times 2.28)$ or $1.55 \le \beta_t \le 3.01$.

(F) Linear Acceleration Models

There are many linear acceleration models [89–95] for leadfree solder alloys. Basically, they are based on (1) frequency and maximum temperature [92, 94], (2) dwell time and maximum temperature [83, 91], and (3) frequency and mean temperature [95]. The one proposed by Lall et al. [94] will be adopted in the current section.

$$\alpha = \left(\frac{\Delta T_t}{\Delta T_o}\right)^{2.3} \left(\frac{f_o}{f_t}\right)^{0.3} \exp\left[4562\left(\frac{1}{T_{\max,o}} - \frac{1}{T_{\max,t}}\right)\right]$$

In this equation, $T_{\max,t}$, f_t, ΔT_t, and $T_{\max,o}$, f_o, ΔT_o are the maximum temperature during cycling (in degrees Kelvins), the temperature cycling frequency, and the temperature range (in degree Celsius), respectively, at testing condition and at operating condition. In our case, $\Delta T_t = 180$ °C, $T_{\max,t} = 125$ °C, and $f_t = 28.8$ cycle/day, and we know that the solder joint can survive up to 300 cycles (Table 4.11 and Fig. 4.141). However, we also need to know if this value of 300 cycles is sufficient to meet the 5 year operating condition given by 0–80 °C with 1 cycle per day ($\Delta T_o = 80$ °C, $T_{\max,o} = 80$ °C, and $f_o = 1$ cycle/day).

$$\alpha = \left(\frac{180}{80}\right)^{2.3} \left(\frac{1}{28.8}\right)^{0.3} \exp\left[4562\left(\frac{1}{273+80} - \frac{1}{273+125}\right)\right]$$

$$\alpha = 10.76$$

The solder joints will survive 10.76 × 300 cycles per day = 3,228 ÷ 365 = 8.84 years > 5 years (Fig. 4.145).

4.8.8 Thermal Cycling Simulation of the Package PCB Assemble

The simulation of the heterogeneous integration PCB assembly has been reported in Sect. 4.8.4. The accumulated creep strain time history at various solder joint locations is shown in Fig. 4.144. It can be seen that the maximum accumulated creep strain occurs at the corner solder joint A underneath the Chip 1 and the second maximum, accumulated creep strain occurs at the corner solder joint F underneath Chip 2A. Thus, any solder joint failure should initiate at these locations. This is confirmed with the failure analysis of thermal cycling test results, as shown in Figs. 4.146 and 4.147.

The accumulated creep strain contours at solder joint A and solder joint F at 450 s (85 °C) and 2,250 s (−40 °C) are shown in Figs. 4.146 and 4.147, respectively. It can be seen that for both solder joints and at 85 and −40 °C, the maximum accumulated creep strain occurs near the interface between the RDL-substrate and the bulk solders. Thus, the failure mode of the solder joints is the cracking of solder near the interface between the RDL and the bulk solders. Again, this failure mode confirms with the thermal cycling test results (Figs. 4.146 and 4.147).

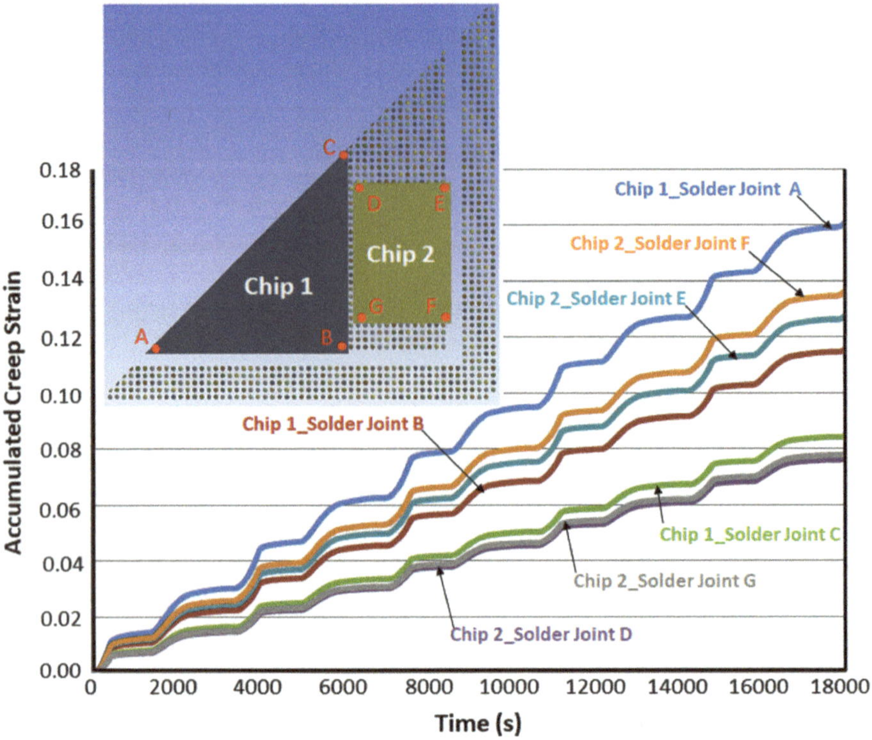

Fig. 4.145 Accumulated creep strain versus time at various locations of solder joints

4.8.9 *Summary*

Some important results and recommendations are summarized as follows.

- A very high-throughput fan-out chip-last panel-level packaging for heterogeneous integration has been demonstrated. Emphasis has been placed on the design, materials, process, fabrication, and reliability of the heterogeneous integration of one large chip (10 mm × 10 mm) and two small chips (7 mm × 5 mm) by a fan-out method on an RDL-first substrate.
- The test chips have been designed and the test-chip wafers have been fabricated. Also, the heterogeneous integration of the three-chip package has been designed and fabricated.
- The panel size for making the RDL-first substrate is 515 mm × 510 mm and has been divided into 12 strips. Each strip (240 mm × 74 mm) has 33 packages. Thus, in one shot, it can make the RDL-first substrate for 1188 chips in 396 heterogeneous integration packages.
- There are three RDLs of the RDL-first substrate. The target of the metal linewidth and spacing of the first RDL1 is 2 μm, the second RDL2 is 5 μm, and the

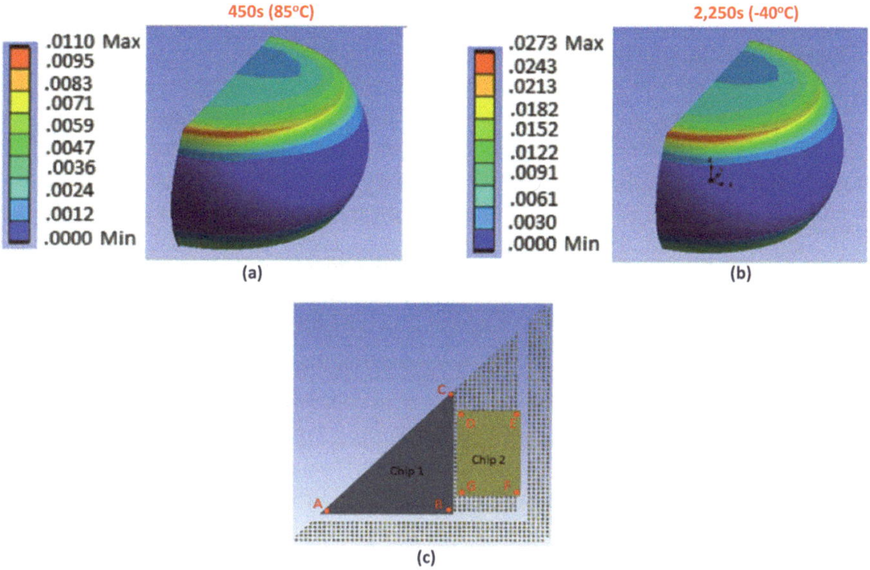

Fig. 4.146 Accumulated creep strain at solder joint A at 450 s (85 °C) and 2,250 s (240 °C)

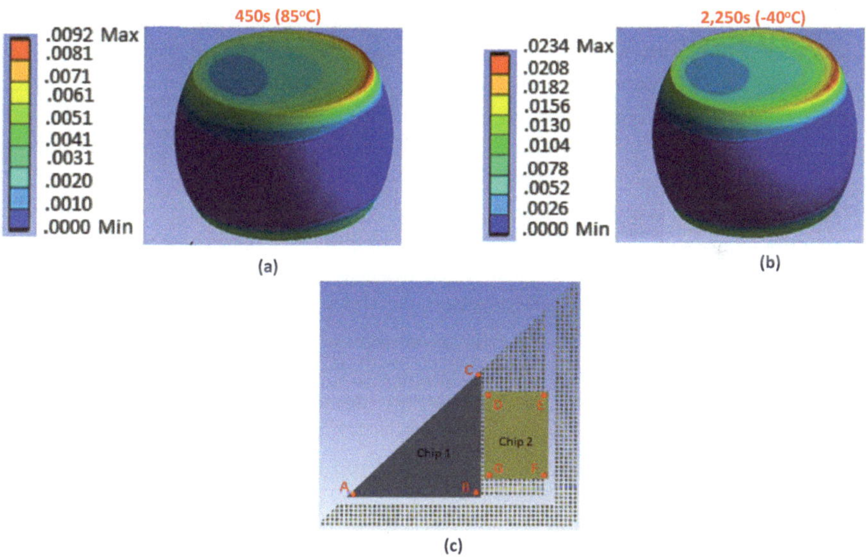

Fig. 4.147 Accumulated creep strain at solder joint F at 450 s (85 °C) and 2,250 s (240 °C)

third RDL3 is 10 μm. However, the fabricated metal linewidth of the first RDL1 was equal to 2.5 μm and the metal line spacing of the first RDL1 was equal to 1.69 μm. Thus, there are rooms for improvement, such as a better estimation of the compensation of photoresist, LDI, ECD Cu, Cu etching, etc. The fabricated metal linewidth and spacing of the RDL2 and RDL3 were close to the targets.

- The method of picking and placing all chips on the substrate and then mass reflowed had low yield. It is because of the warpage of the substrate and the thermal expansion mismatch between the chips and the substrate.
- TCB the chip-to-substrate worked. First, a flux was applied to the RDL-first substrate. Then, TCB the large (10 mm × 10 mm) chip and TCB the small chips (Chips 2A and 2B). The substrate temperature was at 120 °C and the bond-head temperature was at 250 °C.
- The bonded chip-to-substrate assembly has been underfilled with a curing condition equal to 165 °C for 2 h.
- The underfilled assembly has been laminated with an EMC with the following conditions: first stage: 120 °C/vacuum for 30 s and press (0.68 MPa) for 30 s and second stage: 100 °C and press (0.58 MPs) for 60 s.
- The fixtures for solder ball mounting of the heterogeneous integration of three-chip packages on a strip have been designed and fabricated. The fixtures consisted of the vacuum, holders, and two stencils: one is for fluxing and the other is for solder ball mounting.
- Reliability assessments, such as the thermal cycling of the heterogeneous integration of three-chip package PCB assembly, have been performed by a nonlinear temperature- and time-dependent finite-element simulation. The maximum accumulated creep strain per cycle at the corner solder joint is equal to 0.0164, which is too small to create a solder joint thermal-fatigue reliability problem, especially for mobile products such as smartwatches, smartphones, and tablets, for which the life is less than three years. The thermal cycling test has just begun, and it will take a few months to complete the test and failure analysis. The results will be presented in another future publication.
- Thermal cycling test of the heterogeneous integration of three-chip package PCB assembly has been performed. There were 49 channels, and each covered certain outer rows of solder joints. The test condition was −55 °C ↔ 125 °C at 50 min cycle.
- The failure criterion was when the resistance of the daisy chain of the heterogeneous integration PCB assembly increased by 50%. The cycle at which the first solder joint of a channel failed is considered as the cycle-to-failure of the heterogeneous integration package.
- For the median rank, the Weibull slope and the characteristic life of the solder joints of the package assembly are, respectively, 2.28 and 1,702 cycles.
- For 90% confidence (in nine of 10 cases), the true Weibull slope is $1.55 \leq \beta_t \leq 3.01$ and the true 10% life is $387 \leq x_t \leq 817$ cycles.
- The typical failure location of the heterogeneous integration package PCB assembly occurs near the solder joints underneath the corner (A) of Chip 1 and the corner (F) of Chip 2A.
- The typical failure mode of the heterogeneous integration package PCB assembly is the cracking of solder joint near the interface between the RDL-substrate and the bulk solders.

- For an operating condition (0–80 °C with 1 cycle per day), the solder joint of the heterogeneous integration of 3-chip package can survive 8.84 years. A nonlinear, time- and temperature-dependent 3-D finite element simulation of the heterogeneous integration PCB assembly confirmed the failure location and failure mode with the thermal cycling test results.

4.9 Hybrid Substrate for Heterogeneous Integration

Since the invention of surface laminar circuit (SLC) by IBM in Japan at Yasu, Tsukada et al. [96], Tsukada and Tsuchida [97], and Tsukada [98] in the early 1990s, the build-up package substrate (carrier) has been used substantially for semiconductor advanced packaging [1] such as that schematically shown in Fig. 4.148a. It can be seen that the package substrate is supporting multiple chips with controlled collapse chip connection (C4) bumps or Cu pillar with solder cap (μbumps) and underfill (2D IC integration). Usually, the package substrate consists of a core and build-up layers on both sides of the core. For example, the second-generation extreme-performance yield computing (EPYC), and 7002-series shipped by AMD in mid-2019 has a 9-2-9 (nine upper build-up layers, two core-layer, and nine bottom build-up layers) package substrate [99, 100].

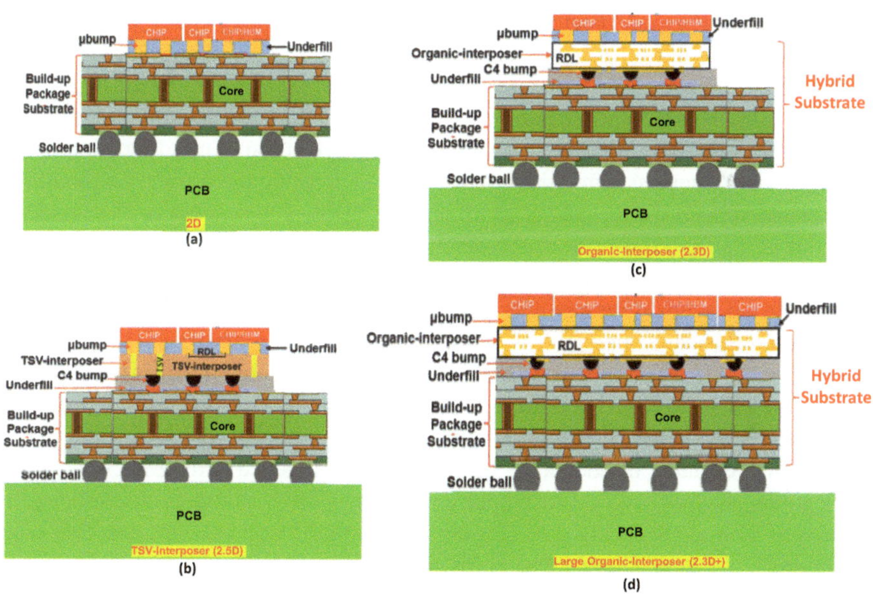

Fig. 4.148 **a** 2-D IC integration. **b** 2.5-D IC integration. **c** 2.3-D IC integration. **d** 2.3-D+ IC integration

In the past decade, because of the very high-density, high I/Os, high-performance, and ultrafine pitch requirements such as the field-programmable gate array (FPGA), most package substrates are not enough to support the chips and a through-silicon via (TSV) interposer [13, 14] such as that schematically shown in Fig. 4.148b is needed (2.5D IC integration). For example, the Virtex-7 HT family shipped by Xilinx/TSMC in 2013 [101] has a TSV interposer (10 μm diameter and 100 μm deep) with four top redistribution layers (RDLs): three Cu damascene layers and one aluminum layer. The lateral interconnections between the sliced FPGA chips are connected mainly by the 0.4 μm-pitch (minimum) RDLs of the interposer [102, 103]. In this case, because the TSV interposer does most of the work, a 5-2-5 build-up package substrate is adequate. However, the TSV interposer is very expensive due to its large size which leads to high yield loss, and most users are paying for its performance.

The very first 2.3D IC integration paper was published by STATSChipPac [33] at IEEE/ECTC 2013. Their motivation is to save costs by replacing the TSV interposer with a fan-out [33] fine metal L/S RDL substrate (or organic interposer). The hybrid substrate structure consists of a build-up package substrate, C4 solder joints with underfill, and an organic interposer, which is schematically shown in Fig. 4.148c. Since then, there are many publications [104–123]. Works [33, 104–107] are with fan-out chip-first packaging process, while works [108–123] are with fan-out chip-last (RDL-first) packaging process. In this chapter, only hybrid substrates fabricated with fan-out chip-last (or RDL-first) will be discussed. The minimum metal L/S of the organic interposer is usually 2 μm. High-volume manufacturing with this hybrid substrate technology is expected by the end of 2023.

Usually, the build-up package substrate and the fan-out organic interposer are fabricated separately. Then, there are at least two different assembly processes shown in Figs. 4.149 and 4.150. Most people use Fig. 4.149; they first bond the chips to the organic interposer. It is followed by underfill and EMC (epoxy molding compound) molding. Then, they assemble the module (chip + organic interposer) on the build-up package substrate with solder bump and underfill. The process in Fig. 4.150 is first combining the build-up package substrate and the organic interposer with solder bump and underfill (the hybrid substrate) and then testing it and making sure it is a known-good hybrid substrate. Finally, bond the chips on the known-good hybrid substrate. In this case, there is less chance of losing the known-good die. Also, the logistic is simpler; after receiving the known-good hybrid substrate from the substrate houses, the OSAT (outsourced semiconductor assembly and test) houses just bond the chips/HBMs on the known-good hybrid substrate.

Recently, because of the drive of artificial intelligence (AI) and 5G/6G, the applications, such as high-performance computing (HPC), autonomous vehicle, the Internet of Things (IoT), big data (for cloud computing), and instant data (for edge computing), are demanding more chips in a package [124]. Thus, it is logical to enlarge the organic interposer to the size of the package substrate, so it can support more chips, as schematically shown in Fig. 4.148d (2.3D+ IC integration).

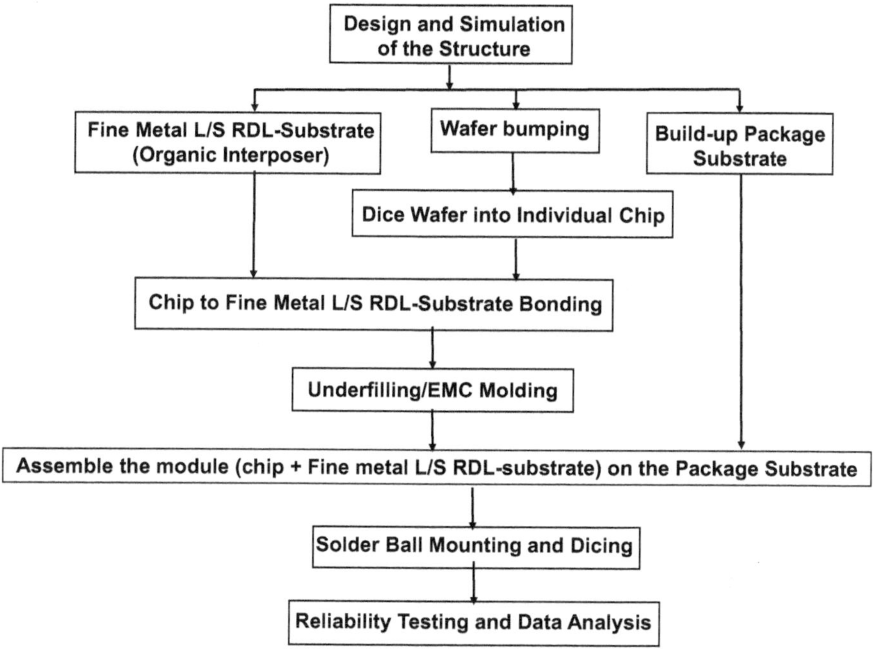

Fig. 4.149 Process flow for chip-bond on organic interposer first and then combined with build-up package substrate

4.10 Hybrid Substrate with PID by RDL-First Panel-Level Packaging

4.10.1 Test Chips

The test chips are shown in Fig. 4.151. The dimensions of the large chip are 10 mm × 10 mm × 150 μm, and there are 3592 daisy-chained pads. The dimensions of the small chip are 5 mm × 5 mm × 150 μm with 1072 daisy-chained pads. For both chips, the pitch of the outer peripheral pads is 50 μm and of all the inner pads is 200 μm [21, 22].

For both chips, the wafer bumping is with the standard physical vapor deposition (PVD) for seed layer and electrochemical deposition (ECD) for Cu-pillar and solder. The wafers are then diced into individual bumped chips.

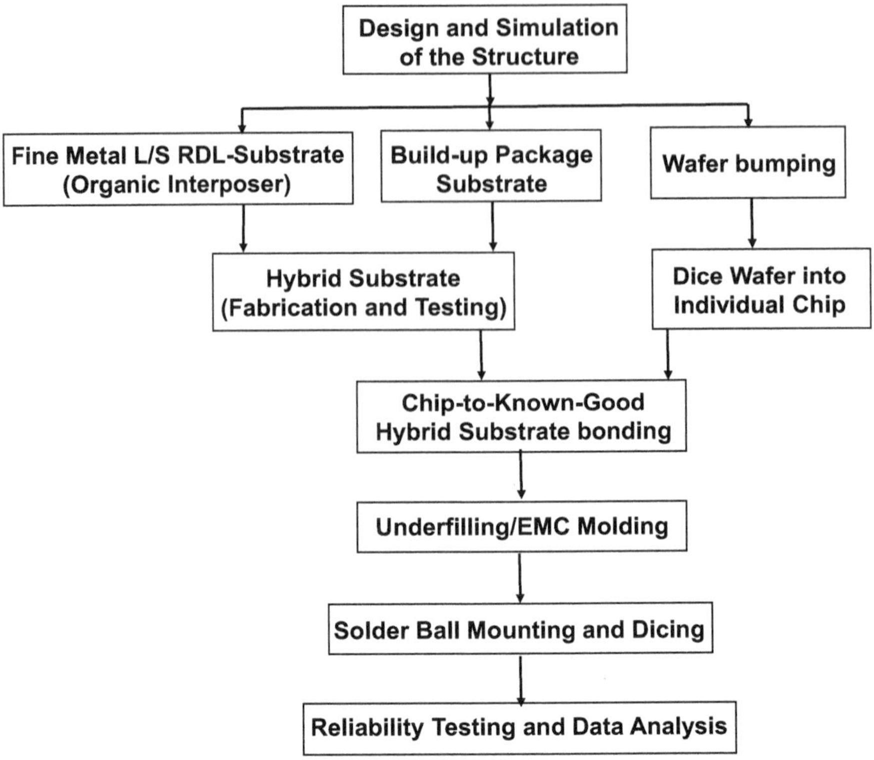

Fig. 4.150 Process flow for combining the organic interposer and build-up package substrate first and then perform chip bonding

4.10.2 Test Package

Figures 4.152, 4.153, 4.154, 4.155 and 4.156 schematically show the test package. It can be seen that there are two chips: Chip 1 (10 mm × 10 mm) and Chip 2 (5 mm × 5 mm). The gap (spacing) between Chip 1 and Chip 2 is 100 μm. The chips are bonded on the hybrid substrate which consists of the fine metal L/S RDL substrate (20 mm × 15 mm × 32 μm) (organic interposer) and build-up substrate (23 mm × 23 mm × 1.3 mm), as shown in Figs. 4.152, 4.153, 4.154, 4.155 and 4.156.

For both chips, the Cu pad dimensions are 40 μm × 40 μm, the Ti/Cu (0.1 μm/0.2 μm) under bump metallurgy (UBM) pad size is 32 μm diameter, and the PI2 (passivation) opening is 20 μm diameter. The diameter of the Cu-pillar is 32 μm and its height is 22 μm, and the height of the SnAg solder cap is 15 μm plus a 3 μm Ni-barrier.

There are three RDLs in the RDL substrate and each RDL consists of one metal layer (ML) and one dielectric layer (DL). Figure 4.154 shows the definition and value of RDLx, MLx, DLxy, Vxy, and L/S. The solder resist opening, and thickness are, respectively, 80 and 2 μm.

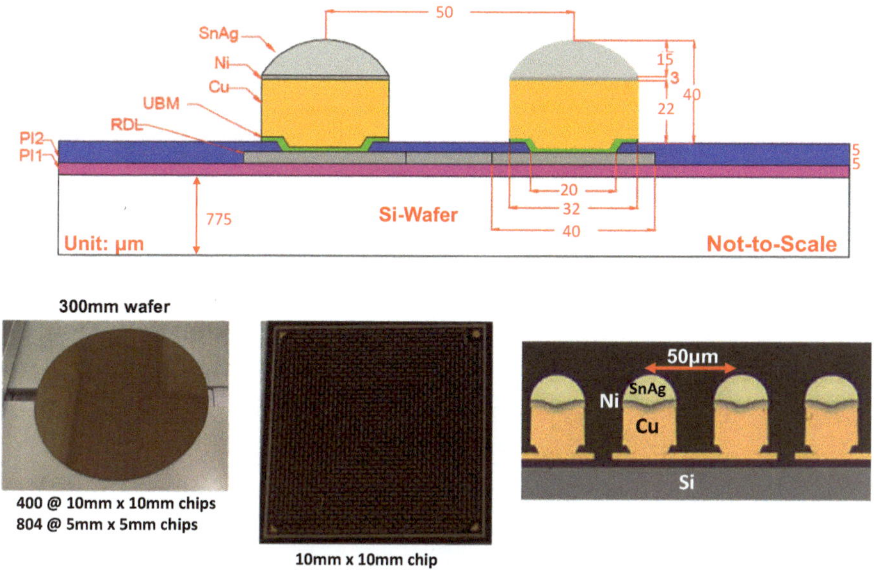

Fig. 4.151 Test chips with 50 μm pitch

The temporary panel for fabricating the RDL substrate is shown in Fig. 4.155. It can be seen that the panel size is 515 mm × 510 mm × 1.1 mm and is made of glass with a coefficient of thermal expansion (CTE) equals to 8.5 × 10^{-6}/°C. The panel is divided into four strips and each strip (257.5 mm × 255 mm) has 154 (20 mm × 15 mm) RDL substrates. Thus, in one shot, it can make RDL substrates for 1232 chips in 616 heterogeneous integration packages. Figure 4.156 shows the top side and bottom side of the hybrid substrate. (1) On the top side, there are 3592 + 1072 = 4664 pads, Fig. 4.156a, which are for bonding the chips to the substrate. (2) On the bottom side, there are 4039 solder ball pads (105 μm diameter) on a 0.225 mm pitch, Fig. 4.156c, which are solder mask defined (SMD) and its opening is 80 μm. These pads are for Sn3Ag0.5Cu lead-free C4 solder bump mounting, and the solder ball diameter is 80 μm.

4.10.3 Organic Interposer Fabrication

A released film (sacrificial layer) is slit coating on a glass panel and then a Ti/Cu seed layer is deposited on the top, as shown in Fig. 4.157. The thickness of the sacrificial layer is 1 μm and it is a light-to-heat conversion film. The contact pad can be obtained by photoresist, laser direct imaging (LDI), development, ECD Cu, and stripping off the photoresist. Then, a photoimageable dielectric (PID) is slit coating and the first DL DL01 of RDL1 is obtained by LDI. It is followed by PVD the Ti/Cu, photoresist,

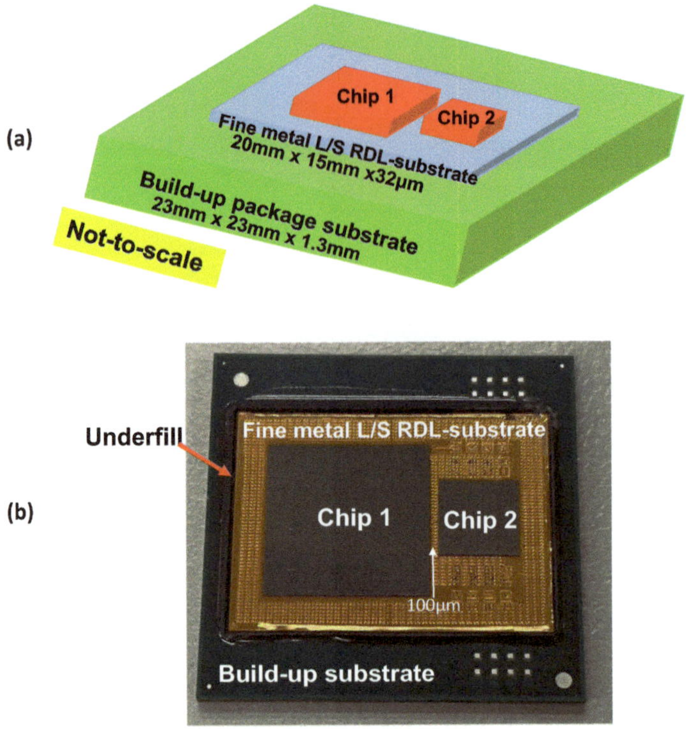

Fig. 4.152 **a** Schematic of the test package. **b** Assembled test package. Chips are bonded on the hybrid substrate which consists of the (20 mm × 15 mm) fine metal L/S RDL substrate and (23 mm × 23 mm) build-up substrate

LDI and development, ECD the Cu, strip off the photoresist, and etch off the TiCu to obtain the first ML ML1 of RDL1. Figure 4.156b shows the image of the RDL1 through the glass carrier. DL12 and ML2 of RDL2, and DL23 and ML3 of RDL3 can be obtained by repeating the same process steps.

The solder mask (passivation) can be obtained by slit coating a PID and LDI. Then, the surface is finishing with electroless nickel electroless palladium immersion gold (ENEPIG). The fabrication of the fine metal L/S RDL substrate is completed. The bottom side of the RDL-substrate is shown in Fig. 4.156d.

The SEM images of the RDL substrate are shown in Fig. 4.158. Table 4.12 shows the L, S, and thickness (H). For RDL1, L = 2.07 and 2.08 μm (not exactly 2 μm), S = 1.74 and 1.81 μm (not 2 μm), and H = 2.06 and 2.17 μm (not 3 μm). For RDL2, L = 5.6 and 6.03 μm (not 5 μm), S = 3.91 and 4.5 μm (not 5 μm), and H = 4.35 and 4.48 μm (not 5 μm). For RDL3, L = 10.24 and 10.69 μm (not 10 μm), S = 9.46 and 10.18 μm (not 10 μm), and H = 3.39 and 5.11 μm (not 5 μm). Thus, there are rooms for improvement such as using better PID, higher resolution LDI, or a stepper.

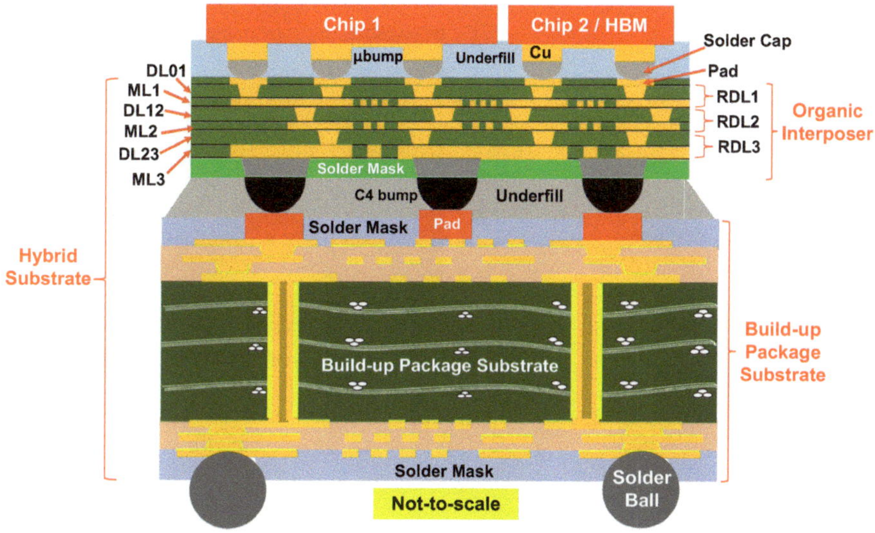

Fig. 4.153 Schematic of the cross section of the heterogeneous integration test package

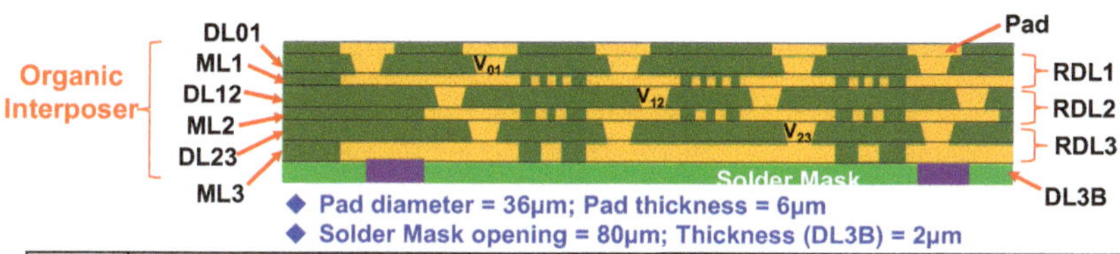

RDLs	Key Elements of RDL	Value	Line width (L) / Spacing (S)
RDL1	ML1 (Metal layer 1)	3μm	2/2μm
	DL01 (Dielectric layer between the contact pad and ML1)	3μm	NA
	V_{01} (Via opening between contact pad and ML1)	10μm	NA
RDL2	ML2 (Metal layer 2)	5μm	5/5μm
	DL12 (Dielectric layer between ML1 and ML2)	6μm	NA
	V_{12} (Via opening between ML1 and ML2)	20μm	NA
RDL3	ML3 (Metal layer 3)	5μm	10/10μm
	DL23 (Dielectric layer between ML2 and ML3)	4μm	NA
	V_{23} (Via opening between ML2 and ML3)	18μm	NA

Fig. 4.154 Fine metal L/S RDL (organic) substrate

4.10.4 *Build-Up Substrate Fabrication*

The package substrate (23 mm × 23 mm × 1.3 mm) is a 2-2-2 structure, as schematically shown in Fig. 4.159a. There are 4039 pads on a 0.225 mm pitch on each

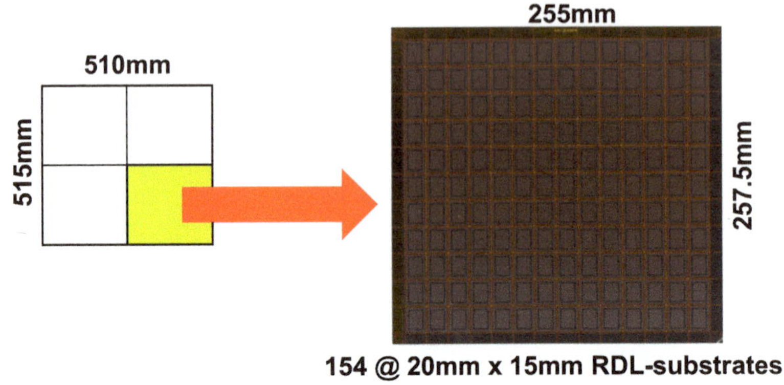

Fig. 4.155 Panel for fabricating the RDL-substrate and the fabricated one

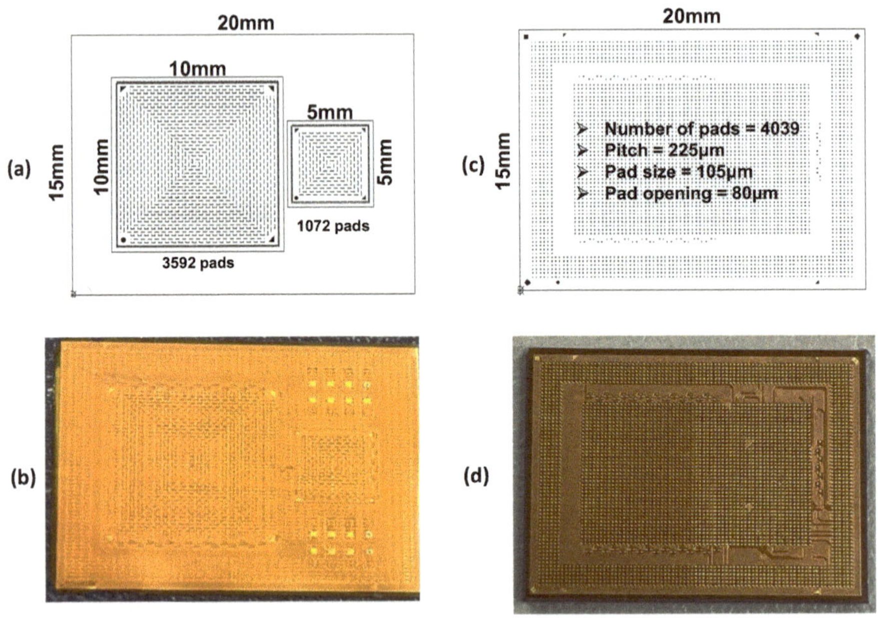

Fig. 4.156 **a** Schematic of the top view of RDL substrate. **b** Top view of the fabricated RDL substrate. **c** Schematic of the bottom view of RDL substrate. **d** Bottom view of the fabricated RDL substrate

substrate, which is made on a 510 mm × 510 mm panel, Fig. 4.159b, by a conventional method. The image of the 2-2-2 build-up substrate is shown in Fig. 4.159c. It can be seen that for the through via, the open size is 200 μm and the depth is 820 μm, and for the stacked via, the open size is 60 μm and the depth is 30 μm. The top side and bottom side of the fabricated build-up package substrate panel are shown in Fig. 4.159d, e, respectively.

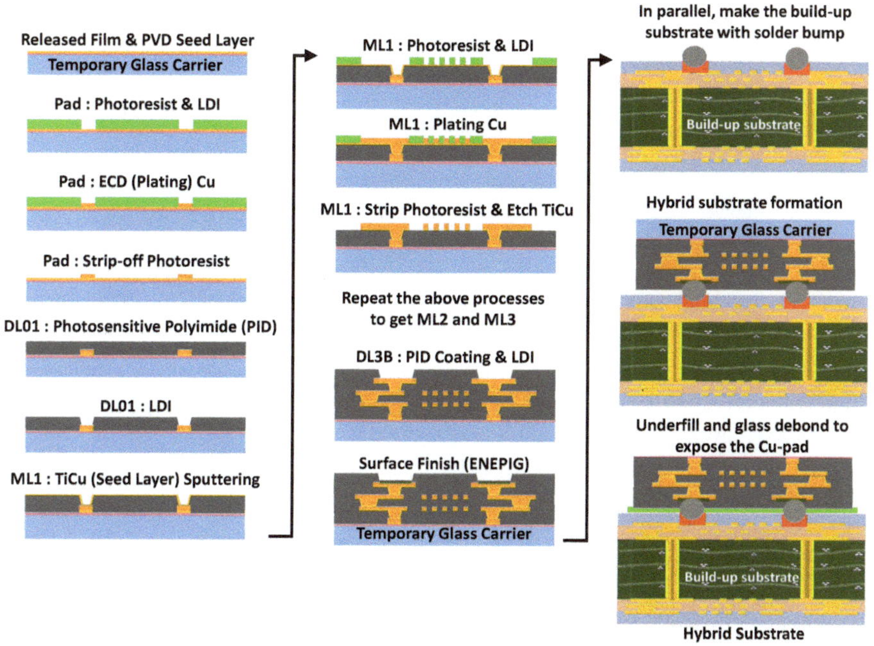

Fig. 4.157 Key steps to fabricate the hybrid substrate (fine metal L/S RDL substrate + build-up substrate)

Table 4.12 RDLs: design versus measurement values

RDLs	Items	Design (μm)	Measured (μm)
Metal of RDL1 (ML1)	Linewidth (L)	2	2.07, 2.08, …
	Spacing (S)	2	1.74, 1.81, …
	Thickness (H)	3	2.06, 2.17, …
Metal of RDL2 (ML2)	Linewidth (L)	5	5.6, 6.03, …
	Spacing (S)	5	3.91, 4.5, …
	Thickness (H)	5	4.35, 4.48, …
Metal of RDL3 (ML3)	Linewidth (L)	10	10.24, 10.69, …
	Spacing (S)	10	9.46, 10.18, …
	Thickness (H)	5	3.39, 5.11, …

Figure 4.160 shows the individual build-up package substrate. Figure 4.160a, b show, respectively, the schematic and fabricated top side of the package, while Fig. 4.160c, d show, respectively, the schematic and fabricated bottom side of the package. There are 475 pads with 1 mm pitch on the bottom side of the package. The pad size is 500 μm and is SMD with the solder mask opening = 300 μm.

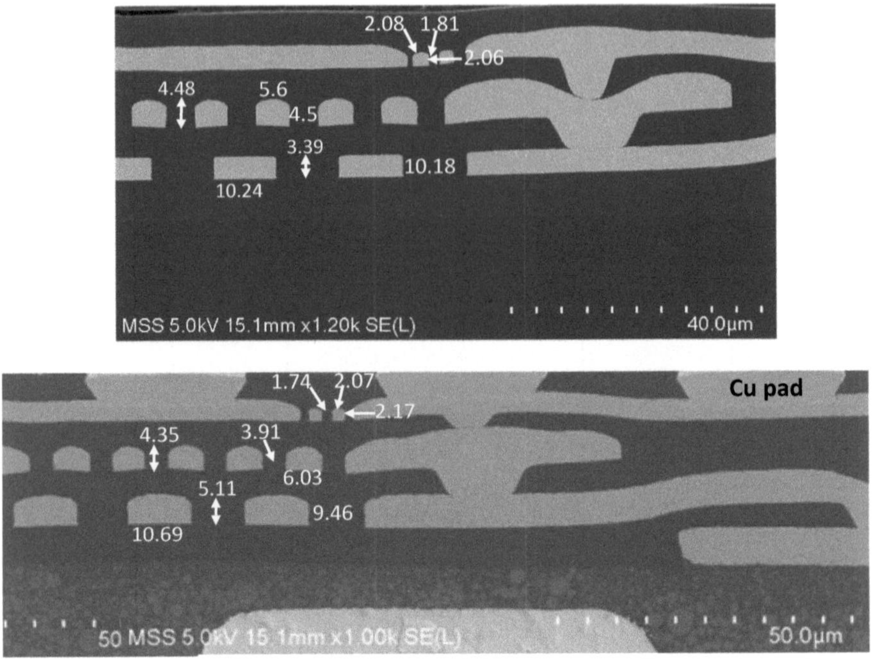

Fig. 4.158 Cross sections of SEM image of RDL substrate

4.10.5 C4 Bump Formation

The C4 bumps are fabricated by stencil printing a Sn3Ag0.5Cu solder paste with a 29 μm-thick stainless-steel stencil on the build-up package substrate. During solder-reflow process, due to the surface tension of the molten solder, which creates smooth truncated spherical 30 μm diameter solder bumps, as shown in Fig. 4.161.

4.10.6 Warpage Measurements

The warpage of the build-up package substrate (BU), fine metal L/S RDL substrate with glass carrier RDL (G), and fine metal L/S RDL substrate with organic carrier RDL (O) at various temperatures has been measured by the shadow Moiré method via the TherMoire Platform. The results are shown in Fig. 4.162. It can be seen that the warpage of BU and RDL (G) is very small. However, the warpage of RDL (O) is very large compared with the others. (This is because the thermal expansion mismatch between the glass carrier and the RDL substrate is smaller than that between the organic carrier and the RDL substrate.) Thus, in this section, the hybrid substrate

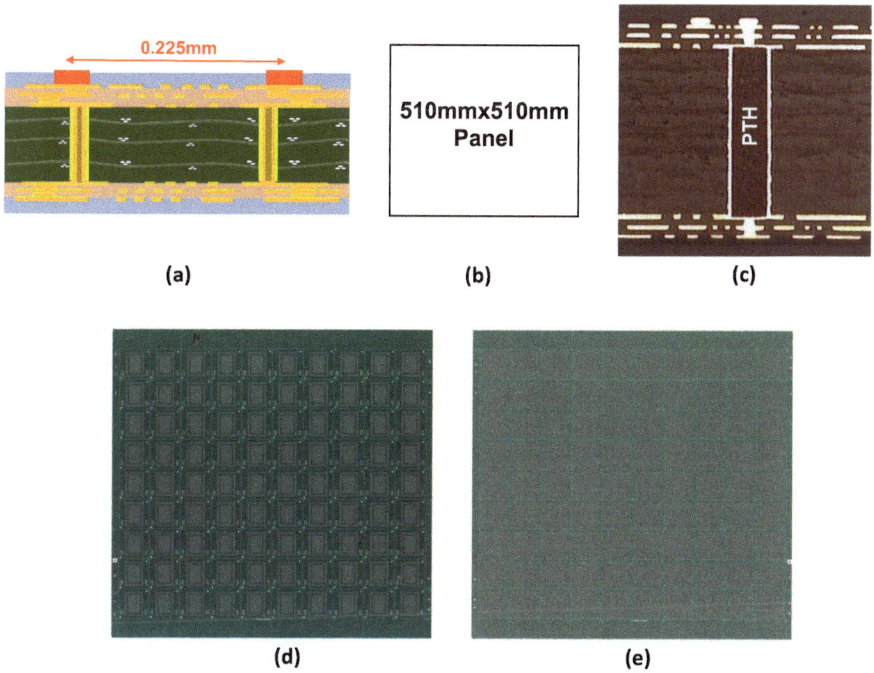

Fig. 4.159 2-2-2 Panel. **a** Schematic. **b** Panel size. **c** Cross section image. **d** Top view of the fabricated build-up substrate. **e** Bottom view of the fabricated build-up substrate

will be formed by the combination of the build-up package substrate and the fine metal L/S RDL substrate with the temporary glass carrier.

4.10.7 Hybrid Substrate Formation

Now that we have the fine metal L/S RDL substrate with glass carrier and the build-up package substrate with solder bumps ready, it is time to fabricate the hybrid substrate, as shown in Fig. 4.157. First, identify the location of the bumps on the package substrate and the pads on the RDL substrate with glass carrier by using a lookup and lookdown camera; second, apply flux on both the bumps and pads; and third, pick and place the RDL substrate with glass carrier on the package substrate and then reflow. The top view of an assembled hybrid substrate is shown in Fig. 4.163. A typical SEM image of the hybrid substrate is shown in Fig. 4.164 and close-up images are shown in Fig. 4.165. It can be seen that the fine metal L/S RDL substrate, solder joints, underfill, and the build-up package substrate are properly assembled.

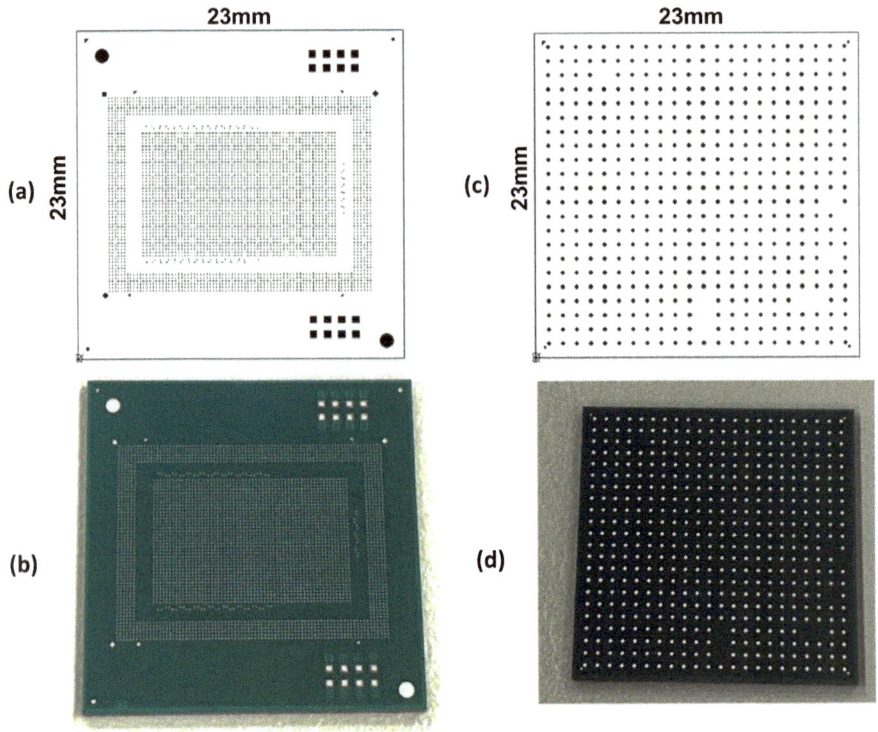

Fig. 4.160 Individual build-up package substrate. **a** Top side. **b** Fabricated top side. **c** Bottom side. **d** Fabricated bottom side

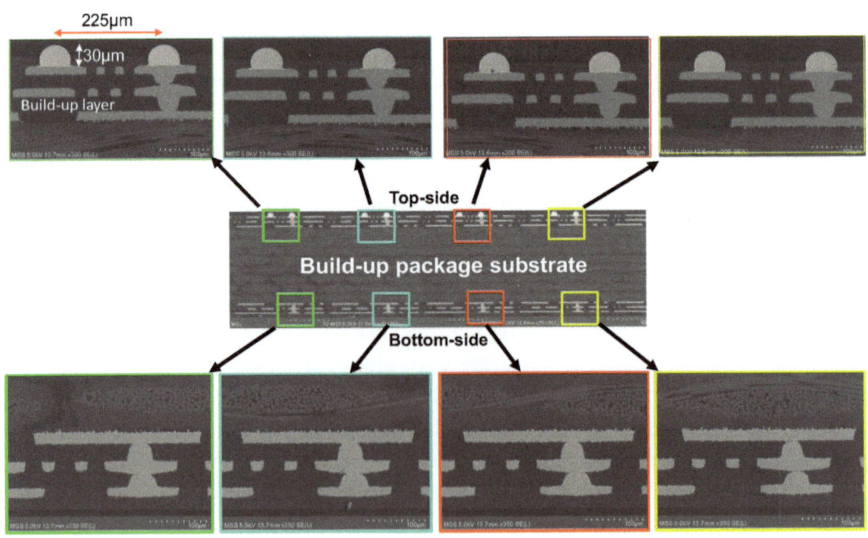

Fig. 4.161 SEM images of the fabricated package substrate with C4 bumps

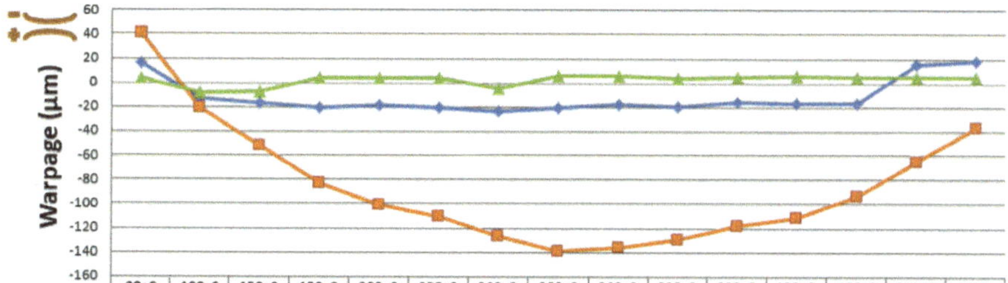

	30oC	100oC	150oC	180oC	200oC	220oC	240oC	260oC	240oC	220oC	200oC	180oC	150oC	100oC	30oC
BU	16.9	-12.8	-16.2	-19.7	-17.8	-20.2	-23.1	-20	-17.1	-18.8	-15.5	-16.5	-16.8	16.1	19.1
RDL(O)	41.7	-19.9	-51.5	-82.8	-101	-110.3	-125.9	-138.3	-135.6	-129.2	-117.4	-110.8	-92.9	-64.9	-35.8
RDL(G)	4	-8	-7	4	4	4	-4	6	6	4	5	6	5	5	5

Fig. 4.162 Shadow Moire warpage measurement of build-up substrate (BU), fine metal L/S RDL substrate on glass carrier RDL (G), and fine metal L/S RDL substrate on organic carrier RDL (O)

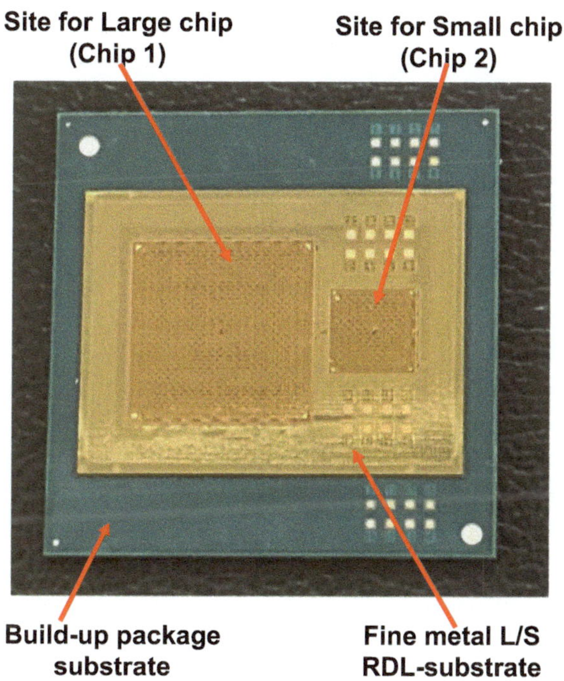

Fig. 4.163 Hybrid substrate = fine metal L/S RDL substrate with glass carrier soldered on the build-up package substrate

4.10.8 Remove the Temporary Glass Panel

Before the final package assembly, we have to debond the glass panel to expose the bonding pads on the fine metal L/S RDL substrate, as shown in Fig. 4.157. By

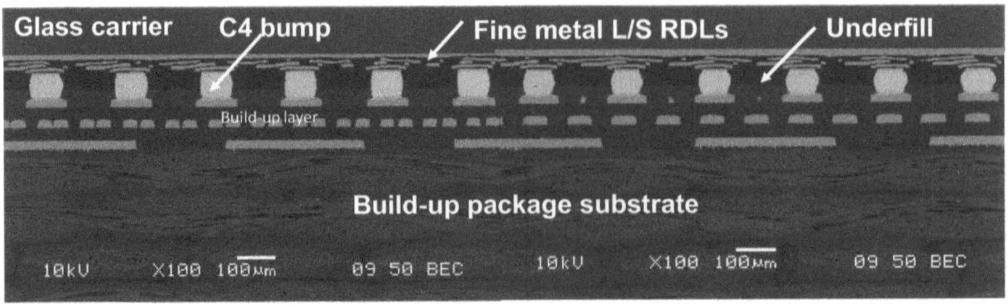

Fig. 4.164 Cross section image of the hybrid substrate

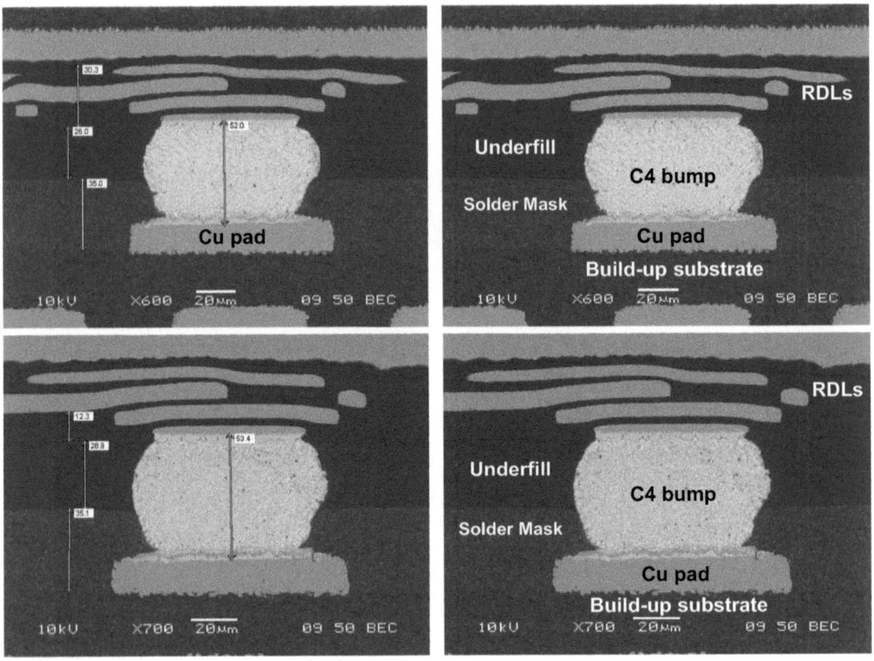

Fig. 4.165 Close-up look at the cross-section SEM images of the hybrid substrate

scanning a laser with 355 nm of wavelength on the glass, the sacrificial layer will become a powder and the glass carrier is removed. Then, chemicals clean the surface.

4.10.9 Chips-to-Hybrid Substrate Bonding

Now that we have the microbumped chips (Fig. 4.151) and the hybrid substrate (Figs. 4.163, 4.164 and 4.165) ready, it is time to do the final assembly. First, identify

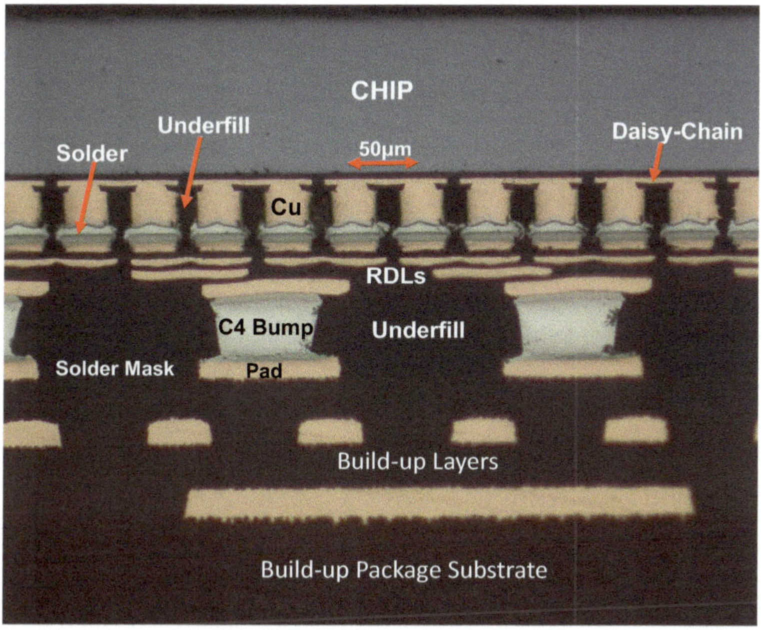

Fig. 4.166 Heterogeneous integration of two-chip package on the hybrid substrate which is the fine metal L/S RDL substrate and build-up substrate

the location of the microbumps on the chips and the bonding pads on the hybrid substrate by using a lookup and lookdown camera; second, apply flux on both the bumps and pads; and third, pick and place the chips on the hybrid substrate and then reflow. The assembled two-chip package is shown in Fig. 4.152b.

4.10.10 Underfilling

The curing temperature, time, and CTE of the underfill are, respectively, 160 °C, 2 h, and 22×10^{-6}/°C (<Tg = 115 °C). Figure 4.166 shows a typical cross section, which can be seen that the chip, microbumps, underfill, RDL substrate, C4-bumps, and build-up package substrate are properly done.

4.10.11 Thermal Cycling Simulation and Results

Please see Sect. 4.11.8 of this book.

4.10.12 Summary

Some important results are summarized as follows.

- The feasibility of the design, materials, process, and fabrication of a heterogeneous integration of two chips with 50 μm pitch on a hybrid substrate with PID by a fan-out RDL-first panel-level packaging has been demonstrated.
- In order to increase throughput, the fine metal L/S (2 μm/2 μm) RDL substrate (organic interposer) has been fabricated on a temporary glass panel.
- The hybrid substrate has been fabricated by soldering the fine metal L/S RDL substrate on a build-up package substrate and underfilled.
- The 50 μm pitch microbumped chips are bonded to the hybrid substrate and underfilled. There are not any carriers transferred.

4.11 Hybrid Substrate with ABF by RDL-First Panel-Level Packaging

4.11.1 The Structure and Material Properties

The structure of this section is exactly the same as that in Sect. 4.10, except the dielectric material for making the organic interposer has been changed from PID to ABF (Ajinomoto Build-Up Film) [23]. The dimensions of the organic interposer fabricated with the PID and ABF are shown in Fig. 4.167. It can be seen that for both cases, the minimum L/S are 2 μm/2 μm. However, metal line thickness for the ABF is thicker than that for the PID.

The material properties of the PID and ABF for fabricating the organic interposers as well as for the build-up package substrate are shown in Table 4.13. The dimensions of the organic interposer for both cases are given in Sect. 4.10.2.

4.11.2 Fabrication Process of the Organic Interposer

The fabrication process of the organic interposer with PID has been reported in Fig. 4.157 of Sect. 4.10. Figure 4.168 shows the fabrication process of the organic interposer with ABF. The scanning electron microscope (SEM) cross section images from the PID and ABF dielectric materials that the metal lines with the ABF dielectric (Fig. 4.169a) are much flatter than those with the PID dielectric (Fig. 4.169b). Table 4.14 shows the metal L, S, and H from design and fabrication. It can be seen that, for both dielectric materials, the fabricated values of the L/S/H are not exactly the same as those from the design values (the smaller the L/S/H the larger the discrepancy). Thus, there are rooms for improvements, e.g., better estimation of compensation of photoresist, LDI, ECD Cu, Cu etching, etc.

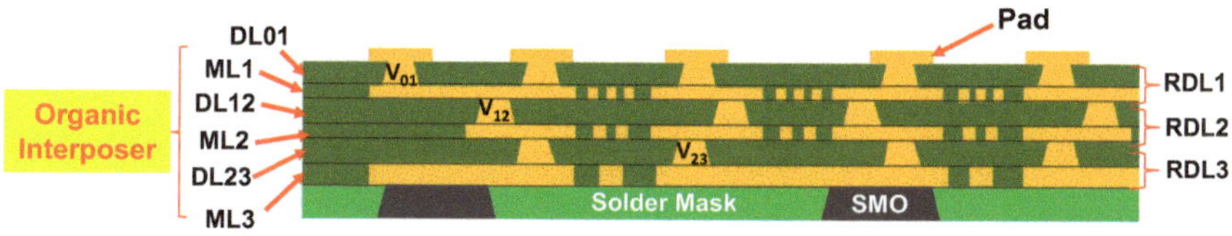

RDLs	Key Elements of RDL	Line width (L) / Spacing (S) (PID and ABF)	Thickness (H)	
			PID	ABF
RDL1	ML1 (Metal layer 1)	2/2μm	3μm	3μm
	DL01 (Dielectric layer between the contact pad and ML1)	NA	3μm	3.5μm
	V_{01} (Via opening between contact pad and ML1)	NA	10μm	20μm
RDL2	ML2 (Metal layer 2)	5/5μm	5μm	8μm
	DL12 (Dielectric layer between ML1 and ML2)	NA	6μm	10μm
	V_{12} (Via opening between ML1 and ML2)	NA	20μm	20μm
RDL3	ML3 (Metal layer 3)	10/10μm	5μm	8μm
	DL23 (Dielectric layer between ML2 and ML3)	NA	4μm	7.5μm
	V_{23} (Via opening between ML2 and ML3)	NA	18μm	20μm

Fig. 4.167 Dimensions of the organic interposer with PID and ABF

Table 4.13 PID and ABF material properties

Items	PID	ABF (Organic interposer)	ABF (Build-up substrate)
Curing condition, °C for min	200 for 60	200 for 90	190 for 90
CTE (25–150 °C), 10^{-6}/°C	60	37	39
CTE (150–240 °C), 10^{-6}/°C	–	98	117
Tg, °C	300	156	153
Dielectric constant (Dk)	3 (1 MHz)	3.2 (5.8 GHz)	3.2 (5.8 GHz)
Loss tangent (Df)	–	0.011 (5.8 GHz)	0.017 (5.8 GHz)
Young' modulus (23 °C), GPa	2.1	7.5	5.0
Tensile strength (23 °C), GPa	0.125	125	98
Elongation (23 °C), %	60	5.4	5.6

4.11.3 Warpage

Shadow Moiré method via the TherMoire Platform is used to measure the warpage of the build-up package substrate (Build-up) and the organic interposers with RDL (PID type) and RDL (ABF type). The results are shown in Fig. 4.170. It can be seen that the warpage of both organic interposers (with the temporary glass on) are

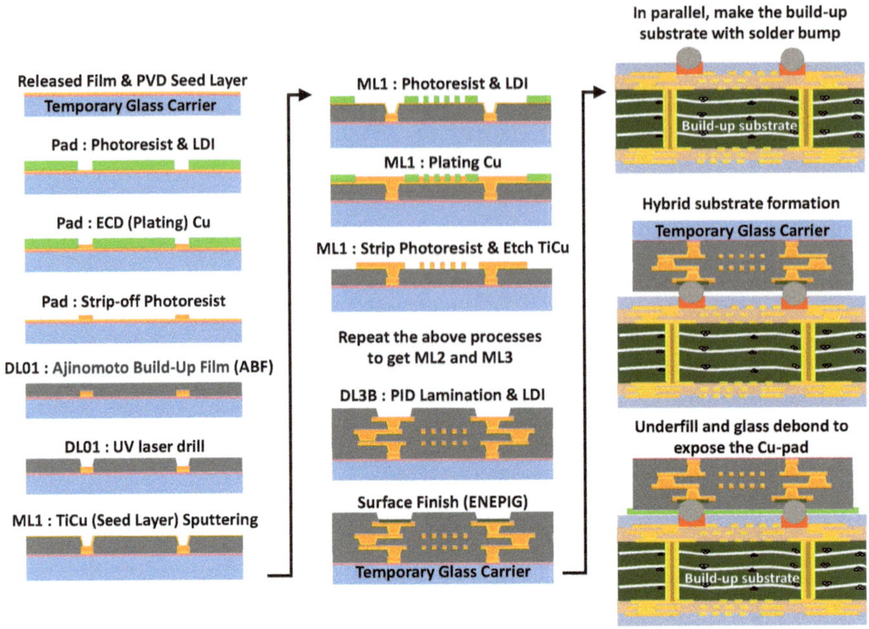

Fig. 4.168 Organic interposer with ABF process flow

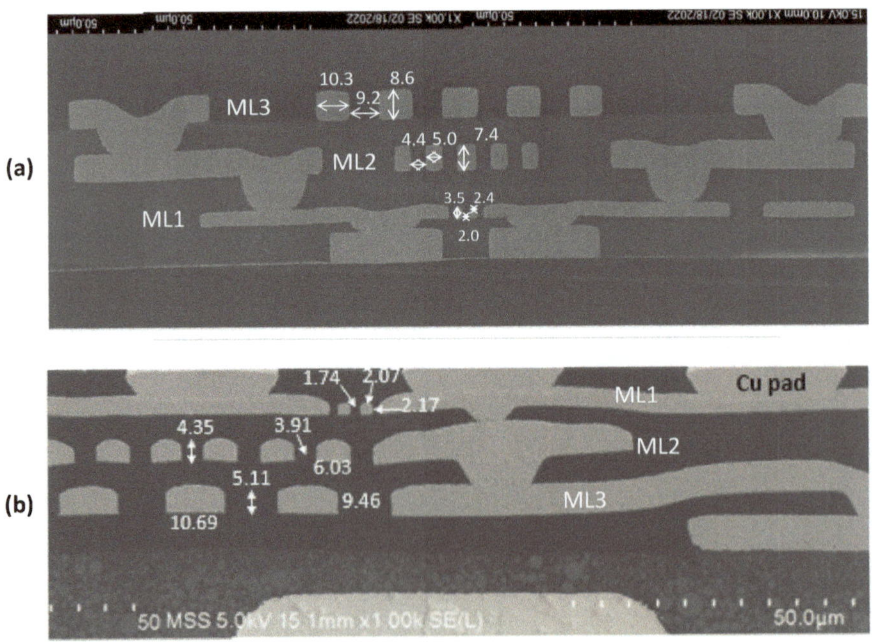

Fig. 4.169 **a** Metal lines (MLs) in RDLs with ABF dielectric and **b** metal lines (MLs) in RDLs with PID dielectric

Table 4.14 RDLs made from PID and ABF: design versus measurement values

RDLs	Items	Design (μm)	Measured (μm)	
			PID	ABF
Metal of RDL1 (ML1)	Linewidth (L)	2	2.1, 2.1, …	2.4, 2.4, …
	Spacing (S)	2	1.7, 1.8, …	1.8, 2.0, …
	Thickness (H)	3	2.1, 2.2, …	3.2, 3.5, …
Metal of RDL2 (ML2)	Linewidth (L)	5	5.6, 6.0, …	5.1, 5.0, …
	Spacing (S)	5	3.9, 4.5, …	4.8, 4.4, …
	Thickness (H)	5 (PID), 8 (ABF)	4.4, 4.5, …	7.2, 7.4, …
Metal of RDL3 (ML3)	Linewidth (L)	10	10.2, 10.7, …	9.9, 10.3, …
	Spacing (S)	10	9.5, 10.2, …	9.4, 9.2, …
	Thickness (H)	5 (PID), 8 (ABF)	3.4, 5.1, …	8.2, 8.6, …

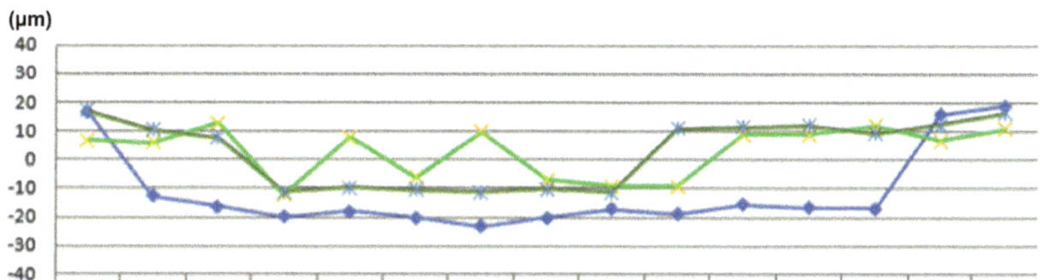

	30oC	100oC	150oC	180oC	200oC	220oC	240oC	260oC	240oC	220oC	200oC	180oC	150oC	100oC	30oC
Build-up	16.9	-12.8	-16.2	-19.7	-17.8	-20.2	-23.1	-20	-17.1	-18.8	-15.5	-16.5	-16.8	16.1	19.1
RDL (PID)	7	6	13	-12	8	-6	10	-7	-9	-9	9	9	12	7	11
RDL (ABF)	17.2	10.4	7.6	-11.3	-9.7	-10.5	-11.4	-10.1	-11.3	11.1	11.6	12.1	9.4	12.4	16.4

Fig. 4.170 Warpage for the build-up package substrate and organic interposers with PID and ABF

smaller than that of the build-up package substrate. Also, the warpage of the organic interposer with ABF is slightly larger than that with PID.

4.11.4 Fabrication Process of the Build-Up Package Substrate

The fabrication process of the build-up package substrate is the same as that in Sect. 4.10.

4.11.5 Fabrication Process of the Hybrid Substrate

The hybrid substrate is fabricated by combining the build-up package substrate and the organic interposer with the C4 bumps and underfill. Figure 4.171 shows the cross sections of the hybrid substrate with ABF and PID. It can be seen that the metal lines of the RDLs with ABF, Fig. 4.171a, are much flatter than those with PID, Fig. 4.171b.

4.11.6 Chips-to-Hybrid Substrate Bonding

The chip-to-hybrid substrate bonding profiles are shown in Fig. 4.172a for the larger chip 1 and Fig. 4.172b for the smaller Chip 2. It can be seen that for Chip 1: (a) force control is used for bonding, (b) 300 gf contact force with a searching speed of 2 mm/s, and (c) contact of BH (bond head)/BS (bond stage) at 120 °C/120 °C with a BH peak temperature of 295 °C, and a ramping rate of 80 °C/s. For Chip 2: (a) position control is used for bonding to maintain the same vertical height as Chip 1, (b) 300 gf contact fore with a searching speed of 1 mm/s, and (c) contact of BH/BS at 120 °C/120 °C with a BH peak temperature of 295 °C, and a ramping rate of 80 °C/s.

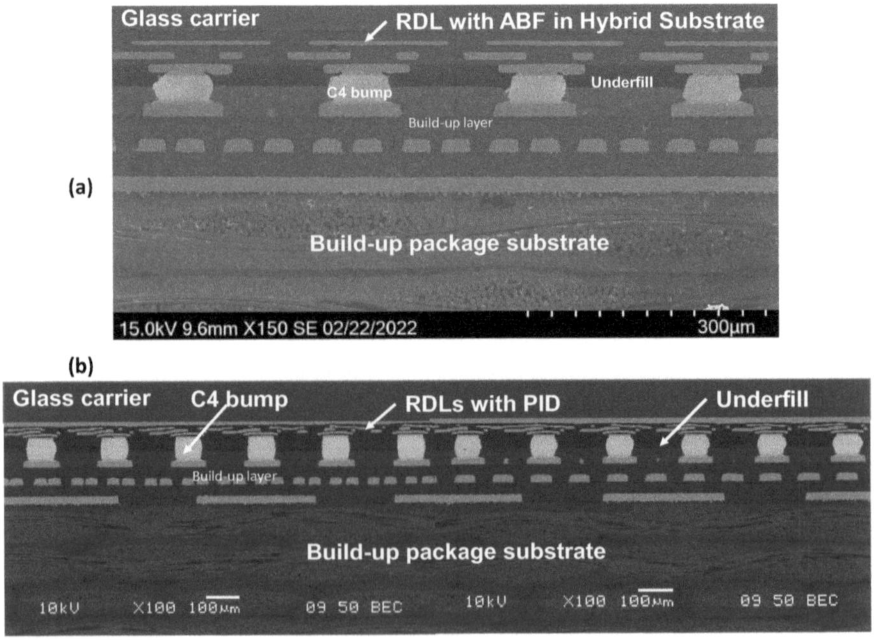

Fig. 4.171 **a** Hybrid substrate with ABF and **b** hybrid substrate with PID

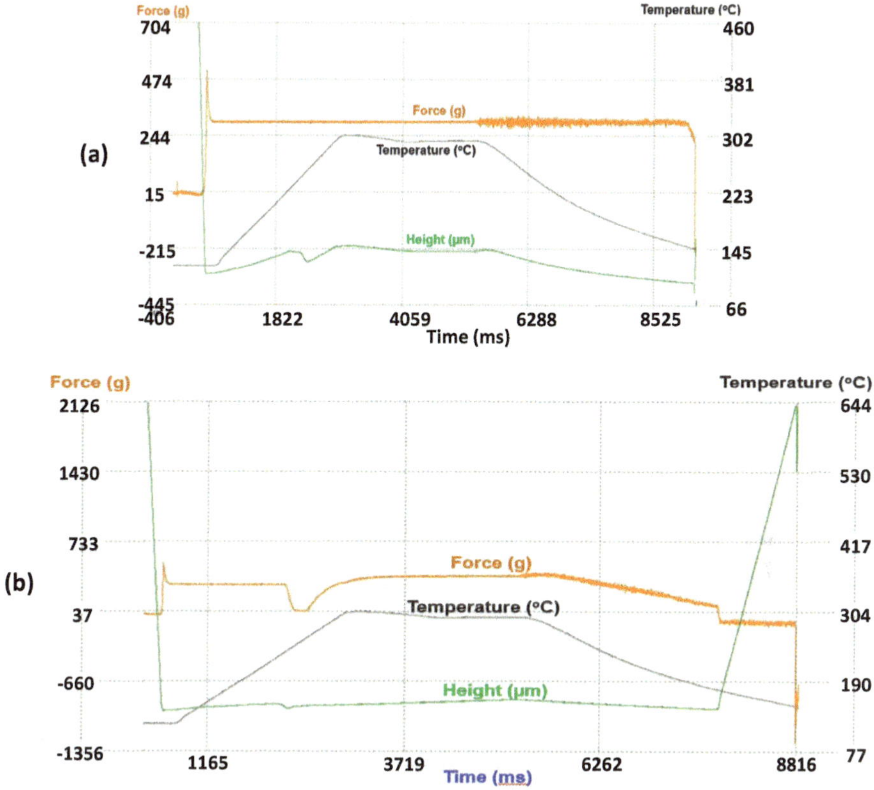

Fig. 4.172 Chip-to-hybrid substrate bonding profiles (ABF): **a** Chip 1 (10 mm × 10 mm) and **b** Chip 2 (5 mm × 5 mm)

4.11.7 Quality of Chips-to-Hybrid Substrate Bonding

For both hybrid substrates, a typical cross section image of the μbump solder joints is shown in Fig. 4.173. It can be seen that the microbump solder joints look normal (there is not any failure).

For the hybrid substrate with PID, the continuity check is still OK after chip-to-hybrid substrate bonding. However, for the hybrid bonding with ABF, the continuity check is not all OK after chip-to-hybrid substrate bonding.

First, a typical good cross section of hybrid substrate with ABF under Chip 2 is shown in Fig. 4.174a, b for a close-up view. It can be seen that the μbump solder joints and the C4 solder joints are normal and there is not any obvious failure except there are some voids in the C4 solder joints.

Some poor quality of solder joints has been observed for hybrid substrate with ABF. A typical cross section is shown in Fig. 4.175 for the smaller Chip 2. It can be seen that there is one C4 solder joint failure (separation between the organic

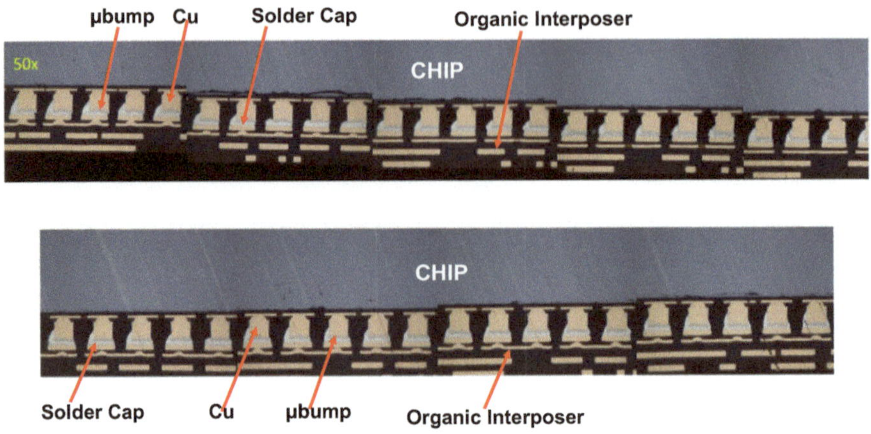

Fig. 4.173 Microbump solder joints in the final assembly

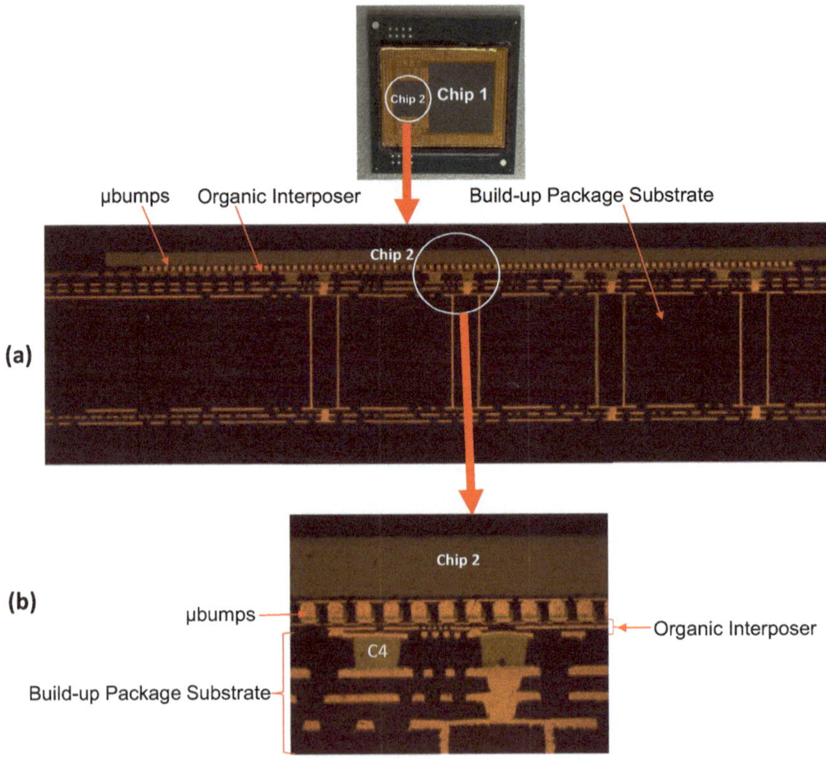

Fig. 4.174 **a** Cross section image of hybrid substrate with ABF and **b** close-up view

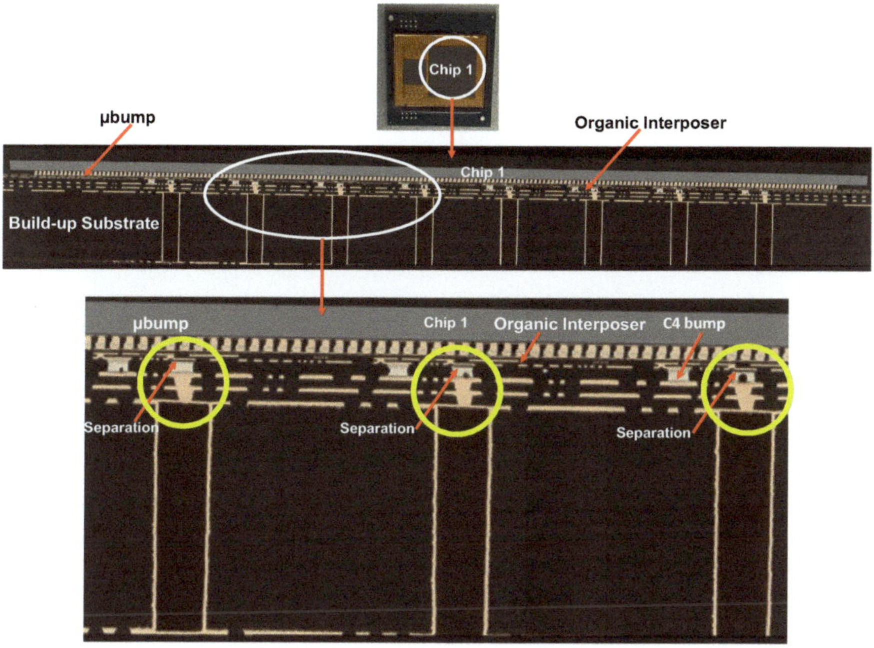

Fig. 4.175 Cross section image of hybrid substrate with ABF of Chip 2

interposer and the C4 solder joint). This failure occurs at the solder joint on top of one of the stacked vias of the build-up package substrate (circled with yellow color). The other solder joints are fine.

A typical cross section of the hybrid substrate with ABF is shown in Fig. 4.176 for the larger Chip 1. In this case, all the solder joints on top of the stacked vias of the build-up package substrate (circled with yellow color) are failed (separation between the organic interposer and the C4 solder joint). However, the solder joints right next to those on top of the stacked vias are normal (no separation).

The possible reasons for there are failures (separation between the organic substrate and the C4 solder joint right on top of the stacked via) for the hybrid substrate with ABF after chip-to-substrate bonding and there is not failure for the hybrid substrate with PID are: (a) the warpage of the organic substrate with ABF is slightly larger than that with PID, so the bonding between the hybrid substrate with ABF is not as strong (adhesive) as the one with PID, and (b) the effect of stress concentration from the stacked vias of the build-up package substrate during bonding may promote the tendence of failure.

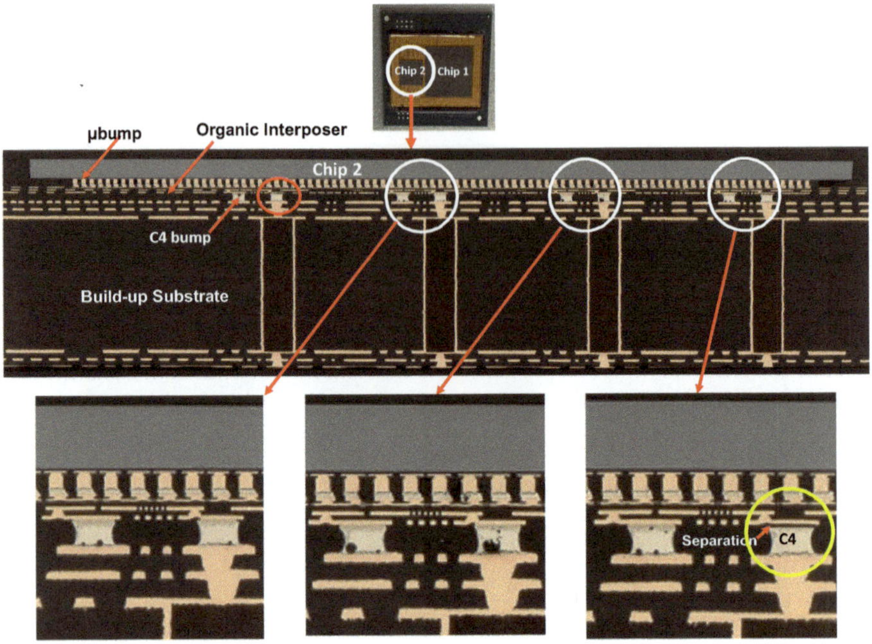

Fig. 4.176 Cross section image of hybrid substrate with ABF of Chip 1

4.11.8 Thermal Cycling Simulation and Results

The thermal-fatigue reliability of the solder joints in the hybrid substrate for the heterogeneous integration of two chips with the PID and ABF dielectric materials is performed by a nonlinear temperature and time-dependent finite element analysis. The cross section (Fig. 4.177) for finite element model (Fig. 4.178) for both final hybrid substrate assemblies are the same except for their material properties. Finer meshes are used for higher stress/strain areas. Again, since the focus of the present simulation is on the joint reliability, in order to simplify the modeling and save the computing time, those three RDLs (organic interposer) are modeled as an equivalent block as shown in Figs. 4.167 and 4.177 with the following effective material properties:

Hybrid substrate with PID

$$\text{Eeff} = \frac{121 \times (3+3+5+8) + 2.1 \times (3+6+4) + 2.1 \times 2}{(3+3+5+8) + (3+6+4) + 2} = 68.54\ \text{GPa}$$

$$\text{CTEeff} = \frac{16.3 \times (3+3+5+8) + 60 \times (3+6+4) + 60 \times 2}{(3+3+5+8) + (3+6+4) + 2} = 35.58 \times 10^{-6}/^{\circ}\text{C}$$

$$v\text{eff} = \frac{0.34 \times (3+3+5+8) + 0.3 \times (3+6+4) + 0.3 \times 2}{(3+3+5+8) + (3+6+4) + 2} = 0.32$$

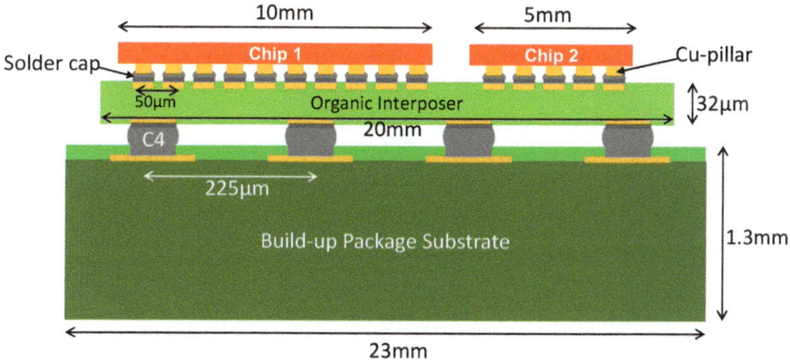

Fig. 4.177 Cross section for finite element modeling

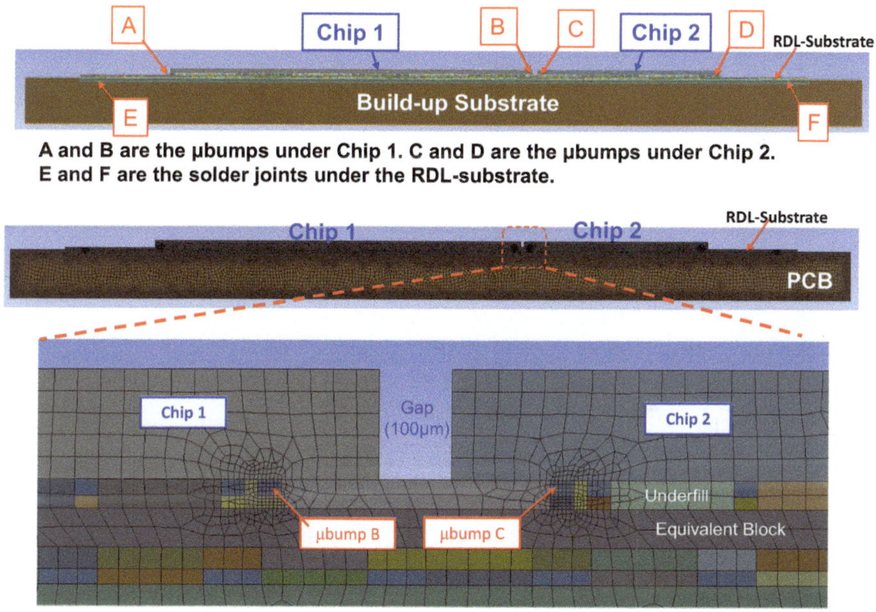

Fig. 4.178 Finite element model for determining the accumulated creep strain in solder joints

Hybrid substrate with ABF

$$\text{Eeff} = \frac{121 \times (8 + 3 + 8 + 8) + 7.5 \times (3.5 + 10 + 7.5) + 7.5 \times 5}{(8 + 3 + 8 + 8) + (3.5 + 10 + 7.5) + 5} = 65.32\ \text{GPa}$$

$$\text{CTEeff} = \frac{16.3 \times (8 + 3 + 8 + 8) + 37 \times (3.5 + 10 + 7.5) + 37 \times 5}{(8 + 3 + 8 + 8) + (3.5 + 10 + 7.5) + 5} = 26.45 \times 10^{-6}/^{\circ}\text{C}$$

$$\upsilon eff = \frac{0.34 \times (8 + 3 + 8 + 8) + 0.3 \times (3.5 + 10 + 7.5) + 0.3 \times 5}{(8 + 3 + 8 + 8) + (3.5 + 10 + 7.5) + 5} = 0.32$$

where Eeff, CTEeff, and veff are, respectively, the effective Young's modulus, effective coefficient of thermal expansion, and effective Poisson's ratio of the RDL-first substrate equivalent block. Figure 4.179 shows the temperature boundary condition for simulation. The temperature is −40 °C ↔ 85 °C. The cycle time is 60 min and the ramp-up, ramp-down, dwell-at-hot, and dwell-at-cold are each 15 min. Stress-free is at room temperature. Table 4.15 shows the material properties for modeling. All the materials are assumed to be constant except solder (Sn3Ag0.5Cu), which is time and temperature dependent and obeys the Anand viscoplasticity constitute equation.

The Anand flow equation that governs the inelastic strain ε has the following form [48]:

Table 4.15 Material properties

Materials	Young's modulus (GPa)	Poisson's ratio	CTE (10^{-6}/°C)
Silicon	131	0.278	2.8
Copper	121	0.34	16.3
Solder	49 − 0.07 T (°C)	0.3	21 + 0.017 T (°C)
Underfill	4.5	0.35	50
PID	2.1	0.3	60
ABF	7.5	0.3	37
Equivalent block (PID)	68.54	0.32	35.38
Equivalent block (ABF)	65.32	0.32	26.45
Solder mask (PCB)	4.1	0.3	39
PCB	$E_x = E_y = 22$, $E_z = 10$	0.28	$\alpha_x = \alpha_y = 18$, $\alpha_z = 70$

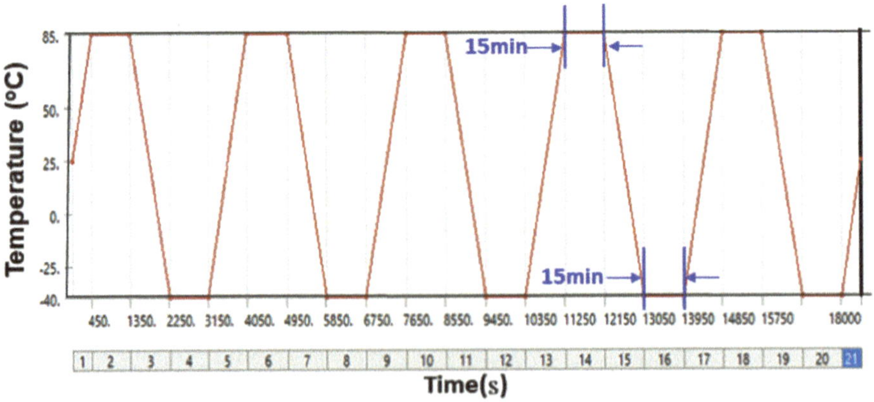

Fig. 4.179 Temperature boundary condition

$$\frac{d\varepsilon}{dt} = A\left[\sinh\left(\frac{\xi\sigma}{s}\right)\right]^{1/m}\exp\left(-\frac{Q}{RT}\right)$$

where $d\varepsilon/dt$ is the inelastic strain rate, A is the pre-exponential factor, Q is the activation energy, R is the universal gas constant (8.314 J K^{-1} mol^{-1}), T is the absolute temperature (Kelvin), ζ is the multiplier of stress, σ is the equivalent stress, and m is strain rate sensitivity. The evolution equation that defines an internal variable s of the model and its saturation value is

$$\frac{ds}{dt} = \left\{h_0(|B|)^{\alpha}\frac{B}{|B|}\right\}\frac{d\varepsilon}{dt}$$

where

$$B = 1 - \frac{s}{s*}$$

and

$$s^* = \hat{s}\left[\frac{d\varepsilon_p/dt}{A}\exp\left(\frac{Q}{RT}\right)\right]^n$$

where h_0 is the hardening/softening constant, α is the strain rate sensitivity of hardening/softening. The quantity s^* represents a saturation value of deformation resistances. $\hat{s}$ is a coefficient, and n is the strain rate sensitivity for the saturation value of deformation resistance, respectively. The evolution equation has an extra initial condition of $s(0) = s_0$, where s_0 is the initial value of s at initial time $t = 0$. The Anand model does not have an explicit yield criterion or a loading–unloading criterion and viscoplastic flow occurs for any non-zero stress.

There are a total of nine (9) constants (parameters) of the Anand equation (model), namely A, Q, ξ, m, s, n, h_0, α, and s_0. For Sn3Ag0.5Cu, these constants are shown in Table 4.16 [48].

Table 4.16 Anand's parameters for Sn3Ag0.5Cu

Parameters	Definition in Anand's model	Sn3Ag0.5Cu
A (1/s)	Pre-exponential factor	3,484
Q/R (°K)	Activation energy/universal gas constant	9,096
ξ	Stress multiplier	4
m	Strain rate sensitivity of stress	0.2
$\hat{s}$ (MPa)	Coefficient for deformation resistance saturation value	26.4
n	Strain rate sensitivity of saturation value	0.01
h_o (MPa)	Hardening constant	144,000
a	Strain rate sensitivity of hardening/softening	1.9
s_o (MPa)	Initial value of deformation resistance	18.07

The deformed shape and undeformed shape (dark lines) of the PID assembly are shown in Fig. 4.180. At 450 s (85 °C), the hybrid substrate expands more than the chips, and the structure is deformed in a concave shape (smiling face), Fig. 4.180a. At 2250 s (−40 °C), the hybrid substrate shrinks more than the chips and the structure is deformed into a convex shape (crying face), Fig. 4.180b. The deformed shape of the ABF assembly (Fig. 4.181a, b) is very similar to the PID assembly.

Accumulated equivalent inelastic strain contour distributions at the critical microbump solder joints A–D (between the organic interposer and the chips), and the critical C4-bump solder joints E and F (between the organic interposer and the build-up package substrate) with the PID are shown in Fig. 4.182. It can be seen that (a) the maximum accumulated equivalent inelastic strain occurs near the corner and in a very small area and (b) the values of the maximum accumulated equivalent inelastic strain are very small. The accumulated maximum equivalent inelastic strain

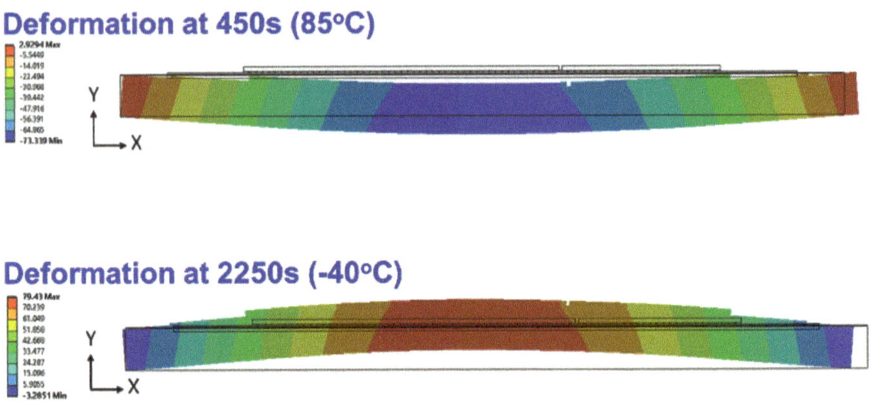

Fig. 4.180 Displacement of the structure (with PID) at 450 s (85 °C) and 2250 s (240 °C)

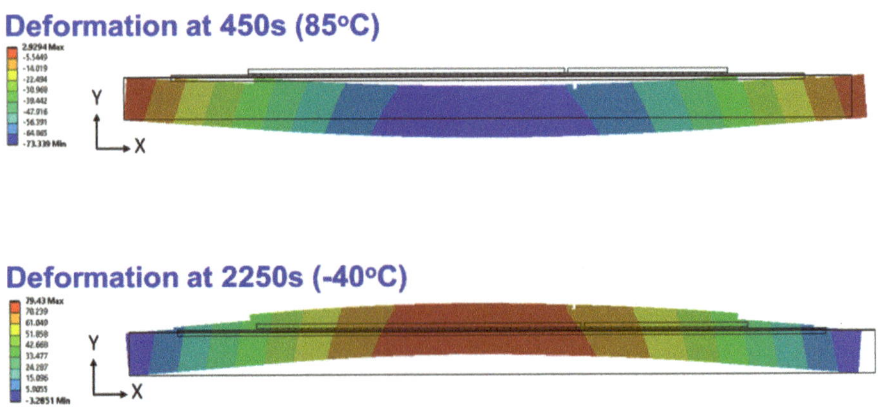

Fig. 4.181 Displacement of the structure (with ABF) at 450 s (85 °C) and 2250 s (240 °C)

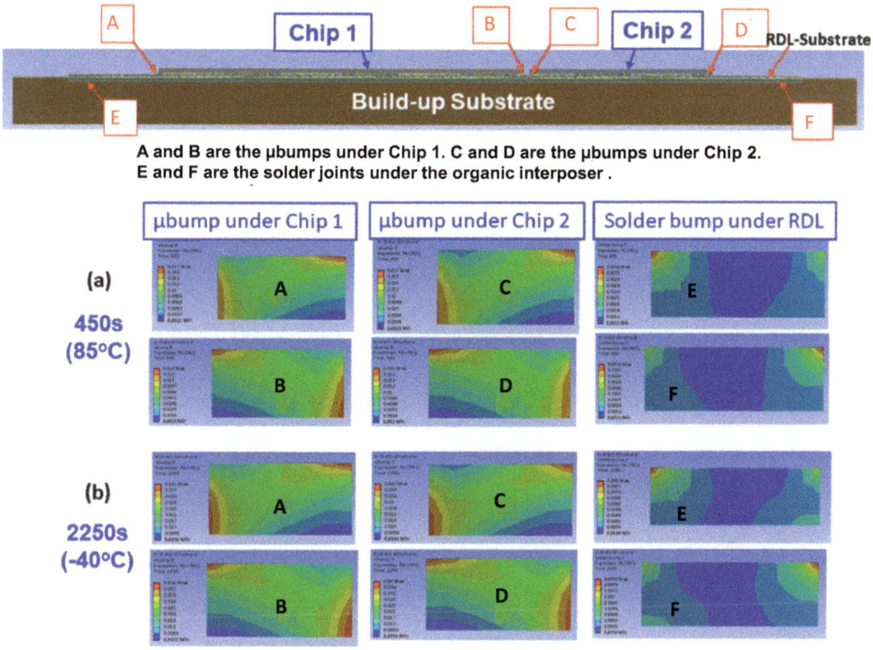

Fig. 4.182 Accumulated equivalent inelastic strain contours in solder joints on the hybrid substrate with PID

time-histories at microsolder joints A, B, C, and D and C4 solder joints E and F are shown in Fig. 4.183. It can be seen that the largest accumulated equivalent inelastic strain per cycle (10.7%) occurs at microsolder joints A and C. The largest accumulated equivalent inelastic strain per cycle in the C4 solder joints E and F is ~1.5%/cycle. All these values are too small and too localized to create solder joint reliability.

Figures 4.184 and 4.185 show, respectively, the accumulated equivalent inelastic strain contours and the accumulated equivalent inelastic strain time-histories in microsolder joints A, B, C, and D and C4 solder joints E and F with ABF. It can be seen that the behavior of the accumulated equivalent inelastic strain courteous and the accumulated equivalent inelastic strain time-histories in the solder joint A, B, C, D, E, and F between the PID hybrid substrate assembly and the ABF hybrid substrate assembly is very similar. Also, the maximum values are almost the same. Thus, the heterogeneous integration of these two chips on the hybrid substrates with PID and ABF dielectrics should have the same level of thermal fatigue reliability solder joints which can survive most operating conditions.

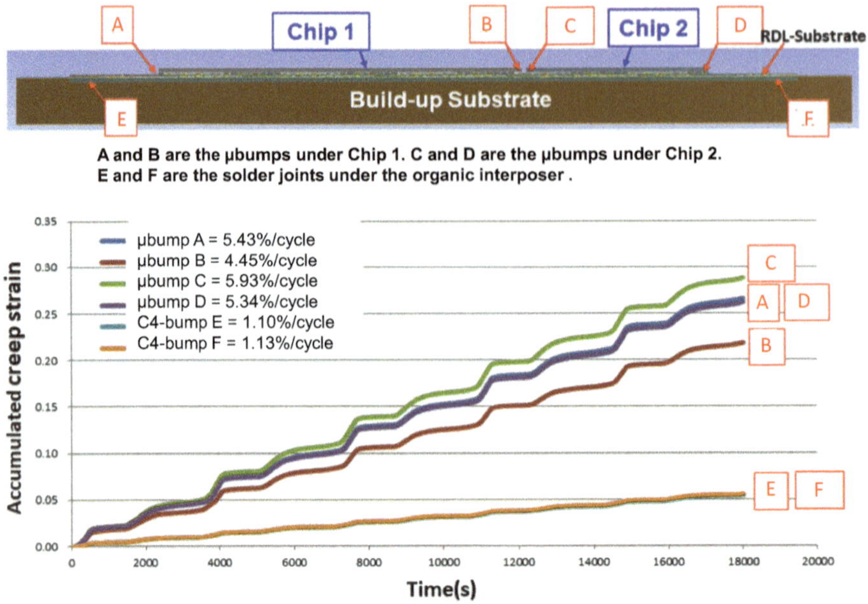

Fig. 4.183 Accumulated equivalent inelastic strain time-history in solder joints on the hybrid substrate with PID

4.11.9 Summary

Some important results and recommendations are summarized as follows:

- The comparisons (such as interposer, process, performance, and applications) between the 2.5D (TSV interposer) and 2.3D (organic interposer) have been shown in Table 4.17.
- The hybrid substrate (organic interposer + build-up package substrate + solder joints + underfill) of an organic interposer for the heterogeneous integration of two chips has been developed.
- The RDLs with ABF are thicker than those with PID.
- Before chip-to-hybrid substrate bonding, the hybrid substrate has been tested to make sure it is a known-good hybrid substrate.
- The metal lines of the RDLs in the organic interposer with ABF are flatter than those with PID. However, the thickness of the RDLs with ABF is thicker than those with PID.
- The warpage of the organic interposer with ABF is slightly larger than that with PID.
- After chip-to-hybrid substrate bonding, for hybrid substrates with both PID and ABF, the micro solder joints between the chips and the known-good hybrid substrates look normal (there is not any failure).

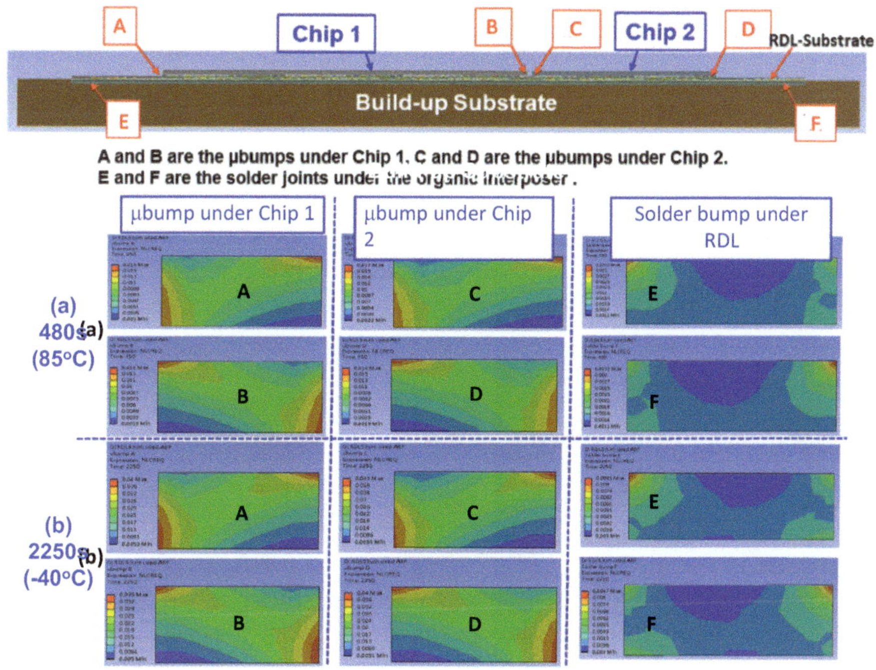

Fig. 4.184 Accumulated equivalent inelastic strain contours in solder joints on the hybrid substrate with ABF

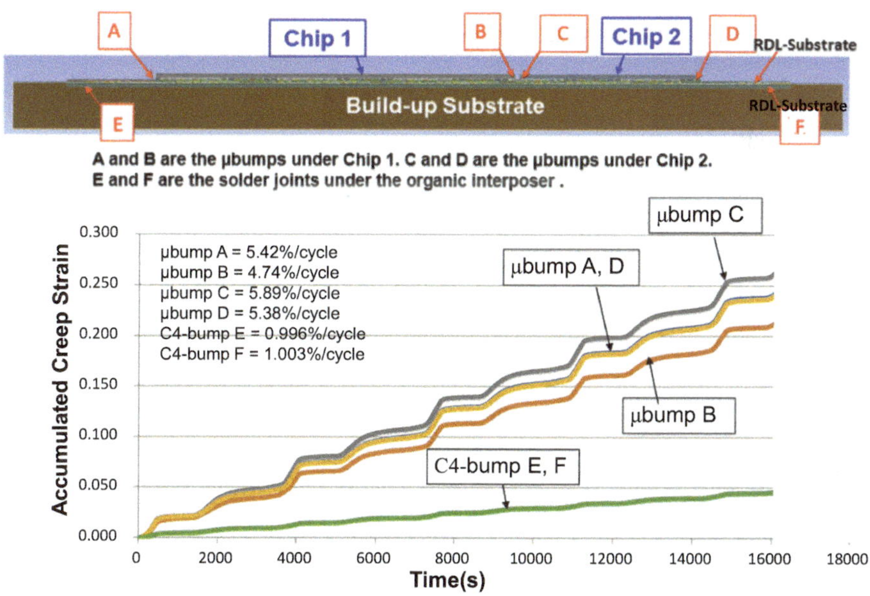

Fig. 4.185 Accumulated equivalent inelastic strain time-history in solder joints on the hybrid substrate with ABF

Table 4.17 Compression between TSV-interposer (2.5D) and organic-interposer (2.3D)

	2.3D (chip-last)	2.5D
SoC size (mm × mm)	≤20 × 20	≤25 × 25
No. of HBMs	6	8
RDL interposer	Cu traces in polymer	Cu traces in SiO_2
RDL (metal L/S) interposer (μm)	≥2	<1
RDL (layer) interposer	≤6	≤8 (usually 4 is enough)
Vertical interconnect	Via in RDL interposer	TSV
C4 bump	Yes	Yes
Underfill (chip/interposer)	Yes	Yes
Underfill (interposer/sub.)	Yes	Yes
Total package profile	Tall	Tall
Package substrate size	Large	Large
Process steps	More	Most
Density	High	Higher
Performance	High	Higher
Decoupling capacitor	Integrated passive devices	Deep trench capacitors
Cost	High	Higher
Applications	HPC, data center	HPC, high bandwidth data center

- After chip-to-hybrid substrate (with PID) bonding, the C4 solder joints between the organic interposer and the build-up package substrate look normal (there is not any failure).
- After chip-to-hybrid substrate (with ABF) bonding, there is one failure (separation between the organic interposer and the solder joint) of the C4 solder joint under the smaller Chip 2 and the failure location is right above the stacked vias of the build-up package substrate. Also, there are many failures (separations) under the larger Chip 1 and the failure locations are on top of the stacked vias. All the other C4 solder joints (not directly on top of a stacked via) look normal (no failure).
- The thermal-fatigue reliability of the solder joints in the hybrid substrate for the heterogeneous integration of two chips with PID and ABF dielectric materials has been performed by a nonlinear temperature and time dependent finite element analysis.
- The accumulated creep strain contours and time-history at the micro solder joints between the organic interposer and the chips and the C4 solder joints between the organic interposer and the build-up package substrate for both PID and ABF hybrid substrates are very similar. Also, the maximum values occur at a very small local volume.

4.12 Hybrid Substrate with Ultralarge Organic Interposer for Heterogeneous Integration

4.12.1 The Structure

The structure (hybrid substrate) under consideration (85 × 80 × 2.13 mm) is shown in Fig. 4.186. It can be seen that it consists of a 5-2-5 build-up package substrate (2 mm thick), an ultralarge organic interposer (0.05 mm thick) with four RDLs, and the C4 Sn0.7Cu solder joints (~80 μm thick) with underfill. Each RDL layer consists of the dielectric layer (P) and metal layer (ML). The linewidth, spacing, and thickness of P1, …, P5, and ML1, …, ML4 are shown in the table of Fig. 4.187, where the via (V) size between MLs is also provided [24].

4.12.2 Finite Element Analysis

Due to the double symmetry, only one quarter of the hybrid substrate is analyzed. There are more than 10,000 solder joints in the quarter model. Most of the solder joints

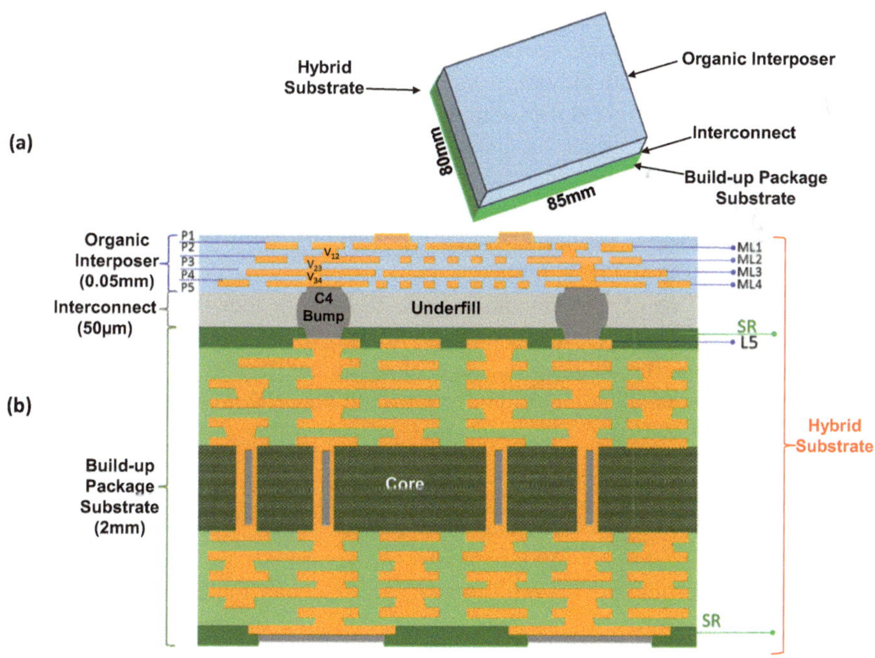

Fig. 4.186 Hybrid substrate with ultralarge organic interposer. **a** Three-dimensional view. **b** Cross-sectional view

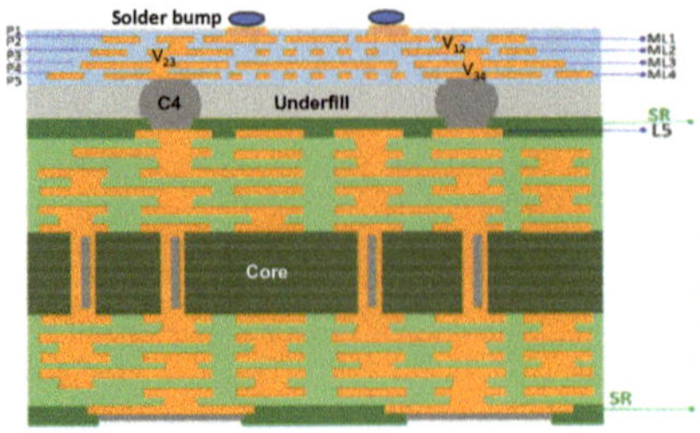

Package Substrate:
- 5-2-5 with double-side solder resist
- Surface finishing: IT
- Thickness: 2mm
- Size: 98mm x 89mm

Organic Interposer:
- 4 RDLs: 4 metal layers and 5 dielectric layers
- Surface finishing: ENEPIG
- Thickness:0.05mm
- Size: 87mm x 82mm

Hybrid Substrate:
- Organic interposer + package substrate
- Solder joints: ~ 50,000
- Size: 85mm x 80mm

RDLs	Key Elements of RDL	Line width/ Spacing/ Thickness (μm)
RDL1	ML1 (Metal layer 1)*	12/29/8
	P1 (Dielectric between pad and ML1)	NA/NA/6
RDL2	ML2 (Metal layer 2)	8/8/4
	P2 (Dielectric between ML1 and ML2)	NA/NA/5
	V_{12} (Via between ML1 and ML2)	20
RDL3	ML3 (Metal layer 3)	60/30/4
	P3 (Dielectric between ML2 and ML3)	NA/NA/5
	V_{23} (Via between ML2 and ML3)	20
RDL4	ML4 (Metal layer 4)	8/8/4
	P4 (Dielectric between ML3 and ML4)	NA/NA/5
	V_{23} (Via between ML3 and ML3)	20
	P5 (Dielectric between ML4 and pad)	NA/NA/5

*There are some ML1 with L/S = 2/2μm near the corners of the organic interposer

Fig. 4.187 Key features and dimensions of the hybrid substrate with ultralarge organic interposer

are dummy (mechanical) joints used to absorb the strain and stress due to the thermal expansion between the organic substrate and Fig. 4.187. Key features and dimensions of the hybrid substrate with ultralarge organic interposer (Fig. 4.188). Finite element modeling of 1/4 of the hybrid substrate with ultralarge organic interposer. Build-up package substrate, and they are distributed around the corners of the hybrid substrate. The farthest DNP (distance to neutral point) signal solder joint is located 38.2 mm (in the x-direction) and 33.8 mm (in the y-direction) from the center of symmetry. Finer meshes are used to the corner solder joints and the farthest DNP signal solder joint, as shown in Fig. 4.188.

The key objective of the simulation is to make sure that the solder joints of the hybrid substrate with ultralarge organic interposer are reliable during temperature cycling. Thus, the build-up package substrate and the organic interposer are approximated by equivalent blocks with equivalent material properties.

4.12.3 Anand Viscoplasticity Constitute Equation

The material properties of the hybrid substrate are shown in Table 4.18. Figure 4.189 shows the equivalent block model and the calculation of the material properties of the package substrate equivalent block and the organic interposer equivalent block. The final equivalent material properties are shown at the top of Table 4.19.

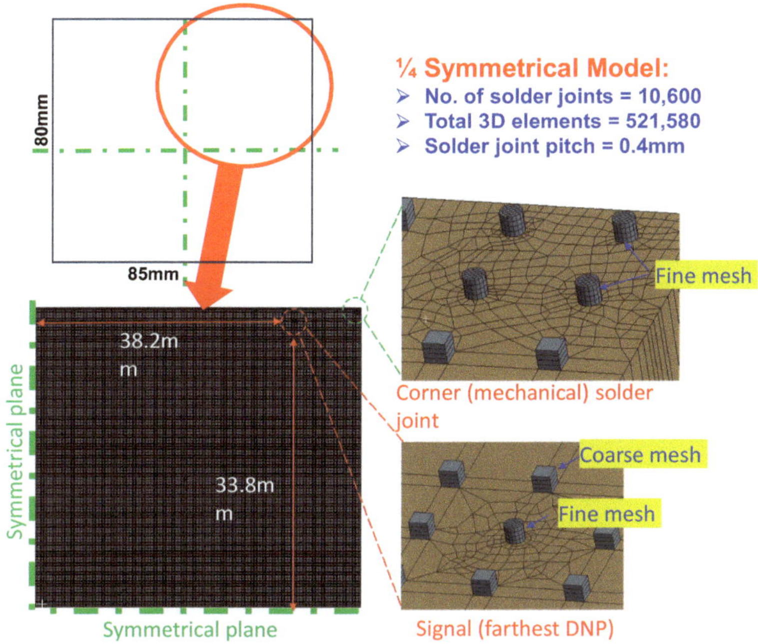

Fig. 4.188 Finite element modeling of 1/4 of the hybrid substrate with ultralarge organic interposer

Table 4.19 shows the material properties for modeling. All the materials are assumed to be constant except solder (Sn0.7Cu), which is temperature dependent and obeys the Anand viscoplasticity constitute equation [125]. The Anand flow equation given in Sect. 4.11 for SnAgCu will be used again for the Sn0.7Cu with different material parameters as shown in the bottom of Table 4.19.

4.12.4 Boundary Condition

Figure 4.190 shows the temperature boundary condition for simulation. The temperature is −40 °C ↔ 85 °C. The cycle time is 60 min and the ramp-up, ramp-down, dwell-at-hot, and dwell-at-cold are each 15 min. Stress-free is at room temperature.

4.12.5 Accumulated Equivalent Inelastic Strain

The deformed shape (color contours) and undeformed shape (dark lines) of the hybrid substrate are shown in Fig. 4.191. At 450 s (85 °C), the organic interposer expands more than the package substrate and the structure is deformed in a convex shape

Table 4.18 Material properties and structural elements for equivalent models

Structure	Materials	Thickness for each layer (μm)	CTE (10^{-6}/°C)	Young's modulus (GPa)	Poisson's ratio
Organic interposer	PID	10	60 @ 25–300 °C	2.0	0.3
	Cu	4 @ RDL 2 & RDL 3; 8 @ RDL 1 & RDL 4 (connected with solder)	16.3	121	0.34
Package substrate	ABF	30	20 @ 25–150 °C 49 @ 150–240 °C	13	0.3
	Cu	15	16.3	121	0.34
	SR	20	38 @ 25–130 °C	7	0.3
	Core	1440	5–7 @ 30–120 °C	32–34	0.3
Solder bump	Sn0.7Cu	80	21 + 0.017 T	49 − 0.07 T	0.3
Underfill	Underfill	80	30 @ 25–90 °C 125 @ >90 °C	7 @ 25–90 °C 0.07 @ >90 °C	0.3

Table 4.19 Equivalent material properties for equivalent blocks. Parameters for (Sn0.7Cu) Anand's model

Materials	CTE (10^{-6}/°C)	Young's modulus (GPa)	Poisson's ratio
Organic interposer	43.22	47.7	0.32
Package substrate	9.74	37.49	0.3
Sn0.7Cu	21 + 0.017 T	49 − 0.07 T	0.3
Underfill	30	7	0.3

Parameters	Definition in Anand's model	Sn0.7Cu
A (1/s)	Pre-exponential factor	3,484
Q/R (°K)	Activation energy/universal gas constant	9,096
ξ	Stress multiplier	4
m	Strain rate sensitivity of stress	0.2
$\hat{s}$ (MPa)	Coefficient for deformation resistance saturation value	26.4
n	Strain rate sensitivity of saturation value	0.01
h_o (MPa)	Hardening constant	144,000
a	Strain rate sensitivity of hardening/softening	1.9
s_o (MPa)	Initial value of deformation resistance	18.07

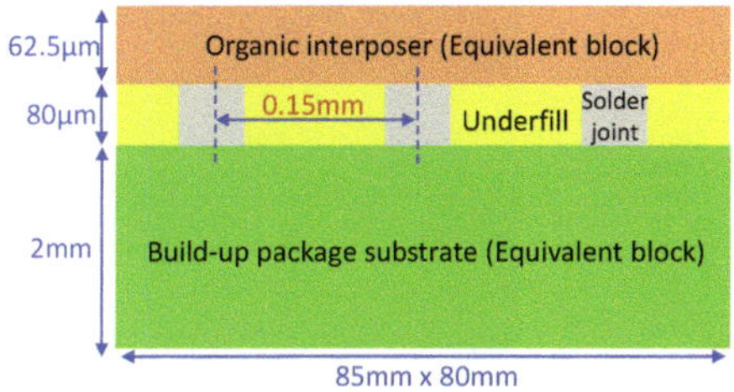

$$M_{eq} = \frac{\sum_{i=1}^{n} M_i * T_i}{T_i}$$

- M_{eq} = Equivalent material property
- M_i = i-Layer material property
- T_i = i-Layer thickness
- n = Total number of layers

TABLE 2A EQUIVALENT MATERIAL PROPERTIES FOR ORGANIC INTERPOSER

Material	Thickness (μm)	Young's modulus (GPa)	Poisson's ratio	CTE(10^{-6}/°C)
PID	6.5	2	0.3	60
RDL 1	8	121	0.34	16.3
PID	10	2	0.3	60
RDL 2	4	121	0.34	16.3
PID	10	2	0.3	60
RDL 3	4	121	0.34	16.3
PID	10	2	0.3	60
RDL 4	8	121	0.34	16.3
PID	2	2	0.3	60
Equivalent		47.70	0.32	43.22

TABLE 2B EQUIVALENT MATERIAL PROPERTIES FOR PACKAGE SUBSTRATE

Material	Thickness (μm)	Young's modulus (GPa)	Poisson's ratio	CTE(10^{-6}/°C)
SR	20	7	0.3	38
L1	15	121	0.34	16.3
ABF	30	13	0.3	20
L2	15	121	0.34	16.3
ABF	30	13	0.3	20
L3	15	121	0.34	16.3
ABF	30	13	0.3	20
L4	15	121	0.34	16.3
ABF	30	13	0.3	20
L5	15	121	0.34	16.3
ABF	30	13	0.3	20
L6	15	121	0.34	16.3
Core	1440	33	0.3	6
L7	15	121	0.34	16.3
ABF	30	13	0.3	20
L8	15	121	0.34	16.3
ABF	30	13	0.3	20
L9	15	121	0.34	16.3
ABF	30	13	0.3	20
L10	15	121	0.34	16.3
ABF	30	13	0.3	20
L11	15	121	0.34	16.3
ABF	30	13	0.3	20
L12	15	121	0.34	16.3
SR	20	7	0.3	38
Equivalent		37.49	0.30	9.74

Fig. 4.189 Equivalent material properties of the package substrate and ultralarge organic interposer

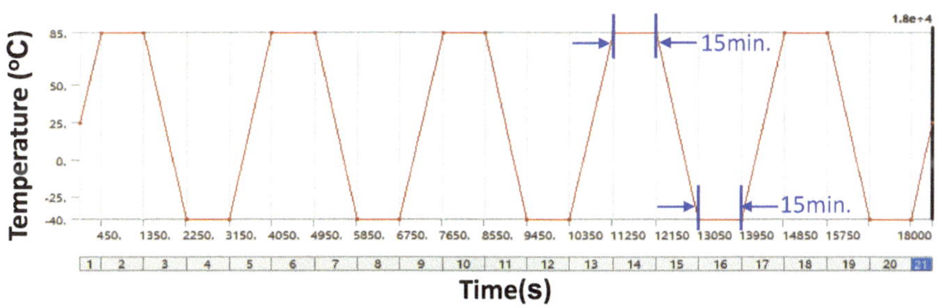

Fig. 4.190 Temperature cycling boundary condition

(crying face), see Fig. 4.191a. At 2250 s (−40 °C), the organic interposer shrinks more than the package substrate and the structure is deformed into a concave shape (smiling face), see Fig. 4.191b.

The accumulated equivalent inelastic strain contour distributions at the corner (mechanical) solder joint at 450 s (85 °C) and at 2250 s (−40 °C) are shown in Fig. 4.192a, b, respectively. It can be seen that the maximum inelastic strain at the corner solder joint occurs at a small local area and the value is 0.46% at 450 s (85 °C) and 1.23% at 2250 s (−40 °C).

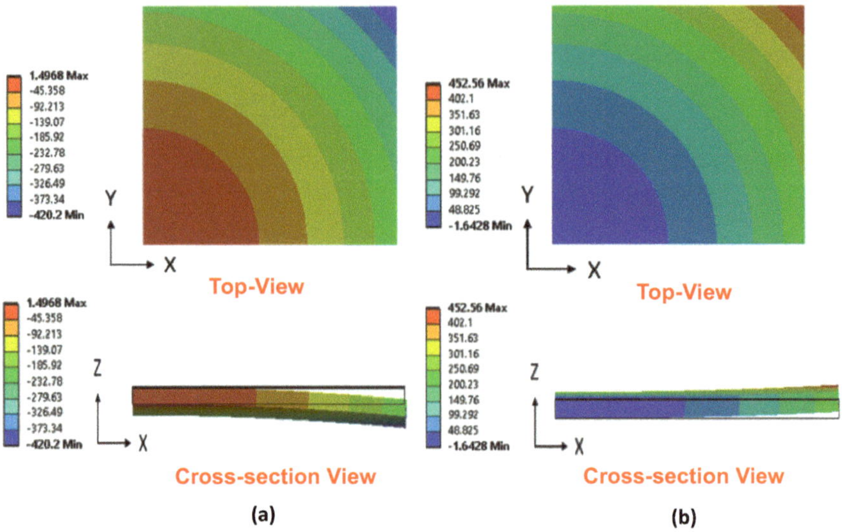

Fig. 4.191 Undeformed (black lines) and deformed (color contours) shapes of the hybrid substrate. **a** At 450 s (85 °C). **b** At 2250 s (−40 °C)

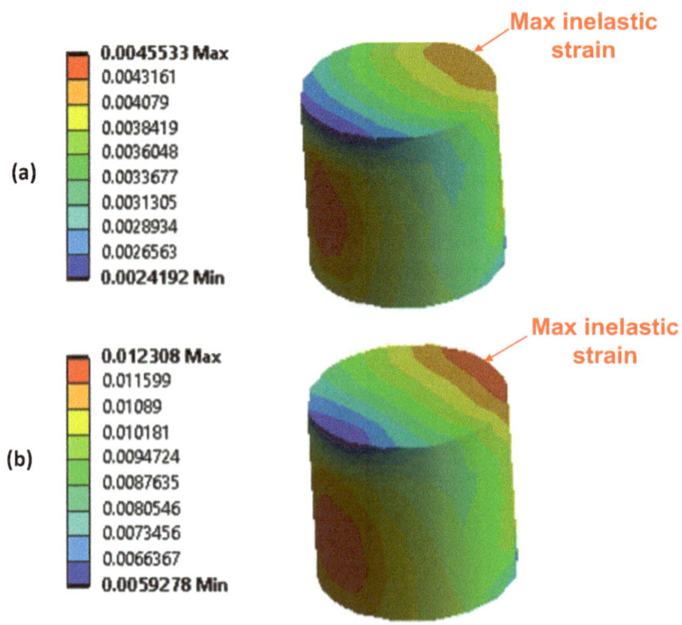

Fig. 4.192 Accumulated equivalent inelastic strain contours at corner (mechanical) solder joint. **a** At 450 s (85 °C). **b** At 2250 s (−40 °C)

The accumulated equivalent inelastic strain contour distributions at the farthest DNP signal solder joint at 450 s (85 °C) and at 2250 s (−40 °C) are shown in Fig. 4.193. It can be seen that the maximum inelastic strain at the farthest DNP signal solder joint is 0.13% at 450 s (85 °C) and only 0.3% at 2250 s (−40 °C). The accumulated maximum equivalent inelastic strain at 18,000 s (25 °C) and the time histories at the corner (mechanical) solder joint and the farthest DNP signal solder joint are shown in Figs. 4.194 and 4.195, respectively. It can be seen that: (1) the largest accumulated equivalent inelastic strain at 18 000 s (25 °C) is 8.8% for the corner solder joint and 1.8% for the farthest DNP signal solder joint and (2) the largest accumulated equivalent inelastic strain per cycle at the corner (mechanical) solder joint is only 1.8% per cycle and that at the farthest DNP signal solder joint is only 0.36% per cycle. All these values are too small to create solder joint reliability problem.

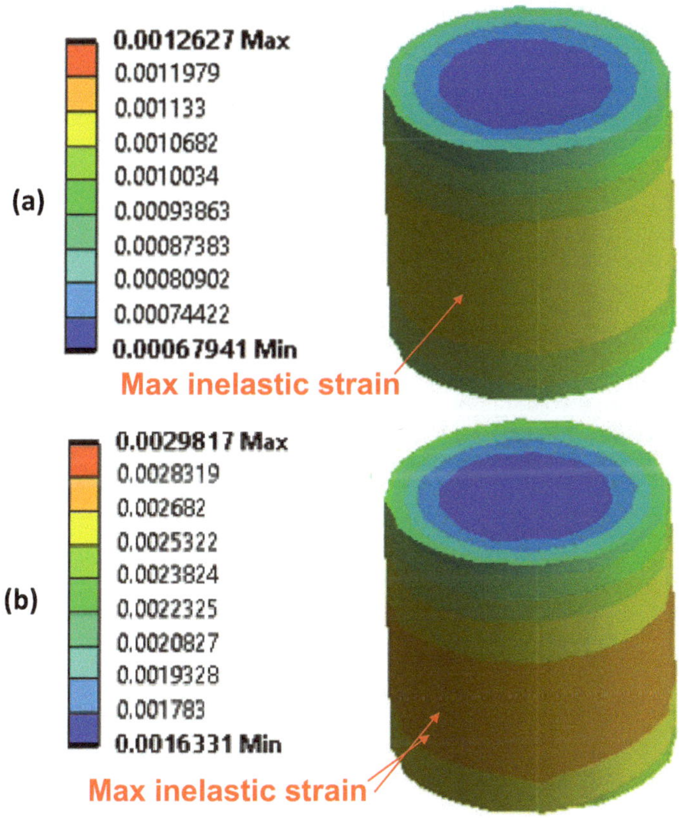

Fig. 4.193 Accumulated equivalent inelastic strain contours at the farthest DNP signal solder joint. **a** At 450 s (85 °C). **b** At 2250 s (−40 °C)

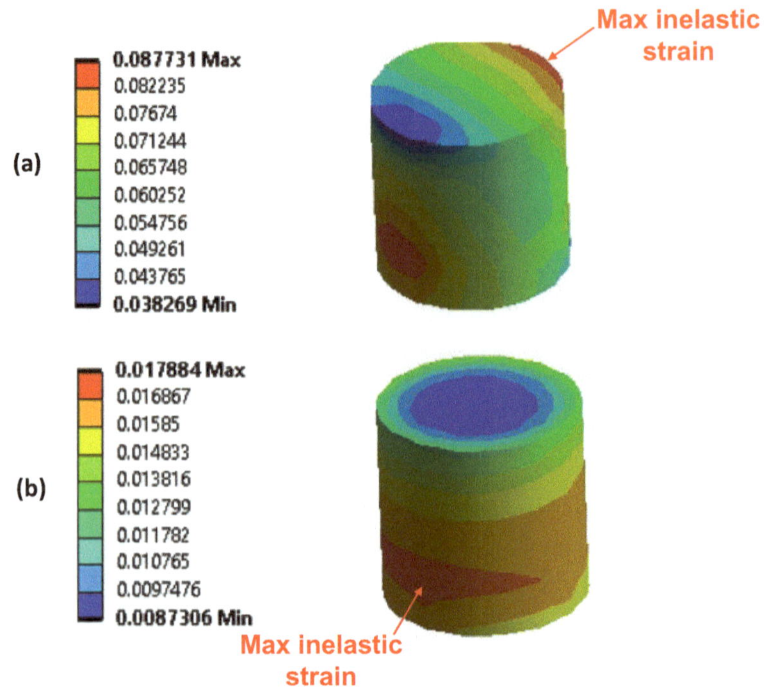

Fig. 4.194 Accumulated equivalent inelastic strain contours at 18 000 s (25 °C). **a** At corner (mechanical) solder joint. **b** At farthest DNP signal solder joint

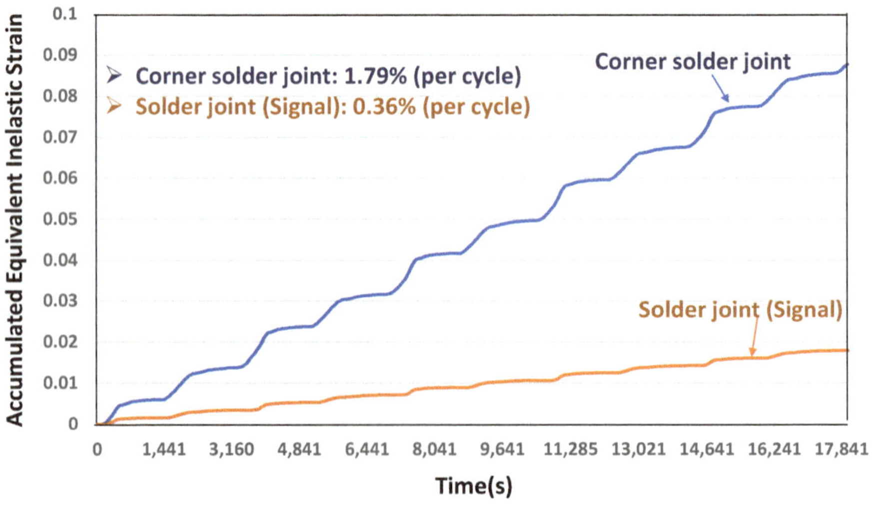

Fig. 4.195 Accumulated equivalent inelastic strain time history at the corner (mechanical) solder joint and farthest DNP signal solder joint

4.12.6 Hybrid Substrate with Ultralarge Organic Interposer

The hybrid substrate of the current section consists of an ultralarge organic interposer, a build-up package substrate, and C4 solder bumps with underfill (Fig. 4.186). The key process steps in fabricating the hybrid substrate are shown in Fig. 4.196.

(A) Ultralarge Organic Interposer

The four-RDL ultralarge organic interposer (87 × 82 × 0.05 mm) is fabricated with the following process. A released film (sacrificial layer) is slit coating on a temporary glass panel (515 mm × 510 × 1.1 mm), and then, a Ti/Cu seed layer is deposited by physical vapor deposition (PVD) on the top, as shown in Fig. 4.196. The thickness of the sacrificial layer is 1 μm and it is a light-to-heat conversion (LTHC) film. The contact pad can be obtained by photoresist, laser direct imaging (LDI), development, electrochemical deposition (ECD) Cu, and stripping off the photoresist. Then, a photoimageable dielectric (PID) is slit coating and the first dielectric layer (P1) of RDL1 is obtained by LDI. It is followed by PVD the Ti/Cu, photoresist, LDI and development, ECD the Cu, strip off the photoresist, and etch off the TiCu to obtain the first ML (ML1) of RDL1. P2 and ML2 of RDL2, P3 and ML3 of RDL3, and P4 and ML4 of RDL4 can be obtained by repeating the same process steps. The solder mask (passivation, P5) can be obtained by slit coating a PID and LDI. Then, the surface is finished with ENEPIG (electroless nickel electroless palladium immersion gold). The C4 solder bumps at 0.15 mm pitch are formed by stencil Sn0.7Cu solder ball mounting and then reflowed with the temperature profile shown in Fig. 4.197.

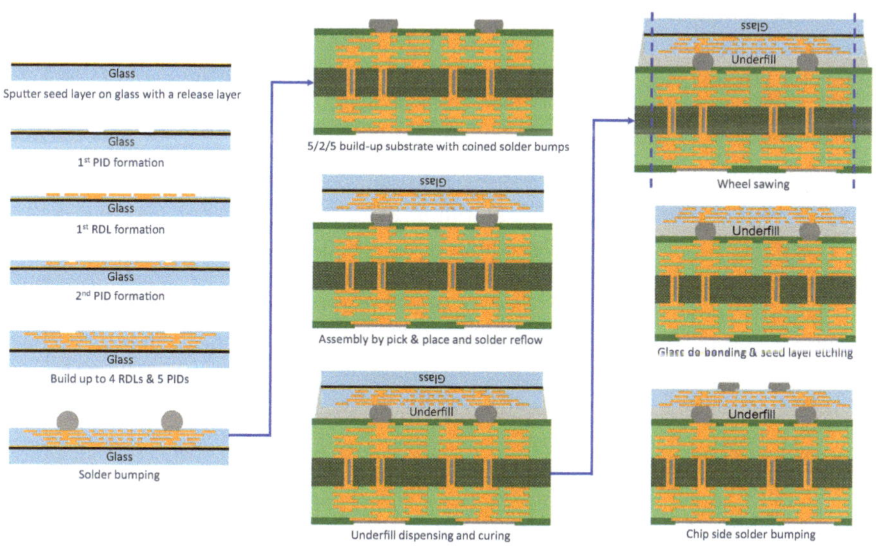

Fig. 4.196 Key fabrication steps of the hybrid substrate with ultralarge organic interposer

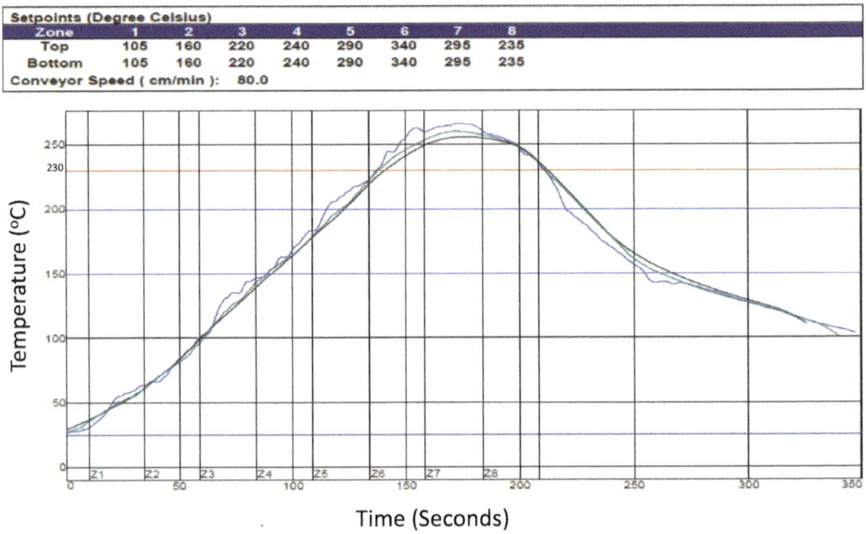

Fig. 4.197 Reflow temperature profile of the solder bump

The top view and cross-sectional view of the organic interposer are shown in Fig. 4.198. It can be seen from Fig. 4.198a that there are 16 units of ultralarge organic interposer on the glass panel. A typical cross section of the organic interposer is shown in Fig. 4.198b. It can be seen that the RDLs and C4 solder bumps are properly done. Figure 4.199 shows the RDLs with L/S = 2/2 μm near the corners of the organic interposer. It can be seen that the average linewidth is not exactly equal to 2 but 2.2 μm and the line spacing is not exactly equal to 2 but 1.6 μm. Thus, there are rooms for improvement, e.g., a better estimation of the compensation of photoresist, LDI, ECD Cu, and Cu etching.

(B) Build-Up Package Substrate

The 5-2-5 build-up package substrate (98 × 89 × 2 mm) is fabricated on a 510 × 510 mm panel by a conventional method, see Fig. 4.200a. It can be seen that there are 16 units (98 × 89 mm) of build-up package substrate on the 510 × 510 mm panel. The dielectric material is Ajinomoto Build-Up Film (ABF). The coined C4 solder bumps are formed by stencil Sn0.7Cu solder ball mounting and then reflowed and pressed, see Fig. 4.200b. There are solder resists on both sides of the substrate and the surface finishing is immersion tin (IT).

(C) Hybrid Substrate

The final assembly of the hybrid substrate is by picking and placing of the individual (87 × 82 mm) organic interposer with the temporary glass carrier on the individual (98 × 89 mm) build-up package substrate and then reflows and underfills. The lead-free reflow temperature profile is shown in Fig. 4.201.

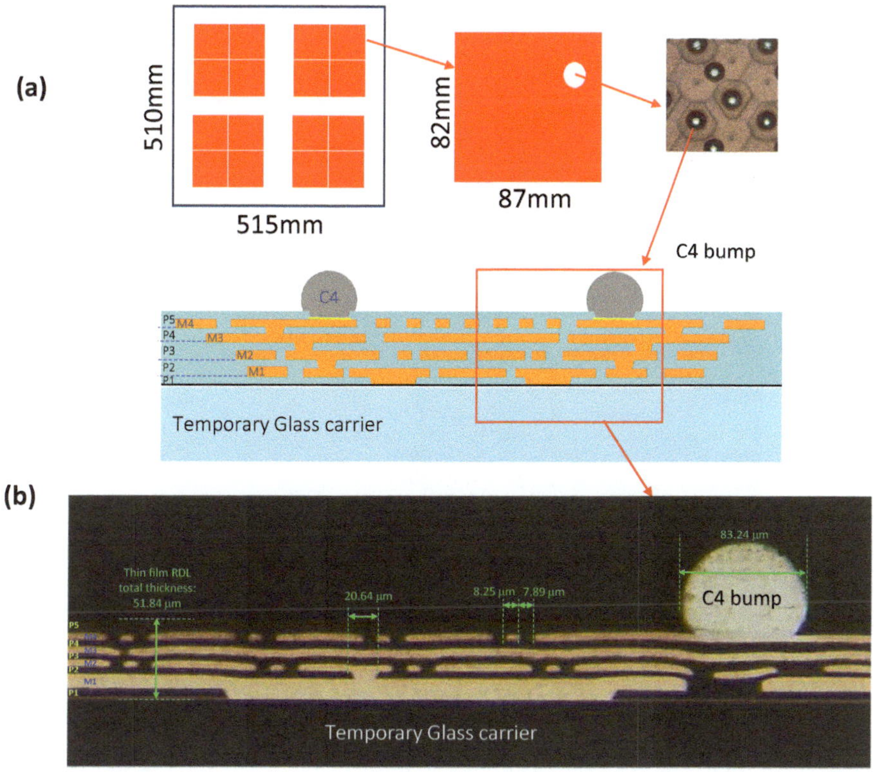

Fig. 4.198 Typical images of the ultralarge organic interposer with glass. **a** Top view. **b** Cross-sectional view

The equipment and the optimal process steps for the underfill dispensing are shown in Fig. 4.202. It can be seen that for such a large structure, the total dispensing time is 14 min long for an individual hybrid substrate. The final hybrid substrate is obtained by wheel sawing its four sides to 85 × 80 mm, as shown in Figs. 4.196, 4.203, 2.204, 4.205, 4.206 and 4.207. The temporary glass carrier is debonded by scanning a laser.

Figure 4.203 shows the X-ray image of the optimal hybrid substrate. The cross sections of the optima hybrid substrate are shown in Fig. 4.204. It can be seen that the center solder joints are a little "pressed" and the corner solder joints are a little "starched."

The warpages of the optimal hybrid substrate without the temporary glass carrier are shown in Fig. 4.205. It can be seen that at 30 °C, the warpage is −123.1 μm (smiling face); at 159 °C, the warpage is 46.8 μm (crying face); at 260 °C, the warpage is 126.2 μm (crying face); at 160 °C, the warpage is 72.5 μm (crying face); and at 31 °C, the warpage is −128.4 μm (smiling face).

The value of the warpage depends on the hybrid substrate structural geometry and material. Usually, the thicker the substrate geometry, the smaller the warpage.

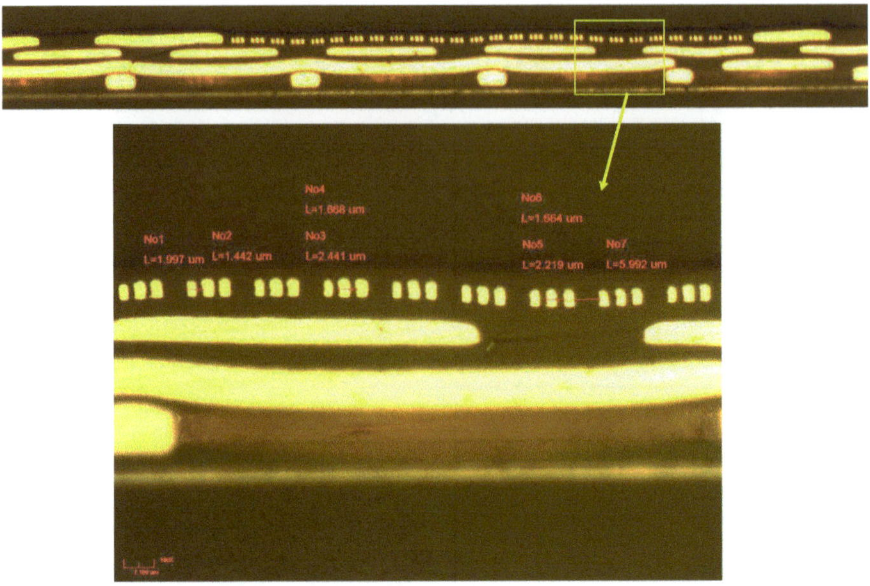

Fig. 4.199 Average L/S = 2.2/1.6 μm

The coefficient of thermal expansion (CTE) of material also plays a very important role on warpage. The CTE of the electroplated Cu is 16.3–17.5 × 10^{-6}/°C. Usually, the CTE of dielectric materials is higher than that of Cu. Thus, the lower the CTE of the PID and ABF dielectric materials, the lower the thermal expansion mismatch between the Cu and the dielectrics, which leads to lower warpage.

Before the optimal hybrid substrate, there are some assembly defects observed. Figure 4.206 shows some open solder joints. Figure 4.207 shows some voids in the underfill.

4.12.7 Summary

Some important results and recommendations are summarized as follows.

- The feasibility of design, materials, process, fabrication, and reliability of a hybrid substrate (85 × 80 mm), which consists of an ultralarge organic interposer, buildup package substrate, and solder joints with underfill, has been demonstrated.
- The ultralarge organic interposer consists of four RDLs with L/S = 2 μm (minimum) and a thickness of 50 μm. The PID is with a CTE = 60 × 10^{-6}/°C (25–300 °C). The organic interposer (87 × 82 mm) has been fabricated by a fan-out RDL-first process on a large glass temporary carrier (515 × 510 × 1.1 mm). The measured minimum L/S is not equal to 2 μm, and thus, there are rooms for

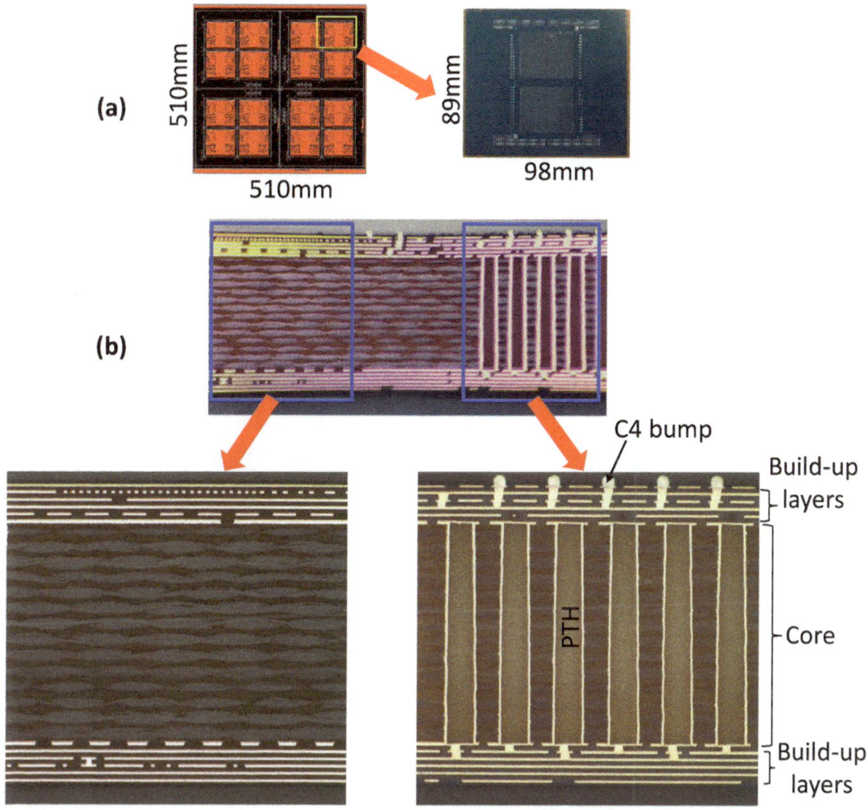

Fig. 4.200 Typical images of the build-up package substrate. **a** Top view. **b** Cross-sectional view

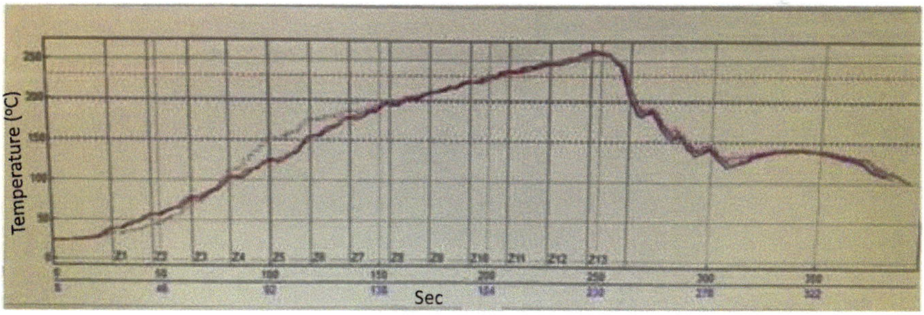

Fig. 4.201 Reflow temperature profile of the final assembly of the hybrid substrate

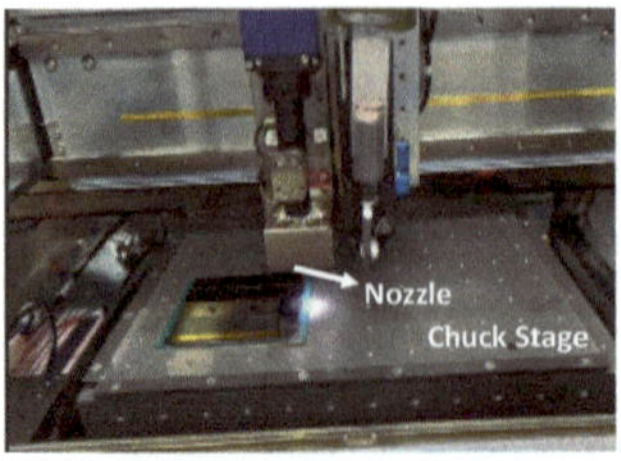

Underfill dispensing equipment

Item	Condition
Dispensing Nozzle	Φ100 μm
Dispensing Frequency	500 Hz
On/Off Time	1 ms/1 ms
Nozzle Heating	60 °C
Chuck Stage Heating	90 °C
Dispensing speed	5.4 mm/sec
Process time	~ 18 min
Gel Weight	1.2 g
Curing condition	1st step: 90 °C/ 100 mins 2nd step: 165 °C/ 120 mins

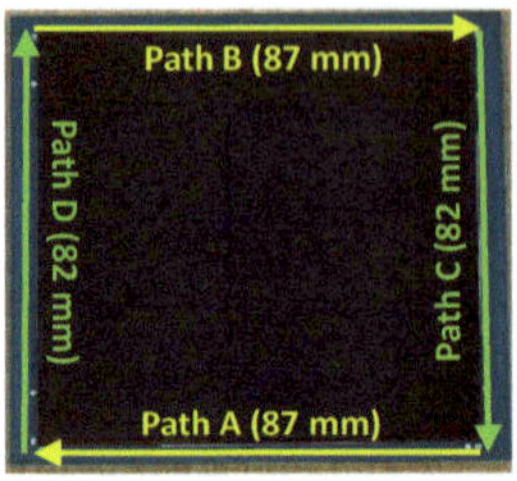

Step	Condition	
1	Path A (150 g) + Path B (150 g)	Waiting 50 sec
2	Path A (150 g) + Path B (150 g)	Waiting 210 sec
3	Path A (150 g) + Path B (150 g)	Waiting 300 sec
4	Path A (75 g) + Path B (75 g)	Waiting 120 sec
5	Path C (75 g) + Path D (75 g)	Waiting 120 sec

5 steps for underfill dispensing (total time = 4 mins)

Fig. 4.202 Underfill dispensing equipment and optimal process steps

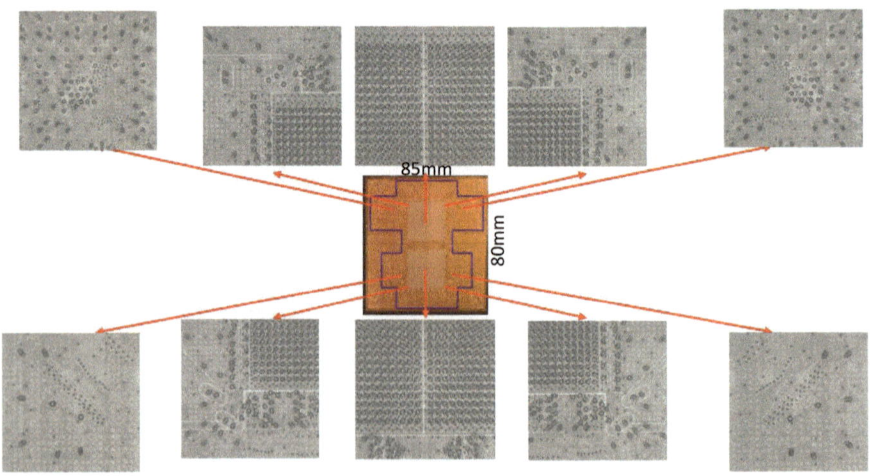

Fig. 4.203 X-ray images of the optimal hybrid substrate

improvement, e.g., a better estimation of the compensation of photoresist, LDI, ECD Cu, Cu etching, and so on.

- The build-up package is 5-2-5. The ABF is with a CTE = 20 × 10^{-6}/°C (25–150 °C) and 49 × 10^{-6}/°C (150–240 °C). The build-up package substrate (98 × 89 × 2 mm) has been fabricated by an ordinary process on a 510 × 510 × 2 mm panel.

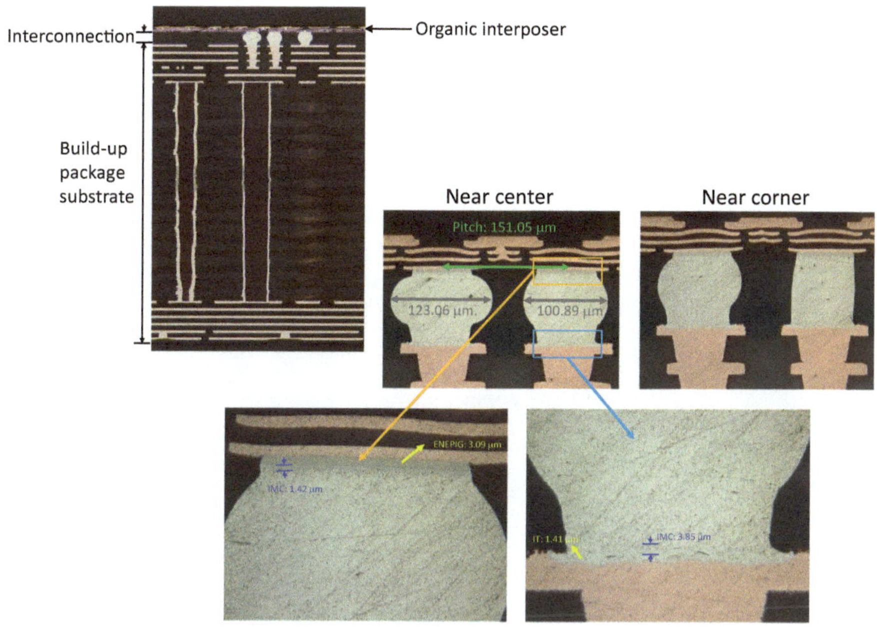

Fig. 4.204 Cross sections of the optimal hybrid substrate

Hybrid substrate without Glass (85 mm x 80 mm)			
Temp. (ºC)	Warpage (µm)	Variation with RT. (%)	Moiré image
30	-123.1	N/A	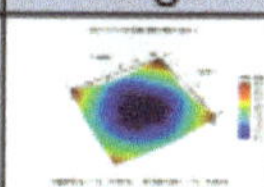
159	46.8	-138.02	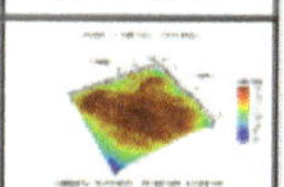
260	126.2	-202.52	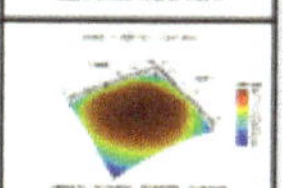
160	72.5	-158.90	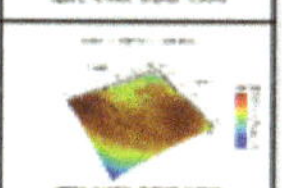
31	-128.4	4.31	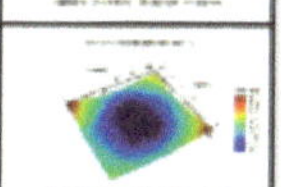

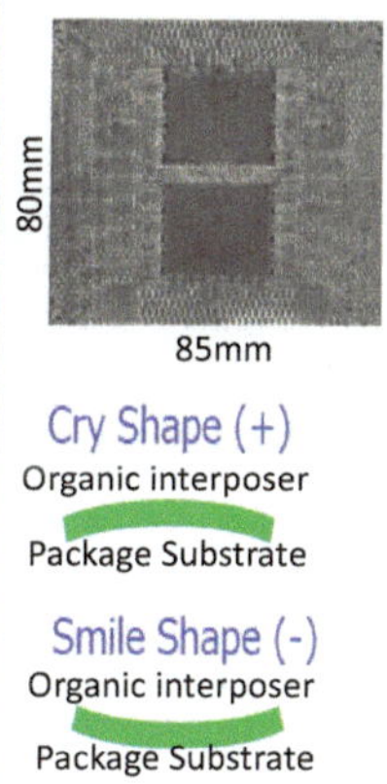

Fig. 4.205 Warpage of the optimal hybrid substrate without glass carrier

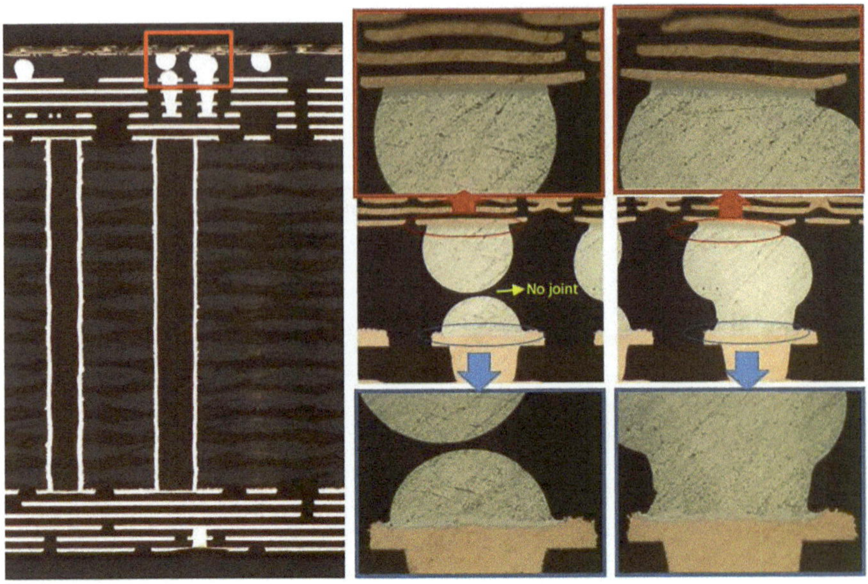

Fig. 4.206 Failed hybrid substrate—open solder joint

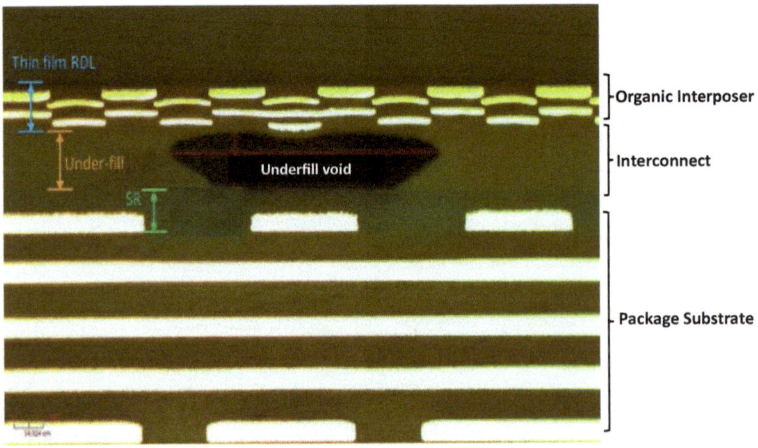

Fig. 4.207 Failed hybrid substrate—void in underfill

- The final assembly of the hybrid substrate is by picking and placing the organic interposer on the build-up package substrate and then fluxing and reflowing the >50,000 Sn0.7Cu solder bumps.
- Underfill dispensing of the gap between the organic interposer and build-up package substrate with >50,000 solder joints of such a large hybrid substrate

have been a big challenge. In this section, it took five steps and 14 min for the dispensing of the underfill. Thus, there are rooms for improvement.

- After underfill, it is wheel sawing the four sides into an 85 × 80 mm hybrid substrate.
- The warpage of the hybrid substrate (without the temporary glass carrier) has been measured by the shadow Moiré method.
- The value of the warpage depends on the hybrid substrate thickness and material. Usually, the thicker the substrate, the smaller the warpage. The CTE of the dielectric material also plays a very important role on the warpage. The CTE of the electroplated Cu is ~17 × 10^{-6}/°C. Usually, the CTE of dielectric materials is higher than that of Cu. Thus, it is recommended that the lower the CTE of the PID and ABF, the lower the thermal expansion mismatch between the Cu and the dielectrics, which leads to lower warpage.
- The reliability of the Sn0.7Cu lead-free solder joint is demonstrated by a nonlinear and time- and temperature-dependent finite element simulation. The accumulated maximum inelastic strains at the solder joint are too small to create solder joint reliability issues.

4.13 Hybrid Substrate with Interconnect-Layer for Heterogeneous Integration

In this section, the fine metal L/S RDL-substrate (organic interposer) and the build-up package substrate or HDI substrate are also fabricated separately, and then they are combined through an interconnect-layer. This is very similar to 2.3D IC integration [20–24] (Sects. 4.10, 4.11, 4.12 and 4.13 of this book) except the solder joint and underfill are eliminated, which are replaced by an interconnect layer [25].

The interconnect layer is about 60 μm thick, which consists of the prepreg and vias (100 μm diameter on top and 80 μm diameter at the bottom) filled with conductive paste and are in β-stage. The fine metal L/S coreless substrate (37 μm thickness) is fabricated by a photoimageable dielectric (PID), laser direct imaging (LDI), and development, physical vapor deposition (PVD), photoresist and LDI, Cu-plating, etc. on a temporary glass carrier. The minimum metal L/S, pad pitch, and pad diameter for the chips are, respectively, 2/2, 60, and 35 μm. The HDI is an eight-layer substrate with 1 mm thickness. The pad pitch and pad diameter are, respectively, 350 and 300 μm. The final assembly of these three separate substrates is by thermal compression and the interconnect layer is then fully cured (C-stage).

Characterizations, such as continuity check is performed to make sure at least the design and fabrication are correct. Optical microscope (OM), X-ray, and scanning electron microscope (SEM) of the cross section of the hybrid substrate is performed to check the quality of metal and dielectrics of the fine metal L/S coreless substrate, the integrity of the conductive paste in the vias, and the delamination of the interconnect-layer. Six times a reflow test and a drop test are performed. A nonlinear finite element analysis is performed to determine the state of strain and stress at the vias filled with

Fig. 4.208 Top-view and cross section view of chiplets heterogeneous integration on hybrid substrate

conductive paste of the interconnect layer and critical solder joints subjected to a thermal cycling condition.

4.13.1 The Structure

Figures 4.208 and 4.209 schematically show the top-view and cross section view of a heterogeneous integration of chips on a high-density organic hybrid substrate. It consists of four major parts: (a) the chips with microbumps, (b) the fine metal L/S RDL-substrate or organic interposer (~37 μm), (c) the interconnect-layer (~60 μm), and (d) the HDI printed circuit board (PCB) (~1 mm).

4.13.2 Test Chips

The test chips for this section are shown in Fig. 4.210. It can be seen that the size of the large chip (Chip 1) is 10 mm × 10 mm × 150 μm with 3760 area array pads on 90 μm pitch and are daisy chained. The Cu pad size is 50 μm × 50 μm. The size of the small chips (Chip 2A and 2B) is 7 mm × 5 mm × 260 μm with 1512 area array daisy chained pads on 60 μm pitch and the Cu pad size is 44 μm × 44 μm. For all the chips, the Ti/Cu (0.1/0.2 μm) under bump metallurgy (UBM) pad size is 35 μm diameter, the passivation (PI2) opening is 20 μm diameter, the Cu-pillar is 35 μm diameter, and 37 μm tall, the SnAg solder cap is 15 μm with a barrier (Ni = 3 μm).

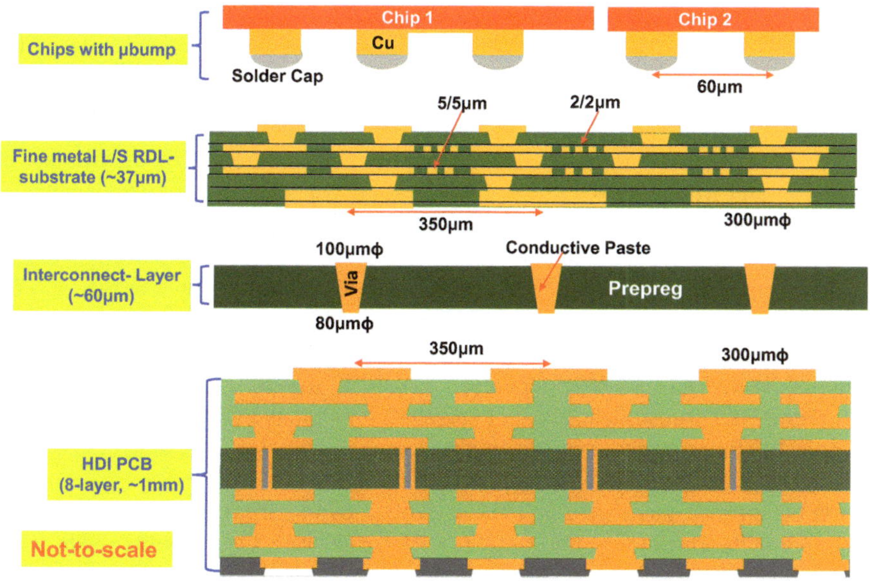

Fig. 4.209 Key parts of the test package

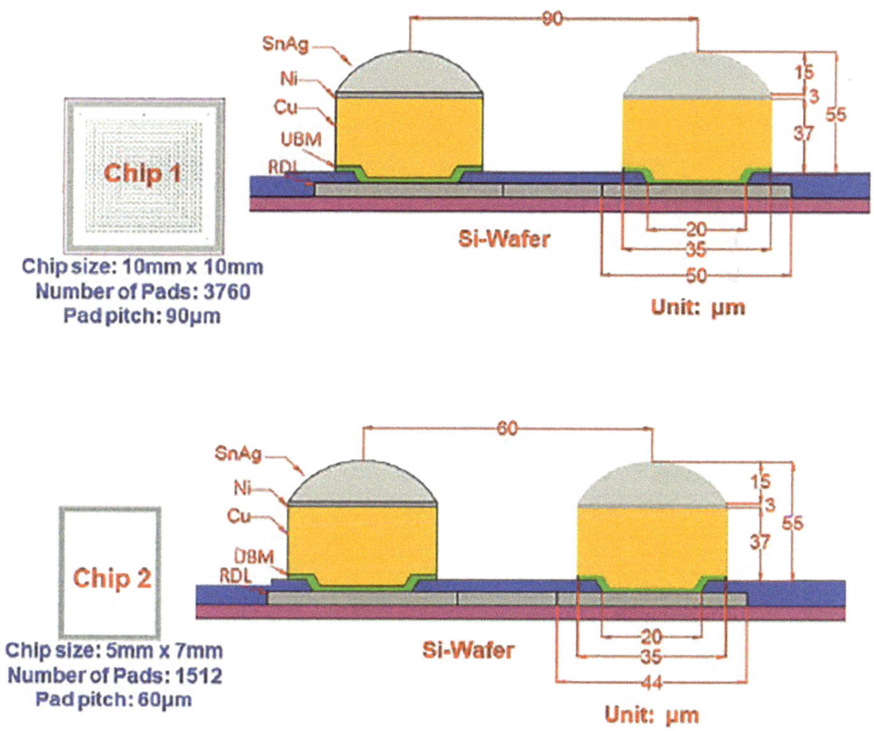

Fig. 4.210 Schematic of the test chips

4.13.3 Fine Metal L/S RDL (Organic Interposer)

There are three RDLs in the RDL-substrate and each RDL consists of one metal layer (ML) and one dielectric layer (DL). Figure 4.211 shows the definition and value of RDLx, MLx, DLxy, Vxy, and L/S/H. It can be seen that the L/S/H of ML1 are 2/2/2.5 μm and ML2 are 5/5/3.5 μm. ML3 is the contact PAD with a diameter = 300 μm and a thickness = 5 μm. The thickness of DLs DL01, DL12, and DL23 are, respectively, 7.5, 6.5, and 5 μm.

The temporary panel for fabricating the RDL-substrate is shown in Fig. 4.212a. It can be seen that the panel size is 515 mm × 510 mm × 1.1 mm and is made of glass with a coefficient of thermal expansion (CTE) equals 8.5×10^{-6}/°C. The panel is divided into 18 strips and each strip (132 mm × 77 mm) has 8 (20 mm × 20 mm) RDL-substrates. Thus, in one shot, it can make RDL-substrates for 432 chips in 144 heterogeneous integration packages. Figure 4.212b, c show the topside and bottom-side of an individual hybrid substrate.

On the topside, there are 2 × 1512 + 3760 = 6784 pads, Fig. 4.212b, which are for bonding the chips to the fine metal L/S RDL-substrate. On the bottom-side, there are 2780 Cu pads (300 μm diameter) on 350 μm pitch, Fig. 4.212c. These pads are for the interconnection of the interconnect layer.

The key process steps in fabricating the fine metal L/S RDL-substrate are shown in Fig. 4.213. First, a released film (sacrificial layer) is slit coated on a temporary glass carrier (515 mm × 510 mm) and then a Ti/Cu seed layer is formed by PVD. It

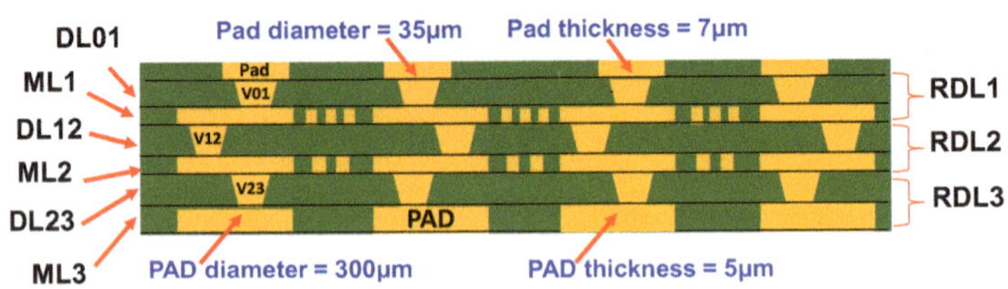

RDLs	Key Elements of RDL	Line width/ Spacing (L/S)	Thickness (H)	Via Dia. (T/B)
RDL1	ML1 (Metal layer 1)	2/2μm	2.5μm	NA
	DL01 (Dielectric layer between Pad and ML1)	NA	7.5μm	NA
	V01 (Via opening between contact Pad and ML1)	NA	NA	17/10μm
RDL2	ML2 (Metal layer 2)	5/5μm	3.5μm	NA
	DL12 (Dielectric layer between ML1 and ML2)	NA	6.5μm	NA
	V12 (Via opening between ML1 and ML2)	NA	NA	32/25μm
RDL3	ML3 (Metal layer 3) (PAD)	NA	5μm	NA
	DL23 (Dielectric layer between ML2 and ML3)	NA	5μm	NA
	V23 (Via opening between ML2 and ML3)	NA	NA	32/25μm

Fig. 4.211 Fine metal L/S 3-layer RDL—(organic) substrate

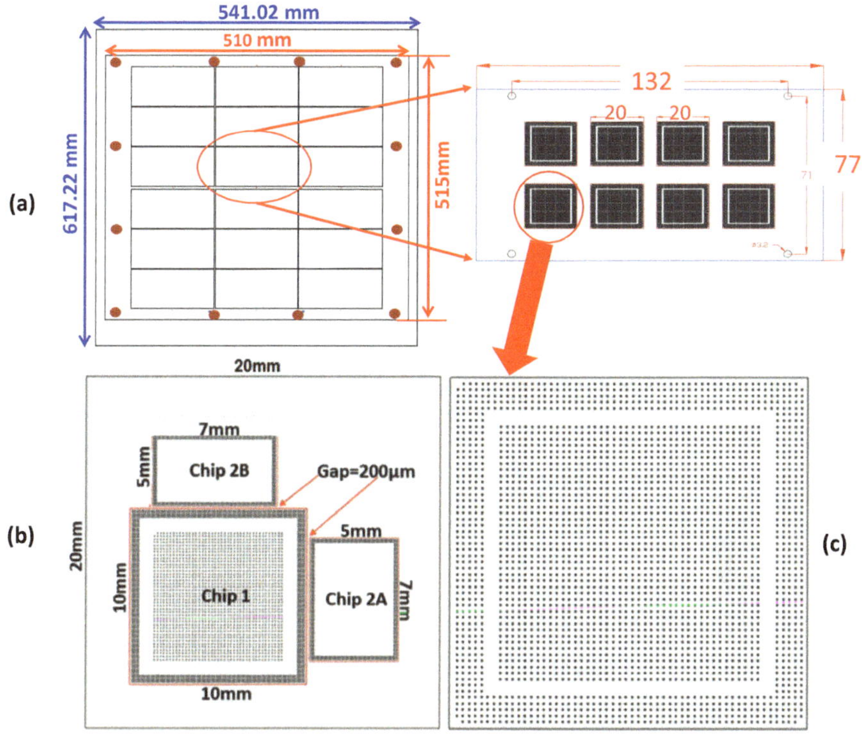

Fig. 4.212 **a** Panel for making the RDL-substrates. RDL-substrate: **b** Top- and **c** bottom-view

is followed by photoresist, LDI, and development. Then, electrochemical deposition (ECD) Cu and strip off the photoresist and etch off the Ti/Cu to obtain the ML (ML3 or PAD) of RDL3. It is followed by slit coating a PID and LDI to get the DL DL23 of RDL3. Then, sputter the Ti/Cu, photoresist, LDI, develop, and ECD the Cu. It is followed by stripping off the photoresist and etching off the TiCu to get the ML ML2 of RDL2. Repeat the same process steps to obtain the ML ML1 of RDL1 and DLs DL12 and DL01 of RDL2 and RDL1, respectively. Then, sputter the Ti/Cu, photoresist, LDI, and develop, and ECD the Cu. It is followed by stripping off the photoresist and etching off the TiCu to get the bonding pad (lead) for the chips. Figure 4.214 shows the panel with 144 three layer (20 mm × 20 mm) RDL-substrates.

Figures 4.215 and 4.216 show, respectively, the OM images of the top side of the (L/S = 2/2 μm) RDL-substrate at 500 and 1000 times of magnifications and the typical cross section. Figure 4.217 shows a typical SEM image of the cross section of the RDL-substrate. It can be seen that (a) for ML1, the line widths (L) are 1.91, 1.98, 1.91, and 1.78 μm, which are close to the target (2 μm), the line spacing (S) are 2.31, 2.37, and 2.37 μm, which are reasonably close to the target (2 μm), and the thicknesses (H) are 3.1 and 3.1 μm, which are close to the target (2.5 μm), (b) for ML2, the L are 7.92 and 7.72 μm, which are not close to the target (5 μm), the

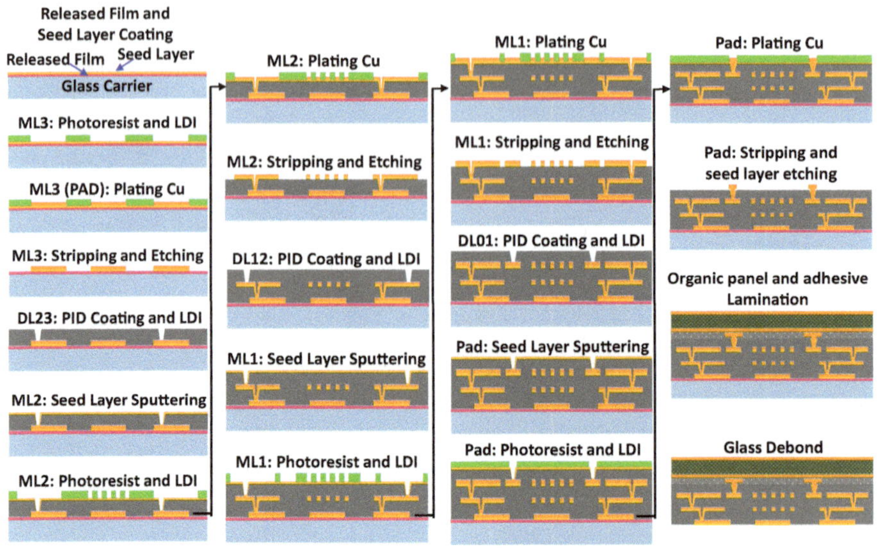

Fig. 4.213 Key process steps in making the fine metal L/S RDL—(organic) substrate

Fig. 4.214 Glass panel for making the RDL-substrate

S are 2.83 and 2.84 μm, which are far from the target (5 μm), and the H is 5.61, which is far from the target (3.5 μm), and (c) for ML3, the H is 8.31 μm, which is also far from the target (5 μm). Thus, there are rooms for improvement, e.g., a better estimation of the compensation of photoresist, LDI, ECD Cu, Cu etching, etc.

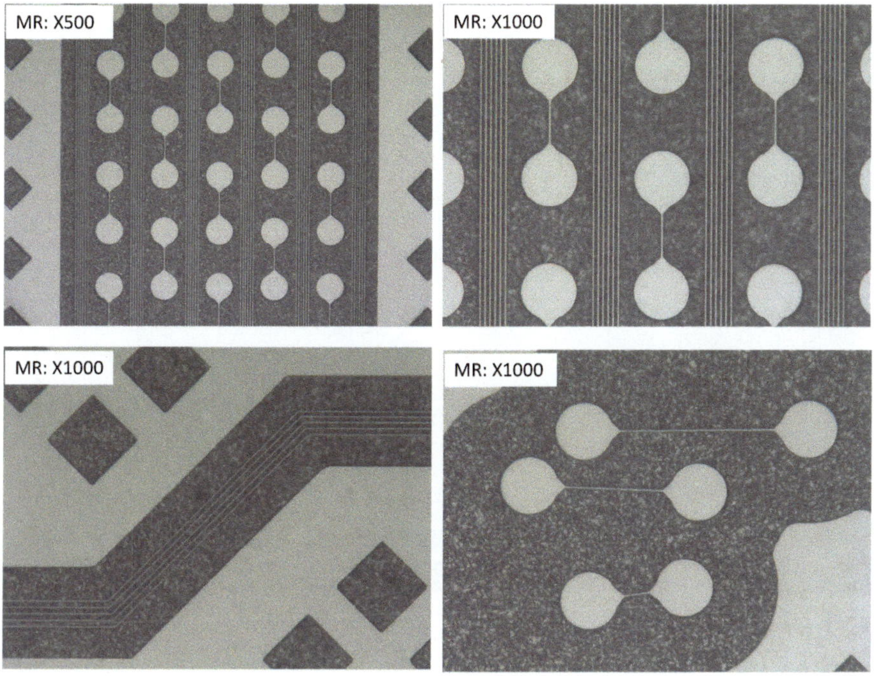

Fig. 4.215 OM images of the RDL-substrate (topside)

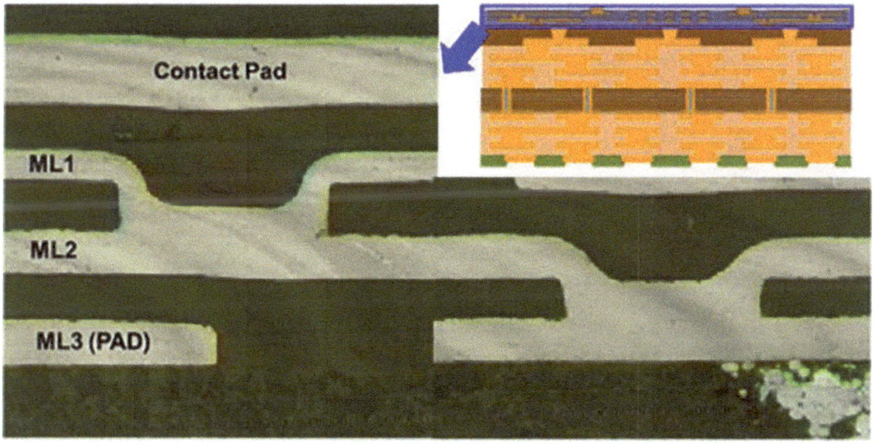

Fig. 4.216 Typical OM image of the RDL-substrate

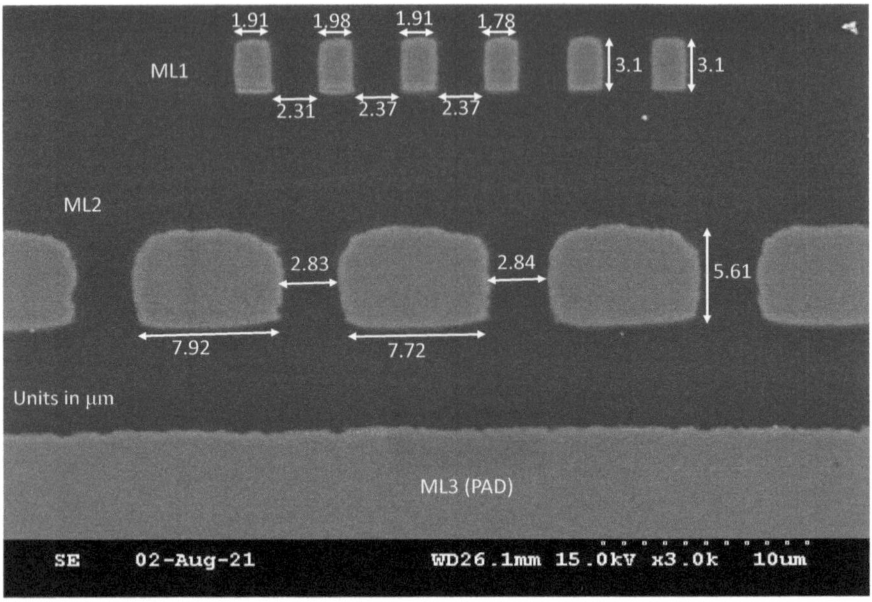

Fig. 4.217 SEM image of the RDL-substrate

4.13.4 *Interconnect-Layer*

Figure 4.218 shows the interconnect layer. First, laminate polyester (PET) on both sides of a prepreg (PP). Then, laser drill vias and print conductive paste into vias. It is followed by stripping off the PET. Both the paste and PP of the interconnect-layer are in β-stage.

4.13.5 *HDI PCB*

The 975 mm-thick HDI PCB has eight layers (Table 4.20) and is fabricated by the conventional process.

4.13.6 *Final Assembly of the Hybrid Interposer*

First, attach the fabricated RDL-substrate with the glass carrier to an organic panel with an adhesive, and then debond the glass carrier as shown in Fig. 4.219. The key final assembly process steps (Fig. 4.220) of these three substrates are by thermal compression. The alignments and correct positions of these three substrates are fixed

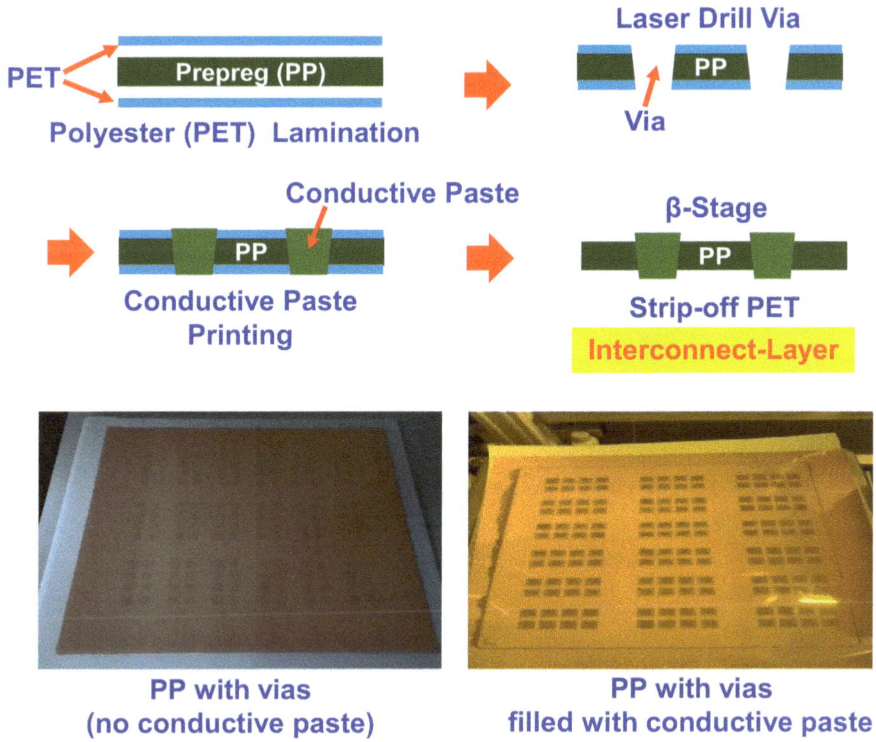

Fig. 4.218 Interconnect-layer (PP and paste are in β-stage)

by more than ten nails around the four sides of these three panel substrates. After thermal compression (lamination), the PP and conductive paste of the interconnect layer are fully cured (C-stage) and the nails are removed. It is followed by dry film lamination at the bottom of the hybrid substrate and then organic panel debonding. It is followed by Cu foil etching, dry film stripping, adhesive plasma etching, and surface finishing.

4.13.7 X-Ray and OM

Figures 4.221 and 4.222 show the OM and X-ray images of the fabricated hybrid substrate. Figure 4.221 shows a typical cross section of the hybrid substrate which consists of the fine metal L/S RDL-substrate, the interconnect-layer, and the HDI PCB. Figure 4.221 also shows the vias filled with the conductive paste nicely and there are not any obvious large void and delamination. The thickness of the interconnect layer is 60 μm, which is thinner than the C4 solder joint (usually 100 μm) of the 2.3-D IC integration.

Table 4.20 HDI PCB

Category	Layer	Material	Item	Diameter/width (μm)	Thickness (μm)
HDI	L1	Cu	Pad Dia.	300	22
		PP 1067	Bottom Dia.	100	55
	L2	Cu	Pad Dia.	200/400	25
		PP 1067	Bottom Dia.	250	60
	L3	Cu	Pad Dia.	400	17
		Core	MTH Dia.	250	254
	L4	Cu	Pad Dia.	400	17
		PP 1067	Bottom Dia.	250	55
	L5	Cu	Pad Dia.	400	17
		Core	MTH Dia.	250	254
	L6	Cu	Pad Dia.	400	17
		PP 1067	Bottom Dia.	250	60
	L7	Cu	Pad Dia.	200/400	25
		PP 1067	Bottom Dia.	100	55
	L8	Cu	Pad Dia.	200	22
		SR	Bottom Dia.	600	20
	HDI total thickness				975

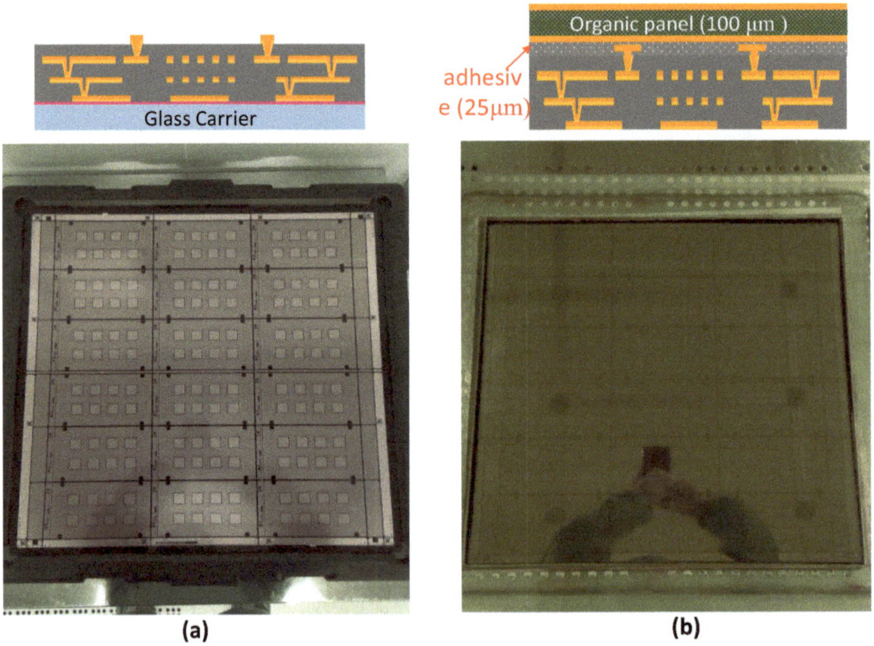

Fig. 4.219 Attach an organic panel to the RDL-substrate and debond the temporary glass carrier

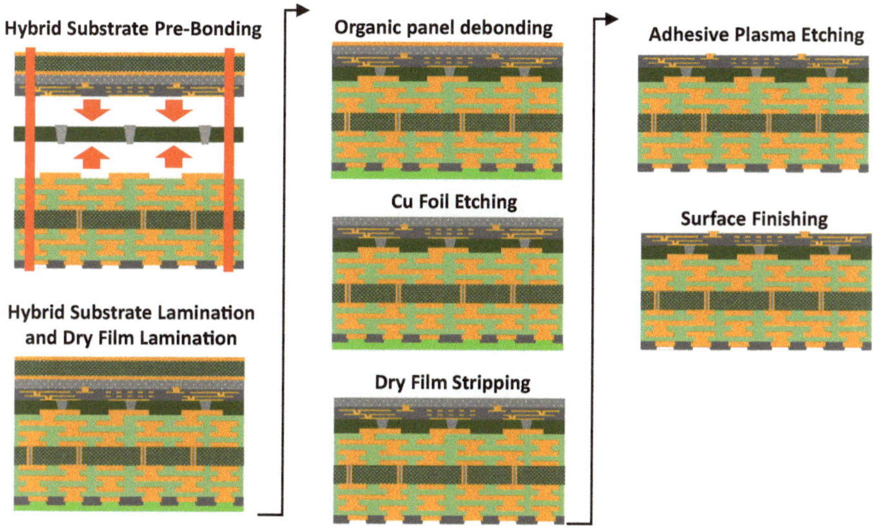

Fig. 4.220 Process steps in making the hybrid substrate

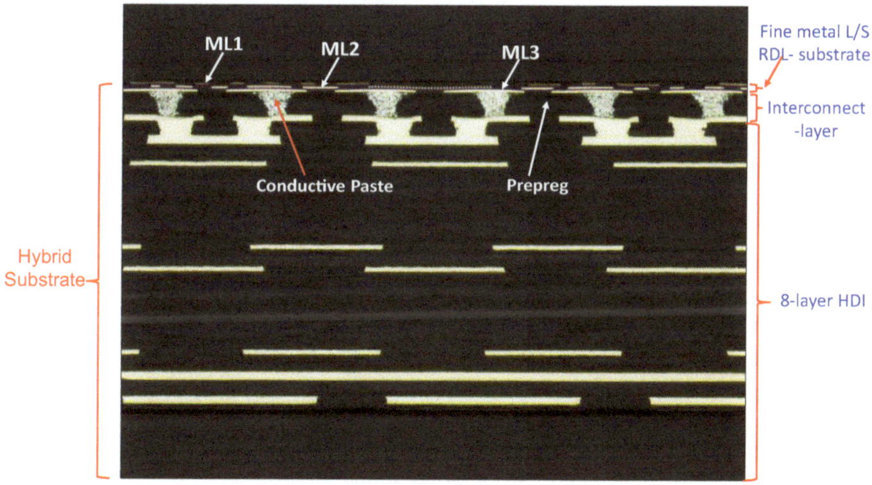

Fig. 4.221 Typical cross section of the hybrid substrate

Figure 4.222 shows the pad for chip bonding, the pads for interconnect layer, the conductive paste, and ML1, ML2, and ML3 with their targets. Figure 4.222 also shows the X-ray image of the conductive paste, the contact pads, and the daisy-chain on PCB. All these demonstrated the assembly is properly done.

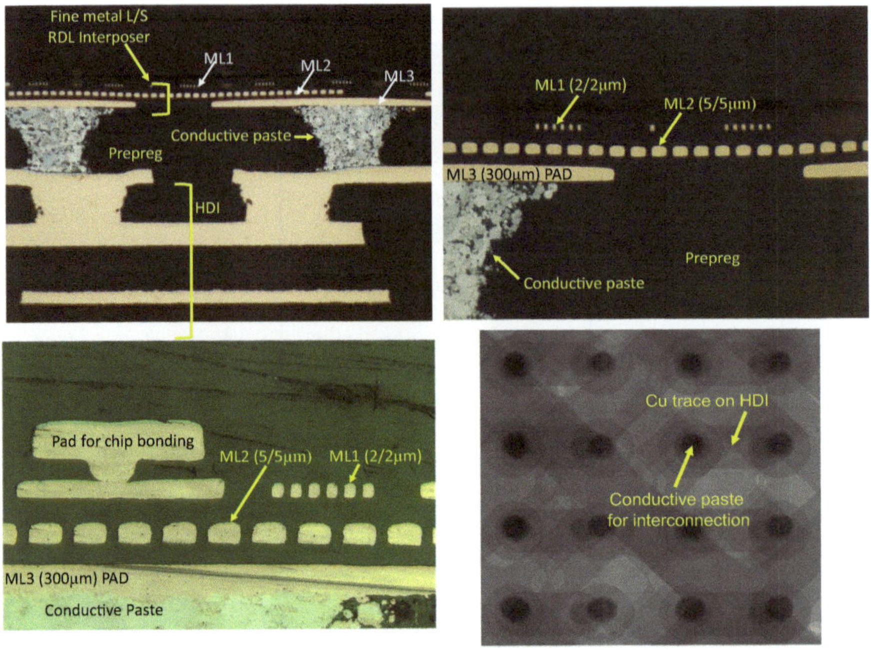

Fig. 4.222 Close-up views of the hybrid substrate

4.13.8 Continuity Check

Figure 4.223 shows the nets for continuity checks. There are 21 hybrid substrates that passed all the nets.

4.13.9 Final Assembly

After the hybrid substrate is ready, it is time to perform chips to hybrid substrate bonding and underfilling. A typical heterogeneous integration of three chips on the hybrid substrate is shown in Fig. 4.224.

4.13.10 Six-Time Reflow

Six-Time Reflow A total of 15 hybrid substrates (without chips) with 26 nets and eight hybrid substrates (with chips) with 37 nets are subjected to six-time reflow with the temperature profile shown in Fig. 4.225. Except for a few failed at the first time

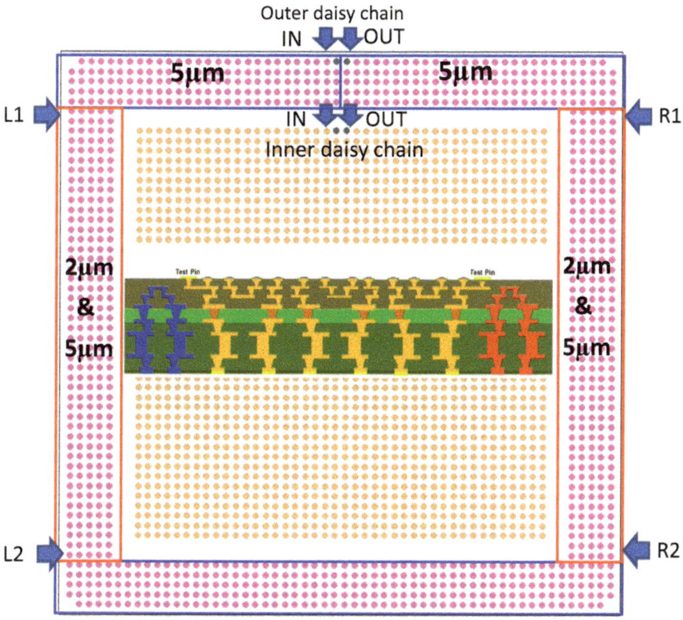

Fig. 4.223 Schematic nets for continuity measurement

of reflow (earlier failures), most passed the six-time reflow test. The failure criterion is when the open circuit is found in a two-wire resistance measurement.

4.13.11 Drop Test

The drop test setup is according to JEDEC Standard JESD22-B111, as shown in Fig. 4.226. After a few tries, the right height (1120 mm) of the drop table is obtained, which yields the drop spectrum with 1500 G/ms (1500 Gs, 0.5 ms half-sine pulse). The sample size is eight packages (hybrid substrates and chips) with 37 nets. After more than 30 drop tests, there are two failures: one failed at 20 drops and the other at 30 drops. The measurement method is a four-wire resistance measurement, and the failure criterion is when the resistance change (R-shift) is 50%. A failure analysis reveals that the failure mode is the cracking of the solder joint as shown in Fig. 4.227 and not at the hybrid substate. This could be due to the poor assembly of the chips to hybrid substrate bonding. Thus, the fabricated hybrid substrate in this article is considered reliable for dropping mobile products.

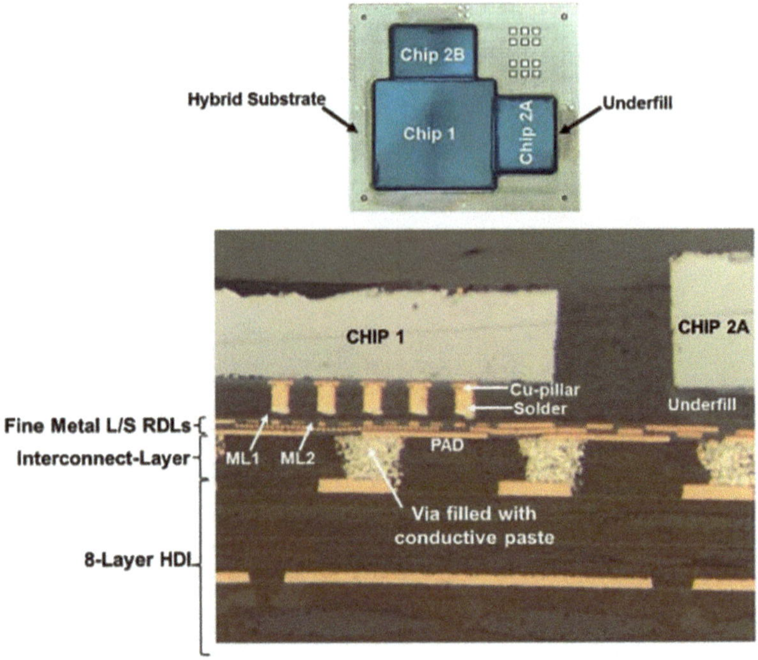

Fig. 4.224 Heterogeneous integration of three chips on the hybrid substrate (top- and cross-sectional view)

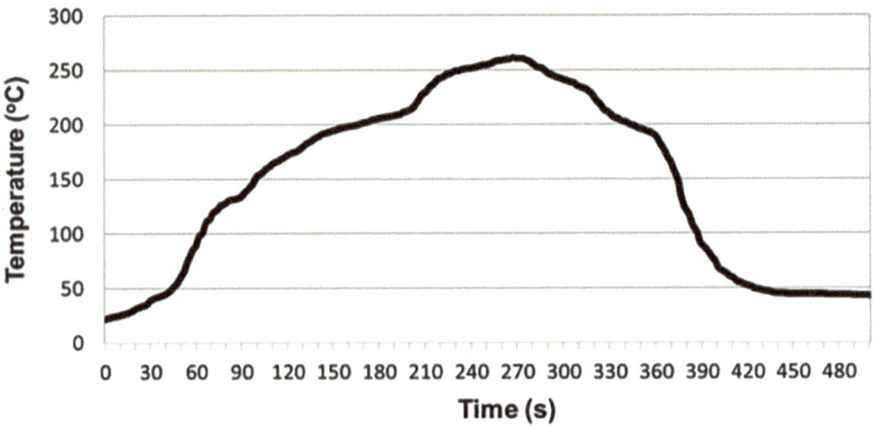

Fig. 4.225 Reflow temperature profile

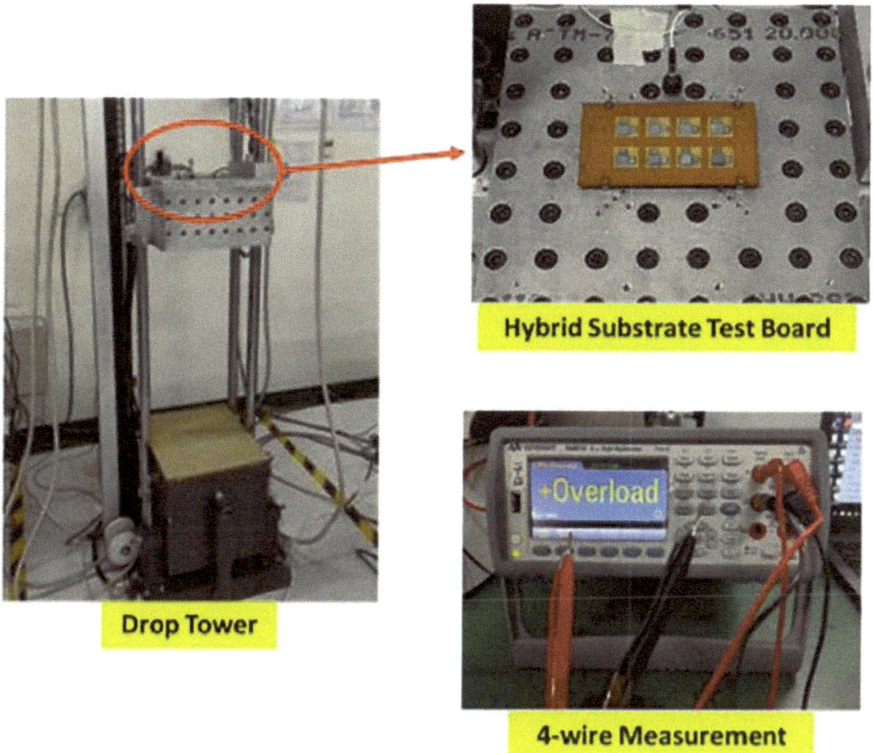

Fig. 4.226 Drop test set-up and measurement

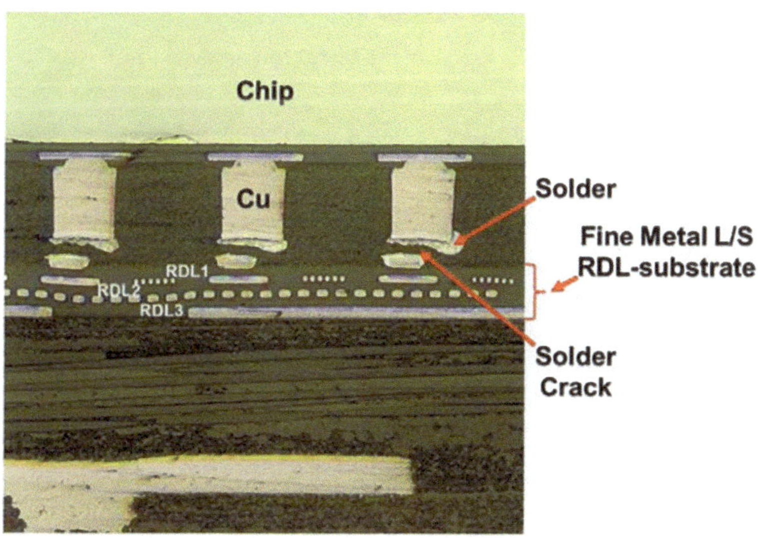

Fig. 4.227 Failure mode due to drop test

4.13.12 Thermal Cycling Simulation

The structural elements and their dimensions of the heterogeneous integration of three chips on the hybrid substrate are shown in Figs. 4.228 and 4.229 and their finite element model is shown in Fig. 4.230. Since the conductive paste filled via and the solder joint are the focus point of interest, much finer meshes are used in the paste filled via areas and the Cu pillars and the solder caps near the chip corners. Due to symmetry about the AA′-axes only half of the structure is modeled.

The material properties of the structural elements are shown in Table 4.21. The electroplated Cu is assumed as an elastic–plastic (bilinear kinematic hardening) material with the stress–strain relation shown in Fig. 4.231 (first Young's modulus = 121 GPa, second Young's modulus = 1.2 GPa, and yield strength = 173 MPa). The HDI is assumed to be an anisotropic material. The SAC (Sn3Ag0.5Cu) solder is assumed to be a temperature dependent material; the Young's modulus = (40–0.07 T) GPa and the CTE = $(21 + 0.017\ \mathrm{T}) \times 10^{-6}/°\mathrm{C}$. The constitutive equation for the SAC is assumed to obey the Garofalo hyperbolic sine creep equation [48]

$$d\gamma/dt = 500,000[\sinh(0.01\sigma)]5 \exp -5,800\ \mathrm{T}$$

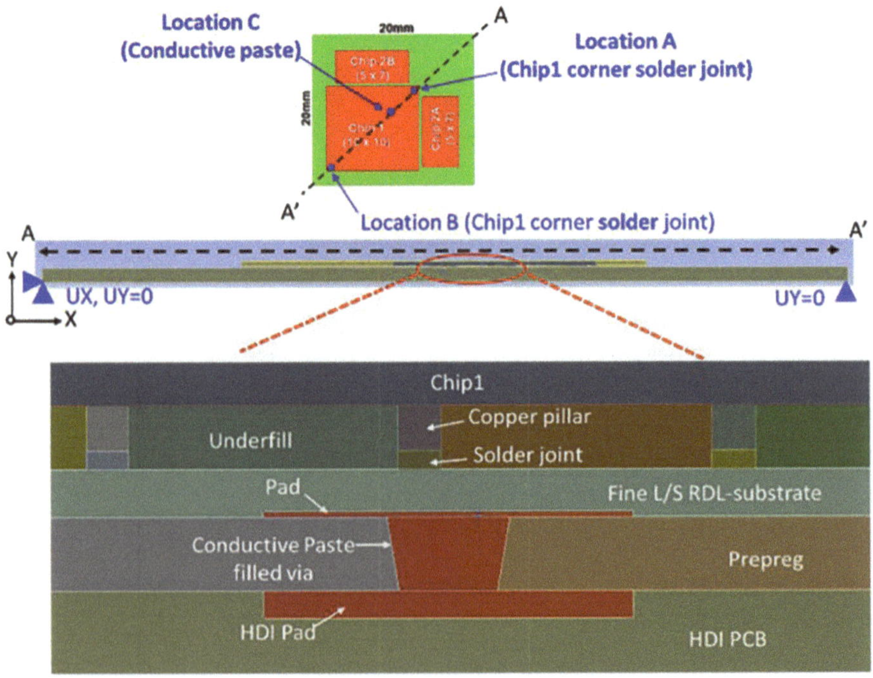

Fig. 4.228 The structure

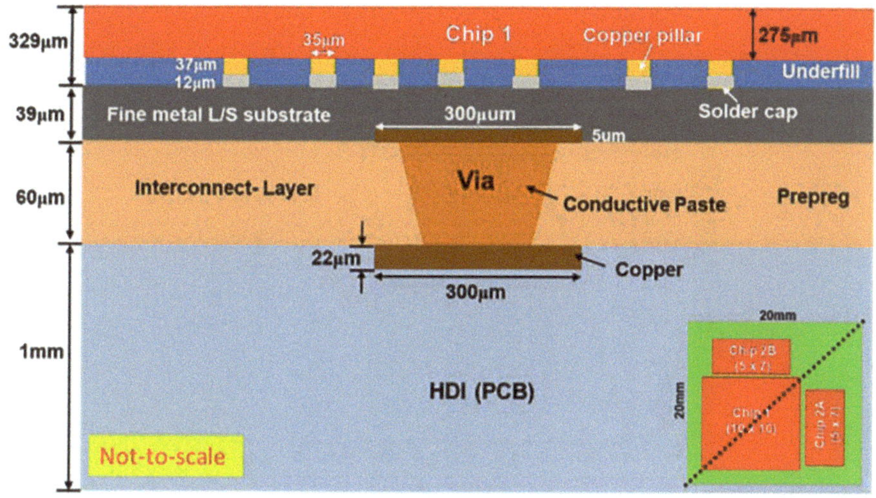

Fig. 4.229 Structural elements for finite element modeling

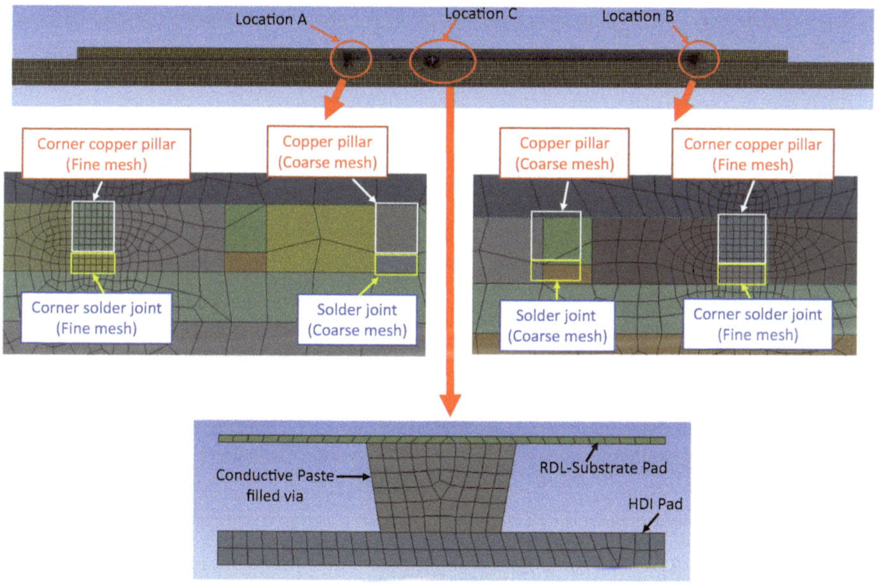

Fig. 4.230 Finite element modeling

The temperature boundary condition is −40 °C ↔ 85 °C, and the dwell-time at hot, dwell-time at cold, ramp-up time, and ramp-down time are 15 min each. Stress free is at room temperature.

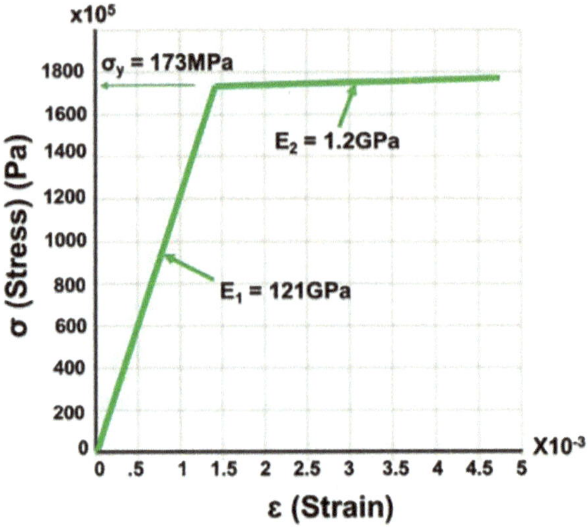

Fig. 4.231 Stress–strain relation for electroplated Cu

Table 4.21 Material properties

Materials	CTE (10^{-6}/°C)	Young's modulus (GPa)	Poisson's ratio
Copper	16.3	121	0.34
HDI PCB	$\alpha_x = \alpha_y = 18$, $\alpha_z = 70$	$E_x = E_y = 22$, $E_z = 10$	0.28
Silicon	2.8	131	0.278
EMC	10 (<150 °C)	19	0.25
Solder	21 + 0.017 T (°C)	49 − 0.07 T (°C)	0.3
RDL substrate	27.9	47.8	0.3
Conductive paste	19.01	20.33	0.3
Prepreg (interconnect-layer)	15	26	0.39
Underfill	50	4.5	0.35

Where *T* is in Celcius

The deformed shape and undeformed shape of the heterogeneous integration of three chips on the hybrid substrate are shown in Fig. 4.232. At 450 s (85 °C), the hybrid substrate expands more than the chips and the whole structure is deformed in a concave shape (smiling face), Fig. 4.232a. At 2250 s (−40 °C), the hybrid substrate shrinks more than the chips and the structure is deformed into a convex shape (crying face), Fig. 4.232b. This is because the CTE of silicon chip is smaller than that of the hybrid substrate.

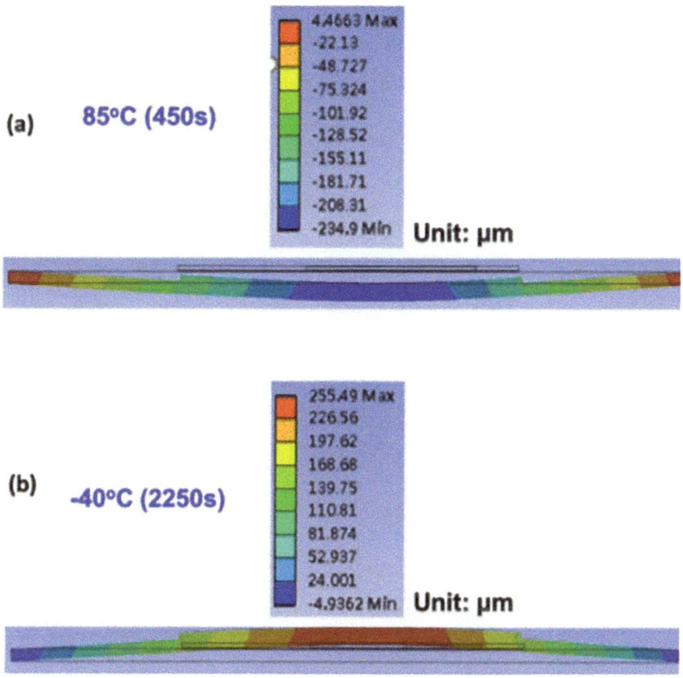

Fig. 4.232 Structural deformation during the first thermal cycle. **a** At 85 °C (450 s). **b** At −40 °C (2250 s)

The maximum warpage of the individual package (20 mm × 20 mm) is 255 μm. According to [12], the allowable warpage for a 13.42 mm × 13.42 mm individual package is 200 μm. Thus, the warpage of the present package should be acceptable.

Figure 4.233 shows the Mises (equivalent) stress acting at the via filled with conductive paste (Location C in Fig. 4.230). Figure 4.233a shows the Mises contours at 85 °C (450 s), while Fig. 4.233b at −40 °C (2250 s). It can be seen that: (a) for both times (or temperatures) the maximum Mises stress occurs near the interface between the interconnect-layer and the fine metal L/S RDL-interposer (this is due to the expansion mismatch between the RDL-interposer and the interconnect layer is larger than that between the interconnect-layer and the HDI PCB) and (b) the Mises stress contour at both times is not much different, and (c) the maximum Mises stress (~20 MPa) is very small to create reliability issues, such as cracking and delamination.

The accumulated creep strain contour distributions at the critical microbump solder joint A (at Location A in Fig. 4.230) is shown in Fig. 4.234 (the contour distributions for solder joint B at Location B in Fig. 4.230 are very similar). The accumulated creep strain time history in solder joints A and B is shown in Fig. 4.235. It can be seen that the maximum accumulated creep strain occurs near the corner of the solder joints. Also, the maximum accumulated creep strain per cycle in the

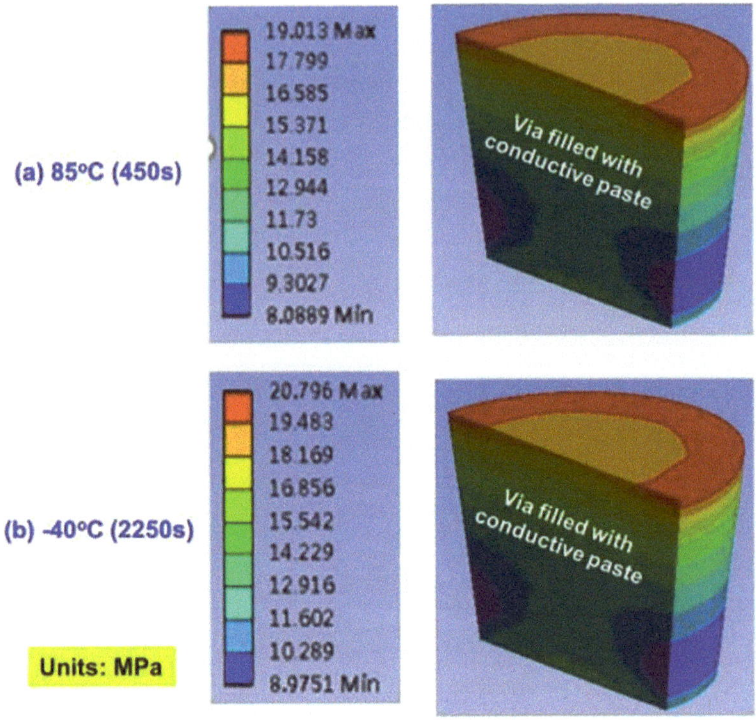

Fig. 4.233 Von Mises stress acting at the via filled with conductive paste. **a** 85 °C (450 s). **b** –40 °C (2250 s)

microbump solder joints A and B is almost the same. The maximum accumulated creep strain per cycle in the microbump solder joints A and B is 3.8% and it occurs in a very small area. Thus, this structure should be reliable for most operating conditions.

The creep strain energy density contour distributions at the microbump solder joint A are shown in Fig. 4.236 (the contour distributions for solder joint B are very similar) and the maximum creep strain energy density time history at solder joints A and B are shown in Fig. 4.237. It can be seen that the maximum creep strain energy density occurs near the corner of the solder joints. The maximum creep strain energy density per cycle in the microbump solder joints A and B is only 1.53 MPa and it occurs at a very small area. Again, this structure should be reliable for most operating conditions.

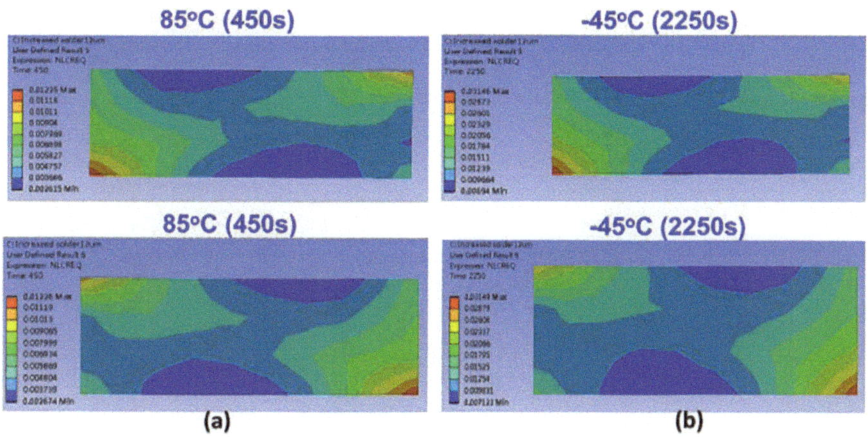

Fig. 4.234 Accumulated creep strain at solder joint A during the first thermal cycle. **a** At 85 °C (450 s). **b** At −40 °C (2250 s)

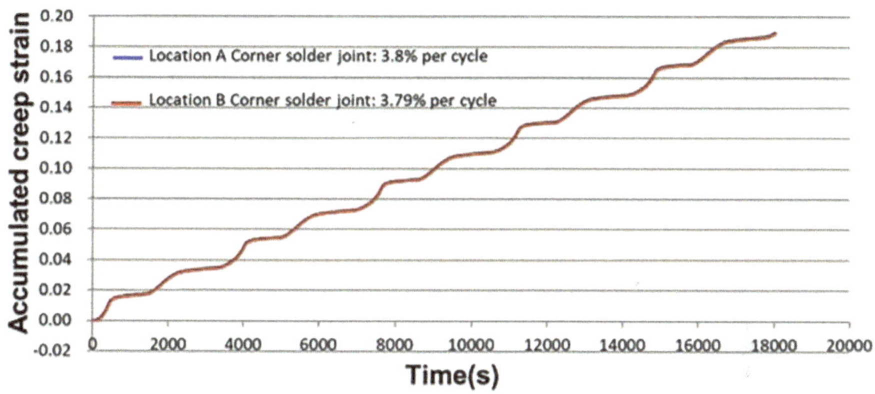

Fig. 4.235 Accumulated creep strain time-history at solder joints A and B

4.13.13 Summary

Some important results are summarized as follows.

- A high-density organic hybrid substrate for heterogeneous integration has been developed. This hybrid substrate consists of three major parts, namely the fine metal L/S RDL-substrate, the interconnect-layer substrate, and the HDI substrate.
- The fine metal L/S RDL-substrate with a minimum metal L/S = 2/2 μm has been fabricated by a fan-out panel level RDL-first process.
- The interconnect layer has been fabricated by the PP and vias filled with conductive paste in β-stage.
- The eight-layer HDI PCB has been fabricated by the conventional process.

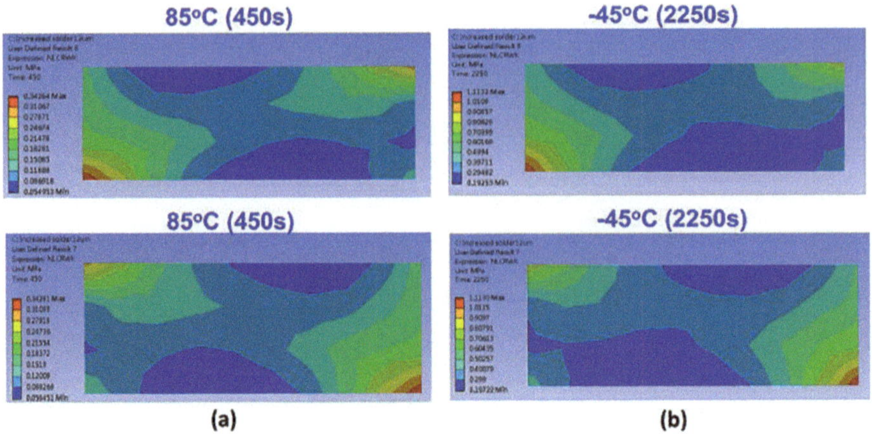

Fig. 4.236 Creep strain energy density at solder joint A during the first thermal cycle. **a** At 85 °C (450 s). **b** At −40 °C (2250 s)

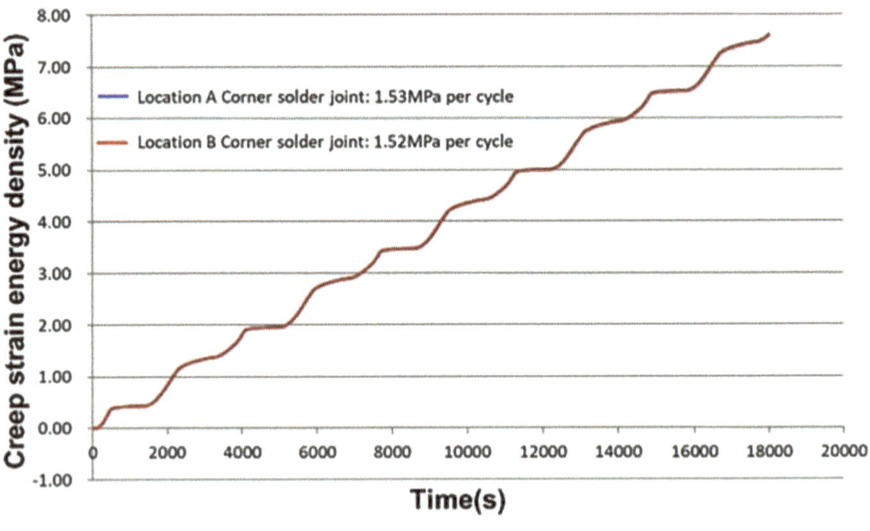

Fig. 4.237 Creep strain energy density time-history at solder joints A and B

- The hybrid substrate has been formed by thermocompression of the RDL-substrate, the interconnect layer substrate, and the HDI substrate and the interconnect-layer become in C-stage.
- Hybrid substrate characterizations, such as OM, X-ray, and SEM demonstrated that the interconnect-layer (both conductive paste and prepreg), MLs, pad for chip bonding, daisy-chains on PCB, etc. are properly fabricated. Continuity checks of the hybrid substrate have been passed.
- The hybrid substrate has passed the reflow test and drop test.

- Nonlinear finite element analysis and result show that the stress state acting at the conductive paste filled via and the surrounding structural elements is very small to create reliability issues, such as delamination and cracking. Also, the accumulated creep strain and the creep strain energy density are too small to create solder joint reliability problems, such as cracking of solder joints.

References

1. Lau, J.H. 2021. *Semiconductor advanced packaging*. New York: Springer.
2. Kauffman, J. 1969. Method of packaging integrated circuits. US Patent 3,436,810, Filed on July 17, 1967, Patented on April 8, 1969.
3. Lin, P., M. McShane, and P. Wilson. 1993. Semiconductor device having a pad array carrier package. US Patent 5,216,278, Filed on March 2, 1992, Patented on June 1, 1993.
4. Elenius, P., and H. Hollack. 2001. Method for forming chip-scale package. US Patent 6,287,893, Filed on July 13, 1998, Patented on September 11, 2001.
5. Elenius, P. 1998. Flip-chip bumping for IC packaging contractors. In *Proceedings of NEPCON west*, 1403–1407.
6. Hedler, H., T. Meyer, and B. Vasquez. 2004. Transfer wafer-level packaging. US Patent 6,727,576, Filed on October 31, 2001, Patented on April 27, 2004.
7. Brunnbauer M., et al. 2006. An embedded device technology based on a molded reconfigured wafer. In *Proceedings of IEEE/ECTC*, 547–551, May 2006.
8. Meyer, T., G. Ofner, S. Bradl, M. Brunnbauer, and R. Hagen. 2006. Embedded 396 wafer level ball grid array (eWLB). In *Proceedings of IEEE/EPTC*, vol. 397, 1–5.
9. Eichelberger, C., and R. Wojnarowski. 1991. High-density interconnect with high volumetric efficiency. US Patent 5,019,946, Filed on September 27, 1988, Patented on May 28, 1991.
10. Fillion, R., R. Wojnarowski, M. Gdula, H. Cole, E. Wildi, and W. Daum. 1994. Method for fabricating an integrated circuit module. US Patent 5,353,498, Filed on July 9, 1993, Patented on October 11, 1994.
11. Eichelberger, C. 1998. Single-chip modules, repairable multi-chip modules, and methods of fabrication thereof. US Patent 5,841,193, Filed on May 20, 1996, Patented on November 24, 1998.
12. Lau, J.H. 2018. *Fan-out wafer-level packaging*. New York: Springer.
13. Lau, J.H. 2013. *Through-silicon vias for 3D integration*. New York: McGraw-Hill Book Company.
14. Lau, J.H. 2016. *3D IC integration and packaging*. New York: McGraw-Hill Book Company.
15. Lau, J.H., N. Fan, and M. Li. 2016. Design, material, process, and equipment of embedded fan-out wafer/panel-level packaging. *Chip Scale Review* 20: 38–44.
16. Kurita, Y., et al. 2010. Fan-out wafer-level packaging with highly flexible design capabilities. In *Proceedings of IEEE ESTC*, 1–6.
17. Motohashi, N., et al. 2011. System in wafer-level package technology with RDL-first process. In *Proceedings of IEEE ECTC*, 59–64.
18. Lau, J.H., C. Ko, K. Yang, C. Peng, T. Xia, P. Lin, J. Chen, P. Huang, H. Liu, T. Tseng, E. Lin, and L. Chang. 2020. Panel-level fan-out RDL-first packaging for heterogeneous integration. *IEEE Transactions on CPMT* 10 (7): 1125–1137.
19. Lau, J.H., C. Ko, C. Peng, K. Yang, T. Xia, P. Lin, J. Chen, P. Huang, T. Tseng, E. Lin, L. Chang, C. Lin, Y. Fan, H. Liu, and W. Lu. 2021. Thermal cycling test and simulation of fan-out chip-last panel-level packaging for heterogeneous integration. *IMAPS Transactions, Journal of Microelectronics and Electronic Packaging* 18: 29–39.

20. Lau, J.H., C. Ko, T. Peng, K. Yang, T. Xia, P. Lin, J. Chen, P. Huang, T. Tseng, E. Lin, L. Chang, C. Lin, and W. Lu. 2020. Chip-last (RDL-first) fan-out panel-level packaging (FOPLP) for heterogeneous integration. *IMAPS Transactions, Journal of Microelectronics and Electronic Packaging* 17(3): 89–98.
21. Lau, J.H., G. Chen, J. Huang, R. Chou, C. Yang, N. Liu, and T. Tseng. 2021. Hybrid substrate by fan-out RDL-first panel-level packaging. *IEEE Transactions on CPMT* 11 (8): 1301–1309.
22. Chou, R., J.H. Lau, G. Chen, J. Huang, C. Yang, N. Liu, and T. Tseng. 2022. Heterogeneous integration on 2.3D hybrid substrate using solder joint and underfill. *IMAPS Transactions, Journal of Microelectronics and Electronic Packaging* 19: 8–17.
23. Lau, J.H., G. Chen, C. Yang, V. Teng, A. Peng, J. Huang, N. Liu, Y. Chen, T. Tseng, and M. Li. 2023. Hybrid substrate for heterogeneous integration. *ASME Transactions, Journal of Electronic Packaging*, published online.
24. Lau, J.H., C. Lin, N. Liu, H. Yang, T. Xia, C. Ko, B. Lin, Y. Chuang, R. Chen, M. Ma, T. Tseng, M. Li, and K. Leung. 2023. Hybrid substrate with ultra-large organic interposer for heterogeneous integration. *IEEE Transactions on CPMT* 13 (9): 1371–1379.
25. Peng, C., J.H. Lau, C. Ko, P. Lee, E. Lin, K. Yang, P. Lin, T. Xia, L. Chang, N. Liu, C. Lin, T. Lee, J. Wang, M. Ma, and T. Tseng. 2022. High-density hybrid substrate for heterogeneous integration. *IEEE Transactions on CPMT* 12 (3): 469–478.
26. Lau, J.H. 2023. Recent advances and trends in multiple system and heterogeneous integration with TSV-interposers. *IEEE Transactions on CPMT* 13 (1): 3–25.
27. Lau, J.H. 2022. Recent advances and trends in multiple system and heterogeneous integration with TSV-less interposers. *IEEE Transactions on CPMT* 12 (9): 1271–1281.
28. Lau, J.H., G. Chen, C. Yang, A. Peng, J. Huang, C. Peng, C. Ko, H. Yang, Y. Chen, and T. Tseng. 2022. Hybrid Substrate for chiplet design and heterogeneous integration packaging. In *Proceedings of IEEE/IEDM*, 3.5.1–3.5.4.
29. Keser, B., et al. 2007. The redistributed chip package: A breakthrough for advanced packaging. In *Proceedings of IEEE ECTC*, 286–291.
30. Sharma, G., et al. 2011. Design and development of multi-die laterally placed and vertically stacked embedded micro-wafer-level packages. *IEEE Transactions on Components, Packaging, and Manufacturing Technology* 1(1): 52–59.
31. Lim, Y., et al. 2010. High quality and low loss millimeter wave passives demonstrated to 77-GHz for SiP technologies using embedded wafer level packaging platform (EMWLP). *IEEE Transactions on Advanced Packaging* 33 (4): 1061–1071.
32. Yoon, S., J.A. Caparas, Y. Lin, and P.C. Marimuthu. 2012. Advanced low profile PoP solution with embedded wafer level PoP (eWLB-PoP) technology. In *Proceedings of IEEE/ECTC*, 1250–1254.
33. Yoon, S., P. Tang, R. Emigh, Y. Lin, P.C. Marimuthu, and R. Pendse. 2013. Fanout flipchip eWLB (embedded wafer level ball grid array) technology as 2.5D packaging solutions. In *Proceedings of IEEE ECTC*, 1855–1860.
34. Che, F., D. Ho, M.Z. Ding, and X. Zhang. 2015. Modeling and design solutions to overcome warpage challenge for fan-out wafer level packaging (FO-WLP) technology. In *Proceedings of IEEE/EPTC*, 1–8.
35. Lin, Y., et al. 2016. Wafer warpage experiments and simulation for fan-out chip on substrate. In *Proceedings of IEEE ECTC*, 13–18.
36. Lau, J.H., M. Li, M. Li, T. Chen, I. Xu, X. Qing, Z. Cheng, N. Fan, E. Kuah, Z. Li, K. Tan, Y. Cheung, E. Ng, P. Lo, K. Wu, J. Hao, S. Koh, R. Jiang, X. Cao, R. Beica, S. Lim, N. Lee, C. Ko, H. Yang, Y. Chen, M. Tao, J. Lo, and R. Lee. 2018. Fan-out wafer-level packaging for heterogeneous integration. *IEEE Transactions on CPMT* 8 (9): 1544–1560.
37. Ko, C.T., H. Yang, Lau, J.H., M. Li, M. Li, C. Lin, J.W. Lin, T. Chen, I. Xu, C. Chang, J. Pan, H. Wu, Q. Yong, N. Fan, E. Kuah, Z. Li, K. Tan, Y. Cheung, E. Ng, K. Wu, J. Hao, R. Beica, M. Lin, Y. Chen, Z. Cheng, S. Koh, R. Jiang, X. Cao, S. Lim, N. Lee, M. Tao, J. Lo, and R. Lee. 2018. Chip-first fan-out panel level packaging for heterogeneous integration. *IEEE Transactions on CPMT* 8(9): 1561–1572.

38. Lau, J.H., M. Li, Y. Lei, M. Li, I. Xu, T. Chen, Q. Yong, Z. Cheng, W. Kai, P. Lo, Z. Li, K. Tan, Y. Cheung, N. Fan, E. Kuah, C. Xi, J. Ran, R. Beica, S. Lim, N. Lee, C. Ko, H. Yang, Y. Chen, M. Tao, J. Lo, and R. Lee. 2018. Reliability of fan-out wafer-level heterogeneous integration. *IMAPS Transactions, Journal of Microelectronics and Electronic Packaging* 15 (4): 148–162.
39. Ko, C.T., H. Yang, J.H. Lau, M. Li, M. Li, C. Lin, J.W. Lin, C. Chang, J. Pan, H. Wu, Y. Chen, T. Chen, I. Xu, P. Lo, N. Fan, E. Kuah, Z. Li, K. Tan, C. Lin, R. Beica, M. Lin, X. Cao, S. Lim, N. Lee, M. Tao, J. Lo, and R. Lee. 2018. Design, materials, process, and fabrication of fan-out panel-level heterogeneous integration. *IMAPS Transactions, Journal of Microelectronics and Electronic Packaging* 15 (4): 141–147.
40. Lau, J.H., C. Ko, C. Lin, T. Tseng, K. Yang, T. Xia, P. Lin, C. Peng, E. Lin, L. Chang, N. Liu, S. Chiu, and Z. Lee. 2021. Fan-out panel-level packaging of mini-LED RGB display. *IEEE Transactions on CPMT* 11 (5): 739–747.
41. Yu, D. 2014. Wafer-level system integration (WLSI) technologies for 2D and 3D system-in-package. In *Proceedings of SEMIEUROPE*.
42. Lin, J., J.-P. Hung, N.-W. Liu, Y.-C. Mao, W.-T. Shih, and T.-H. Tung. 2015. Packaged semiconductor device with a molding compound and a method of forming the same. US Patent 9 000 584 B2, Apr. 7, 2015.
43. Tseng, C., C.-S. Liu, C.-H. Wu, and D. Yu. 2016. InFO (wafer level integrated fan-out) technology. In *Proceedings of IEEE/ECTC*, 1–6.
44. Hsieh, C., C.-H. Wu, and D. Yu. 2016. Analysis and comparison of thermal performance of advanced packaging technologies for state-of-the-art mobile applications. In *Proceedings of IEEE/ECTC*, 1430–1438.
45. Lau, J.H., M. Li, N. Fan, E. Kuah, Z. Li, K. Tan, T. Chen, I. Xu, M. Li, Y.M. Cheung, Wu Kai, Ji Hao, R. Beica, T. Taylor, C. Ko, H. Yang, Y. Chen, S. Lim, N. Lee, J. Ran, K. Wee, Q. Yong, C. Xi, M. Tao, J. Lo, and R. Lee. 2017. Fan-out wafer-level packaging (FOWLP) of large chip with multiple redistribution layers (RDLs). *IMAPS Transactions, Journal of Microelectronics and Electronic Packaging* 14(4): 123–131.
46. Lau, J.H., M. Li, Q. Li, I. Xu, T. Chen, Z. Li, K. Tan, X. Qing, C. Zhang, K. Wee, R. Beica, C. Ko, S. Lim, N. Fan, E. Kuah, K. Wu, Y. Cheung, E. Ng, X. Cao, J. Ran, H. Yang, Y. Chen, N. Lee, M. Tao, J. Lo, and R. Lee. 2018. Design, materials, process, and fabrication of fan-out wafer-level packaging. *IEEE Transactions on CPMT* 8 (6): 991–1002.
47. Lau, J.H., M. Li, Y. Li, M. Li, I. Au, T. Chen, S. Chen, Q. Yong, J. Madhukumar, K. Wu, N. Fan, E. Kuah, Z. Li, K. Tan, W. Bao, S. Lim, R. Beica, C. Ko, and X. Cao. 2018. Warpage measurements and characterizations of FOWLP with large chips and multiple RDLs. *IEEE Transactions on CPMT* 8 (10): 1729–1737.
48. Lau, J.H., and N.C. Lee. 2020. *Assembly and reliability of lead-free solder joints*. New York: Springer.
49. Lau, J., R. Lee, M. Yuen, and P. Chan. 2010. 3D LED and IC wafer level packaging. *MicroElectronics International* 27 (2): 98–105.
50. Chen, Y. 2020. *Latest mini LED and micro LED definition and technology analysis*. Taiwan: LED Inside.
51. Han, S. 2020. *MSI to launch gaming notebook backlit by Lextar-made mini-LED*. Taipei, Taiwan: Digitimes.
52. Chen, M. 2020. *Taiwan suppliers to play major role in Apple mini-LED offerings in 2021*. Taipei, Taiwan: Digitimes.
53. Yoon, G. 2020. *Samsung electronics preparing for mass-production of mini-LED TVs*. South Korea: Korea IT News.
54. Lau, J.H. 1994. *Handbook of fine pitch surface mount technology*. New York: Van Nostrand Reinhold.
55. Lau, J.H. 1994. *Chip on board technologies for multichip modules*. New York, NY, USA: Van Nostrand Reinhold.
56. Eslampour, H., S. Lee, S. Park, T. Lee, S. Yoon, and Y. Kim. 2010. Comparison of advanced PoP package configurations. In *Proceedings of IEEE/ECTC*, 1946–1950.

57. Yoon, S., J. Caparas, Y. Lin, and P. Marimuthu. 2012. Advanced low profile PoP solution with embedded wafer level PoP (eWLB-PoP) technology. In *Proceedings of IEEE/ECTC*, 1250–1254.
58. Tseng, C.-F., C.-S. Liu, C.-H. Wu, and D. Yu. 2016. InFO (wafer level integrated fan-out) technology. In *Proceedings of IEEE/ECTC*, 1–6.
59. Hsieh, C.-C., C.-H. Wu, and D. Yu. 2016. Analysis and comparison of thermal performance of advanced packaging technologies for state-of-the-art mobile applications. In *Proceedings of IEEE/ECTC*, 1430–1438.
60. Kurita, Y., K. Soejima, K. Kikuchi, M. Takahashi, M. Tago, M. Koike, et al. 2006. A novel "SMAFTI" package for inter-chip wide-band data transfer. In *Proceedings of ECTC*, 289–297.
61. Kawano, M., S. Uchiyama, Y. Egawa, N. Takahashi, Y. Kurita, K. Soejima, et al. 2006. A 3D packaging technology for 4 Gbit stacked DRAM with 3 Gbps data transfer. In *Proceedings of IEMT*, 581–584.
62. Kurita, Y., S. Matsui, N. Takahashi, K. Soejima, M. Komuro, M. Itou, et al. 2007. A 3D stacked memory integrated on a logic device using SMAFTI technology. In *Proceedings of ECTC*, 821–829.
63. Kawano, M., N. Takahashi, Y. Kurita, K. Soejima, M. Komuro, S. Matsui. 2008. A 3-D packaging technology for stacked DRAM with 3 Gb/s data transfer. *IEEE Transactions on Electron Devices* 1614–1620.
64. Motohashi, N., Y. Kurita, K. Soejima, Y. Tsuchiya, M. Kawano. 2009. SMAFTI package with planarized multilayer interconnects. In *Proceedings of ECTC*, 599–606.
65. Kurita, M., S. Matsui, N. Takahashi, K. Soejima, M. Komuro, M. Itou, et al. 2009. Vertical integration of stacked DRAM and high-speed logic device using SMAFTI technology. *IEEE Transactions on Advanced Packaging* 657–665.
66. Kurita, Y., N. Motohashi, S. Matsui, K. Soejima, S. Amakawa, K. Masu, et al. 2009. SMAFTI packaging technology for new interconnect hierarchy. In *Proceedings of IITC*, 220–222.
67. Huemoeller, R., and C. Zwenger. 2015. Silicon wafer integrated fan-out technology. *Chip Scale Review* 34–37.
68. Bu, L., F. Che, M. Ding, S. Chong, and X. Zhang. 2015. Mechanism of moldable underfill (MUF) process for fan-out wafer level packaging. In *IEEE/EPTC proceedings*, 1–7.
69. Che, F., D. Ho, M. Ding, and D. Woo. 2016. Study on process induced wafer level warpage of fan-out wafer level packaging. In *IEEE/ECTC proceedings*, 1879–1885.
70. Rao, V., C. Chong, D. Ho, D. Zhi, C. Choong, S. Lim, D. Ismael, and Y. Liang. 2016. Development of high density fan out wafer level package (HD FOWLP) with multi-layer fine pitch RDL for mobile applications. In *IEEE/ECTC proceedings*, 1522–1529.
71. Chen, Z., F. Che, M. Ding, D. Ho, T. Chai, and V. Rao. 2017. Drop impact reliability test and failure analysis for large size high density FOWLP package on package. In *IEEE/ECTC proceedings*, 1196–1203.
72. Lim, T., and D. Ho. 2018. Electrical design for the development of FOWLP for HBM integration. In *IEEE/ECTC proceedings*, 2136–2142.
73. Ho, S., H. Hsiao, S. Lim, C. Choong, S. Lim, and C. Chong. 2019. High density RDL build-up on FO-WLP using RDL-first approach. In *IEEE/EPTC proceedings*, 23–27.
74. Boon, S., D. Wee, R. Salahuddin, and R. Singh. 2019. Magnetic inductor integration in FO-WLP using RDL-first approach. In *IEEE/EPTC proceedings*, 18–22.
75. Hsiao, H., S. Ho, S.S. Lim, W. Ching, C. Choong, S. Lim, H. Hong, and C. Chong. 2019. Ultra-thin FO package-on-package for mobile application. In *IEEE/ECTC proceedings*, 21–27.
76. Lin, B., F. Che, V. Rao, and X. Zhang. 2019. Mechanism of moldable underfill (MUF) process for RDL-1st fan-out panel level packaging (FOPLP). In *IEEE/ECTC proceedings*, 1152–1158.
77. Sekhar, V., V. Rao, F. Che, C. Choong, and K. Yamamoto. 2019. RDL-1st fan-out panel level packaging (FOPLP) for heterogeneous and economical packaging. In *IEEE/ECTC proceedings*, 2126–2133.
78. Ma, M., S. Chen, P.I. Wu, A. Huang, C.H. Lu, A. Chen, C. Liu, and S. Peng. 2016. The development and the integration of the 5 μm to 1 μm half pitches wafer level Cu redistribution layers. In *IEEE/ECTC proceedings*, 1509–1614.

79. Kim, Y., J. Bae, M. Chang, A. Jo, J. Kim, S. Park, D. Hiner, M. Kelly, and W. Do. 2017. SLIM™, high density wafer level fan-out package development with submicron RDL. In *IEEE/ECTC proceedings*, 18–13.
80. Hiner, D., M. Kolbehdari, M. Kelly, Y. Kim, W. Do, J. Bae, M. Chang, and A. Jo. 2017. SLIM™ advanced fan-out packaging for high performance multi-die solutions. In *IEEE/ECTC proceedings*, 575–580.
81. Lin, B., C. Ko, W. Ho, C. Kuo, K. Chen, Y. Chen, and T. Tseng. 2017. A comprehensive study on stress and warpage by design, simulation and fabrication of RDL-first panel level fan-out technology for advanced package. In *IEEE/ECTC proceedings*, 1413–1418.
82. Suk, K., S. Lee, J. Youn, K. Lee, H. Kim, S. Lee, P. Kim, D. Kim, D. Oh, and J. Byun. 2018. Low cost Si-less RDL interposer package for high performance computing applications. In *IEEE/ECTC proceedings*, 64–69.
83. Hwang, T., D. Oh, E. Song, K. Kim, J. Kim, and S. Lee. 2018. Study of advanced fan-out packages for mobile applications. In *IEEE/ECTC proceedings*, 343–348.
84. Lee, C, J. Su, X. Liu, Q. Wu, J. Lin, P. Lin, C. Ko, Y. Chen, W. Shen, T. Kou, S. Huang, A. Lin, Y. Lin, and K. Chen. 2018. Optimization of laser release process for throughput enhancement of fan-out wafer level packaging. In *IEEE/ECTC proceedings*, 1818–1823.
85. Cheng, W., C. Yang, J. Lin, W. Chen, T. Wang, and Y. Lee. 2018. Evaluation of chip-last fan-out panel level packaging with G2.5 LCD facility using FlexUPTM and mechanical de-bonding technologies. In *IEEE/ECTC proceedings*, 386–391.
86. Cheng, S., C. Yang, W. Cheng, S. Cheng, W. Chen, H. Lai, T. Wang, and Y. Lee. 2019. Application of fan-out panel level packaging techniques for flexible hybrid electronics systems. In *IEEE/ECTC proceedings*, 1877–1822.
87. Chang, K., C. Huang, H. Kuo, M. Jhong, T. Hsieh, M. Hung, and C. Wan. 2019. Ultra high density IO fan-out design optimization with signal integrity and power integrity. In *IEEE/ECTC proceedings*, 41–46.
88. Lin, Y., M. Yew, M. Liu, S. Chen, T., Lai, P. Kavle, C. Lin, T. Fang, C. Chen, C. Yu, K. Lee, C. Hsu, P. Lin, F. Hsu, and S. Jeng. 2019. Multilayer RDL interposer for heterogeneous device and module integration. In *EEE/ECTC proceedings*, 931–936.
89. Lau, J.H., C. Ko, T. Peng, K. Yang, T. Xia, P. Lin, J. Chen, P. Huang, T. Tseng, E. Lin, L. Chang, C. Lin, and W. Lu. 2020. Chip-last (RDL-first) fan-out panel-level packaging (FOPLP) for heterogeneous integration. *IMAPS Transactions, Journal of Microelectronics and Electronic Packaging* 17: 89–98.
90. Lau, J., W. Dauksher, and P. Vianco. 2003. Acceleration models, constitutive equations and reliability of lead-free solders and joints. In *IEEE/ECTC proceedings*, 229–236.
91. Pan, N., G. Henshall, F. Billaut, S. Dai, M. Strum, R. Lewis, E. Benedetto, and J. Rayner. 2005. An acceleration model for Sn-Ag-Cu solder joint reliability under various thermal cycle conditions. In *SMTA international conference proceedings*, 876–883.
92. Vasudevan, V., and X. Fan. 2008. An acceleration model for lead-free (SAC) solder joint reliability under thermal cycling. In *IEEE/ECTC proceedings*, 139–145.
93. Miremadi, J., G. Henshall, A. Allen, E. Benedetto, and M. Roesch. 2009. Lead-free solder-joint-reliability model enhancement. In *IMAPS proceedings*, 316–323.
94. Lall, P., A. Shirgaokar, and D. Arunachalam. 2012. Norris-Landzberg acceleration factor and Goldmann constants for SAC305 lead-free electronics. *ASME Transactions, Journal of Electronic Packaging* 134: 1–8.
95. Osterman, M. 2018. Modeling temperature cycle fatigue life of select SAC solders. In *SMTA international conference.*
96. Tsukada, Y., S. Tsuchida, and Y. Mashimoto. 1992. Surface laminar circuit packaging. In *Proceedings of IEEE/ECTC*, San Diego, CA, USA, 22–27.
97. Tsukada, Y., and S. Tsuchida. 1992. Surface laminar circuit, a low cost high density printed circuit board. In *Proceedings of surface mount international conference*, San Jose, CA, USA, 537–542.
98. Tsukada, Y. 1994. Solder bumped flip chip attach on SLC board and multichip module. In *Chip on board: Technology for multichip modules*, ed. J.H. Lau, 410–443. New York, NY, USA: Van Nostrand.

99. Naffziger, S., K. Lepak, M. Paraschou, and M. Subramony. 2020. AMD chiplet architecture for high-performance server and desktop products. In *IEEE international solid-state circuits conference (ISSCC) digest technical papers*, 44–45.
100. Naffziger, S. 2020. Chiplet meets the real world: Benefits and limits of chiplet designs. In *Proceedings of symposia on VLSI technology and circuits*, 1–39.
101. Xilinx and TSMC. 2013. TSMC reach volume production on all 28 nm CoWoS based all programmable 3D IC families. http://press.xilinx.com/2013-10-20-Xilinx-and-TSMCReach-Volume-Productionon-all-28nm-CoWoSbased-All-Programmable-3D-IC-Families. Accessed 1 Nov 2022.
102. Banijamali, B., S. Ramalingam, K. Nagarajan, and R. Chaware. 2011. Advanced reliability study of TSV interposers and interconnects for the 28 m technology FPGA. In *Proceedings of IEEE/ECTC*, 285–290.
103. Banijamali, B., et al. 2011. Ceramics vs. low-CTE organic packaging of TSV silicon interposers. In *Proceedings of IEEE/ECTC*, 573–576.
104. Yip, L., R. Lin, C. Lai, and C. Peng. 2022. Reliability challenges of high-density fan-out packaging for high-performance computing applications. In *Proceedings of IEEE/ECTC*, 1454–1458.
105. Yu, D. 2018. Advanced system integration technology trends. In *Proceedings of SiP Global Summit (SEMICON)*, Taiwan.
106. Lau, J.H. 2022. Recent advances and trends in multiple system and heterogeneous integration with TSV-less interposers. *IEEE Transactions on Components, Packaging, and Manufacturing Technology* 12(8): 1271–1281.
107. Jayaraman, S. 2022. Advanced packaging: HDFO for next generation devices. In *Proceedings of IWLPC*, 1–28.
108. Suk, K., et al. 2018. Low-cost Si-less RDL interposer package for high performance computing applications. In *Proceedings of IEEE/ECTC*, 64–69.
109. You, S.-H., et al. 2018. Advanced fan-out package Si/PI/thermal performance analysis of novel RDL packages. In *Proceedings of IEEE/ECTC*, 1295–1301.
110. Lin, Y.-H., et al. 2019. Multilayer RDL interposer for heterogeneous device and module integration. In *Proceedings of IEEE/ECTC*, 931–936.
111. Lin, P.-Y., et al. 2021. Reliability performance of advanced organic interposer (CoWoS-R) packages. In *Proceedings of IEEE/ECTC*, 723–728.
112. Lin, M.L., et al. 2022. Organic interposer CoWoS-R+ (plus) technology. In *Proceedings of IEEE/ECTC*, 1–6.
113. Chang, K.T., et al. 2019. Ultra high density IO fan-out design optimization with signal integrity and power integrity. In *Proceedings of IEEE/ECTC*, 41–46.
114. Lai, W.-H., et al. 2020. A comparative study of 2.5D and fan-out chip on substrate: Chip first and chip last. In *Proceedings of IEEE/ECTC*, 354–360.
115. Fang, J.-K., M.-L. Huang, H.-J. Tu, W.-L. Lu, and P. Yang. 2020. A production-worthy fan-out solution—ASE FOCoS chip last. In *Proceedings of IEEE/ECTC*, 290–295.
116. Cao, L. 2022. Advanced FOCOS (fanout chip on substrate) technology for chiplets heterogeneous integration. In *Proceedings of IWLPC*, 1–6.
117. Cao, L., et al. 2022. Advanced fanout packaging technology for hybrid substrate integration. In *Proceedings of IEEE/ECTC*, 1362–1370.
118. Lee, T.-C., S.-H. Yang, H.-Y. Wu, and Y.-J. Lin. 2022. Chip last fanout chip on substrate (FOCoS) solution for chiplets integration. In *Proceedings of IEEE/ECTC*, 1970–1974.
119. Yin, W.-J., et al. 2022. Mechanical and thermal characterization analysis of chip-last fan-out chip on substrate. In *Proceedings of IEEE/ECTC*, 1711–1719.
120. Lia, J., et al. 2021. Large size multilayered fan-out RDL packaging for heterogeneous integration. In *Proceedings of IEEE/ECTC*, 239–243.
121. Miki, S., H. Taneda, N. Kobayashi, K. Oi, K. Nagai, and T. Koyama. 2019. Development of 2.3D high density organic package using low temperature bonding process with Sn-bi solder. In *Proceedings of IEEE/ECTC*, 1599–1604.

122. Murayama, K., S. Miki, H. Sugahara, and K. Oi. 2020. Electro-migration evaluation between organic interposer and build-up substrate on 2.3D organic package. In *Proceedings of IEEE/ECTC*, 716–722.
123. Kim, J., et al. 2021. Cost effective 2.3D packaging solution by using fanout panel level RDL. In *Proceedings of IEEE/ECTC*, 310–314.
124. Lau, J.H. 2023. *Chiplet design and heterogeneous integration packaging*. New York: Springer.
125. Pang, J. 2012. *Lead free solder: Mechanics and reliability*. New York: Springer.

Chapter 5
Chiplet Communications (Bridges)

5.1 Introduction

There are many advanced packaging technologies [1–4] listed in Fig. 5.1 along with their performance and density rankings. Figure 5.2 shows the groups of advanced packaging such as 2D, 2.1D, 2.3D, 2.5D, and 3D IC integration. The focus of this chapter is on chiplets and heterogeneous integration with emphasis on chiplet communications. The major difference between chiplet and heterogeneous integration is that a "chiplet" is a chip design method, while a "heterogeneous integration" is a chip packaging method [1, 2].

There are at least five different chiplet design and heterogeneous integration packaging, as shown in Figs. 5.3, namely, (1) chip partition and heterogeneous integration (driven by cost and technology optimization), Fig. 5.3a, (2) chip split and heterogeneous integration (driven by cost and semiconductor manufacturing yield), Fig. 5.3b, (3) multiple system and heterogeneous integration with thin-film layer directly on top of a build-up package substrate (2.1D IC integration), Fig. 5.3c, (4) multiple system and heterogeneous integration with TSV-less interposer (2.3D IC integration), Fig. 5.3d, and (5) multiple system and heterogeneous integration with TSV interposers (2.5D and 3D IC integration), Fig. 5.3e. (3)–(5) are driven by formfactor and performance.

In chip partition and heterogeneous integration, Fig. 5.3a, the SoCs, such as the logic and I/Os, are partitioned into functions (chiplets): logic and I/O. These chiplets can be stacked (integrated) by the front-end CoW (chip-on-wafer) or WoW (wafer-on-wafer) methods and then assembled (integrated) on the same substrate of a single package by using heterogeneous integration techniques (Fig. 5.4) [1–50]. It should be emphasized that the front-end chiplets' integration can yield a smaller package area and better electrical performance but is optional.

In chip split and heterogeneous integration, Fig. 5.3b, the SoC, such as logic, is split into smaller chiplets, such as logic1, logic2, and logic3. These chiplets can be stacked (integrated) by the front-end CoW or WoW methods and then assembled on

J. H. Lau, *Flip Chip, Hybrid Bonding, Fan-In, and Fan-Out Technology*,
https://doi.org/10.1007/978-981-97-2140-5_5

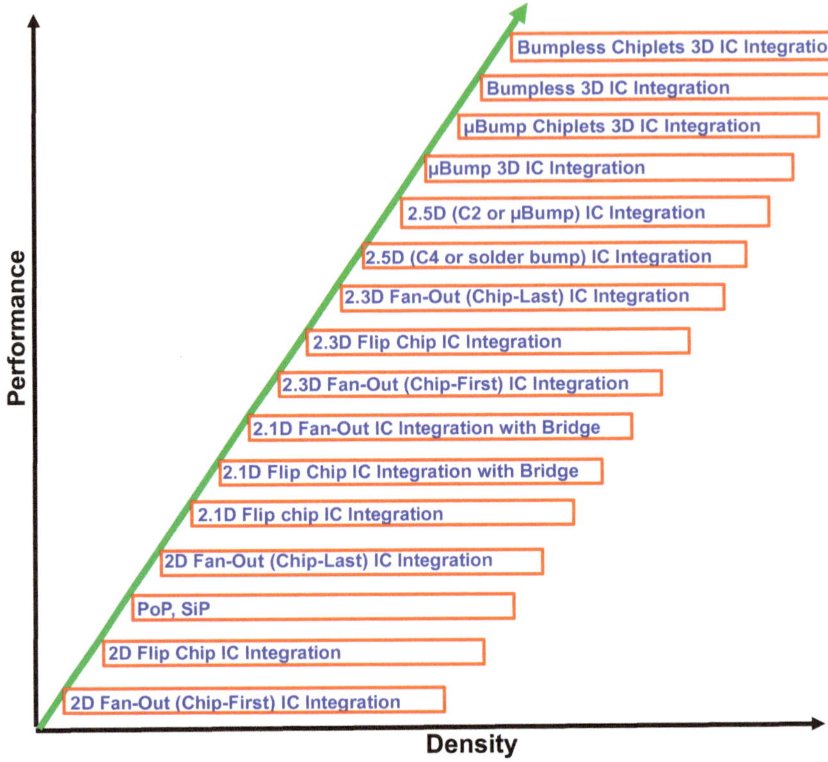

Fig. 5.1 Advanced packaging based on density and performance

the same substrate of a single package by using heterogeneous integration techniques (Fig. 5.4) [1–50]. Again, the front-end integration of chiplets is optional.

In multiple system and heterogeneous integration with thin-film layers directly on top of the build-up package substrate, Fig. 5.3c, the SoC such as the CPU, logic, and HBM are supported by a build-up package substrate with thin-film layers. This is driven by performance and formfactor and for high-density and high-performance applications [51–59]. Because of the flatness of the build-up package substrate, the yield loss of this integration is very high, and it is not in high-volume production today.

In multiple system and heterogeneous integration with TSV-less interposers, Fig. 5.3d, the SoC such as the CPU, logic, and HBM are supported by a fine metal L/S RDL-substrate (organic-interposer) and then on a build-up package substrate [60–85]. This is driven by performance and formfactor and for high density and performance applications. This integration is in small volume production today.

In multiple system and heterogeneous integration with TSV-interposers, Fig. 5.3e, the SoC such as the CPU, logic, and HBM are supported by a passive (2.5D) or active (3D) TSV-interposer and then on a build-up package substrate [86–100]. This

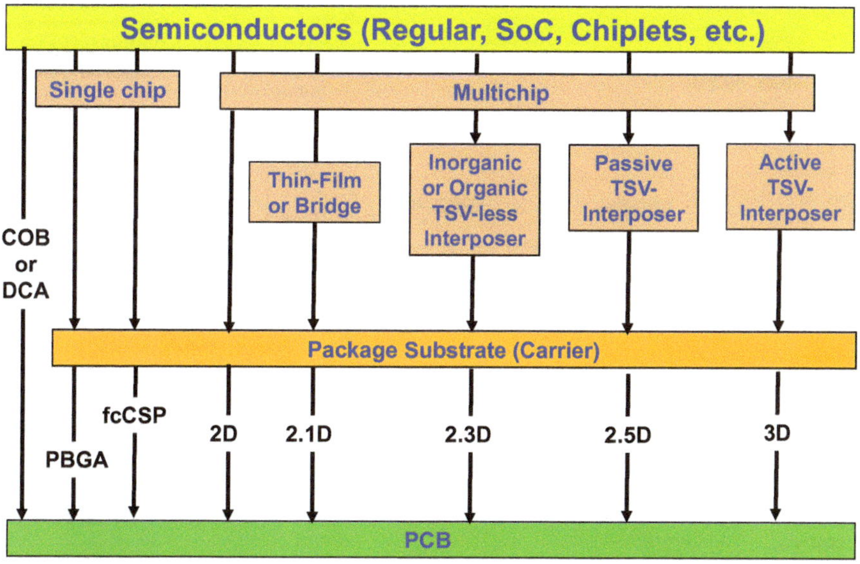

Fig. 5.2 Groups of advanced packaging: 2D, 2.1D, 2.3D, 2.5D, and 3D IC integration

is driven by performance and formfactor and for extremely high-density and high-performance applications. This integration is in volume production today.

The advantages of chiplet design and heterogeneous integration packaging are: (a) yield improvement (lower cost) during semiconductor manufacturing; (b) fast time-to-market; (c) cost reduction during design; (d) better thermal performance; (e) reusable intellectual property (IP); and (f) modularization. The disadvantages (challenges) are: (a) additional area for interfaces; (b) higher packaging costs; (c) more packaging complexity and design effort; and (d) past methodologies are less suitable. Therefore, one of the major focuses of packaging technologists is to reduce the area for interfaces of the communication between chiplets and the packaging cost. In this chapter, various kinds of bridges for the communications of chiplet design and heterogeneous integration packaging such as (a) those embedded on top of an organic package substrate, (b) those embedded in fan-out epoxy molding compound (EMC) with RDLs, and (c) flexible bridge, will be systematically presented and discussed. Some challenges and recommendations will also be provided.

5.2 Background

In the past, the communication of chiplet design and heterogeneous integration packaging is by fine metal line width and spacing (L/S) TSV-interposer, build-up high-density organic substrate, and fan-out RDLs. Figure 5.5 shows the Virtex-7 HT family shipped by Xilinx in 2013. In 2011, Xilinx asked TSMC to fabricate

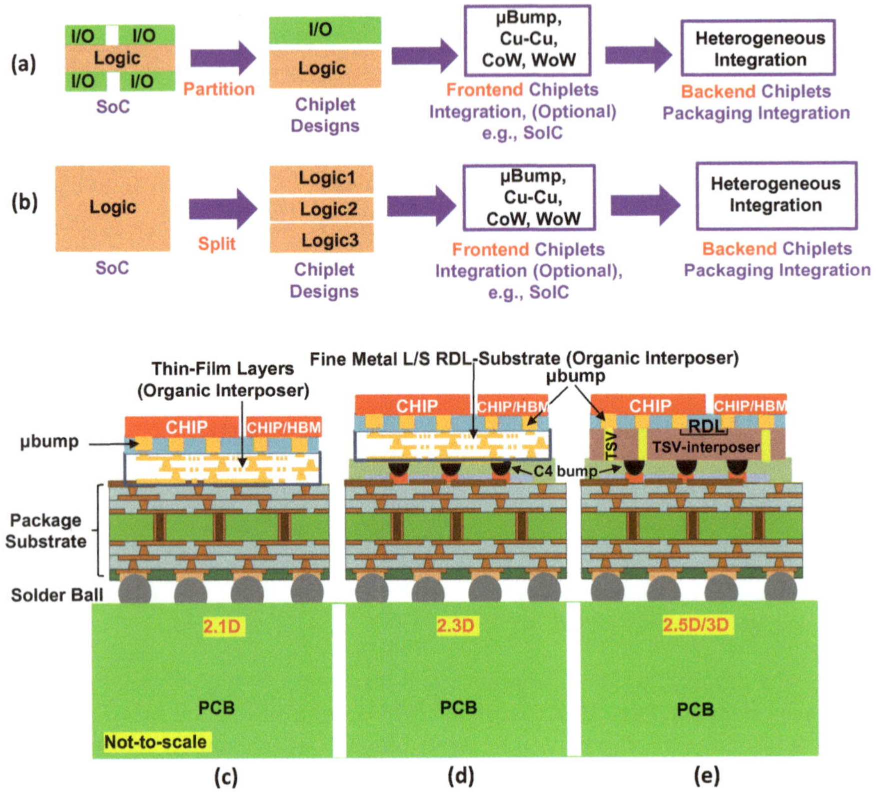

Fig. 5.3 Chiplet design and heterogeneous integration packaging. **a** Chip partition. **b** Chip split. **c** Multiple system and heterogeneous integration with thin-film layers. **d** Multiple system and heterogeneous integration with organic interposer. **e** Multiple system and heterogeneous integration with TSV-interposer

its field-programable gate array (FPGA) system-on-chip (SoC) with 28 nm process technology [94, 101]. Because of the large chip size, the yield was very poor. Then, Xilinx redesigned and split the large FPGA into four smaller chiplets as shown in Fig. 5.5 and TSMC manufactured the chiplets at high yield (with the 28 nm process technology) and packaged them on their chip-on-wafer-on-substrate (CoWoS) technology. CoWoS is a 2.5D IC integration, which is the key structure (substrate) to let those 4 chiplets do lateral communications. The minimum pitch of the four redistribution layers (RDLs) on the TSV-interposer is 0.4 μm. The TSV-interposer is known to have a very high cost due to its manufactured yield loss (the size of the TSV-interposer is very large even it is with a 64 nm process).

Figure 5.6 shows AMD's 2nd-generation extreme-performance yield computing (EPYC) server processors [27, 28], the 7002-series, shipped in mid-2019. One of AMD's solutions is to partition the SoC into chiplets, reserving the expensive leading-edge silicon for the central processing unit (CPU) core while leaving the I/Os and

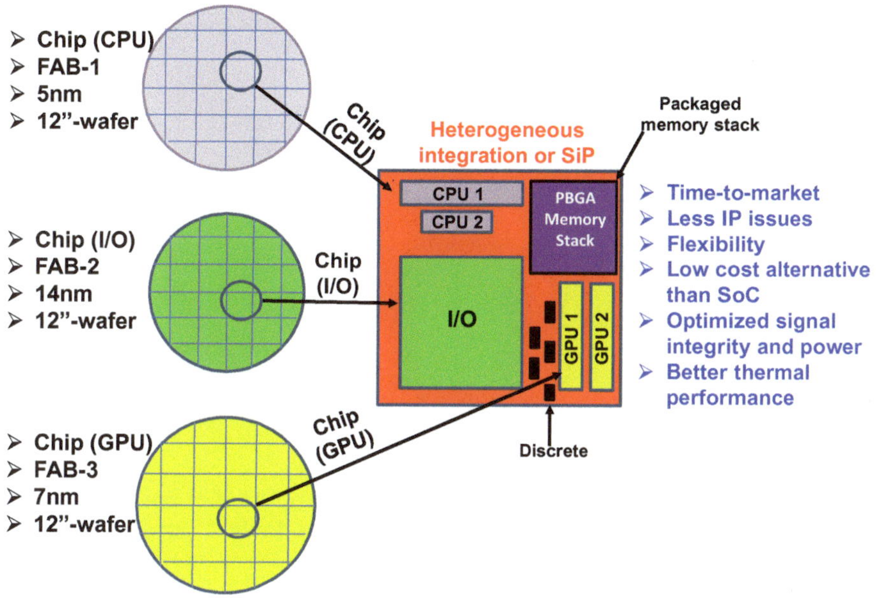

Fig. 5.4 Heterogeneous integration

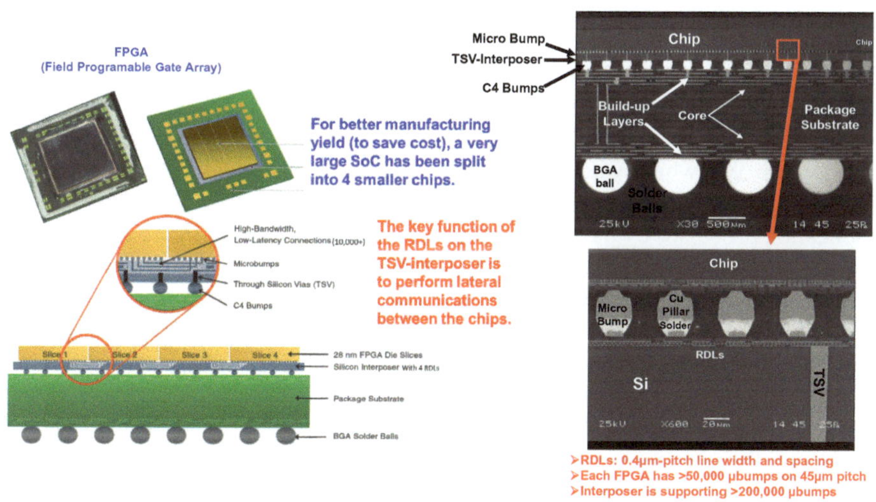

Fig. 5.5 Chip split supported by a passive TSV-interposer

memory interfaces in n-1 generation silicon. Another solution is to split the CPU core into smaller chiplets. In this case, each core complex die (CCD), or CPU compute die, is split into two smaller chiplets. AMD used the expensive 7 nm process technology fabricated by TSMC (in early 2019) for the core CCD chiplets and moved

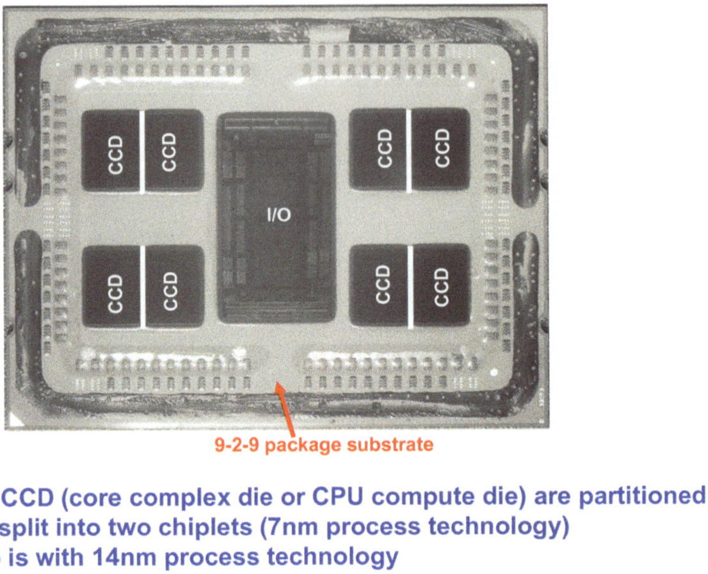

Fig. 5.6 Chip partition and chip split supported by a build-up packaging substrate

the dynamic random-access memory (DRAM) and logic to a mature 14 nm I/O die fabricated by GlobalFoundries. The 2nd-generation EPYC is a 2D chiplets IC integration technology, i.e., all the chiplets are side-by-side on a 9–2-9 build-up package substrate. The 20-layer fine metal L/S organic substrate is not cheap.

Figure 5.7 shows an application processor chipset in the smartphone of HTC (Desire 606 W), which was shipped in 2013. The application processor chipset is SPREADTRUM SC8502, which is a heterogeneous integration of the modem and application processor by the fan-out chip-first process. These chips are supported by the fan-out 2-layer redistribution layers (RDLs) substrate and then solder balled on a printed circuit board (PCB).

It should be noted that the requirement of communications (RDLs) between chipets is fine-metal L/S/H (thickness) and at a very small and local area of the chiplets. There is no reason to use the whole TSV-interposer, the whole organic substrate, or the whole fan-out RDLs to support the communication between chiplets. Therefore, the concept of using small area and a fine-metal L/S/H RDL bridge to connect the chiplets to perform communication (to reduce cost) for chiplet design and heterogeneous integration packaging has been proposed, see for example [102–104], and is a very hot topic today. There are at least two different groups of bridge, namely rigid bridge and flexible bridge.

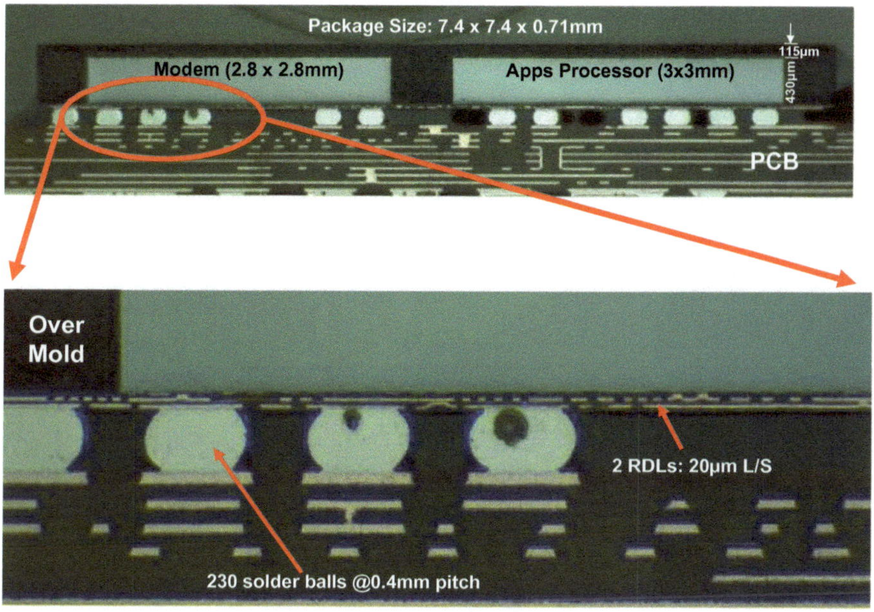

Fig. 5.7 Chiplets supported by a fan-out RDL

5.3 Rigid Bridge Versus Flexible Bridge

The flexible bridge is the RDL itself. The flexible bridge consists of fine-metal L/S/H conductors in a dielectric polymer, such as polyimide film. The very first flexible bridge patent application US 2006/0095639 A1 was filed by SUN Microsystems on November 2, 2004 (Fig. 5.8). For high-speed and high-frequency applications, such as millimeter wave frequencies, the dielectric layer can also be a liquid crystal polymer (LCP) and is called LCP-flexible bridge. Flexible bridge is out of the scope of this book.

Today, most of the products and publications with bridges are rigid bridges. Rigid bridge consists of the RDLs and the substrate. Most rigid bridges are made with silicon substrate and the RDLs are fabricated on a silicon wafer. Some rigid bridges are even with TSVs. There are at least two groups of rigid bridges, namely, (1) rigid bridges with build-up package substrate, and (2) rigid bridges with fan-out RDL substrate.

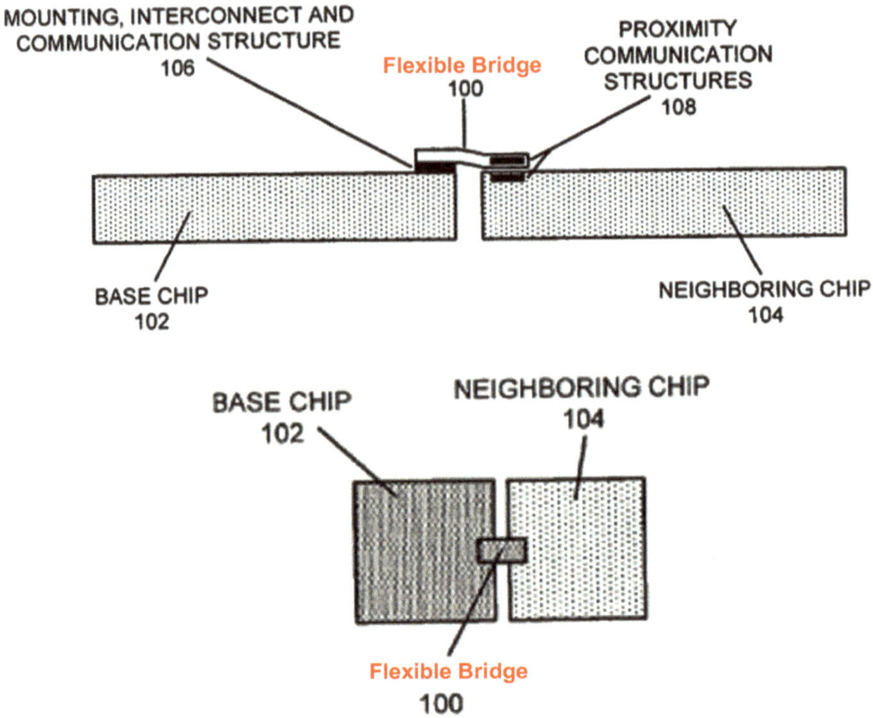

Fig. 5.8 Flexible bridge proposed by Sun Microsystems

5.4 How to Fabricate the Silicon Bridge with TSV?

5.4.1 Fabrication of TSVs

The fabrication process of TSVs is shown in Fig. 5.9. The process starts with a SiN_x/SiO_x insulation layer by either thermal oxidation or PECVD (plasma enhanced chemical vapor deposition) as shown in Fig. 5.9. After photoresist and TSV lithography, the TSV is etched into the Si substrate by Bosch-type DRIE (deep reactive ion etch) to form a high aspect ratio (10.5) via structure. The etched TSV structure is then processed with a SiO_x liner by SACVD (subatmosphere chemical vapor deposition), a Ta barrier layer, and a Cu seed layer by PVD (physical vapor deposition). Cu electroplating is used to fill the TSV structure. The final blind TSV has a top opening of approximately 10 μm in diameter and a depth of about 105 μm, which gives an aspect ratio of 10.5. In such a high aspect ratio via structure, a bottom-up plating mechanism is applied to ensure a seamless TSV with a reasonably low Cu thickness in the field.

The SEM cross-sectional images are shown in Fig. 5.10. It can be seen that the diameter of the TSV is slightly decreased at the bottom, which is expected from the

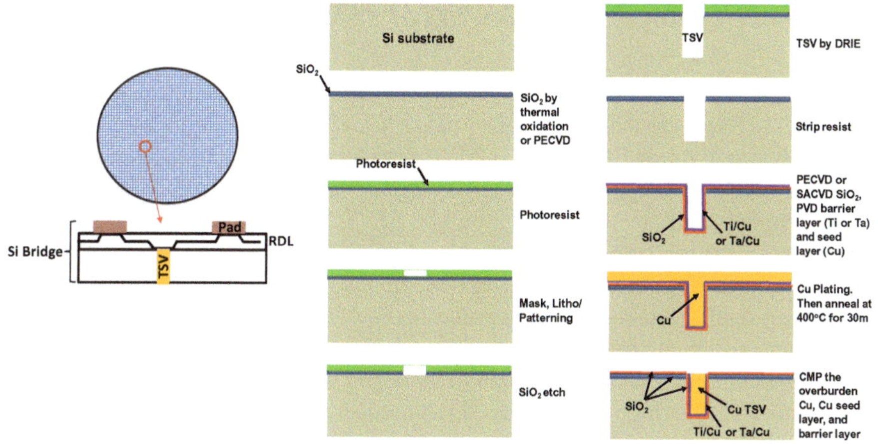

Fig. 5.9 TSV fabrication process

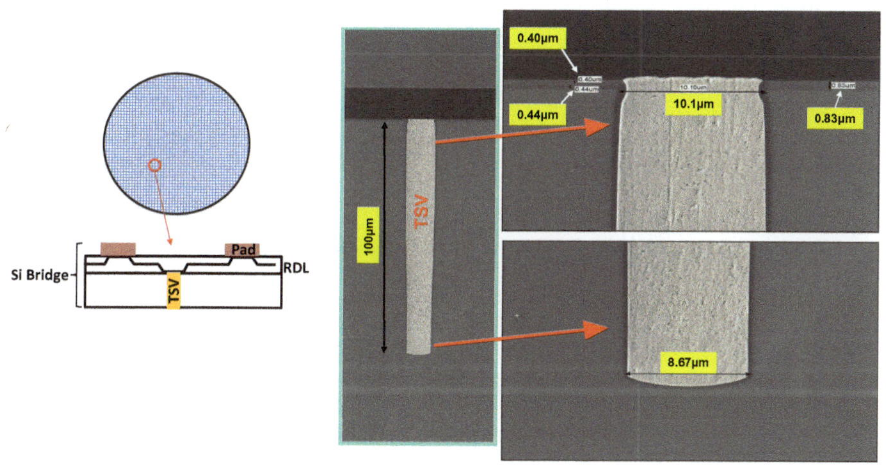

Fig. 5.10 SEM images of TSVs

etching process. The Cu thickness at the field is <5 μm. The post-plating anneal is at 400 °C for 30 min. To complete the TSV process, excess Cu in the field is removed by CMP (chemical–mechanical polishing).

5.4.2 Fabrication of RDLs on Top of TSV

In general, RDL consists of the dielectric layer and the metal conducting layer. There are at least two ways to fabricate RDLs [105, 106]. The first method is by using polymers, such as polyimide (PI) PWDC 1000 (Dow Corning), benzocyclobutene (BCB) cyclotene 4024–40 (Dow Chemical), polybenzo-bisoxazole (PBO) HD-8930 (HD Micro Systems), and the fluorinated aromatic AL-X 2010 (Asahi Glass Corporation) to make the dielectric layer and electroplating (such as Cu) to make the metal layers. The second method is the Cu damascene method, which is primarily modified from the conventional semiconductor back-end-of-line to make the Cu metal RDLs. In general, much thinner structures (both dielectric layers and Cu RDLs), finer pitches, smaller linewidth and line-spacing can be obtained with the Cu damascene method. The polymer/Cu-plating method will be mentioned first.

(A) Fabrication of RDLs: Polymer + Cu-Plating and Etching Method

Continuing with the wafer from Figs. 5.9 and 5.10, the fabrication process for the RDLs using polymers is shown in Fig. 5.11 and also listed as follows. UBM (under bump metallurgy) is included.

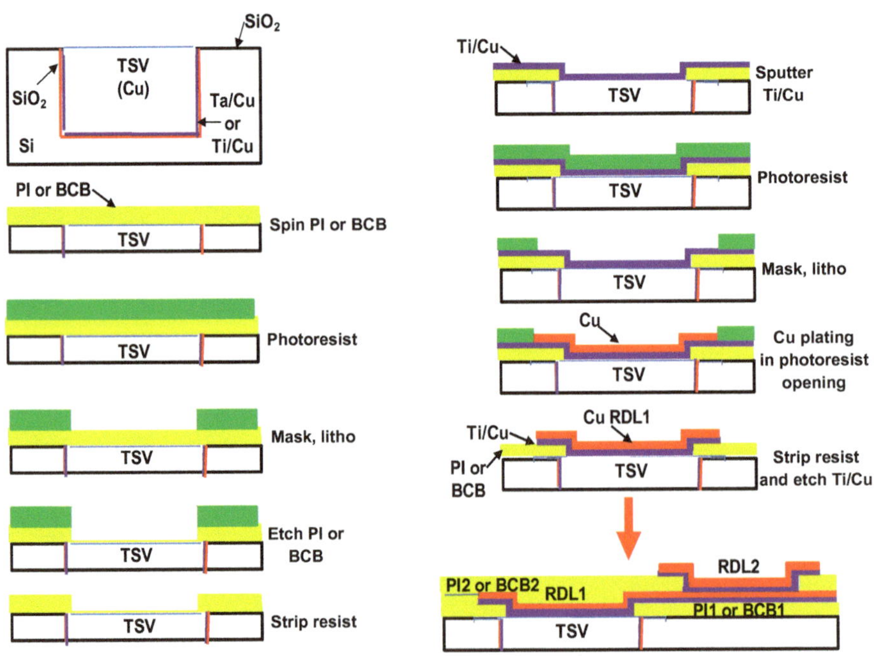

Fig. 5.11 RDL fabrication process by polymer + ECD + etching

1. Spin the polymers such as PI or BCB on the wafer and cure for 1 h. This will form a 4–7-μm-thick layer.
2. Apply photoresist and mask, then use photolithography techniques (align and expose) to open vias on the PI or BCB.
3. Etch the PI or BCB.
4. Strip off the photoresist.
5. Sputter Ti and Cu over the entire wafer.
6. Apply a photoresist and mask, and then use photolithography techniques to open the redistribution-trace locations.
7. Electroplate Cu in photoresist openings.
8. Strip off the photoresist.
9. Etch off the Ti/Cu and RDL1 is completed.
10. Repeat steps 1–9 for RDL2, and so forth.
11. Repeat step 1 (for UBM).
12. Apply photoresist and mask, and then use photolithography techniques (align and expose) to open vias on the PI or BCB for the desired bump pads and to cover the redistribution traces.
13. Etch the desired vias on the PI or BCB.
14. Strip off the photoresist.
15. Sputter Ti and Cu over the entire wafer.
16. Apply photoresist and mask, and then use photolithography techniques to open the vias on the bump pads to expose the areas with UBM.
17. Electroplate the Cu core.
18. Strip off the photoresist.
19. Etch off the Ti/Cu.
20. Electroless Ni and immersion Au. UBM is completed.

A typical cross section of the RDLs with polymers (e.g., BCB) as passivation and Cu plating as metal layers is shown in Fig. 5.12 [107]. It can be seen that the thickness of the passivation layers, BCB1 and BCB2, is about 6–7 μm, and the RDL is about 4 μm. It should be noted that photolithography can be directly applied on PI or BCB for larger dimensions. In that case the first set of photoresists is not needed.

(B) Fabrication of RDLs: SiO_2 + Cu Damascene and CMP Method

Another method of RDL fabrication is by a Cu damascene process. Starting with the wafer from Figs. 5.9 and 5.10, the fabrication process of RDLs with a Cu damascene technique is primarily based on the semiconductor back-end-of-line process. The details are shown in Fig. 5.13 and listed below.

1. SiO_2 layer by PECVD (plasma-enhanced chemical vapor deposition).
2. Apply photoresist and mask, then use photolithography techniques (align and expose) to open vias on the SiO_2.
3. RIE (reactive ion etch) of SiO_2.
4. Strip off the photoresist.

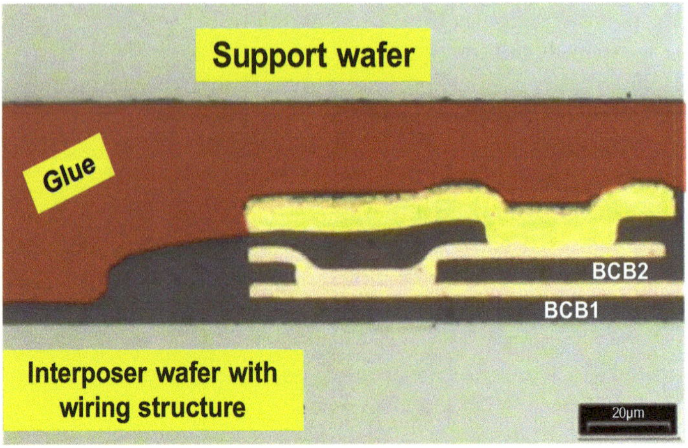

Fig. 5.12 RDLs by polymer + ECD + etching fabrication process

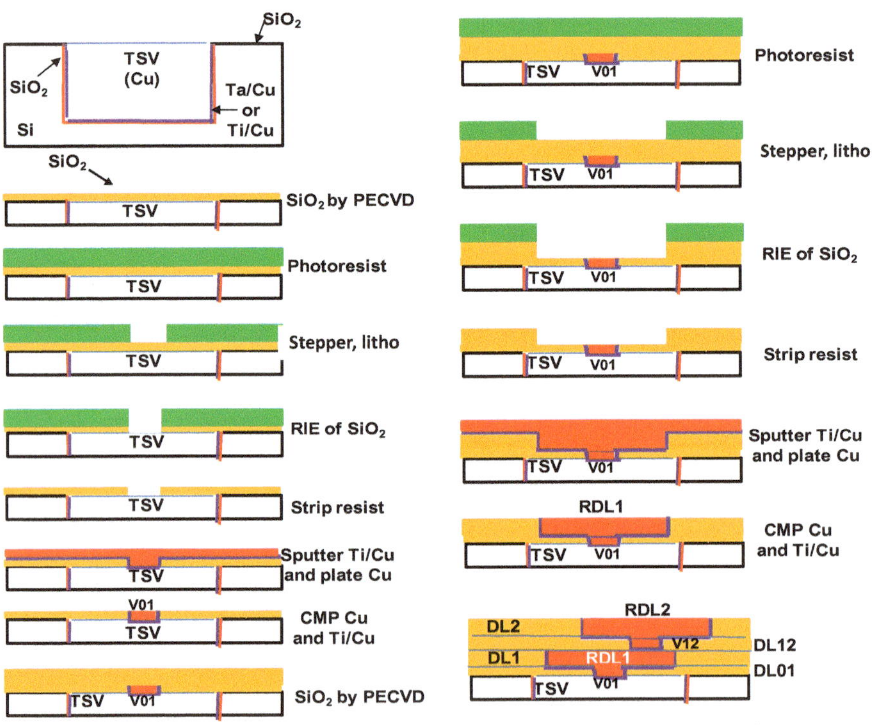

Fig. 5.13 RDL fabrication process by SiO_2 (PECVD) + Cu-damascene + CMP

5. Sputter Ti and Cu and electrochemical deposition (ECD) Cu over the entire wafer.
6. CMP the Cu and Ti/Cu. V01 (the via connecting the TSV to RDL1) is completed.
7. Repeat step 1.
8. Apply photoresist and mask, and then use photolithography techniques to open the redistribution trace locations.
9. Repeat step 3.
10. Repeat step 4.
11. Repeat step 5.
12. CMP the Cu and Ti/Cu. RDL1 is completed.
13. Repeat step 1 through step 6 to complete V12 (the via connecting the RDL1 to RDL2).
14. Repeat step 7 through step 12 to complete RDL2 and any additional layers.
15. (For UBM) Repeat step 1.
16. Apply photoresist and mask, and then use photolithography techniques (align and expose) to open vias on the SiO_2 for the desired bump pads and cover the redistribution traces.
17. Etch the desired vias on the SiO_2.
18. Strip off the photoresist.
19. Sputter Ti and Cu over the entire wafer.
20. Apply photoresist and mask, and then use photolithography techniques to open the vias on the bump pads to expose the areas with UBM.
21. Electroplate Cu core.
22. Strip off the photoresist.
23. Etch off the Ti/Cu.
24. Electroless Ni and immersion Au. UBM is completed.

It should be noted that the RDLs can also be fabricated by the dual Cu damascene method as shown in Fig. 5.14 [105, 106]. SEM images of the RDL cross sections fabricated by a Cu damascene technique are shown in Fig. 5.15. The minimum RDL line width is 3 μm. The thickness of RDL1 and RDL2 is 2.6 μm and of RDL3 is 1.3 μm. The passivation thickness between RDLs is 1 μm.

5.5 Intel's Embedded Multi-die Interconnect Bridge (EMIB)

The most famous rigid bridge is Intel's embedded multi-die interconnect bridge (EMIB) [102–104]. Figure 5.16 shows one of Intel's EMIB patents [102]. It can be seen that the EMIB die is embedded in the cavity of a build-up package substrate, which is supporting the chiplets. Figure 5.17 shows Intel's processor (Kaby Lake) that combines its high-performance × 86 cores with AMD's Radeon Graphics into the same processor package using Intel's own EMIB as well as HBM (2017). Intel canceled all the Kaby Lake-G products in October 2019. Since then, Intel has been

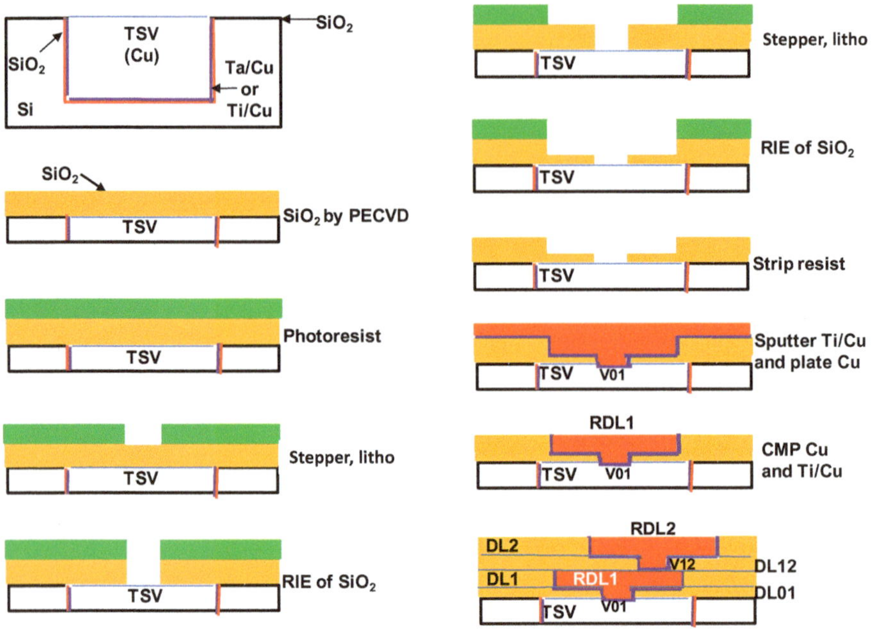

Fig. 5.14 RDL fabrication process by SiO_2 (PECVD) + dual Cu-damascene + CMP

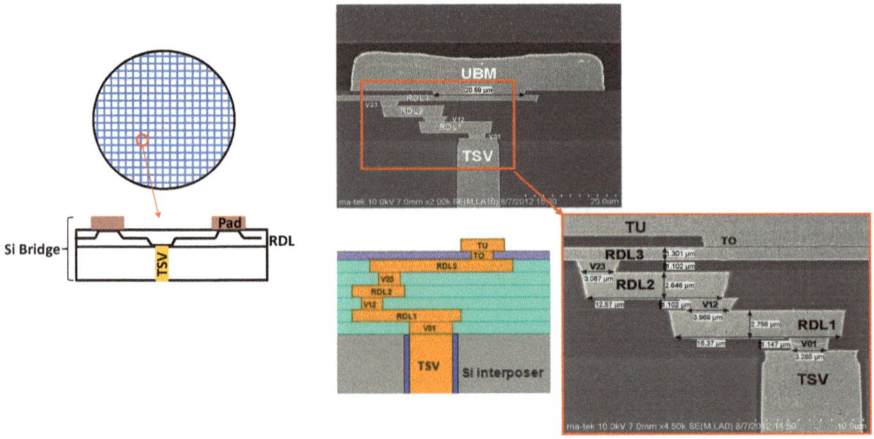

Fig. 5.15 SEM images of RDLs by SiO_2 (PECVD) + dual Cu-damascene + CMP

shipping its Agilex (Fig. 5.18) with two to five EMIBs in 2019, and the Ponte Vecchio GPU (Fig. 5.19) with 11 EMIBs in 2022. Sapphire Rapids is the next-generation of Intel's Xeon scalable processor (Fig. 5.20). It consists of four SoCs and they are connected with 10 EMIBs. There are another four EMIBs that connect the four SoCs

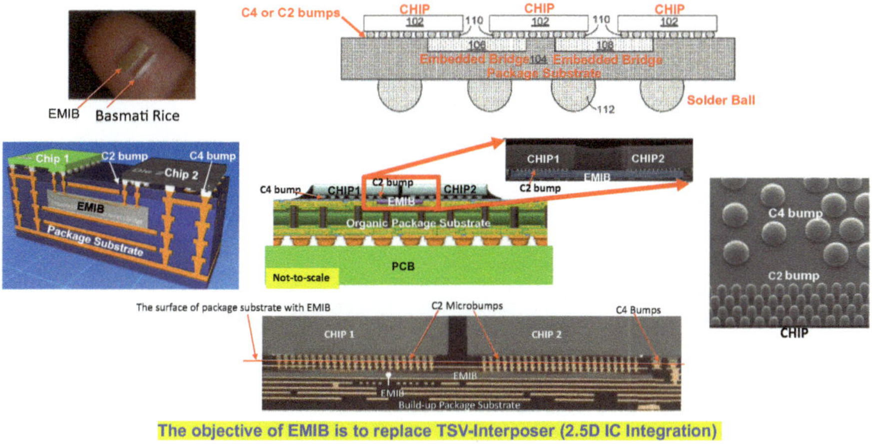

Fig. 5.16 Intel's embedded multi-die interconnect bridge (EMIB)

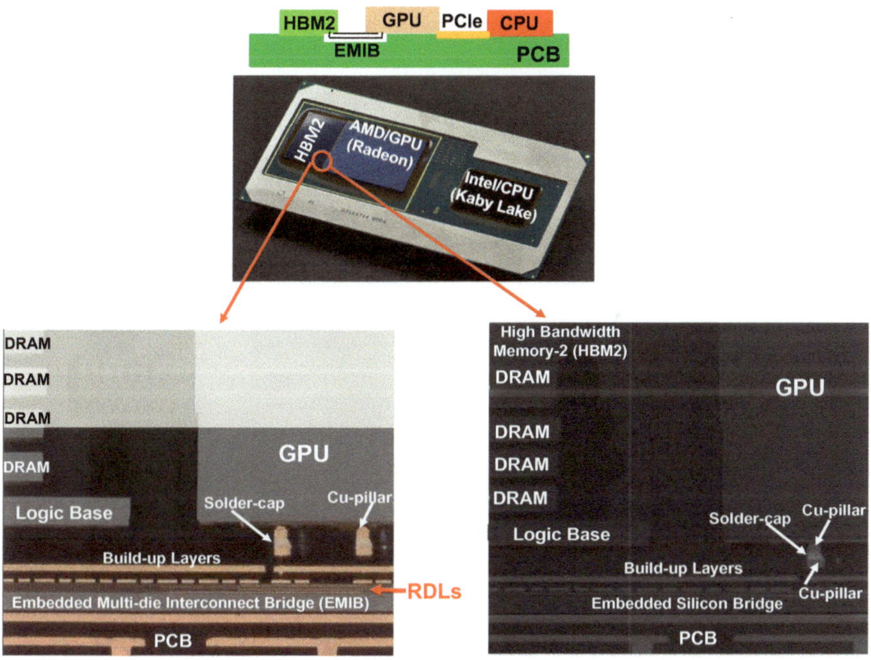

Fig. 5.17 Intel's processor Kaby Lake and AMD's Radeon graphics with one EMIB

and the four HBMs. Lately, Intel has been discussing reducing the number of SoCs to two and shipping a new iteration of the product.

For EMIB, there are at least three important tasks (Fig. 5.16) namely: (1) wafer bumping of two different kinds of bumps on the chiplets wafer (but there are not

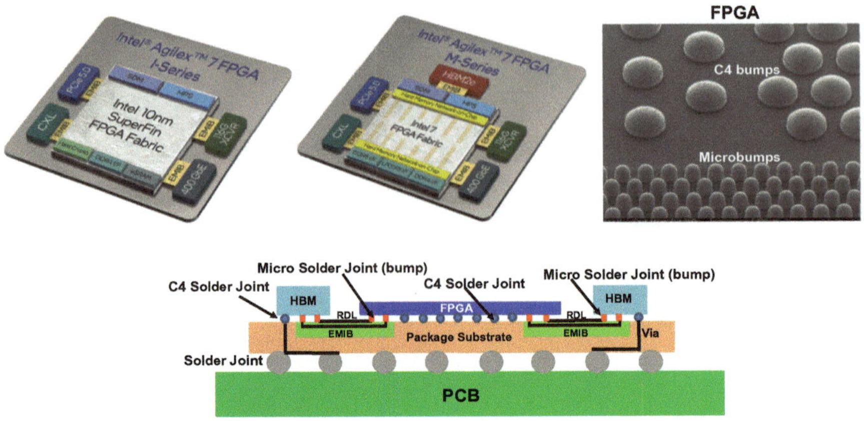

Fig. 5.18 Intel Agilex with two to five EMIBs

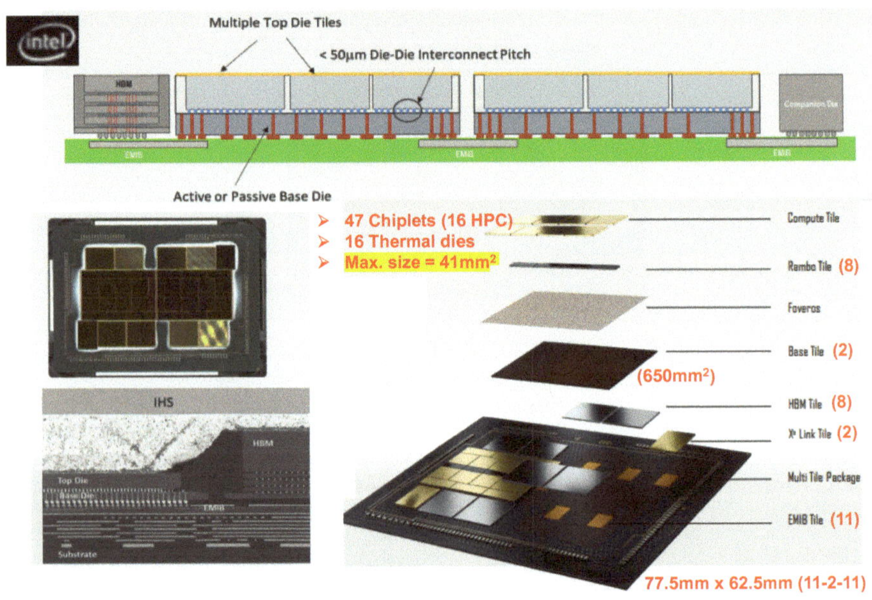

Fig. 5.19 Intel's Ponte Vecchio GPU with 11 EMIBs

bumps on the bridge); (2) embedding the bridge in the cavity of a build-up substrate and then laminating the top surface of the substrate; and (3) bonding the chiplets on the substrate with the embedded bridge.

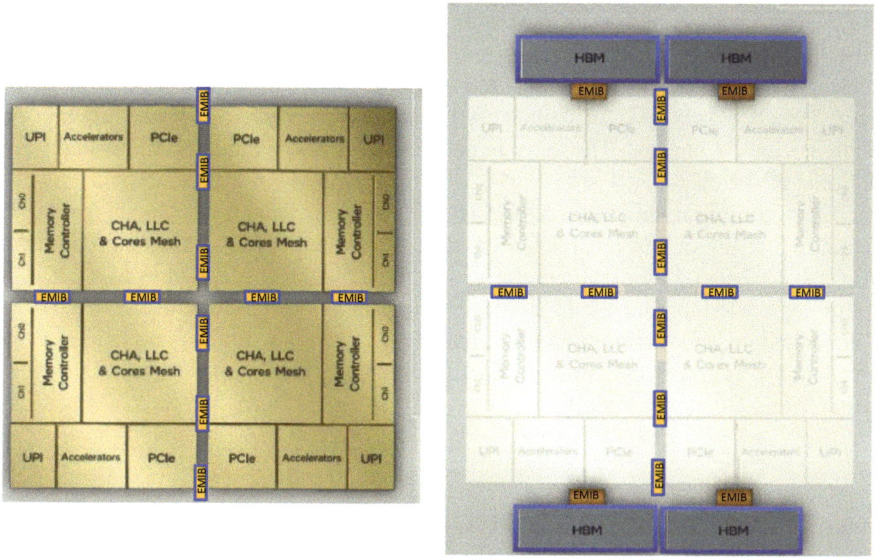

Fig. 5.20 Intel's Sapphire Rapids with 14 EMIBs

5.5.1 Solder Bumps for Embedded Multi-die Interconnect Bridge

It can be seen from Fig. 5.16 that there are two kinds of bumps on the chiplet, namely the C4 (controlled collapse chip connection) bumps and the C2 (chip connection or copper-pillar with solder-cap micro) bumps. Thus, wafer bumping of the chiplets wafer poses a challenge, but Intel has already taken care of this issue.

5.5.2 Fabrication of Embedded Multi-die Interconnect Bridge Substrate

There are two major tasks in fabricating the organic package substrate with EMIB (Fig. 5.16). One is to make the EMIB, and the other is to make the substrate with EMIB. To make the EMIB, one must first build the RDLs (including the contact pads) on a Si-wafer. The way to make the RDLs depends on the line width/spacing of the conductive wiring of the RDLs. Finally, attach the non-RDL side of the Si-wafer to a die-attach film, and then singulate the Si-wafer.

To make the substrate with an EMIB, first place the singulated EMIB with the die-attached film on top of the Cu foil in the cavity of the substrate. It is followed by laminating a dielectric film on the whole organic package substrate. Then, drilling (on the dielectric film) and Cu plating to fill the holes (vias) to make connections to the contact pads of the EMIB. Continue Cu plating to make lateral connections of the substrate. Then, it is followed by laminating another dielectric film on the whole substrate and drilling (on dielectric) and Cu plating to fill the holes and make contact pads. (Smaller pads on a finer pitch are for C2 bumps, while larger pads on a gross pitch are for C4 bumps.) The organic package substrate with an EMIB is ready for bonding of the chips.

Today, the minimum metal L/S/H is 2/2/2 μm and the bridge size is from 2 × 2 mm to 8 × 8 mm, but most are equal and < 5 × 5 mm [103]. The dielectric layer thickness is 2 μm. Usually, there are ≤ 4 RDLs. One of the challenges of the EMIB technology is to fabricate the organic build-up package substrate with cavities for the silicon bridges and then laminate (with pressure and temperature) another build-up layer on top (to meet the substrate surface flatness requirement) for chiplets (with both C2 and C4 bumps) bonding.

5.6 IBM's Direct Bonded Heterogeneous Integration (DBHi)

During IEEE/ECTC 2021 and 2022, IBM presented seven papers on "Direct Bonded Heterogeneous Integration (DBHi) Si Bridge" [108–114] (Fig. 5.21). The major differences between Intel's EMIB and IBM's DBHi are as follows: (a) For Intel's EMIB, there are two different bumps (C4 and C2) on the chiplets (and there are no bumps on the bridge) (Fig. 5.16), while for IBM's DBHi, there are C4 bumps on the chiplets and C2 bumps on the bridge (Fig. 5.21). (b) For Intel's EMIB, the bridge is embedded in the cavity of a build-up substrate with a die-attach material and then laminated with another build-up layer on top. Therefore, the substrate fabrication is very complicated. For IBM's DBHi, the substrate is just a regular build-up substrate with a cavity on top as shown in Fig. 5.22.

The bonding assembly process of DBHi is very simple (Fig. 5.22). First, bond the Chip 1 and the bridge with nonconductive paste (NCP) and thermal compression bonding (TCB). Then, bond Chip 2 and the bridge with NCP and TCB. Those steps are followed by placing the module (Chip 1 + bridge + Chip 2) on the organic substrate with a cavity and then going through the standard flip-chip reflow assembly process. The underfill under the bridge is optional. Figure 5.21 shows the demonstration by IBM.

During I EEE/ECTC 2023, IBM presented a paper on "Direct Bonded Heterogeneous Integration (DBHi): Surface bridge approach for dietiling" [115]. The authors demonstrated that the cavity of a build-up package substrate is not necessary for their Si bridge technology as shown in Fig. 5.23.

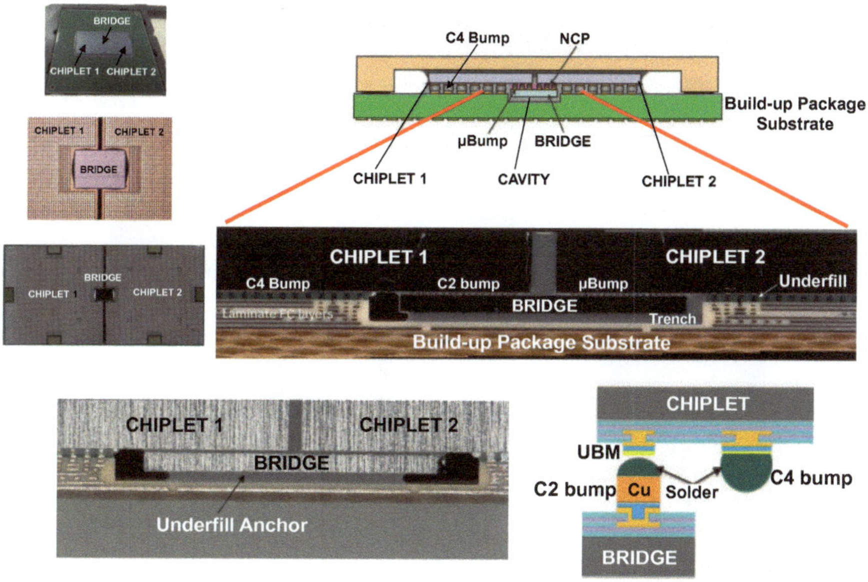

Fig. 5.21 IBM's direct bonded heterogeneous integration (DBHi) Si bridge

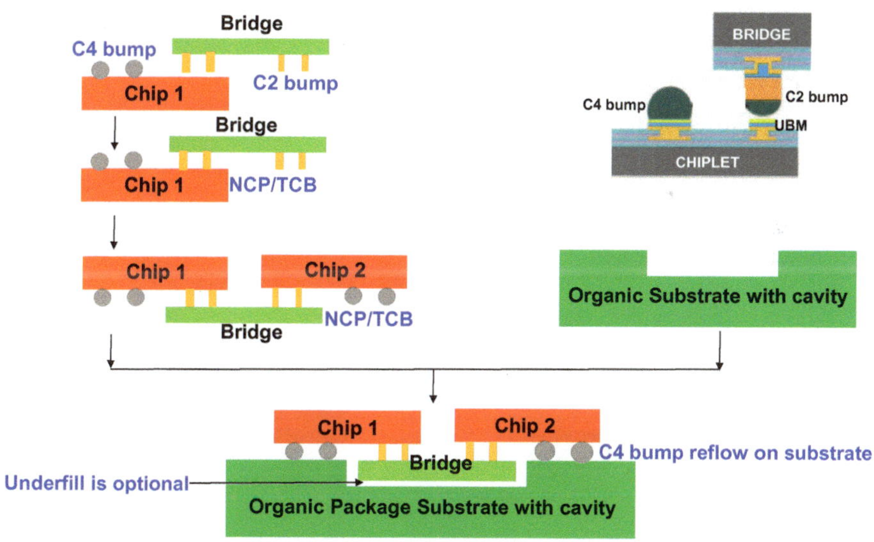

Fig. 5.22 IBM's DBHi on package substrate with cavity

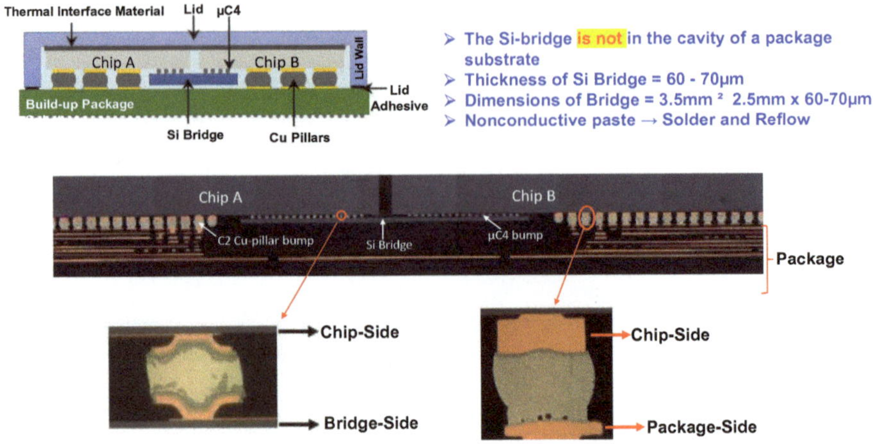

Fig. 5.23 IBM's DBHi on package substrate without cavity

5.7 Apple's UltraFusion

UltraFusion is Apple's innovative packaging architecture that interconnects the die of two M1 Max chips to create a SoC with unprecedented levels of stunningly compact design, extensive connectivity, performance and capabilities. This architecture does not combine two M1 Max dies into a single chip package - it also makes the two dies present themselves as a single chip. The interconnection between the two M1 Max dies is by a "silicon bridge" as shown in Fig. 5.24. The chiplets and the Si-bridge are supported by an ordinary build-up package substrate without any cavity.

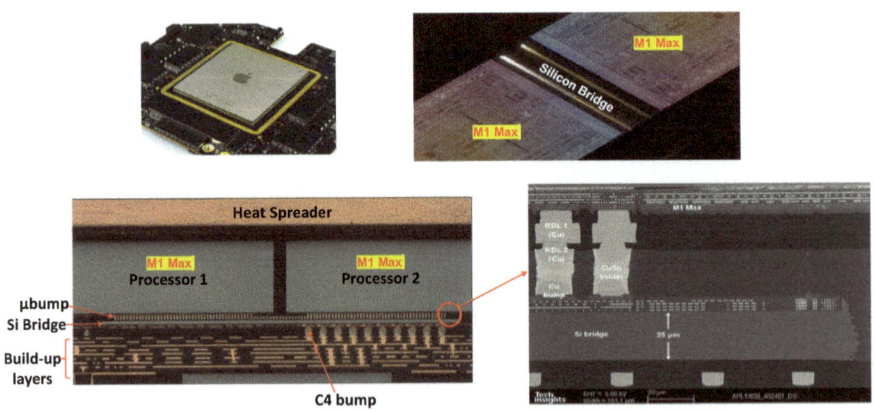

Fig. 5.24 Apple's UltraFusion are supported by a build-up package substrate without cavity

Fig. 5.25 AMD's Instinct MI250X are supported by a build-up package substrate without any cavity

5.8 AMD's Instinct MI250X Compute Accelerator

The AMD Instinct MI250X compute accelerator is shown in Fig. 5.25. It can be seen that there are two GPUs (the second-generation matrix cores for HPC driven by AI) with each having a size of 790 mm^2. Each GPU is connected to four HBM2e with Si bridges on microbumps. The GPU and the HBM2E are supported by a build-up package substrate without any cavity.

5.9 Hybrid Bonding Bridge

Unimicron proposed the use of Cu-Cu hybrid bonding for the bridge between chiplets in chiplet design and heterogeneous integration packaging, (Fig. 5.26). The advantages of this structure are: (1) higher density, (2) better performance, and (3) ordinary package substrate. There are at least two options: one is with C4 bumps on the package substrate, and the other is with C4 bumps on the chiplet wafer.

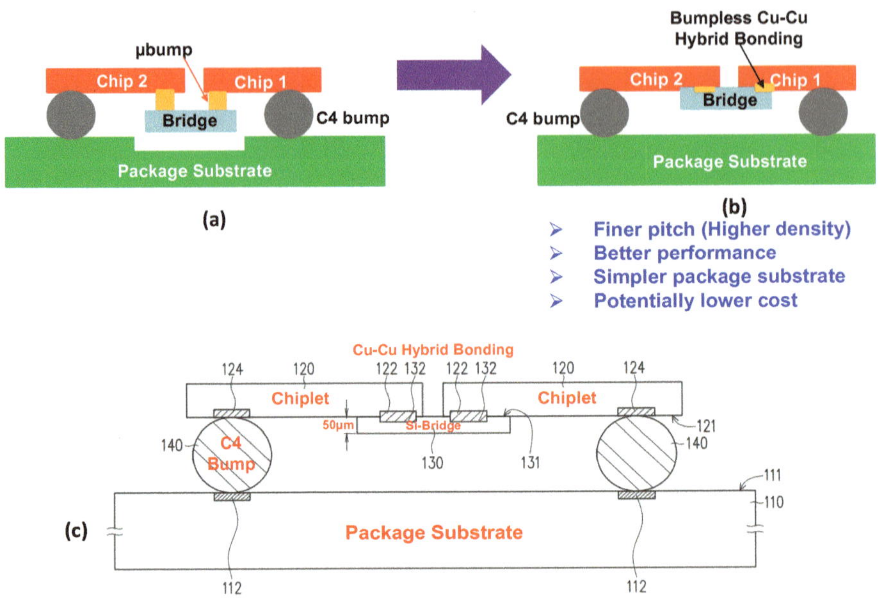

Fig. 5.26 Hybrid boding to connect chiplets and Si bridge

5.9.1 *Hybrid Bonding Bridge with C4 Bumps on the Package Substrate*

Figure 5.27 shows the process flow of a hybrid bonding bridge with C4 bumps on the package substrate. For the bridge wafer, the processing starts off with chemical vapor deposition (CVD) to make a dielectric material such as SiO_2 and then it is planarized by an optimized chemical mechanical polishing (CMP) process to make the Cu dishing. Then, the bridge wafer is diced into individual chips (still on the blue tape of the wafer) after application of a protective coating layer on the wafer surfaces to prevent any particles and contaminants that may cause interface voids during the subsequent bonding process. These steps are followed by activating the bonding surface by using plasma and hydration processes for better hydrophilicity and a higher density of a hydroxyl group on the bonding surface.

To process the chiplet wafer, repeat the CVD process for the SiO_2, CMP for the Cu dishing, and plasma and hydration of the activation of the bonding surface. Then, pick and place the individual bridge chip on the chiplet wafer and perform the SiO_2-to-SiO_2 bonding at room temperature. These steps are followed by annealing to achieve covalent bonding between oxide layers and metallic bonding between Cu-Cu contacts and the diffusion of Cu atoms.

For the package substrate, the process is to stencil print the solder paste on the substrate and then reflow into C4 solder bumps. For the final assembly, the bridge +

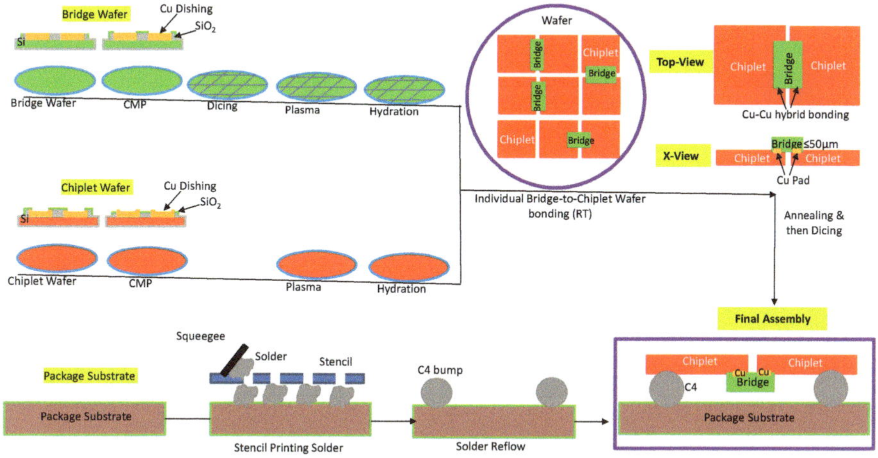

Fig. 5.27 Hybrid bonding bridge–bridge wafer, chiplet wafer, and package substrate process flow. C4 bumps are on package substrate

chiplets module is picked and placed on the package substrate, then the C4 bumps are reflowed.

5.9.2 *Hybrid Bonding Bridge with C4 Bumps on the Chiplet Wafer*

Figure 5.28 shows the process flow of the hybrid bonding bridge with C4 bumps on the chiplet wafer. It can be seen that, compared with the C4 bumps for the package substrate case, the process steps for the bridge wafer and the chiplet wafer are the same up to the bridge-to-chiplet wafer bonding step. After that, the C4 bumps are fabricated by wafer bumping on the chiplet wafer. Then, the chiplet wafer is diced into individual modules (bridge + chiplets with C4 bumps). The final assembly is accomplished by picking and placing the individual module on the package substrate and reflowing the C4 solder bumps.

5.10 Ridge Bridge Embedded in Fan-Out EMC with RDLs

The rigid bridges by Intel, IBM, Apple, AMD etc. are either embedded in or are on an organic build-up package substrate w/o cavity. There is another class of rigid bridge, which is embedded in the fan-out EMC and connected to the fan-out RDL substrate.

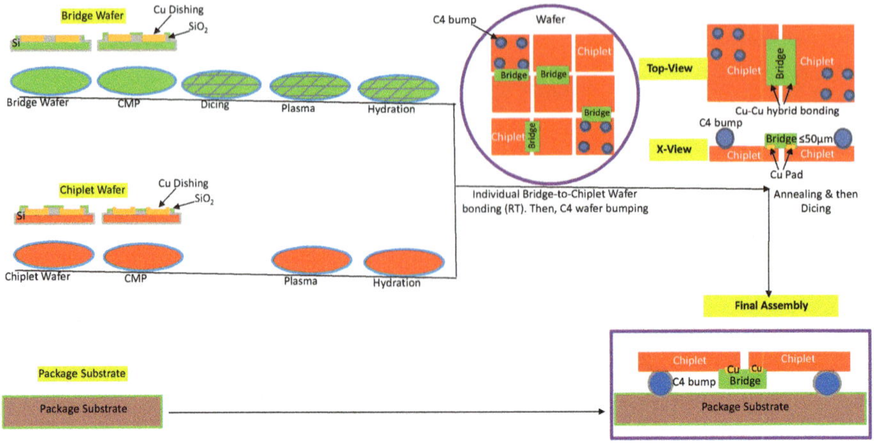

Fig. 5.28 Hybrid bonding bridge—bridge wafer, chiplet wafer, and package substrate process flow. C4 bumps are on chiplet wafer

5.10.1 *Applied Materials' Bridge-First and Face-Up Process*

On May 12, 2020, Applied Materials obtained the U.S. patent 10,651,126 [116]. The company's design embedded the bridge in EMC by the fan-out chip (bridge) first and die face-up process (Fig. 5.29). This could be the very first patent of a rigid bridge embedded in fan-out EMC.

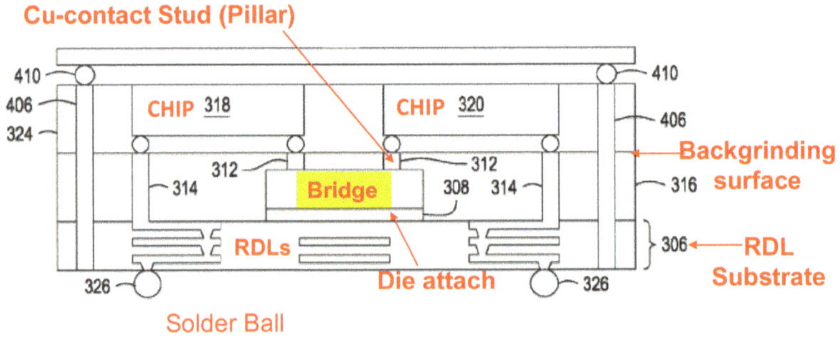

Fig. 5.29 Applied Materials' bridge patent with fan-out chip (bridge) first and die face-up process (US 10,651,126)

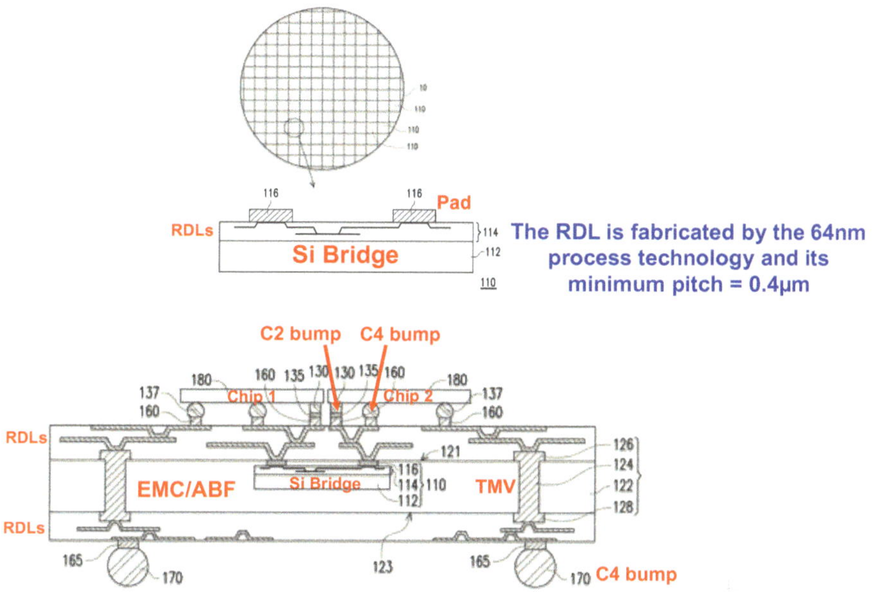

Fig. 5.30 Unimicron's bridge patent with fan-out chip (bridge) first and die facedown process (US 11,410,933)

5.10.2 Unimicron's Bridge-First and Face-Down Process

On May 7, 2021, Unimicron filed and obtained the U.S. patent U.S. 11,410,933 on Aug. 9, 2022 [117] in which the bridge is embedded in the fan-out EMC by the chip (bridge) first and die face-down process (Fig. 5.30).

5.10.3 IME's Bridge-Last Process

On May 25, 2021, IME obtained the U.S. patent U.S. 11,018,080 [118] in which the bridge is embedded in the EMC and connected to the RDL-substrate by the chip (bridge) last or RDL-first fan-out process, (Fig. 5.31).

5.10.4 Publications by TSMC, ASE, Amkor, SPIL, and IME

In the past couple of years, there are many publications in rigid bridges embedded in fan-out EMC with RDLs. For examples, On Aug. 25, 2020, during TSMC's Annual Technology Symposium, the company announced its integrated fanout local silicon interconnect (InFO_ LSI) and chip-on-wafer-on-substrate local silicon interconnect

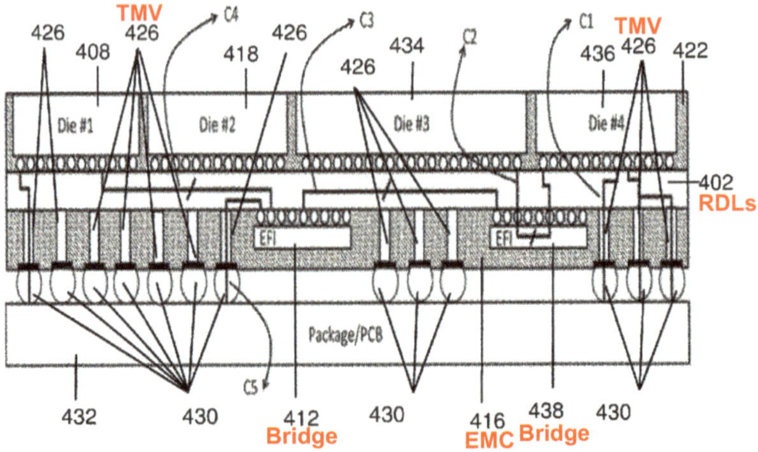

Fig. 5.31 IME's bridge patent with fan-out chip (bridge) last or RDL-first process (US 11,018,080)

(CoWoS_LSI) (Fig. 5.32a and b, respectively) [119]. During IEEE/ECTC (June 2021), there were at least four papers published regarding the application of fan-out packaging technology to embed the rigid bridge in the EMC with RDLs for the chiplets to perform lateral communications. All four of these papers discuss very similar technologies. In Ref. [120], ASE embedded the bridge in the EMC using the fan-out packaging method and called it stacked Si bridge fan-out chip-on-substrate (sFOCoS) (Fig. 5.32 (c)). In Ref. [121], Siliconware Precision Industries Co., Ltd. (SPIL) called its similar technology fan-out embedded bridge (FO-EB) (Fig. 5.32d). In Ref. [122], Amkor referred to its comparable technology as S-Connect fan-out interposer (Fig. 5.32e). In Ref. [123], IME presented its bridge and called it embedded fine interconnect (EFI) (Fig. 5.32f).

5.11 Packaging Technology Driven by Artificial Intelligence (AI)

Recently, because of the drive of AI such as Open AI's GPT, generative AI and conversational AI, and 5G/6G, the applications such as high-performance computing (HPC), autonomous vehicle, internet of things (IoTs), big data (for cloud computing) and instant data (for edge computing) are demanding more advanced semiconductor packaging technology [1–4]. One of the most popular advanced packaging technologies is the 2.5D or 3D integrated circuit (IC) integration [1] as schematically shown in Fig. 5.33. It can be seen that the system on chip (SoC) devices, such as the central processing unit (CPU) and graphics processing unit (GPU), and high-bandwidth memory (HBM) are supported by a passive (2.5D) or active (3D) through-silicon via (TSV)-interposer and then on a build-up package substrate. Finally, the whole

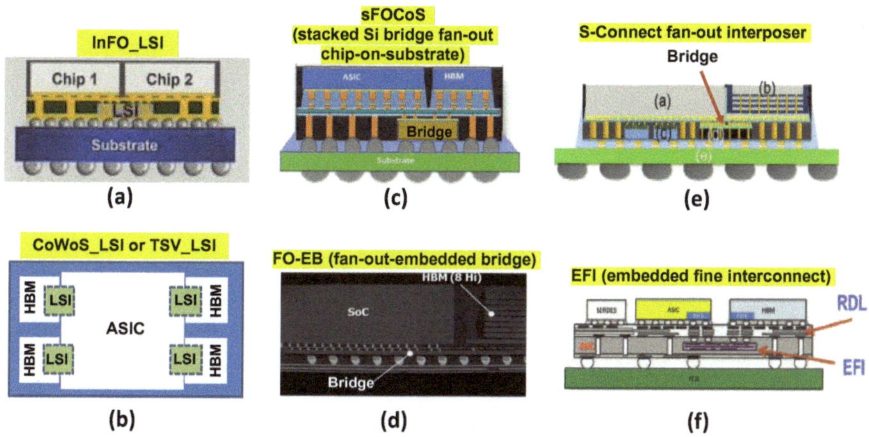

Fig. 5.32 **a** TSMC's InFO-LSI. **b** TSMC's CoWoS-LSI. **c** ASE's sFOCoS. **d** SPIL's FO-EB. **e** Amkor's S-connect with Si-bridge. **f** IME's bridge with EFI

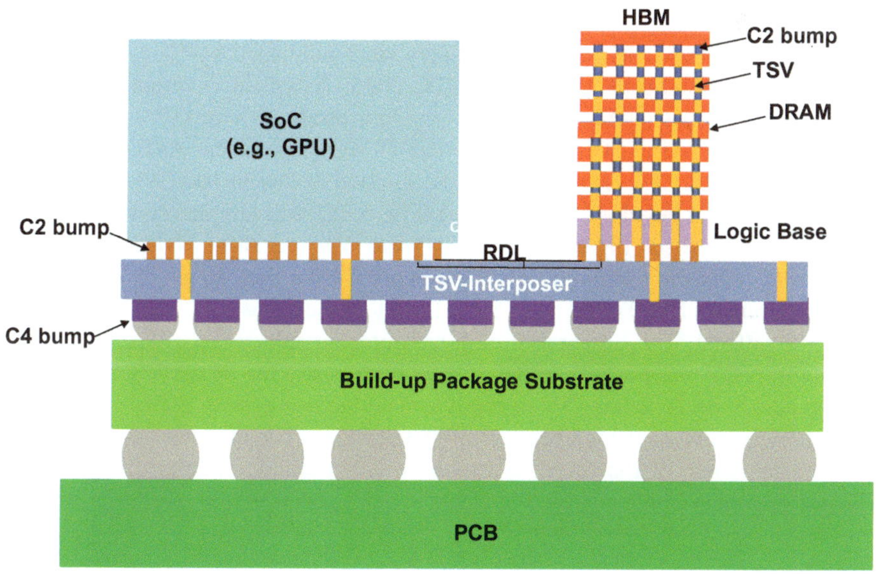

Fig. 5.33 Passive TSV-interposer (2.5D IC integration) is the most popular advanced packaging for applications driven by AI

module is attached to a printed circuit board (PCB) with ball grid array (BGA) solder balls and solder paste. This multiple chiplet system and heterogeneous integration packaging are driven by performance and formfactor and for extremely high-density and high-performance applications [9, 11].

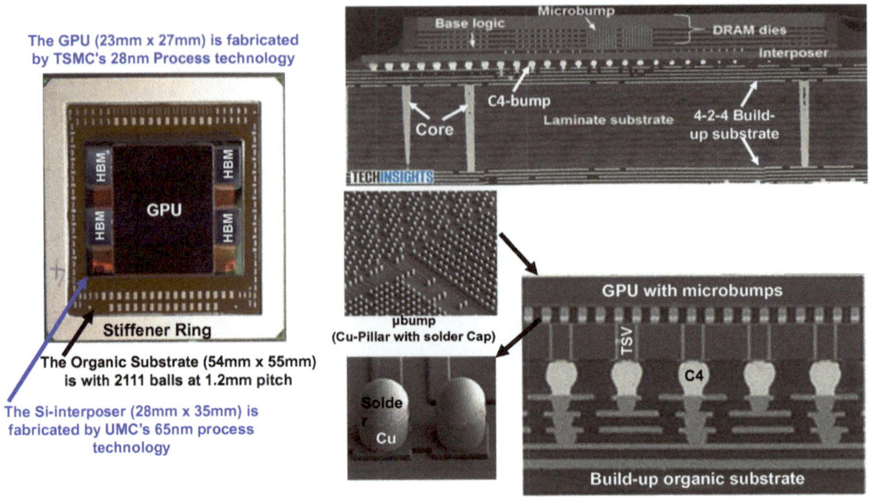

Fig. 5.34 AMD/UMC's GPU (Fiji)

The very first 2.5D IC integration papers were published by CEA-Leti [86] at IEEE/ECTC 2005 and [87] at IEEE/ECTC 2006. A typical example of 2.5D IC integration is shown in Fig. 5.33. The very first product (Virtex™-7 HT family) of 2.5D was shipped in 2013 by Xilinx and TSMC. Since then, AMD shipped its Radeon R9 Fury X GPU (Fig. 5.34), Nvidia shipped its Pascal 100 GPU (Fig. 5.35), Fujitsu shipped its Fugaku (A64FX CPU) (Fig. 5.36), and Graphcore shipped Bow (an intelligence processing unit) (Fig. 5.37), etc. Very recently, Nvidia shipped its A100 GPU (826mm^2) with six HBM2 supported by a very large TSV-interposer (Fig. 5.38) and its H100 GPU (Fig. 5.39) for AI applications.

TSMC called 2.5D IC integration chip-on-wafer-on-substrate (CoWoS). The 2.5D or 3D with TSV-interposer is known for its high cost. One of the key reasons for its high cost is because the TSV-interposer manufacturing (with the 64 nm process technology) yield loss is high because of its large size. The objective of this section is to present some of the recent advances in using silicon bridges to replace the TSV-interposer.

5.12 TSMC's CoWoS-L

During IEEE/ECTC 2023, TSMC published two papers on replacing its TSV-interposer by local silicon interconnect (LSI) or Si-bridge embedded in epoxy molding compound (EMC) with RDLs [124, 125]. The key reason for this technology is to deal with the manufacturing yield loss of the ever-increasing size of the

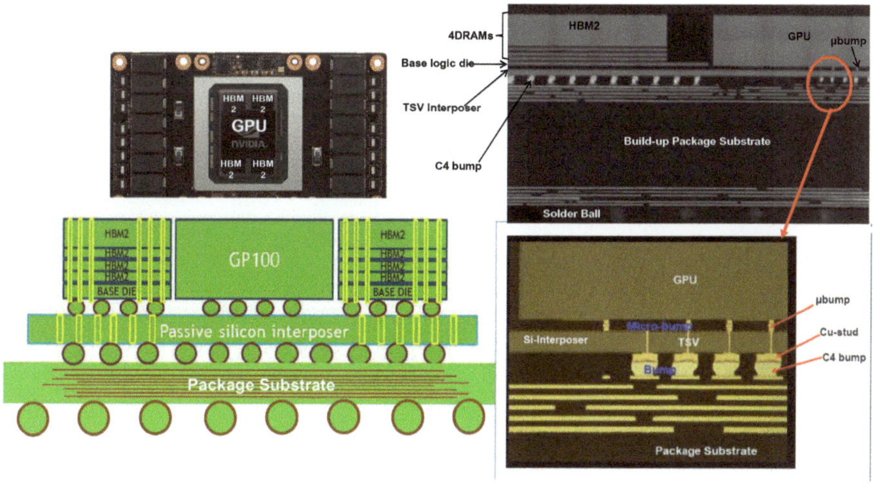

Fig. 5.35 Nvida/TSMC's P100

Fig. 5.36 Fujitsu/TSMC's CPU

TSV-interposer - the yield loss is so high that the cost becomes unbearable. Just look at two examples: TSMC has been using CoWoS with the 1X reticle ($830mm^2$) for Xilinx, and with the 2X reticle ($1660mm^2$) for Nvidia. For a 3X reticle ($2500mm^2$),

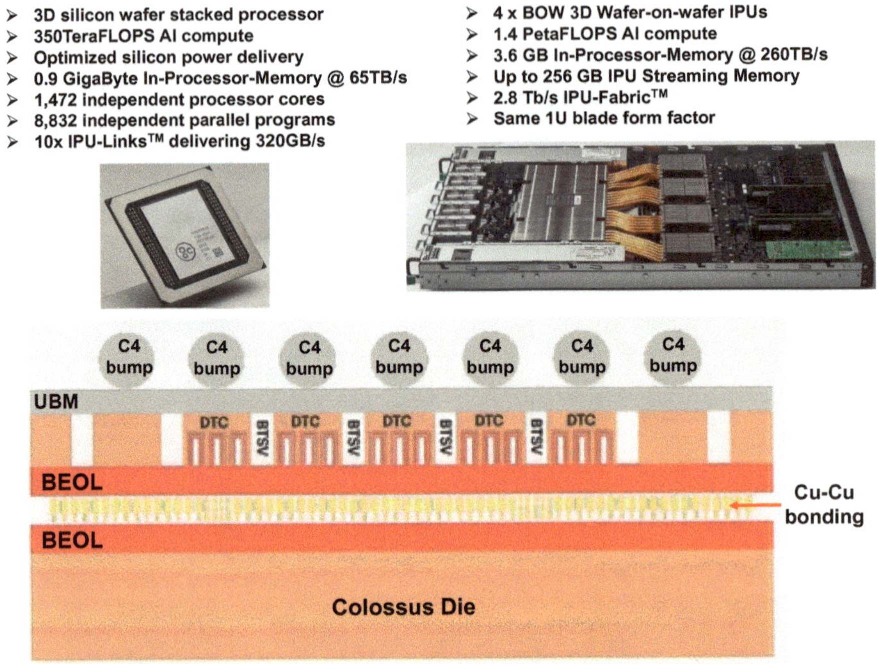

Fig. 5.37 Graphcore/TSMC's IPU (intelligence processing unit) processor

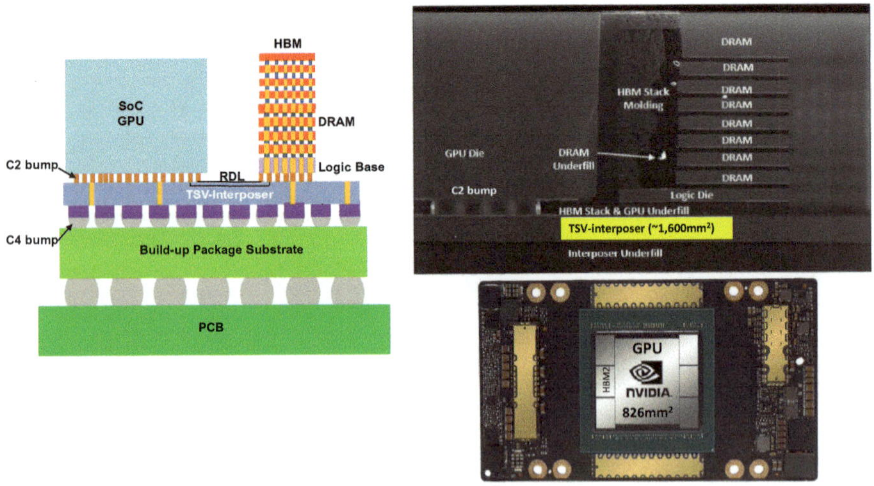

Fig. 5.38 Nvida/TSMC's A000

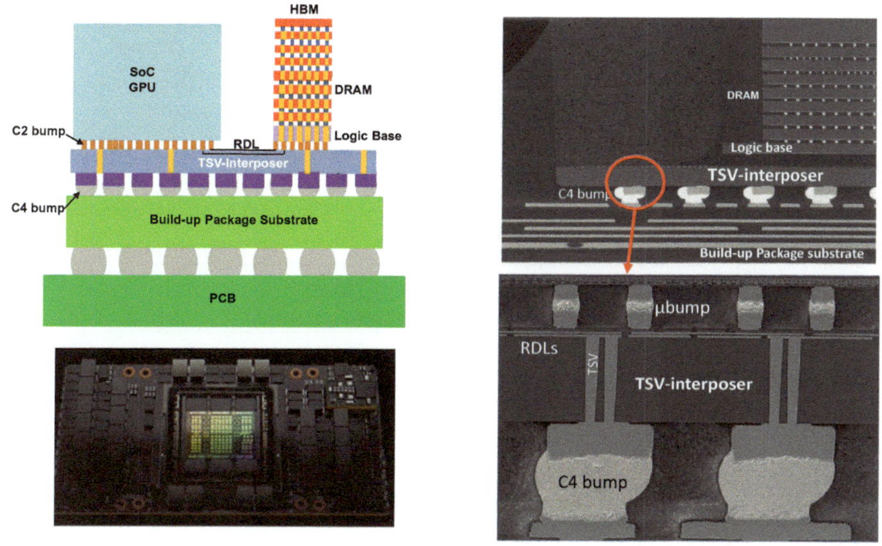

Fig. 5.39 Nvida/TSMC's H100

the yield loss is too high, so TSMC is going to recommend a new chip-on-wafer-on-substrate-LSI interposer (CoWoS-L) as shown in Fig. 5.40a. Figure 5.40b shows the number of physically possible TSV-interposers vs. the size of the TSV-interposer on a 300 mm-wafer. The impact of TSV-interposer size on yield loss per wafer is obvious.

Figure 5.41 shows the replacement of the TSV-interposer (Fig. 5.41a) of CoWoS by the LSI. Figure 5.41b shows the Si-bridge embedded in an EMC with fan-out RDLs (CoWoS-L). CoWoS-L is a new interposer that consists of at least one LSI (or Si-bridge) with or without TSVs and integrated fan-out (InFO) RDLs to form a reconstituted interposer (RI). The small-size LSI inherits all the attractive features of the TSV-inter poser by retaining sub-micron Cu interconnects (RDLs), TSVs, and embedded deep trench capacitors (eDTCs) to ensure good system performance. At the same time, the small-size LSI avoids the issues associated with the large-size TSV-interposer, such as manufacturing yield loss. The metal L/S of the small-size bridge (LSI) has a pitch (minimum) = 0.4 μm. At the end, the large-size TSV-interposer is replaced by the RI, which consists of the EMC, small-size bridge (LSI) and with or without TSV + InFO RDLs. The fabrication of the RDL and TSV is shown in Figs. 5.9–5.15 and 5.42 and the fabrication process of the CoWoS-L is shown in Fig. 5.43.

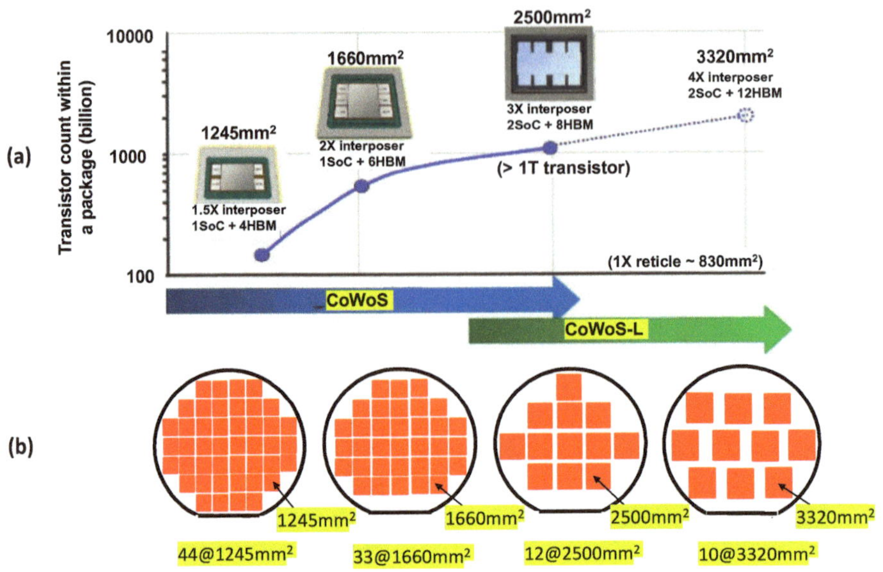

Fig. 5.40 **a** TSMC's roadmap on TSV-interposers. **b** Physical possible number of TSV-interposer vs. size of TSV-interposer

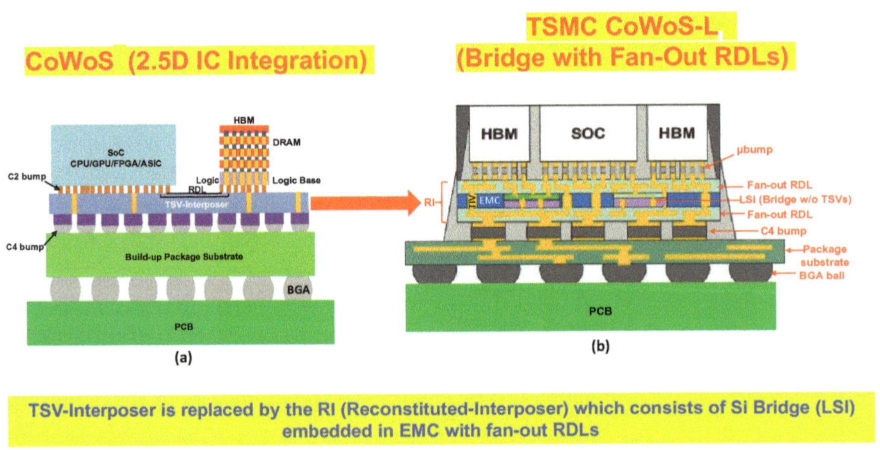

Fig. 5.41 The TSV-interposer in CoWoS (**a**) is replaced by the LSI (Si bridge) embedded in fan-out EMC with RDLs—TSMC's CoWoS-L (**b**)

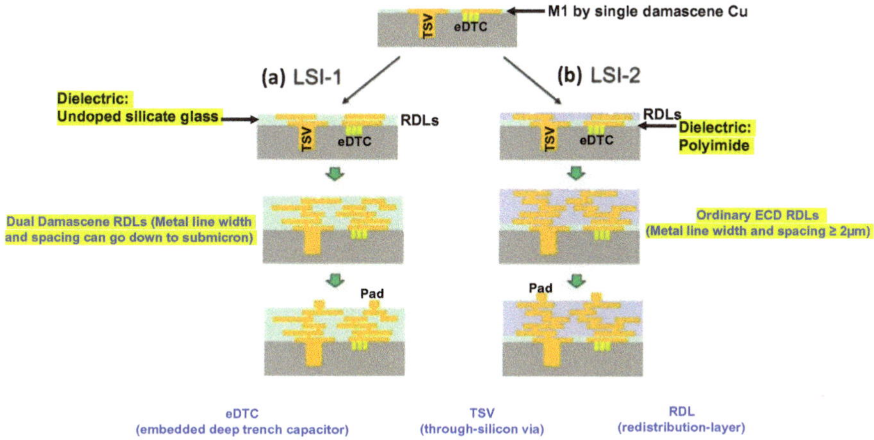

Fig. 5.42 Process in making the Si bridge with TSV and RDLs

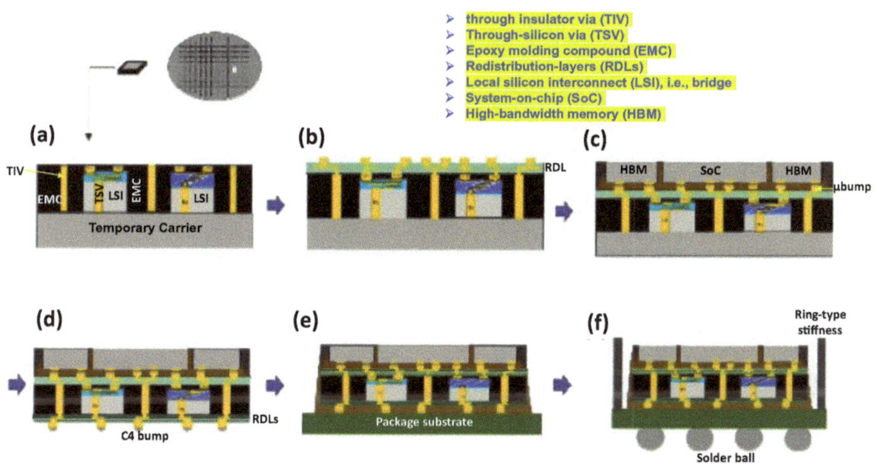

Fig. 5.43 Process in making the CoWoS-L

5.13 SPIL's Fan-Out Embedded Bridge with TSV (FO-EB-T)

During IEEE/ECTC 2023, SPIL presented two papers on FO-EB-T [126, 127]. Very similar to TSMC's CoWoS-L architecture, SPIL replaced the TSV interposer with a Si-bridge with TSVs and RDLs embedded in an EMC with fan-out RDLs as shown in Fig. 5.44. A typical cross section of the FO-EB-T is shown in Fig. 5.45.

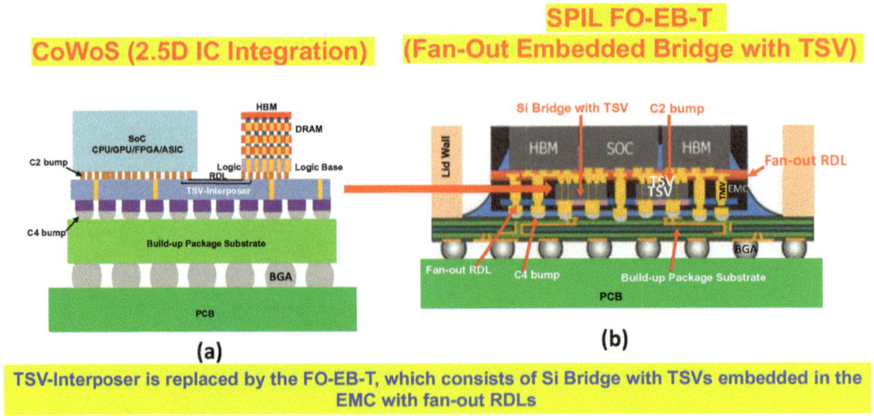

Fig. 5.44 The TSV-interposer in CoWoS (**a**) is replaced by the Si bridge embedded in fan-out EMC with RDLs—SPIL's FO-EB-T (**b**)

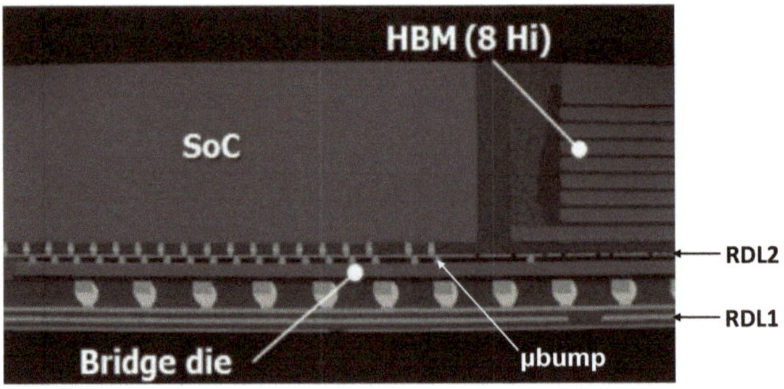

Fig. 5.45 SEM of SPIL's FO-EB-T

5.14 Universal Chiplet Interconnect Express (UCIe) Consortium

An important consortium concerned with bridge technology is the Universal Chiplet Interconnect Express (UCIe), Fig. 5.46. According to the consortium's website, the organization addresses customer requests for a more customizable, package-level integration—combining best-in-class die-to-die interconnect and protocol connections from an interoperable, multivendor ecosystem. This new open industry standard establishes a universal interconnect at the package level. The UCIe board of directors and leadership (promoters) include founding members ASE, AMD, Arm, Google Cloud, Intel Corporation, Meta, Microsoft Corporation, Qualcomm Incorporated, Samsung Electronics, and TSMC, along with newly elected members, Alibaba and NVIDIA. In [128], Intel published the UCIe 1.0 specification, which provides

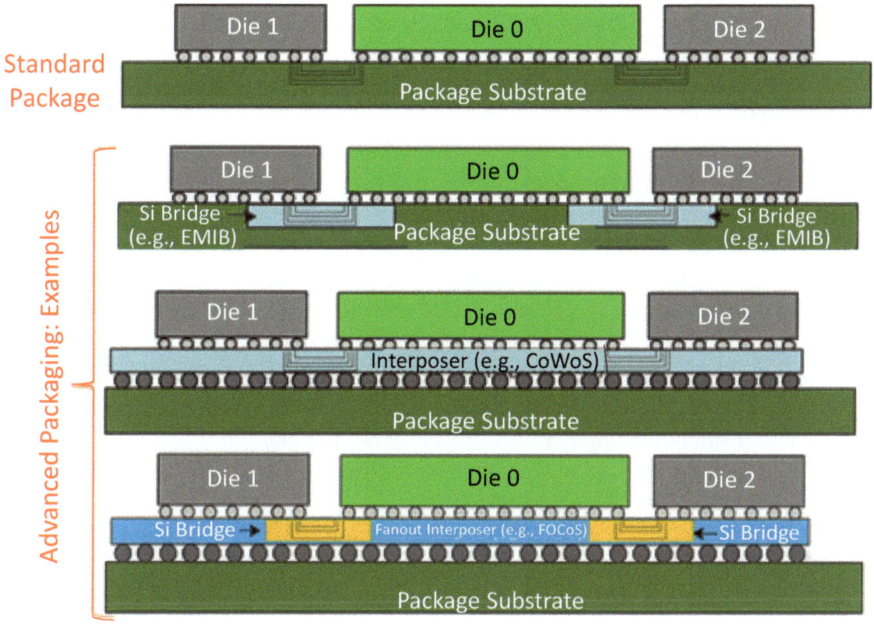

Fig. 5.46 UCIe bridges

a complete standardized die-to-die interconnect with physical layer, protocol stack, software model, and compliance testing. Figure 5.46 shows examples of standard packaging and advanced packaging with chiplet design and heterogeneous integration. It can be seen that there are three different kinds of bridges for advanced packaging: (1) bridge embedded in organic package substrate; (2) bridge embedded in Si-interposer; and (3) bridge embedded in fan-out EMC with RDLs.

5.15 Summary

Some important results and recommendations are summarized as follows.

- A bridge is a small piece of chip without devices but with RDLs to let the chiplets perform horizontal communication.
- Some bridges also perform vertical communication without devices but with RDLs and TSVs.
- There are two groups of bridges: rigid bridge and flexible bridge.
- For rigid bridges, the RDLs are fabricated on a silicon wafer substrate.

- Today, the rigid bridges are embedded on an organic package substrate, such as the EMIB (Intel) and DBHi (IBM) and embedded in fan-out EMC and connected with RDLs, such as those by Applied Materials, TSMC, Unimicron, ASE, Amkor, SPIL, IME, and imec.
- A new rigid bridge called hybrid bonding bridge has been proposed which leads to a very high-performance and very fine pitch package.
- For a flexible bridge, the RDL comprises the conductor layer and the polyimide dielectric layer.
- For 5G millimeter wave high-frequency applications, it is recommended to replace the polyimide with the liquid crystal polymer (LCP), i.e., an LCP-flexible bridge.
- The challenges of various bridges have been provided.
- At present, 2.5D and 3D IC integrations with TSV-interposers are the key advanced semiconductor packaging technologies for HPC, IoTs, autonomous vehicles, big data and instant data applications driven by AI and 5G/6G.
- The manufacturing yield loss of the ever-increasing size of the TSV-interposer is so high that the cost becomes unbearable.
- The TSV-interposer is replaced by the Si-bridge (with RDLs) embedded in the cavity of a build-up packaging substrate, such as the EMIB by Intel.
- The TSV-interposer is replaced by a Si-bridge (with RDLs) that is connecting the chiplets with microbumps such as those developed by IBM, AMD, and Apple. There is no cavity on the build-up package substrate.
- The microbumps discussed in the previous item can be eliminated by using Cu-Cu bump-less hybrid bonding.
- The TSV-interposer is replaced by a Si-bridge (with RDLs and w/o TSVs) embedded in EMC with fan-out RDLs such as those developed by TSMC (CoWoS-L) and SPIL (FO-EB-T).
- Bridge standards are desperately needed. UCIe is the right way to go.

References

1. Lau, J.H. 2023. *Chiplet Design and Heterogeneous Integration Packaging*. New York: Springer.
2. Lau, J.H. 2021. *Semiconductor Advanced Packaging*. New York: Springer.
3. Lau, J.H. 2019. *Heterogeneous Integration*. New York: Springer.
4. Lau, J.H. 2018. *Fan-Out Wafer-Level Packaging*. New York: Springer.
5. Lau, J.H., C. Lin, N. Liu, H. Yang, T. Xia, C. Ko, B. Lin, Y. Chuang, R. Chen, M. Ma, T. Tseng, M. Li, and K. Leung. 2023. Hybrid Substrate with Ultra-Large Organic Interposer for Heterogeneous Integration. *IEEE Transactions on CPMT* 13 (9): 1371–1379.
6. Lau, J. H., G. Chen, C. Yang, V. Teng, A. Peng, J. Huang, N. Liu, Y. Chen, T. Tseng, and M. Li. 2023. Hybrid substrate for heterogeneous integration. *ASME Transactions, Journal of Electronic Packaging*, published online: August 11, 2023.
7. Lau, J. H. 2023. Recent advances and trends in chiplet design and heterogeneous integration packaging. *ASME Transactions, Journal of Electronic Packaging*, published online: June 23, 2023.

8. Lau, J.H. 2023. Recent advances and trends in Cu-Cu hybrid bonding. *IEEE Transactions on CPMT* 13 (3): 399–425.
9. Lau, J.H. 2023. Recent advances and trends in multiple system and heterogeneous integration with TSV-Interposers. *IEEE Transactions on CPMT* 13 (1): 3–25.
10. Lau, J. H., G. Chen, C. Yang, A. Peng, J. Huang, C. Peng, C. Ko, H. Yang, Y. Chen, and T. Tseng. 2022. Hybrid substrate for chiplet design and heterogeneous integration packaging. *Proceedings of IEEE/IEDM*. pp. 3.5.1–3.5.4.
11. Lau, J.H. 2022. Recent advances and trends in multiple system and heterogeneous integration with TSV-less interposers. *IEEE Transactions on CPMT* 12 (9): 1271–1281.
12. Lau, J.H. 2022. Recent advances and trends in advanced packaging. *IEEE Transactions on CPMT* 12 (2): 228–252.
13. Peng, C., J.H. Lau, C. Ko, P. Lee, E. Lin, K. Yang, P. Lin, T. Xia, L. Chang, N. Liu, C. Lin, T. Lee, J. Wang, M. Ma, and T. Tseng. 2022. High-Density hybrid substrate for heterogeneous integration. *IEEE Transactions on CPMT* 12 (3): 469–478.
14. Lau, J.H., G. Chen, J. Huang, R. Chou, C. Yang, N. Liu, and T. Tseng. 2021. Hybrid substrate by Fan-Out RDL-First Panel-Level packaging. *IEEE Transactions on CPMT* 11 (8): 1301–1309.
15. Lau, J.H., C. Ko, C. Lin, T. Tseng, K. Yang, T. Xia, P. Lin, C. Peng, E. Lin, L. Chang, N. Liu, S. Chiu, and Z. Lee. 2021. Fan-Out Panel-Level packaging of Mini-LED RGB display. *IEEE Transactions on CPMT* 11 (5): 739–747.
16. Lau, J.H., C. Ko, K. Yang, C. Peng, T. Xia, P. Lin, J. Chen, P. Huang, H. Liu, T. Tseng, E. Lin, and L. Chang. 2020. Panel-Level Fan-Out RDL-first packaging for heterogeneous integration. *IEEE Transactions on CPMT* 10 (7): 1125–1137.
17. Lau, J.H., C. Ko, T. Tseng, K. Yang, C. Peng, T. Xia, P. Lin, E. Lin, L. Chang, H. Liu, and D. Cheng. 2020. Panel-Level Chip-Scale package with multiple diced wafers. *IEEE Transactions on CPMT* 10 (7): 1110–1124.
18. Lau, J.H. 2019. Recent advances and trends in Fan-Out Wafer/Panel-Level Packaging. *ASME Transactions, Journal of Electronic Packaging* 141: 1–27.
19. Lau, J.H., M. Li, M. Li, T. Chen, I. Xu, X. Qing, Z. Cheng, N. Fan, E. Kuah, Z. Li, K. Tan, Y. Cheung, E. Ng, P. Lo, K. Wu, J. Hao, S. Koh, R. Jiang, X. Cao, R. Beica, S. Lim, N. Lee, C. Ko, H. Yang, Y. Chen, M. Tao, J. Lo, and R. Lee. 2018. Fan-out wafer-level packaging for heterogeneous integration. *IEEE Transactions on CPMT* 8 (9): 1544–1560.
20. Lau, J.H., M. Li, Q. Li, I. Xu, T. Chen, Z. Li, K. Tan, X. Qing, C. Zhang, K. Wee, R. Beica, C. Ko, S. Lim, N. Fan, E. Kuah, K. Wu, Y. Cheung, E. Ng, X. Cao, J. Ran, H. Yang, Y. Chen, N. Lee, M. Tao, J. Lo, and R. Lee. 2018. Design, materials, process, and fabrication of Fan-Out Wafer-Level packaging. *IEEE Transactions on CPMT.* 8 (6): 991–1002.
21. Ko, CT, H. Yang, Lau, J. H., M. Li, M. Li, C. Lin, J.W. Lin, T. Chen, I. Xu, C. Chang, J. Pan, H. Wu, Q. Yong, N. Fan, E. Kuah, Z. Li, K. Tan, Y. Cheung, E. Ng, K. Wu, J. Hao, R. Beica, M. Lin, Y. Chen, Z. Cheng, S. Koh, R. Jiang, X. Cao, S. Lim, N. Lee, M. Tao, J. Lo, and R. Lee. 2018. Chip-First Fan-Out Panel Level Packaging for Heterogeneous Integration. *IEEE Transactions on CPMT*, 8(9) September 2018, pp. 1561–1572.
22. Wuu, J., R. Agarwal, M. Ciraula, C. Dietz, B. Johnson, D. Johnson, R. Schreiber, R. Swaminathan, W. Walker, and S. Naffziger. 2022. 3D V-CacheTM: The implementation of a Hybrid-Bonded 64MB Stacked Cache for a 7nm x86–64 CPU. *Proceedings of IEEE/ISSCC*, pp. 1–36.
23. Agarwal, R., P. Cheng, P. Shah, B. Wilkerson, R. Swaminathan, J. Wuu, and C. Mandalapu. 2022. 3D Packaging for Heterogeneous Integration. *IEEE/ECTC Proceedings*, pp. 1103–1107.
24. Swaminathan, R. 2022. The next frontier: Enabling Moore's Law using heterogeneous integration. *Chip Scale Review*, pp. 11–22.
25. Su, L. 2021. AMD accelerating—The high-performance computing ecosystem. Keynote at Computex, Taipei, Taiwan.
26. Swaminathan, R. 2021. Advanced packaging: enabling moore's Law's next frontier through heterogeneous integration. Proceedings of IEEE Hot Chip Conference, pp. 1–25.
27. Naffziger, S., K. Lepak, M. Paraschour, and M. Subramony 2020. AMD Chiplet Architecture for High-Performance Server and Desktop Products. *IEEE/ISSCC Proceedings*, pp. 44–45.

28. Naffziger, S. 2020. Chiplet meets the real world: benefits and limits of chiplet designs. *Symposia on VLSI Technology and Circuits*. pp. 1–39.
29. Stow, D., Y. Xie, T. Siddiqua, and G. Loh. 2017. Cost-Effective design of scalable High-Performance systems using active and passive interposers. *IEEE/ICCAD Proceedings*, pp. 1–8.
30. Gomes, W., A. Koker, P. Stover, D. Ingerly, S.Siers, S. Venkataraman, C. Pelto, T. Shah, A. Rao, F. O'Mahony, E. Karl, L. Cheney, I. Rajwani, H. Jain, R. Cortez, A. Chandrasekhar, B. Kanthi, and R. Koduri. 2022. Ponte vecchio: a Multi-Tile 3D stacked processor for exascale computing. *Proceedings of IEEE/ISSCC*. pp. 42–44.
31. Sheikh, F., R. Nagisetty, T. Karnik, and D. Kehlet. 2021. 2.5D and 3D Heterogeneous Integration—Emerging applications. *IEEE Solid-State Circuits Magazine*. Fall pp. 77–87.
32. Prasad, C., S. Chugh, H. Greve, I. Ho, E. Kabir, C. Lin, M. Maksud, S. Novak, B. Orr, K. Park, A. Schmitz, Z. Zhang, P. Bai, D. Ingerly, E. Armagan, H. Wu, P. Stover, L. Hibbeler, M. O'Day, and D. Pantuso. 2020. Silicon reliability characterization of intel's foveros 3D integration technology for Logic-on-Logic Die stacking. *Proceedings of IEEE International Reliability Physics Symposium*. pp. 1–5.
33. Wade, M., E. Anderson, S. Ardalan, P. Bhargava, S. Buchbinder, M. Davenport, J. Fini, H. Lu, C. Li, R. Meade, C. Ramamurthy, M. Rust, F. Sedgwick, V. Stojanovic, D. Van Orden, C. Zhang, C. Sun, S. Shumarayev, C. O'Keeffe, T. Hoang, D. Kehlet, R. Mahajan, M. Guzy, A. Chan, and T. Tran. 2020. *TeraPHY: A Chiplet Technology for LowPower, High-Bandwidth In-Package Optical I/O*, 63–71. March/April: IEEE Computer Society.
34. Abdennadher, S. 2021. Testing Inter-Chiplet communication interconnects in a disaggregated SoC design. *Proceedings of IEEE/DTS*, pp. 1–7.
35. Intel Architecture Day, Intel, Santa Clara, CA, USA, Aug. 2020.
36. Gelsinger, P. 2021. Engineering the future. Intel Unleashed Webcast. Santa Clara, CA, USA: Intel.
37. Ingerly, D., S. Amin, L. Aryasomayajula, A. Balankutty, D. Borst, A. Chandra, K. Cheemalapati, C. Cook, R. Criss, K. Enamul1, W. Gomes, D. Jones, K. Kolluru, A. Kandas, G. Kim, H. Ma, D. Pantuso, C. Petersburg, M. Phen-givoni, A. Pillai, A. Sairam, P. Shekhar, P. Sinha, P. Stover, A. Telang, and Z. Zell. 2019. Foveros: 3D integration and the use of Face-to-Face chip stacking for logic devices. *IEEE/IEDM Proceedings*, pp. 19.6.1–19.6.4.
38. Gomes, W., S. Khushu, D. Ingerly, P. Stover, N. Chowdhury, F. O'Mahony, etc. 2020. Lakefield and mobility computer: A 3D stacked 10nm and 2FFL hybrid processor system in $12 \times 12\text{mm}^2$, 1mm Package-on-Package. *IEEE/ISSCC Proceedings*, pp. 40–41.
39. Mahajan, R. and S. Sane. 2021. Advanced packaging technologies for heterogeneous integration. *Proceedings of IEEE Hot Chip Conference*, pp. 1–44.
40. WikiChip. 2020. A look at intel lakefield: A 3D-Stacked Single-ISA heterogeneous Penta-Core SoC. https://en.wikichip.org/wiki/chiplet.
41. Liang, S., G. Wu, K. Yee, C. Wang, J. Cui, and D. Yu. 2022. High performance and energy efficient computing with advanced SoIC^{TM} Scaling. *Proceedings of IEEE/ECTC*. pp. 1090–1094.
42. Chiang, Y., S. Tai, W. Wu, J. Yeh, C. Wang, and D. Yu. 2021. InFO_oS (Integrated Fan-Out on Substrate) technology for advanced chiplet integration. *Proceedings of IEEE/ECTC*, pp. 130–135.
43. Huang, P., C. Lu, W. Wei, C. Chiu, K. Ting, C. Hu, C. Tsai, and S. Hou. 2021. Wafer level system integration of the fifth generation CoWoS®-S with high performance Si Interposer at 2500 mm^2. *Proceedings of IEEE/ECTC*. pp. 101–104.
44. Wu, J., C. Chen, C. Lee, C. Liu, and D. Yu. 2021. SoIC-An ultra large size integrated substrate technology platform for HPC applications. *Proceedings of IEEE/ECTC*. pp. 28–33.
45. Chen, M. F., C. S. Lin, E. B. Liao, W. C. Chiou, C. C. Kuo, C. C. Hu, C. H. Tsai, C. T. Wang and D. Yu. 2020. SoIC for Low-Temperature, Multi-Layer 3D memory integration. *IEEE/ECTC Proceedings*. pp. 855–860.
46. Chen, Y. H., C. A. Yang, C. C. Kuo, M. F. Chen, C. H. Tung, W. C. Chiou, and D. Yu. Ultra high density SoIC with Sub-micron Bond Pitch. *IEEE/ECTC Proceedings*. pp. 576–581.

47. Chen, F., M. Chen, W. Chiou, D. Yu. 2019. System on Integrated Chips (SoIC^{TM}) for 3D heterogeneous integration. *IEEE/ECTC Proceedings*. pp. 594–599.
48. Moore, S. 2022. Graphcore uses TSMC 3D Chip Tech to Speed AI by 40%. *IEEE Spectrum*.
49. Coudrain, P., J. Charbonnier, A. Garnier, P. Vivet, R. Vélard, A. Vinci, F. Ponthenier, A. Farcy, R. Segaud, P. Chausse, L. Arnaud, D. Lattard, E. Guthmuller, G. Romano, A. Gueugnot, F. Berger, J. Beltritti, T. Mourier, M. Gottardi, S. Minoret, C. Ribière, G. Romero, P.-E. Philip, Y. Exbrayat, D. Scevola, D. Campos, M. Argoud, N. Allouti, R. Eleouet, C. Fuguet Tortolero, C. Aumont, D. Dutoit, C. Legalland, J. Michailos, S. Chéramy, and G. Simon. 2019. Active interposer technology for chiplet-based advanced 3D system architectures. *Proceedings of IEEE/ECTC*, pp. 569–578.
50. Vivet, P., E. Guthmuller, Y. Thonnart, G. Pillonnet, C. Fuguet, I. Miro-Panades, G. Moritz, J. Durupt, C. Bernard, D. Varreau, J. Pontes, S. Thuries, D. Coriat, M. Harrand, D. Dutoit, D. Lattard, L. Arnaud, J. Charbonnier, P. Coudrain, A. Garnier, F. Berger, A. Gueugnot, A. Greiner, Q. Meunier, A. Farcy, A. Arriordaz, S. Cheramy, and F. Clermidy. 2021. IntAct: A 96-Core processor with six chiplets 3D-Stacked on an active interposer with distributed interconnects and integrated power management. *IEEE Journal of Solid-State Circuits* 56 (1): 79–97.
51. Shimizu, N., W. Kaneda, H. Arisaka, N. Koizumi, S. Sunohara, A. Rokugawa, and T. Koyama. 2013. Development of organic multi chip package for high performance application. *IMAPS Proceedings of International Symposium on Microelectronics*, pp. 414–419.
52. Oi, K., S. Otake, N. Shimizu, S. Watanabe, Y. Kunimoto, T. Kurihara, T. Koyama, M. Tanaka, L. Aryasomayajula, and Z. Kutlu. 2014. Development of new 2.5D package with novel integrated organic interposer substrate with Ultra-fine Wiring and High-Density bumps. *Proceedings of IEEE/ECTC*. pp. 348–353.
53. Uematsu, Y., N. Ushifusa, and H. Onozeki. 2017. Electrical transmission properties of HBM interface on 2.1-D system in package using organic interposer. *Proceedings of IEEE/ECTC*, pp. 1943–1949.
54. Chen, W., C. Lee, M. Chung, C. Wang, S. Huang, Y. Liao, H. Kuo, C. Wang, and D. Tarng. 2018. Development of novel fine line 2.1 D package with organic interposer using advanced substrate-based process. *IEEE/ECTC Proceedings*. pp. 601–606.
55. Huang, C., Y. Xu, Y. Lu, K. Yu, W. Tsai, C. Lin, C. Chung. 2018. Analysis of warpage and stress behavior in a fine Pitch Multi-Chip interconnection with Ultrafine-Line organic substrate (2.1D). *IEEE/ECTC Proceedings*, pp. 631–637.
56. Islam, N., S. Yoon, K. Tan, and T. Chen. 2019. High density Ultra-Thin organic substrate for advanced flip chip packages. *IEEE/ECTC Proceedings*. pp. 325–329.
57. Kumazawa, Y., S. Shika, S. Katagiri, T. Suzuki, T. Kida, S. Yoshida 2019. Development of novel photosensitive dielectric material for reliable 2.1D package. *Proceedings of IEEE/ECTC*, pp. 1009–1004.
58. Katagiri, S., S. Shika, Y. Kumazawa, K. Shimura, T. Suzuki, T. Kida, S. Yoshida. 2020. Novel photosensitive dielectric material with superior electric insulation and warpage suppression for organic interposers in reliable 2.1D Package. *Proceedings of IEEE/ECTC*. pp. 912–917.
59. Mori, H., and S. Kohara. 2021. Copper Content optimization for warpage minimization of substrates with an asymmetric Cross-Section by Genetic Algorithm. *Proceedings of IEEE/ECTC*, pp. 1521–1526.
60. Yoon, S., P. Tang, R. Emigh, Y. Lin, P. Marimuthu, and R. Pendse. 2013. Fanout flipchip eWLB (Embedded Wafer Level Ball Grid Array) technology as 2.5D packaging solutions. *IEEE/ECTC Proceedings*, pp. 1855–1860.
61. Chen, N. C., T. Hsieh, J. Jinn, P. Chang, F. Huang, J. Xiao, A. Chou, B. Lin. 2016. A novel system in package with Fan-out WLP for high speed SERDES application. *IEEE/ECTC Proceedings*, pp. 1496–1501.
62. Yip, L., R. Lin, C. Lai, and C. Peng. 2022. Reliability challenges of High-Density Fan-out packaging for High-Performance computing applications. *IEEE/ECTC Proceedings*, pp. 1454–1458.

63. Lin, Y., W. Lai, C. Kao, J. Lou, P. Yang, C. Wang, and C. Hseih. 2016. Wafer warpage experiments and simulation for fan-out chip on substrate. *IEEE/ECTC Proceedings*, pp. 13–18.
64. Yu, D. 2018. Advanced system integration technology trends. *SiP Global Summit*, SEMICON Taiwan.
65. Kurita, Y., T. Kimura, K. Shibuya, H. Kobayashi, F. Kawashiro, N. Motohashi, and M. Kawano. 2010. Fan-out wafer level packaging with highly flexible design capabilities. *IEEE/ESTC Proceedings*, pp. 1–6.
66. Motohashi, N., T. Kimura, K. Mineo, Y. Yamada, T. Nishiyama, K. Shibuya, H. Kobayashi, Y. Krita, and M. Kawano. 2011. System in a Wafer-level package technology with RDL-first process. *IEEE/ECTC Proceedings*. pp. 59–64.
67. Huemoeller, R., and C. Zwenger. 2015. Silicon wafer integrated fan-out technology. *Chip Scale Review*. pp. 34–37.
68. Lim, H., J. Yang, and R. Fuentes. 2018. Practical design method to reduce crosstalk for silicon wafer integration Fan-Out technology (SWIFT) packages. *IEEE/ECTC Proceedings*. pp. 2205–2211.
69. Jayaraman. S. 2022. Advanced packaging: HDFO for next generation devices. *Proceedings of IWLPC*, pp. 1–28.
70. Suk, K., S. Lee, J. Kim, S. Lee, H. Kim, S. Lee, P. Kim, D. Kim, D. Oh, and J. Byun. 2018. Low-Cost Si-less RDL interposer package for high performance computing applications. *IEEE/ECTC Proceedings*, pp. 64–69.
71. You, S., S. Jeon, D. Oh, K. Kim, J. Kim, S. Cha, and G. Kim. 2018. Advanced Fan-Out Package SI/PI/Thermal performance analysis of novel RDL packages. *IEEE/ECTC Proceedings*, pp. 1295–1301.
72. Lin, Y., M. Yew, M. Liu, S. Chen, T. Lai, P. Kavle, C. Lin, T. Fang, C. Chen, C. Yu, K. Lee, C. Hsu, P. Lin, F. Hsu, and S. Jeng. 2019. Multilayer RDL interposer for heterogeneous device and module integration. *IEEE/ECTC Proceedings*, pp. 931–936.
73. Lin, P., M. Yew, S. Yeh, S. Chen, C. Lin, C. Chen, C. Hsieh, Y. Lu, P. Chuang, H. Cheng, and S. Jeng. 2021. Reliability performance of advanced organic interposer (CoWoS-R) packages. *Proceedings of IEEE/ECTC*, pp. 723–728.
74. Lin, M., M. Liu, H. Chen, S. Chen, M. Yew, C. Chen, and S. Jeng 2022. Organic interposer CoWoS-R (plus) technology. *Proceedings of IEEE/ECTC*, pp. 1–6.
75. Chang, K., C. Huang, H. Kuo, M. Jhong, T. Hsieh, M. Hung, and C. Wang. 2019. Ultra High-Density IO Fan-Out design optimization with signal integrity and power integrity. *IEEE/ECTC Proceedings*, pp. 41–46.
76. Lai, W., P. Yang, I. Hu, T. Liao, K. Chen, D. Tarng, and C. Hung. 2020. A comparative study of 2.5D and Fan-out chip on substrate: chip first and chip last. *IEEE/ECTC Proceedings*, pp. 354–360.
77. Fang, J., M. Huang, H. Tu, W. Lu, and P. Yang. 2020 A Production-worthy Fan-Out Solution – ASE FOCoS Chip Last. *IEEE/ECTC Proceedings*, pp. 290–295.
78. Cao, L. 2022. Advanced FOCOS (Fanout Chip on Substrate) technology for chiplets heterogeneous integration. *Proceedings of IWLPC*, pp. 1–6.
79. Cao, L., T. Lee, R. Chen, Y. Chang, H. Lu, N. Chao, Y. Huang, C. Wang, C. Huang, H. Kuo, Y. Wu, and H. Cheng. 2022. Advanced fanout packaging technology for hybrid substrate integration. *Proceedings of IEEE/ECTC*, pp. 1362–1370.
80. Lee, T., S. Yang, H. Wu, and Y. Lin. 2022. Chip Last Fanout Chip on Substrate (FOCoS) Solution for Chiplets Integration. *Proceedings of IEEE/ECTC*, pp. 1970–1974.
81. Yin, W., W. Lai, Y. Lu, K. Chen, H. Huang, T. Chen, C. Kao, and C. Hung. 2022. Mechanical and thermal characterization analysis of Chip-last Fan-out Chip on substrate. *Proceedings of IEEE/ECTC*. pp. 1711–1719.
82. Li, J., F. Tsai, J. Li, G. Pan, M. Chan, L. Zheng, S. Chen, N. Kao, D. Lai, K. Wan, and Y. Wang. 2021. Large size multilayered Fan-Out RDL packaging for heterogeneous integration. *IEEE/EPTC Proceedings*. pp. 239–243.
83. Miki, S., H. Taneda, N. Kobayashi, K. Oi, K. Nagai, and T. Koyama. 2019. Development of 2.3D high density organic package using low temperature bonding process with Sn-Bi Solder. *IEEE/ECTC Proceedings*. pp. 1599–1604.

84. Murayama, K., S. Miki, H. Sugahara, and K. Oi. 2020. Electro-migration evaluation between organic interposer and build-up substrate on 2.3D organic package. *IEEE/ECTC Proceedings*, pp. 716–722.
85. Kim, J., J. Choi, S. Kim, J. Choi, Y. Park, G. Kim, S. Kim, S. Park, H. Oh, S. Lee, T. Cho, and D. Kim. 2021. Cost effective 2.3D packaging solution by using fanout panel level RDL. *IEEE/ECTC Proceedings*, pp. 310–314.
86. Souriau, J., O. Lignier, M. Charrier, and G. Poupon. 2005. Wafer level processing Of 3D system in package for RF and data applications. *IEEE/ECTC Proceedings*, pp. 356–361.
87. Henry, D., D. Belhachemi, J-C. Souriau, C. Brunet-Manquat, C. Puget, G. Ponthenier, J. Vallejo, C. Lecouvey, and N. Sillon. 2006. Low electrical resistance silicon through vias: technology and characterization. *IEEE/ECTC Proceedings*, pp. 1360–1366.
88. Selvanayagam, C., J.H. Lau, X. Zhang, S. Seah, K. Vaidyanathan, and T. Chai. 2009. Nonlinear thermal stress/strain analyses of copper filled TSV (Through Silicon Via) and Their Flip-Chip Microbumps. *IEEE Transactions on Advanced Packaging* 32 (4): 720–728.
89. Tang, G.Y., S. Tan, N. Khan, D. Pinjala, J.H. Lau, A. Yu, V. Kripesh, and K. Toh. 2010. Integrated liquid cooling systems for 3-D stacked TSV modules. *IEEE Transactions on CPMT* 33 (1): 184–195.
90. Khan, N., H. Li, S. Tan, S. Ho, V. Kripesh, D. Pinjala, J.H. Lau, and T. Chuan. 2013. 3-D Packaging With Through-Silicon Via (TSV) for electrical and fluidic interconnections. *IEEE Transactions on CPMT* 3 (2): 221–228.
91. Khan, N., V. Rao, S. Lim, H. We, V. Lee, X. Zhang, E. Liao, R. Nagarajan, T.C. Chai, V. Kripesh, and J.H. Lau. 2010. Development of 3-D silicon module With TSV for system in packaging. *IEEE Transactions on CPMT* 33 (1): 3–9.
92. Lau, J.H., and G.Y. Tang. 2012. Effects of TSVs (through-silicon vias) on thermal performances of 3D IC integration system-in-package (SiP). *Journal of Microelectronics Reliability, Vo.* 52 (11): 2660–2669.
93. Lau, J.H., C. Lee, C. Zhan, S. Wu, Y. Chao, M. Dai, R. Tain, H. Chien, J. Hung, C. Chien, R. Cheng, Y. Huang, Y. Lee, Z. Hsiao, W. Tsai, P. Chang, H. Fu, Y. Cheng, L. Liao, W. Lo, and M. Kao. 2014. Low-Cost Through-Silicon hole interposers for 3D IC integration. *IEEE Transactions on CPMT* 4 (9): 1407–1419.
94. Banijamali, B., C. Chiu, C. Hsieh, T. Lin, C. Hu, S. Hou, et al. 2013. Reliability evaluation of a CoWoS-enabled 3D IC package. *IEEE/ECTC Proceedings*, pp. 35–40.
95. Banijamali, B., T. Lee, H. Liu, S. Ramalingam, I. Barber, J. Chang and M. Kim, and L. Yip. 2015 Reliability evaluation of an extreme TSV interposer and interconnects for the 20nm technology CoWoS IC Package. *IEEE/ECTC Proceedings*, pp. 276–280.
96. Lau, J. H. 2014. Overview and outlook of 3D IC packaging, 3D IC integration, and 3D Si Integration. *ASME Transactions, Journal of Electronic Packaging*, 136(4), pp. 1–15.
97. Lau, J.H. 2011. Overview and Outlook of TSV and 3D Integrations. *Journal of Microelectronics International* 28 (2): 8–22.
98. Zhang, X., J. Lin, S. Wickramanayaka, S. Zhang, R. Weerasekera, R. Dutta, K. Chang, K. Chui, H. Li, D. Ho, L. Ding, G. Katti, S. Bhattacharya, and D. Kwong. 2015. Heterogeneous 2.5D Integration on through silicon interposer. *Applied Physics Reviews*, 2, pp. 021308 1–56.
99. Hou, S., W. Chen, C. Hu, C. Chiu, K. Ting, T. Lin, W. Wei, W. Chiou, V. Lin, V. Chang, C. Wang, C. Wu, and D. Yu. 2017. Wafer-Level integration of an advanced Logic-Memory system through the Second-Generation CoWoS technology. *IEEE Transactions on Electron Devices*, pp. 4071–4077.
100. Hsieh, M.C., S.T. Wu, C.J. Wu, and J.H. Lau. 2014. Energy release rate estimation for through silicon vias in 3-D integration. *IEEE Transactions on CPMT* 4 (1): 57–65.
101. http://press.xilinx.com/2013-10-20-Xilinx-and-TSMCReach-VolumeProduction-on-all-28nm-CoWoS-based-All-Programmable-3D-IC-Families.
102. Chiu, C., Z. Qian, and M. Manusharow 2014. Bridge interconnect with air gap in package assembly. US Patent No. 8,872,349.
103. Mahajan, R., R. Sankman, N. Patel, D. Kim, K. Aygun, Z. Qian, et al., "Embedded multi-die interconnect bridge (EMIB) – a high-density, high-bandwidth packaging interconnect," *IEEE/ECTC Proc.*, May 2016, pp. 557–565.

104. Duan, G., Y. Knaoka, and R. McRee. 2021. Die embedded challenges for EMIB advanced packaging technology. *IEEE/ECTC Proc.*, pp. 1–7.
105. Lau, J. H., P. Tzeng, C. Lee, C. Zhan, M. Li, J. Cline, K. Saito, Y. Hsin, P. Chang, Y. Chang, J. Chen, S. Chen, C. Wu, H. Chang, C. Chien, C. Lin, T. Ku, R. Lo, and M. Kao 2023. Redistribution Layers (RDLs) for 2.5D/3D IC Integration. *Proceedings of the 46th IMAPS International Symposium on Microelectronics*, pp. 434–441.
106. Lau, J. H., P. Tzeng, C. Lee, C. Zhan, M. Li, J. Cline, K. Saito, Y. Hsin, P. Chang, Y. Chang, J. Chen, S. Chen, C. Wu, H. Chang, C. Chien, C. Lin, T. Ku, R. Lo, and M. Kao 2014. Redistribution Layers (RDLs) for 2.5D/3D IC Integration. *IMAPS Transactions, Journal of Microelectronic Packaging*, 11: 1, First Quarter pp. 16–24.
107. Zoschke, K., Wolf, J., Lopper, C., Kuna, I., Jürgensen, N., Glaw, V., Samulewicz, K., Röder, J., Wilke, M., Wünsch, O., Klein, M., Suchodoletz, M., Oppermann, H., Braun, T., Wieland, R., and Ehrmann, O. 2011. TSV based silicon interposer technology for wafer level fabrication of 3D SiP modules. *Proceedings of IEEE/ECTC*, Orlando, Florida, pp. 836–842.
108. Sikka, K., R. Bonam, Y. Liu, P. Andry, D. Parekh, A. Jain, M. Bergendahl, et al. 2021. Direct bonded heterogeneous integration (DBHi) Si Bridge. *IEEE/ECTC Proceedings*, pp. 136–147.
109. Matsumoto, K., M. Bergendahl, K. Sikka, S. Kohara, H. Mori, and T. Hisada. 2021. Thermal Analysis of DBHi (Direct Bonded Heterogeneous Integration) Si Bridge. *Proceedings of IEEE/ECTC*, pp. 1382–1390.
110. Jain, A., K. Sikka, J. Gomez, D. Parekh, M. Bergendahl, J. Borkulo, K. Biesheuvel, R. Doll, and M. Mueller. 2021. Laser Versus Blade dicing for direct bonded heterogeneous integration (DBHi) Si Bridge. *Proceedings of IEEE/ECTC*, pp. 1125–1130.
111. Marushima, C., T. Aoki, K. Nakamura, R. Miyazawa, A. Horibe, I. Sousa, and K. Sikka, and T. Hisada. 2022. Dimensional parameters controlling capillary underfill flow for Void-Free encapsulation of a direct bonded heterogeneous integration (DBHi) Si-bridge Package. *Proceedings of IEEE/ECTC*, pp. 586–590.
112. Horibe, A., T. Watanabe, C. Marushima, H. Mori, S. Kohara, R. Yu, M. Bergendahl, T. Magbitang, R. Wojtecki, D. Taneja, M. Godard, C. Cristina, B. Pulido, I. Sousa, K. Sikka, and T. Hisada. 2022. Characterization of Non-Conductive paste materials (NCP) for thermocompression bonding in a direct bonded heterogeneoussly integrated (DBHi) Si-Bridge Package. *Proceedings of IEEE/ECTC*, pp. 625–630.
113. Horbe, A., C. Marushima, T. Watanabe, A. Jian, E. Turcotte, I. Sousa, K. Sikka, and T. Hisada. 2022. Super fine jet underfill dispense technique for robust micro joint in direct bonded heterogeneous integration (DBHi) silicon bridge packages, *Proceedings of IEEE/ECTC*, pp. 631–634.
114. Chowehury, P., K. Sakuma, S. Raghawan, M. Bergendahl, K. Sikka, S. Kohara, T. Hisada, H. Mori, D. Taneja, and I. Sousa. 2022. Thermo-mechanical Analysis of Thermal Compression Bonding Chip-Joint Process. *Proceedings of IEEE/ECTC*, pp. 579–585.
115. Pulido, C., S. Kohara, A. Jain, T. Wassick, D. Taneja, P. McInnes, et al. 2023. Direct-bonded heterogeneous integration (DBHi): Surface bridge approach for die tiling. *Proceedings of IEEE/ECTC*, pp. 1053–1059.
116. Hsiung, C., and A. Sundarrajan. (2020). Methods and apparatus for wafer-level die bridge. US Patent No. 10,651,126.
117. Lau, J. H., C. Ko, P. Lin, T. Tseng, R. Tain, and H. Yang. 2022 Package structure and manufacturing method thereof. US Patent No. 11,410,933.
118. Weerasekera, R., S. Bhattacharys, K. Chang, and V. Rao. 2021. Semiconductor package and method of forming the same. US Patent No. 11,018,080.
119. TSMC, "TSMC's Annual Technology Symposium", Taipei, Taiwan, Aug. 25, 2020.
120. You, J., J. Li, D. Ho, J. Li, M. Zhuang, D. Lai, et al. 2021. Electrical performances of fan-out embedded bridge. *Proceedings of IEEE/ECTC*, pp. 2030–2034.
121. Lee, L., Y. Chang, S. Huang, J. On, E. Lin, O. Yang. 2021 Advanced HDFO packaging solutions for chiplets integration in HPC application. *Proceedings of IEEE/ECTC*, pp. 8–13.
122. Lee, J., G. Yong, M. Jeong, J. Jeon, D. Han, M. Lee, et al. 2021. S-connect fan-out interposer for next-gen heterogeneous integration," *Proceedings of IEEE/ECTC*, June 2021, pp. 96–100.

123. Chong, C., T. Lim, D. Ho, H. Yong, C. Choong, S. Lim, et al. 2021. Heterogeneous integration with embedded fine interconnect. *Proceedings of IEEE/ECTC*, pp. 2216- 2221.
124. Hu, Y., Y. Liang, H. Hu, C. Tan, C. Shen, C. Lee, et al. 2023. CoWoS architecture evolution for next-generation HPC on 2.5D system in package. *IEEE/ECTC Proceedings*, pp. 1022–1026.
125. Hou, S., C. Lee, T. Wang, and H. Hu. 2023. Supercarrier redistribution layers to realize ultra-large 2.5D wafer-scale packaging by CoWoS. *IEEE/ECTC Proceedings*, June 2023, pp. 510–514.
126. Su, P., D. Ho, J. Pu, and Y. Wang. 2023. FO-EB-T package solution for high-performance computing. *IEEE/ECTC Proceedings*, pp. 1038–1042.
127. Lin, V., T. Shih, A. Jang, and Y. Wang. 2023. Fan-out embedded bridge with TSV (FO-EB-T) package characterization and evaluation. *IEEE/ECTC Proceedings*, pp. 1344–1347.
128. Sharma, D., G. Pasdast, Z. Qian, and K. Aygun. 2022. Universal chiplet interconnect express (UCIe): An open industry standard for innovations with chiplets at package level. *IEEE Trans. on CPMT* 122 (9): 1423–1431.

Chapter 6
Co-packaged Optics

6.1 Introduction

Co-packaged optics (CPO) are heterogeneous integration packaging methods to integrate the optical engine (OE) which consists of photonic ICs (PIC) and the electrical engine (EE) which consists of the electronic ICs (EIC) as well as the switch ASIC (application specific IC). The advantages of CPO are: (a) to reduce the length of the electrical interface between the OE/EE (or PIC/EIC) and the ASIC, (b) to reduce the energy required to drive the signal, and (c) to cut the latency which leads to better electrical performance. In this chapter, on-board optics (OBO), near-package optics (NPO), and CPO will be discussed.

Silicon photonics are the semiconductor integration of EIC and PIC on a silicon substrate (wafer) with complementary metal-oxide semiconductor (CMOS) technology, which will be briefly mentioned first and are out of the scope of this book.

6.2 Silicon Photonics

Today, Intel is the only company who is in high volume manufacturing of silicon photonics. As shown in Fig. 6.1, after the IC is fabricated on the silicon wafer, the indium phosphide (InP), a light emitting material, is bonded to the top of the silicon wafer with a thin layer of glass glue. This is what Intel called the hybrid laser technology, in which the IC and laser are manufactured by the CMOS technology on the same silicon wafer. Some of the advantages are: the automate on-wafer optical, electrical, and high-speed test and wafer-level burn-in. A typical example of hybrid laser is shown in Fig. 6.2.

J. H. Lau, *Flip Chip, Hybrid Bonding, Fan-In, and Fan-Out Technology*,
https://doi.org/10.1007/978-981-97-2140-5_6

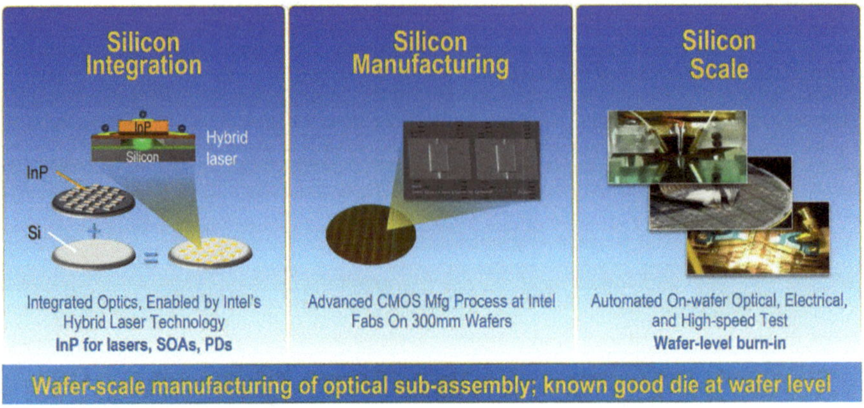

Fig. 6.1 Intel's silicon photonics

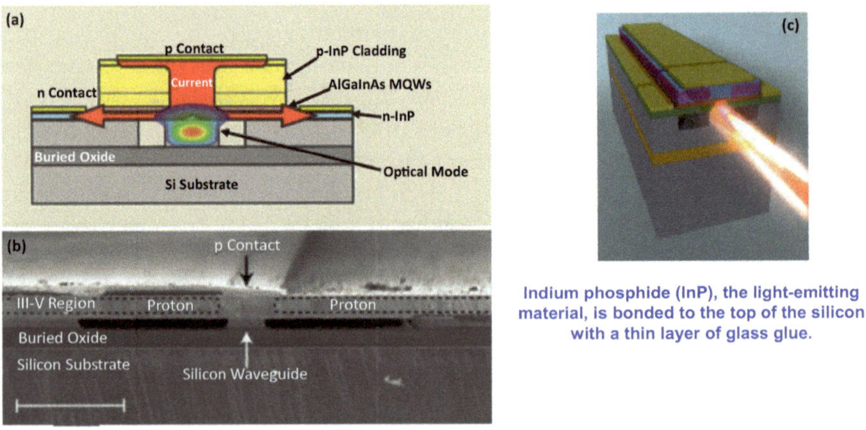

Fig. 6.2 A hybrid silicon laser. **a** Key components. **b** Scanning electron micrograph of the fabricated device. **c** Schematic

6.3 Data Centers

A data center includes switches. routers, storages, etc. such as those shown in Fig. 6.3a. Its functions include the network of computing and storage resources that enable the delivery of shared applications and data. Figure 6.3b shows some of the optical transceivers used in a data center.

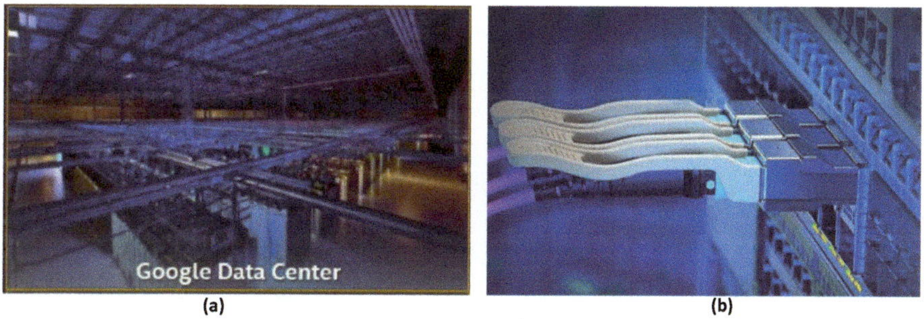

Fig. 6.3 **a** Google data center. **b** Transceivers in a data center

6.4 Optical Transceivers

Optical transceivers are the most important component in a data center as shown in Fig. 6.3b. Optical transceiver is a key component of an optical transmission system that permits coupling of the transmission medium with the active components of the chain, such as switches, routers—optical interfaces or any optical transport equipment, Fig. 6.4.

Optical transceiver is defining the process of converting electric signaling toward the optical transmission with the help of TOSA (transmission optical sub assembly) or Tx module and performing inverse action through the ROSA (receiver optical sub assembly) or Rx module, Figs. 6.4 and 6.5.

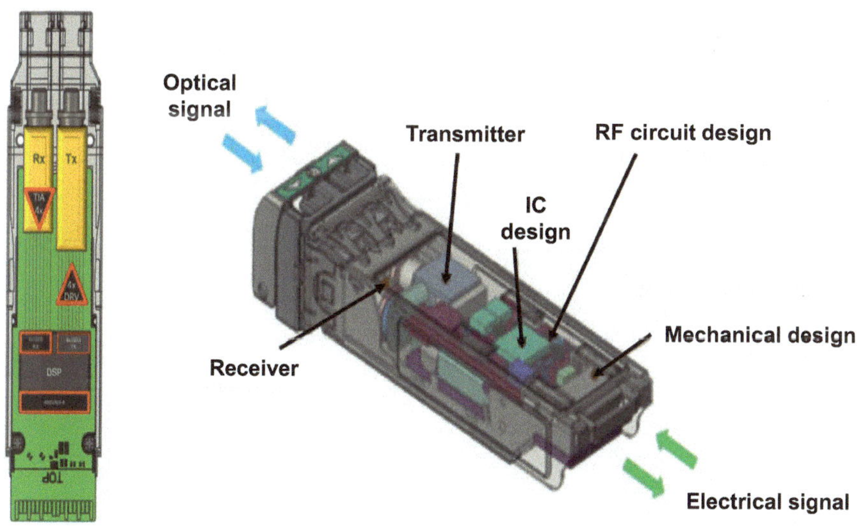

Fig. 6.4 Optical transceivers

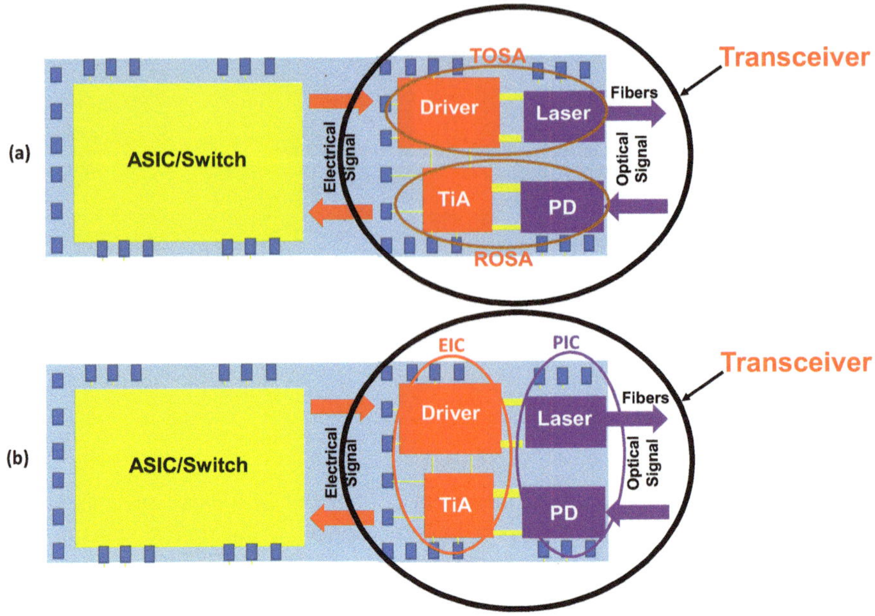

Fig. 6.5 Key components in an optical transceiver **a** TOSA and ROSA **b** PIC and EIC

A TOSA contains a semiconductor laser diode (LD) and laser driver, while a ROSA contains a photodiode (PD), optical interface such as lens, TIA (transimpedance amplifier), and passive electrical interface.

TOSA module converts the electrical signal to the optical transmission light that lands on the fiber. The ROSA is used to receive an optical signal from a fiber and convert it back into an electrical signal, Fig. 6.5. Today, the one used the most is the pluggable transceiver as shown in Figs. 6.3b and 6.6.

Fig. 6.6 Pluggable transceiver

6.5 Optical Engine (OE) and Electrical Engine (EE)

The optical engine and electrical engine are the key components of the optical transceiver for realizing photoelectric signal conversion. For example, the OE of an optical transceiver consists of the PIC such as the PD and laser, etc. as shown in Fig. 6.5b. On the other hand, the EE of an optical transceiver consists of the EIC such as the laser driver and TiA, etc. as shown in Fig. 6.5b.

6.6 OBO, NPO, and CPO

A simple roadmap of pluggable transceivers, OBO, NPO, and CPO is shown in Fig. 6.7. The gigabit interface converter (GBIC) was one of the first standards released in 1995 by the Storage Networking Industry Association (used to be called the Small Form Factor Committee). Since 2000 the small form-factor pluggable (SFP) has been used widely by industry, Fig. 6.7a. Between 2006 and 2010, there were enhanced small-form-factor pluggable (SFP +), quad small-form-factor pluggable (QSFP), and C form-factor pluggable (CFP). Between 2011 and 2015, there were enhanced quad small-form-factor pluggable (QSFP +), QSFP28, SFP28, CFP2, and CFP4. Between 2015 and now, there are the compact size SFP (CSFP), QSFP46 (doubling the data rate achieved by QSFP28), quad small-form-factor pluggable double density (QSFP-DD), octal small form factor pluggable (OSFP), and small form-factor pluggable double density (SFP-DD). It can be seen that the pluggable transceivers have been used by industry since 1995 [1–5] and are used the most today. These pluggable transceivers are mounted on the edge of a printed circuit board (PCB) with the ASIC attached to a package substrate, Fig. 6.7a. The distance from the PIC/EIC of the pluggable transceivers to the ASIC chip is the farthest and thus the power consumption and electrical performance are the worst.

Since 2018, the key components such as OE and EE (PIC/EIC) of the transceivers have been mounted on the same PCB as the packaged ASIC and arranged on the four sides of the packaged ASIC, Fig. 6.7b. This is called OBO (on-board optics), which uses the PCB to do the connection between the packaged ASIC and the OE/EE. The distance between the PIC/EIC and the ASIC is closer than the pluggable transceivers and thus there is an improvement in power and electrical performance.

Since 2020, some people placed the OE/EE on an optional optical substrate side-by-side with the packaged ASIC on the same high-performance substrate as shown in Fig. 6.7c. This is called NPO (near-package optics), which uses the high-performance substrate to do the connection between the packaged ASIC and the OE/EE. There is no high-speed data bandwidth going through the PCB. Better power and electrical performance can be achieved. According to optical internetworking forum (OIF), for NPO lengths between the ASIC and optics can be up to 150 mm with the channel loss constrained to 13 dB. This is why the OIF is developing the extra short reach $(XSR)^{+}$ electrical interface, to expand the XSR's reach for NPO.

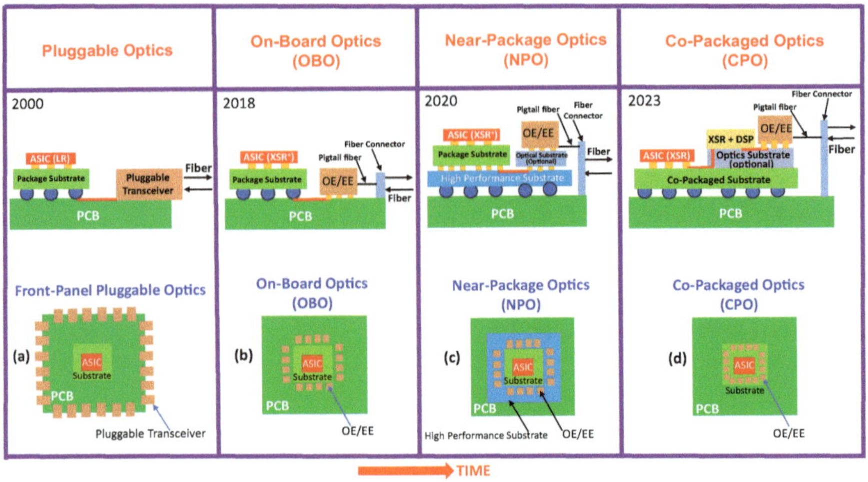

Fig. 6.7 Roadmap of pluggable transceiver, OBO, NPO, and CPO

Since 2023, after the shipment of CPO (co-packaged optics) by Intel (Fig. 6.8) and Broadcom (Fig. 6.9), people are very interested in CPO [6–31]. In CPO, the OE/EE w/o the optics substrate are placed side-by-side along the four edges of the ASIC chip on the same co-packaged substrate, Fig. 6.7d. (It should be emphasis that the optics substrate for the OE/EE is optional.) In this case, the distance between the ASIC and the PIC/EIC is the shortest and thus with the best electrical performance. In short, NPO brings the PIC/EIC closer to the ASIC package on a high-performance substrate, while CPO brings the PIC/EIC and the ASIC chip side-by-side on a co-packaged substrate of the same package. CPO offers superior formfactor, power consumption, and latency benefits compared to NPO. According to OIF, CPO confines the modules OE/EE and host ASIC to 50 mm of each other. The channel loss here is limited to 10 dB. CPO has a lower power consumption because of the shorter spans and 3 dB saving.

6.7 Integration of the PIC and EIC

As mentioned earlier silicon photonic integrates the PIC and EIC on a silicon wafer by CMOS technology which is out of the scope of this book. In this book, the 2D and 3D heterogeneous integration of ASIC, PIC and EIC will be discussed.

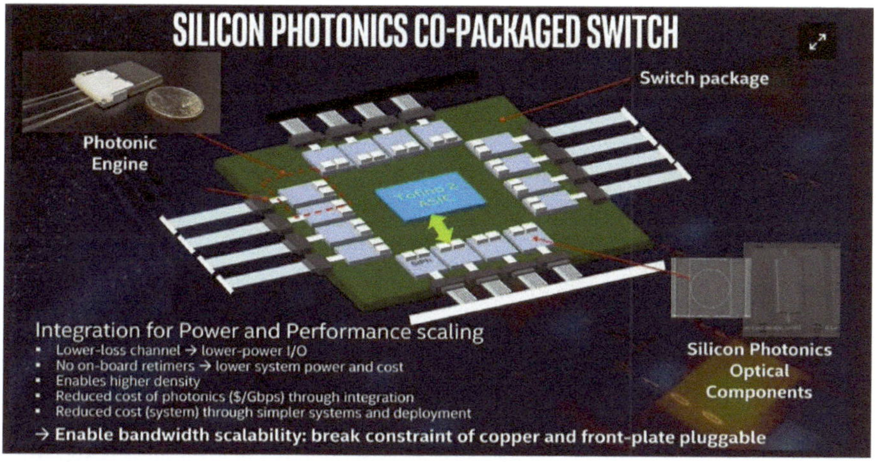

Fig. 6.8 Intel's co-packaged optics switch

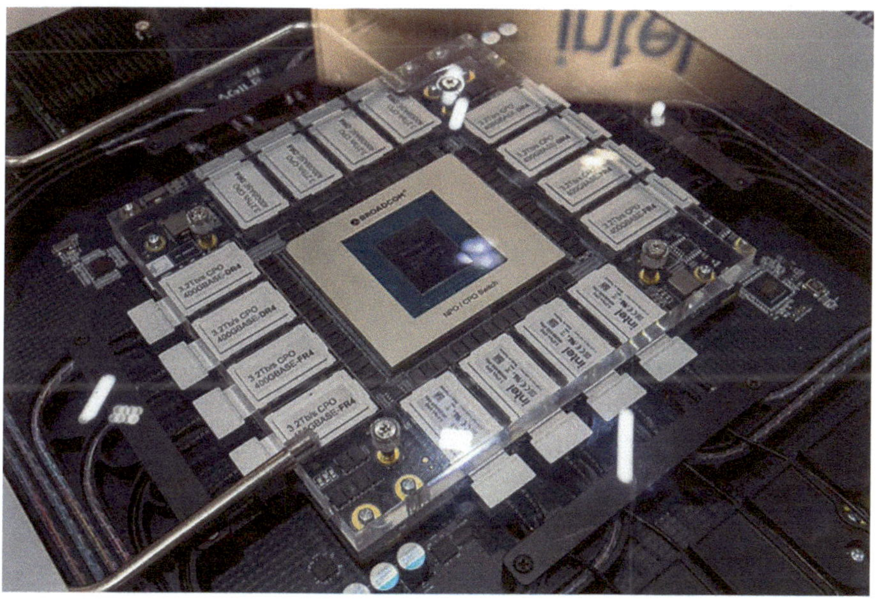

Fig. 6.9 Broadcom's co-packaged optics switch

6.8 2D Heterogeneous Integration of PIC and EIC

Figure 6.10 schematically shows the 2D heterogeneous integration of EIC and PIC on an optical substrate. It can be seen that the EIC and PIC are side-by-side attached to the optional optical substrate with μbumps.

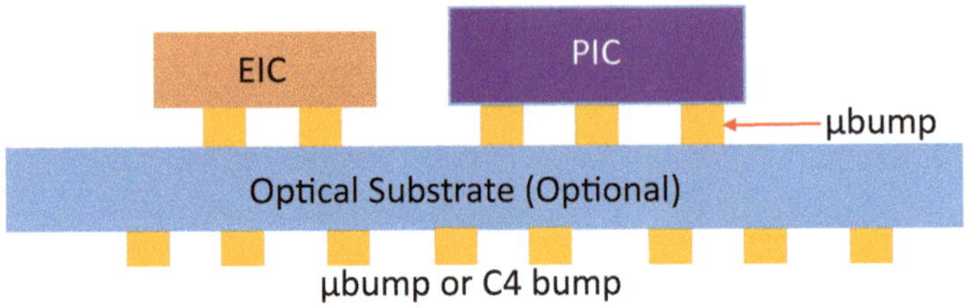

Fig. 6.10 2D heterogeneous integration of EIC and PIC

6.9 2D Heterogeneous Integration of ASIC Switch, PIC and EIC

Figure 6.11 shows various 2D methods of heterogeneous integration of ASIC, PIC, and EIC. Figure 6.11a shows the 2D heterogeneous integration of PIC and EIC on an optional optical substrate side-by-side with the ASIC chip on the same co-packaged substrate with μbumps or C4 (controlled collapse of chip connection) bumps. Then, the co-packaged substrate is attached to a PCB with BGA (ball grid array) [32] solder balls.

Figure 6.11b shows the 2D heterogeneous integration of PIC and EIC on an optional optical substrate side-by-side with the ASIC chip on the same co-packaged

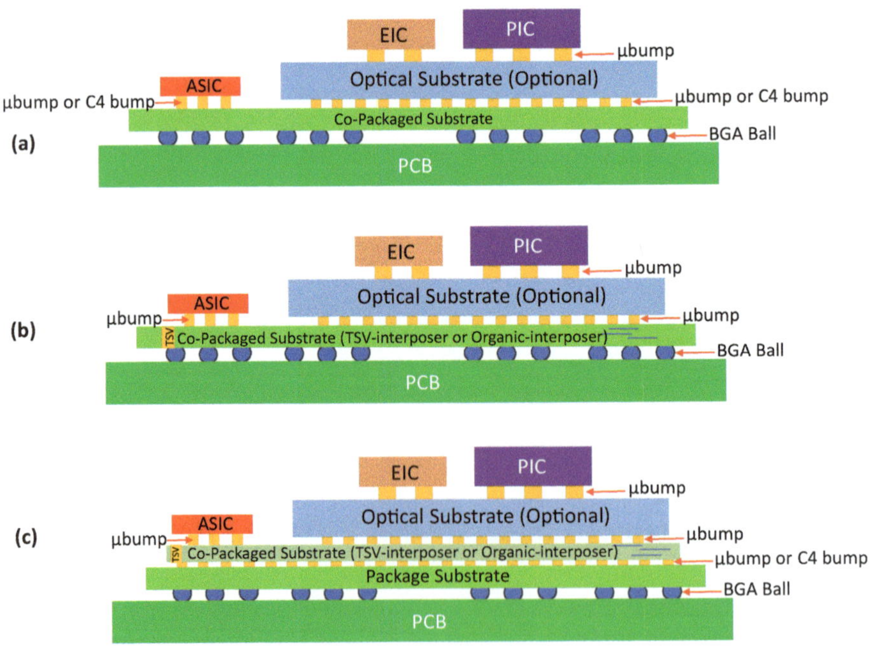

Fig. 6.11 2D heterogeneous integration of ASIC, EIC and PIC

substrate such as the TSV (through-silicon via)-interposer or organic-interposer (Chap. 4) with μbumps. Then, the TSV-interposer or the organic-interposer is attached on a PCB with BGA balls.

Figure 6.11c shows the 2D heterogeneous integration of PIC and EIC on an optional optical substrate side-by-side with the ASIC chip on the same co-packaged substrate such as the TSV-interposer or organic-interposer with μbumps. Then, the TSV-interposer or the organic-interposer is attached on a package substrate with μbumps or C4 bumps. Finally, the package substrate is attached to a PCB with BGA balls. In the industry, this is called 2.5D IC integration if it is with the TSV-interposer or 2.3D IC integration if it is with the organic-interposer. Figure 6.12 shows an example of the 2.5D heterogeneous integration of ASIC, PIC, and EIC [33]. Figure 6.13 shows an example of the 2.3D heterogeneous integration of ASIC, PIC, and EIC [28].

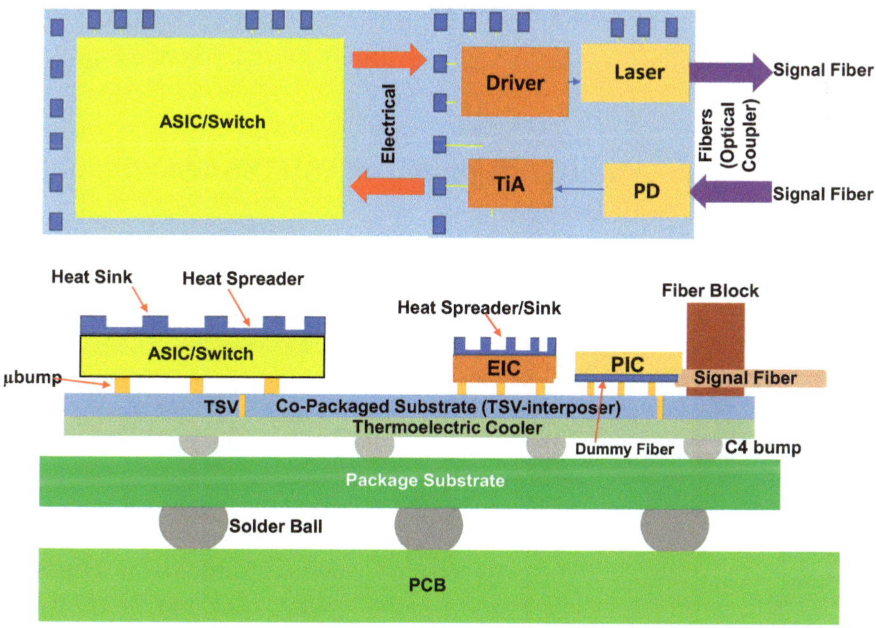

Fig. 6.12 2D heterogeneous integration of ASIC, EIC and PIC on a co-packaged substrate (TSV-interposer)

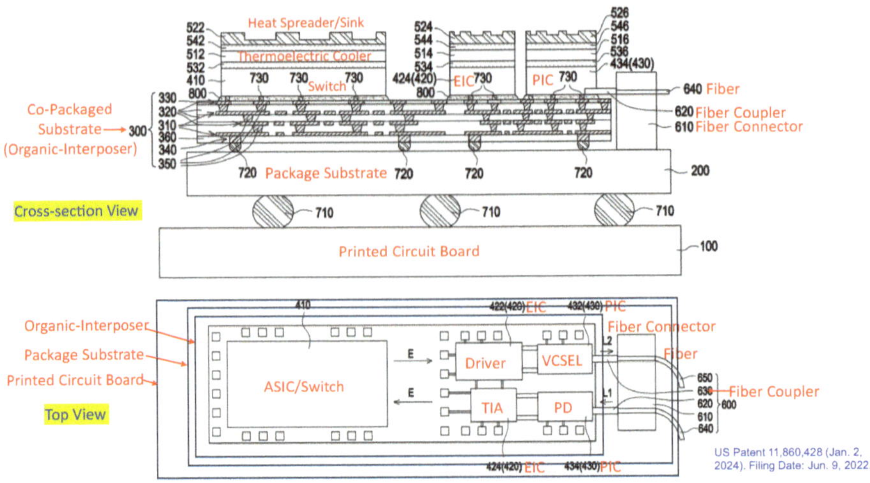

Fig. 6.13 2D heterogeneous integration of ASIC, EIC and PIC on a co-packaged substrate (organic-interposer)

6.10 2D Heterogeneous Integration of ASIC Switch, PIC and EIC with Bridges

Figure 6.14a shows the 2D heterogeneous integration of ASIC, PIC, and EIC with silicon bridges. It can be seen that the chiplets (ASIC, EIC, and PIC) are connected with silicon bridges with μbumps (Chap. 5 of this book). Then, the ASIC, EIC, and PIC are attached on a co-packaged substrate with either μbumps or C4 bumps.

Figure 6.14b shows the 2D heterogeneous integration of ASIC, PIC, and EIC with silicon bridges. It can be seen that the chiplets (ASIC, EIC, and PIC) are connected with silicon bridges with Cu-Cu bumpless hybrid bonding [34]. Then, the ASIC, EIC, and PIC are attached on a co-packaged substrate with either μbumps or C4 bumps.

Figure 6.14c shows the 2D heterogeneous integration of ASIC, PIC, and EIC with Intel's EMIB (embedded multi-die interconnect bridge) [36, 37] in the cavity of a co-packaged substrate. It can be seen that there are two different kinds of bumps (μbumps and C4 bumps) on the chiplets (ASIC, EIC, and PIC).

6.11 3D Heterogeneous Integration of PIC and EIC

There are many different kinds of 3D heterogeneous integration (stacking) of the PIC and EIC as shown in Fig. 6.15. It can be seen from Fig. 6.15a that the EIC and the PIC are stacked (face-to-face) with the μbumps and the signals are fanning out through the wire bonds from the PIC. Figure 6.15b is basically the same as Fig. 6.15a

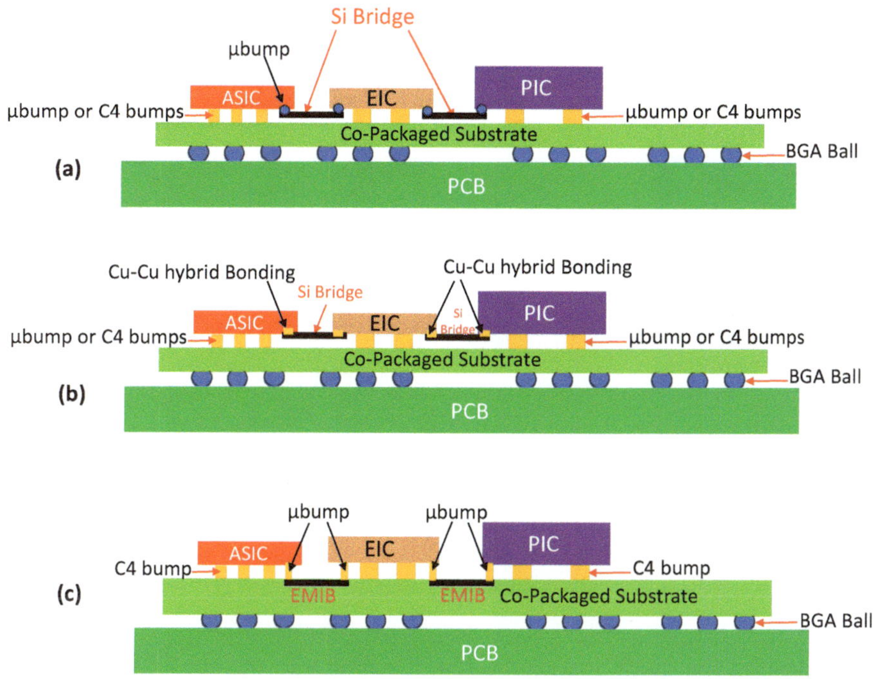

Fig. 6.14 2D heterogeneous integration of ASIC, EIC and PIC with silicon bridges on a co-packaged substrate

except the signals are fanning out through the C4 bumps or μbumps from the PIC. Figure 6.15c shows the EIC and the PIC are stacked (face-to-back) with the μbumps, and the signals are fanning out through the TSVs and the C4 bumps or μbumps from the PIC. Figure 6.15d is basically the same as Fig. 6.15c except the μbumps between the EIC and PIC are eliminated by the bumpless Cu-Cu hybrid bonding. Figure 6.15e shows the EIC and the PIC are stacked (connected) by a TSV-interposer with the μbumps and the signals are fanning out through the RDL and TSV of the TSV-interposer and the TSV and C4 bumps or μbumps from the PIC. Figure 6.15f is basically the same as Fig. 6.15e except the TSV-interposer is replaced by the organic-interposer and the signals are fanning out through the RDLs of the organic-interposer and the TSV and C4 bumps or μbumps from the PIC. Figure 6.15g shows the PIC and EIC are stacked (connected) by a TSV-interposer with μbumps and the signals are fanning out through the μbumps of the TSV-interposer. Figure 6.15h is basically the same as Fig. 6.15g except the TSV-interposer is replaced by the organic-interposer. Figure 6.15i is basically the same as Fig. 6.15g except the μbumps on both sides of the TSV-interposer have been eliminated by the bumpless Cu-Cu hybrid bonding.

Comparing Fig. 6.15c with Fig. 6.15e and f the latter are for extremely high-density and the case where there are difficulties in direct stacking of the EIC and PIC.

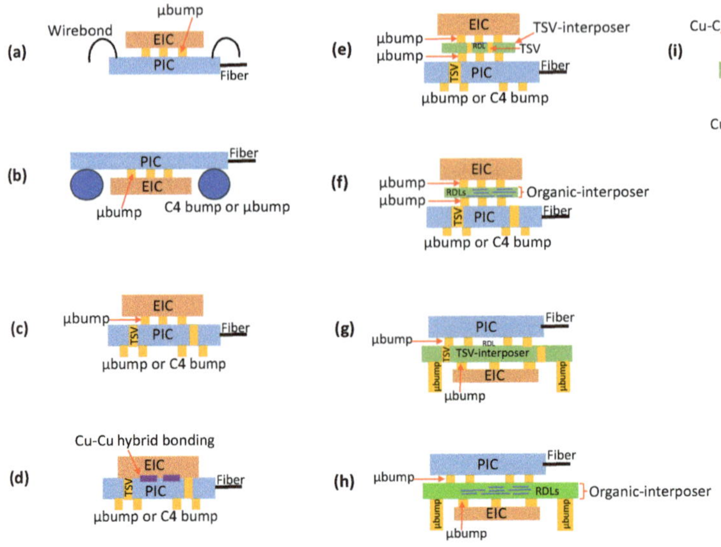

Fig. 6.15 Various 3D heterogeneous integration of EIC and PIC

Comparing Fig. 6.15e with Fig. 6.15g, it can be seen that the TSVs in the PIC as shown in Fig. 6.15e have been eliminated, Fig. 6.15g. Similarly, by comparing Fig. 6.15f with Fig. 6.15h, it can be seen that the TSVs in the PIC as shown in Fig. 6.15f have been eliminated, Fig. 6.15h.

6.12 3D Heterogeneous Integration of ASIC Switch, PIC and EIC

Figure 6.16a shows the 3D heterogeneous integration (stacking) of PIC and EIC (such as those shown in Fig. 6.15) on an optional optical substrate side-by-side with the ASIC chip on the aame co-packaged substrate with µbumps or C4 bumps. Then, the co-packaged substrate is attached to a PCB with BGA balls.

Figure 6.16b shows the 3D heterogeneous integration of PIC and EIC on an optional optical substrate side-by-side with the ASIC chip on the same co-packaged substrate such as the TSV-interposer or organic-interposer with µbumps. Then, the TSV-interposer or the organic-interposer is attached on a PCB with BGA balls.

Figure 6.16c shows the 3D heterogeneous integration of PIC and EIC on an optional optical substrate side-by-side with the ASIC chip on the same co-packaged substrate such as the TSV-interposer or organic-interposer with µbumps. Then, the TSV-interposer or the organic-interposer is attached on a package substrate with µbumps or C4 bumps. Finally, the package substrate is attached to a PCB with BGA balls. Figure 6.17 shows an example of 3D heterogeneous integration of PIC and EIC

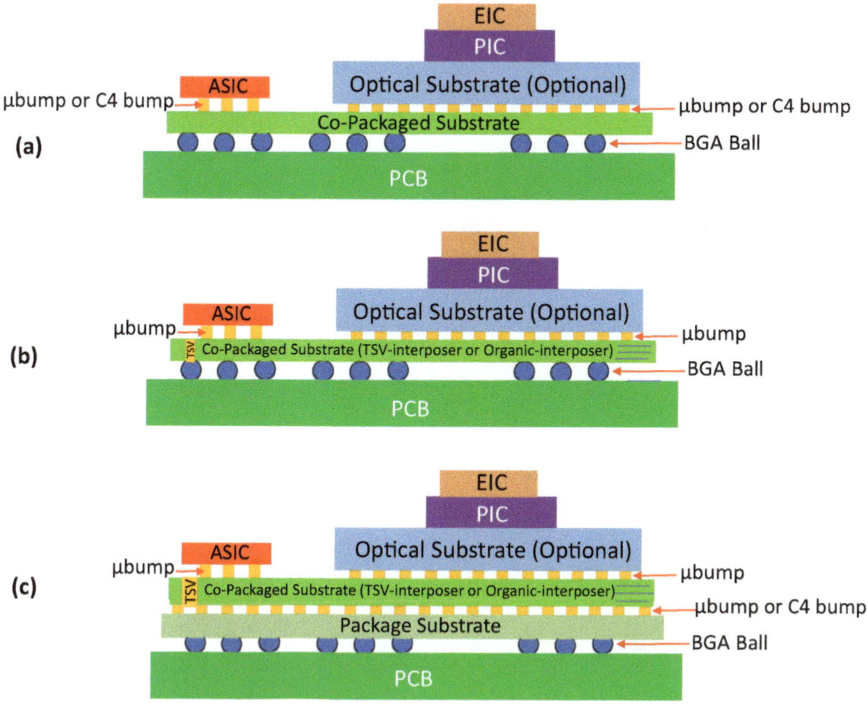

Fig. 6.16 3D heterogeneous integration of ASIC, EIC and PIC

with the technology shown in Fig. 6.15c and 6.16a [24]. Figure 6.18 [35] shows an example of 3D heterogeneous integration of ASIC, PIC and EIC with the technology shown in Figure 6.15c and 6.16c. Figure 6.19 shows one of Nvidia's CPOs, which is the most complicated CPO in the public today.

With co-packaged optics (CPO), the switch chip is typically surrounded by 16 OEs/EEs, all placed on an organic package substrate (see Fig. 6.20). Currently, the 25.6-terabit ethernet switch chip requires 16@1.6 terabits-per-second (1.6Tbps) OEs/EEs while the upcoming 51.2-terabit switch chips will use 3.2Tbps OEs/EEs (the size of EIC and PIC of the OE/EE will be larger). The issue is that the multi-chip module can only be so large. It is challenging with today's packaging technology to surround the 51.2-terabit ASIC with 16@3.2Tbps OEs/EEs.

One way to solve this problem is to stack-up the PIC and EIC as shown in Figs. 6.21 and 6.22. Actually, Fig. 6.21 [29] is a combination of Fig. 6.15d and Fig. 6.16c. Figure 6.22 [30] is a combination of Fig. 6.15h and Fig. 6.16b.

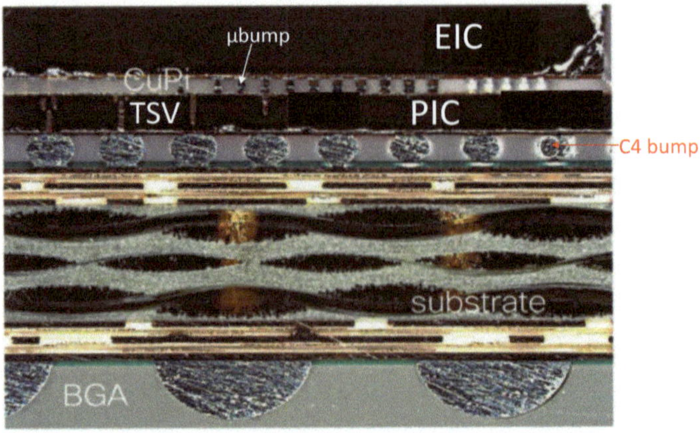

Fig. 6.17 3D heterogeneous integration of EIC and PIC on a co-packaged substrate

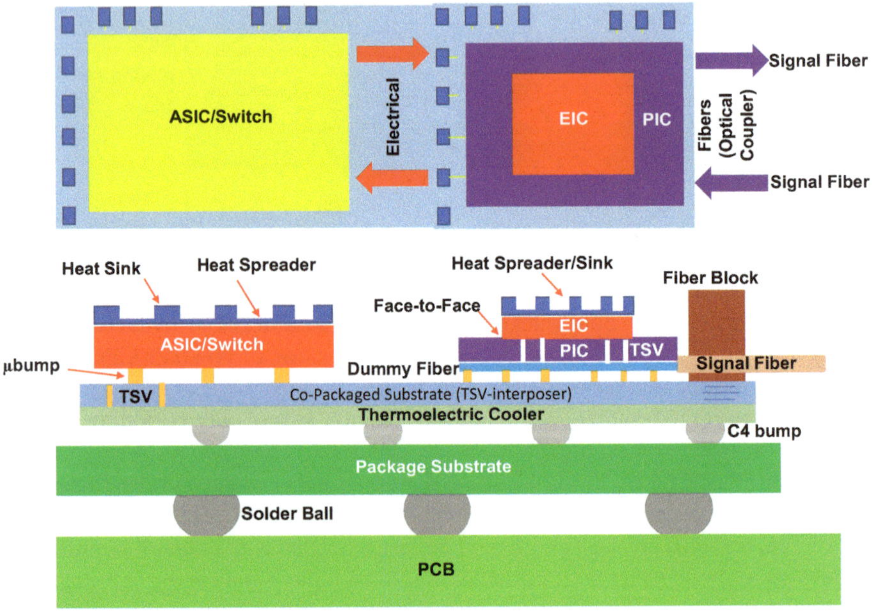

Fig. 6.18 3D heterogeneous integration of ASIC, EIC and PIC on a co-packaged substrate (TSV-interposer)

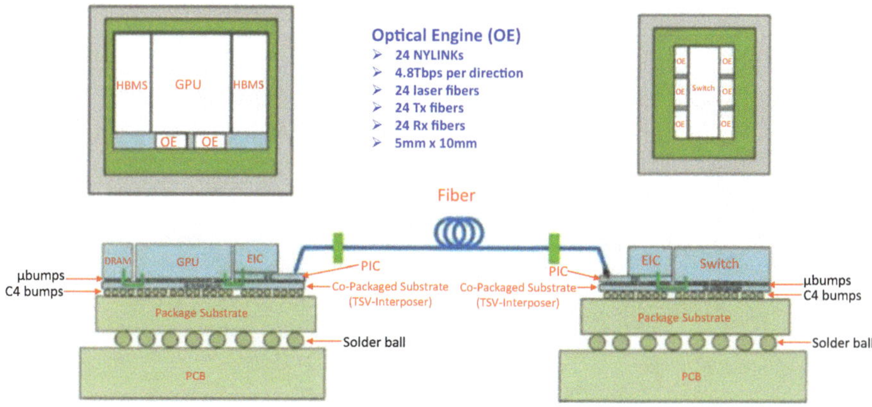

Fig. 6.19 Nvidia's 3D integration of SoC, HBM, EIC and PIC on co-packaged substrates (TSV-interposers)

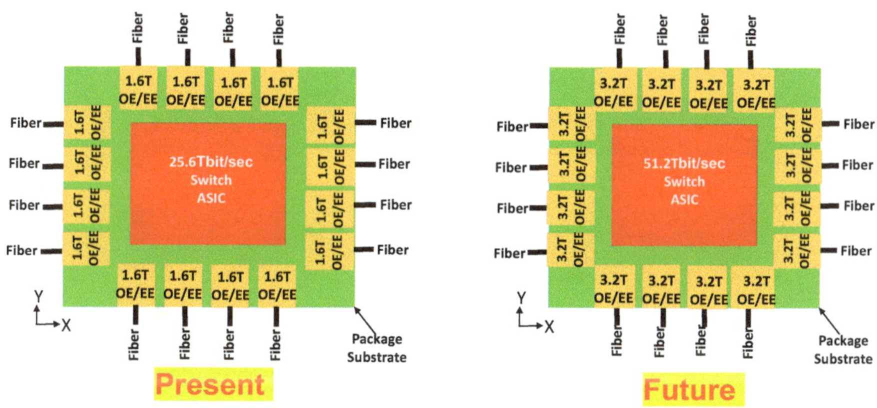

Fig. 6.20 The present switch (25.6Tbit/s) versus the future switch (51.2Tbit/s)

6.13 3D Heterogeneous Integration of ASIC Switch, PIC and EIC with Bridges

Figure 6.23a shows the 3D heterogeneous integration of PIC and EIC (such as those shown in Fig. 6.15), and ASIC with silicon bridges. It can be seen that the ASIC and the 3D stacking of EIC and PIC are connected with a silicon bridge with μbumps. Then, the ASIC and the 3D stacking of EIC and PIC are attached on a co-packaged substrate with either μbumps or C4 bumps.

Figure 6.23b shows the 3D heterogeneous integration of PIC and EIC, and ASIC with Intel's EMIB in the cavity of a package substrate. It can be seen that there are

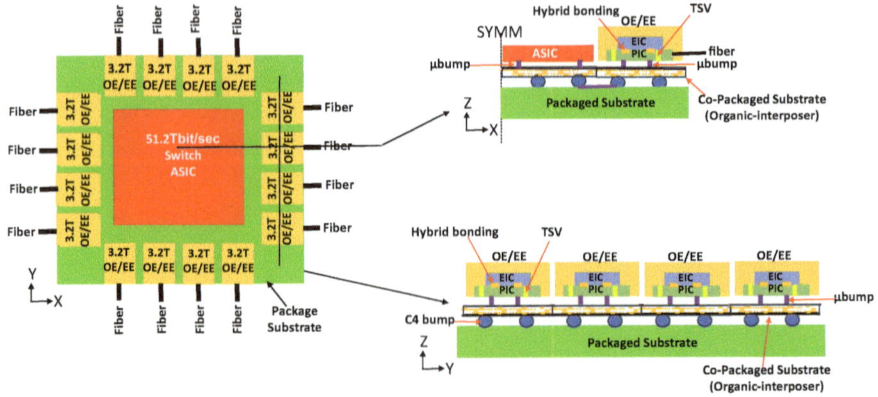

Fig. 6.21 Co-packaged optics method for 51.2Tbit/s switch

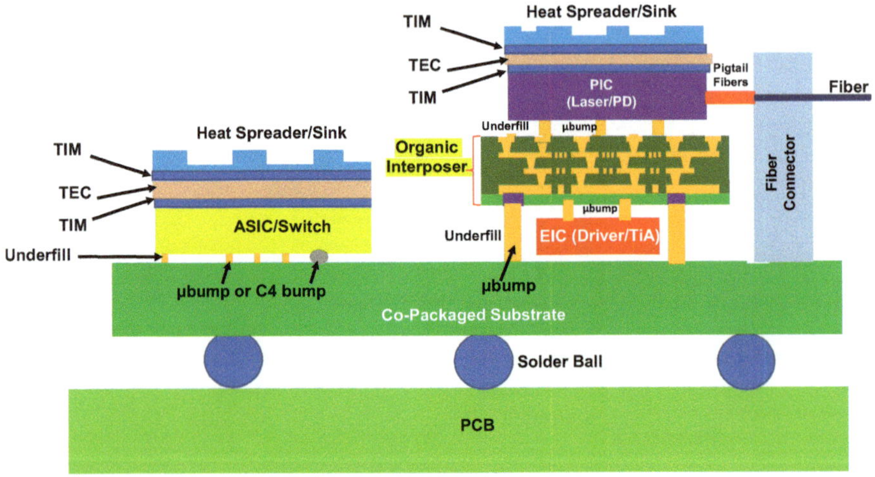

Fig. 6.22 Another Co-packaged optics method for 51.2Tbit/s switch

two different kinds of bumps (μbumps and C4 bumps) on the ASIC and PIC. An example is shown in Fig. 6.24 [34].

Figure 6.23c shows the 3D heterogeneous integration of PIC and EIC, and ASIC with a silicon bridge embedded in a co-packaged substrate such as the fan-out epoxy molding compound with RDLs (fan-out organic-interposer). The bridge will connect the ASIC and the 3D stacking of PID and EIC as well as the co-packaged substrate through the RDLs and through mold via (TMV).

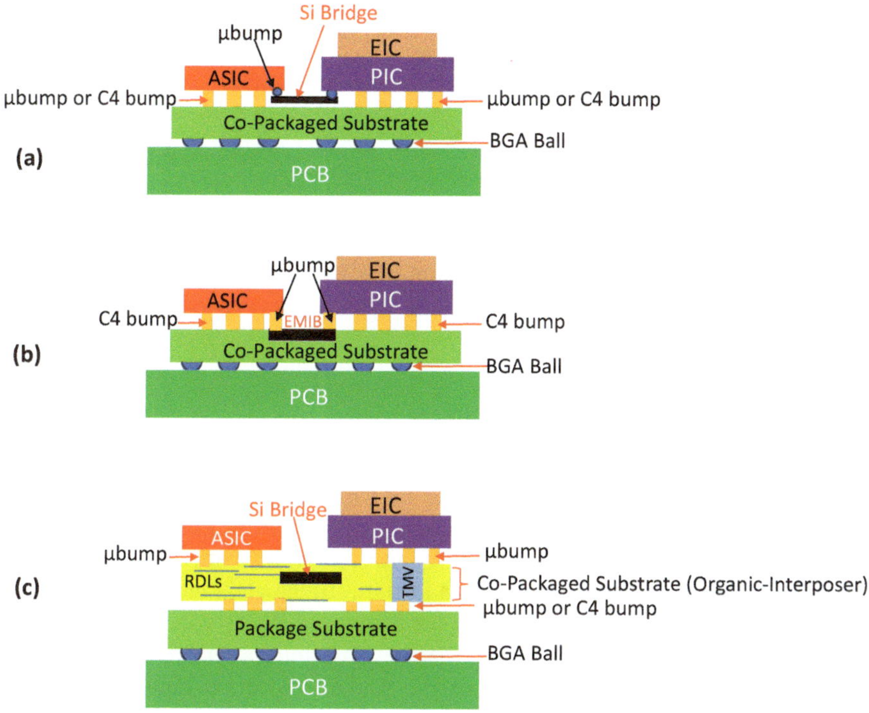

Fig. 6.23 3D heterogeneous integration of ASIC, EIC and PIC with silicon bridges on a co-packaged substrate

Fig. 6.24 An example on 3D heterogeneous integration of ASIC, EIC and PIC with silicon bridges on a co-packaged substrate

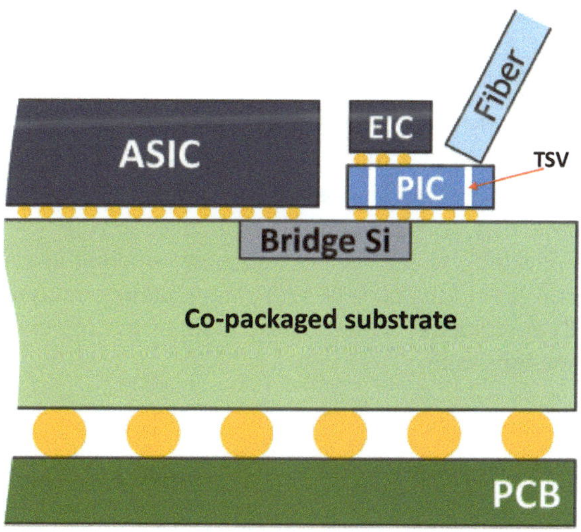

6.14 Heterogeneous Integration of ASIC Switch, PIC and EIC on Glass Substrate

6.14.1 Intel's Announcement of Glass Substrates

Intel has announced a new glass-substrate technology for the next generation of high-power processors. Intel is on the path to delivering 1 trillion transistors on a package by 2030 and its ongoing innovation in advanced packaging including glass substrates will help achieve this goal.

6.14.2 Advantages of Glass Substrate

Since the announcement of glass substrate for supporting (packaging) the chips on September 18, 2023, by Intel, people are very interested in glass substrates. The advantages of glass substrates comparing with the organic substrates are: (a) the surface of glass interposer is smoother and flatter, (b) the glass interposer allows for waveguides which transmit the lights to the devices of PIC, (c) the glass interposer possesses superior optical properties, (d) the glass interposer has better thermal, mechanical, and dimensional stability, (e) the glass interposer has higher interconnect density, and (f) the glass interposer improves signaling speed, power delivery, and design rules.

6.14.3 Challenges of Glass Substrate

On the other hand, the challenges of glass substrate, according to the executive Vice President (Ann Kelleher) are: (a) glass substrates will be more expensive to produce and package than tried-and-tested organic substrates, (b) there will be the yield issues at the start, (c) glass substrates will need to build a viable ecosystem for commercial production, which includes necessary tooling and supply capacity and that's why Intel is working closely with glass-handling equipment and material suppliers, and (d) the company will also have to find ways to outsource test and assembly of these new substrates.

6.14.4 Glass Substrate Examples

Figure 6.25 shows a 3D stacking of PIC and EIC on a glass interposer (substrate) and then side-by-side with the ASIC chip on the same co-packaged substrate. It can be seen that Fig. 6.25 is a combination of Figs. 6.15g and 6.16b, except in Fig. 6.15g, the

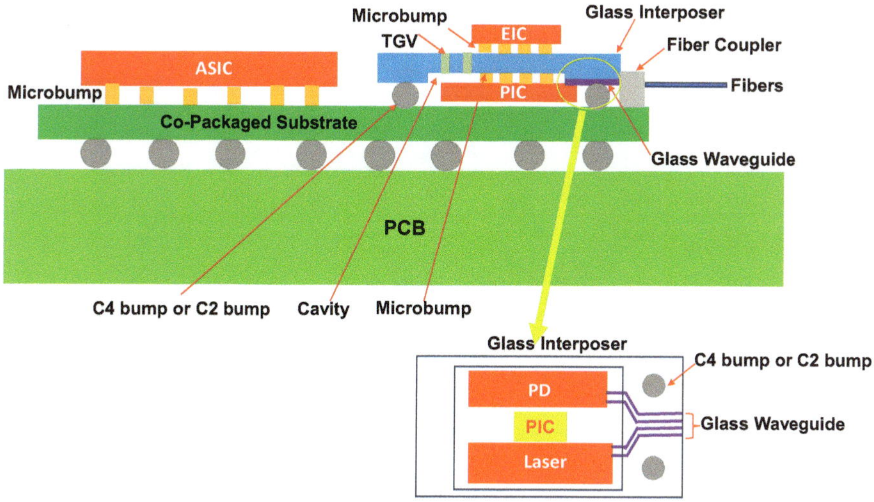

Fig. 6.25 3D heterogeneous integration of EIC and PIC with a glass interposer

TSV-interposer has been replaced by a glass-interposer, Fig. 6.25. Glass waveguides are fabricated on the glass interposer and they are connected to the PIC such as the PD and laser devices. The 3D stacking of the EIC and PIC through the glass interposer is packaged side-by-side with the ASIC chip on the same co-packaged substrate.

Figure 6.26 shows another example of 3D heterogeneous integration of ASIC, PIC, and EIC on a co-packaged substrate (glass interposer). It can be seen that the EIC and PIC are stacking through either μbump or Cu-Cu bumpless hybrid bonding. Then they are arranged side-by-side with the ASIC chip on the same co-packaged substrate (glass interposer). The assembly process of the package is shown in Fig. 6.27. First, etch a cavity and make the TGVs (through glass vias) on the glass substrate. Then place the PIC with a die attach in the cavity with a holding device. The PID is with either C2 bumps (μbumps) or bare Cu-pads with 3 nm recess. Fill the gap with a mold resin. Fabricate the RDLs, the glass waveguide, and then assemble the fiber coupler with fibers. Finally, perform the μbump bonding or the Cu-Cu hybrid bonding between the EIC and PIC.

6.15 Summary

Some important results and recommendations are summarized as follows.

- Silicon photonics are the semiconductor integration of EIC and PIC on a silicon substrate (wafer) with CMOS technology.
- CPO are heterogeneous integration packaging methods to integrate the OE which consists of PIC and the EE which consists of EIC as well as the switch ASIC.

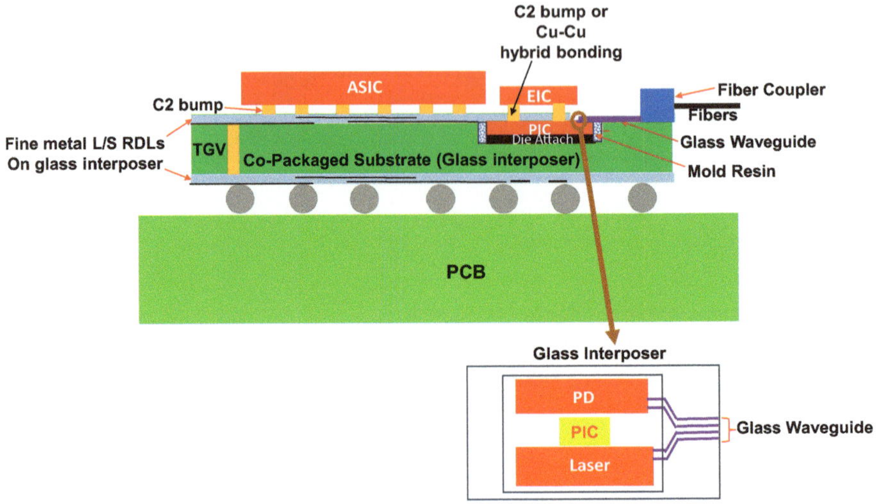

Fig. 6.26 3D heterogeneous integration of ASIC, EIC and PIC on a co-packaged substrate (glass interposer)

- Roadmaps of OBO, NPO, and CPO have been provided.
- Various (9 different) 3D heterogeneous integrations of PIC and EIC have been proposed.
- Various 2D and 3D heterogeneous integrations of ASIC Switch, PIC and EIC (CPO) w/o bridge have been proposed.
- Various heterogeneous integration of ASIC Switch, PIC and EIC (CPO) on glass substrate have been proposed.

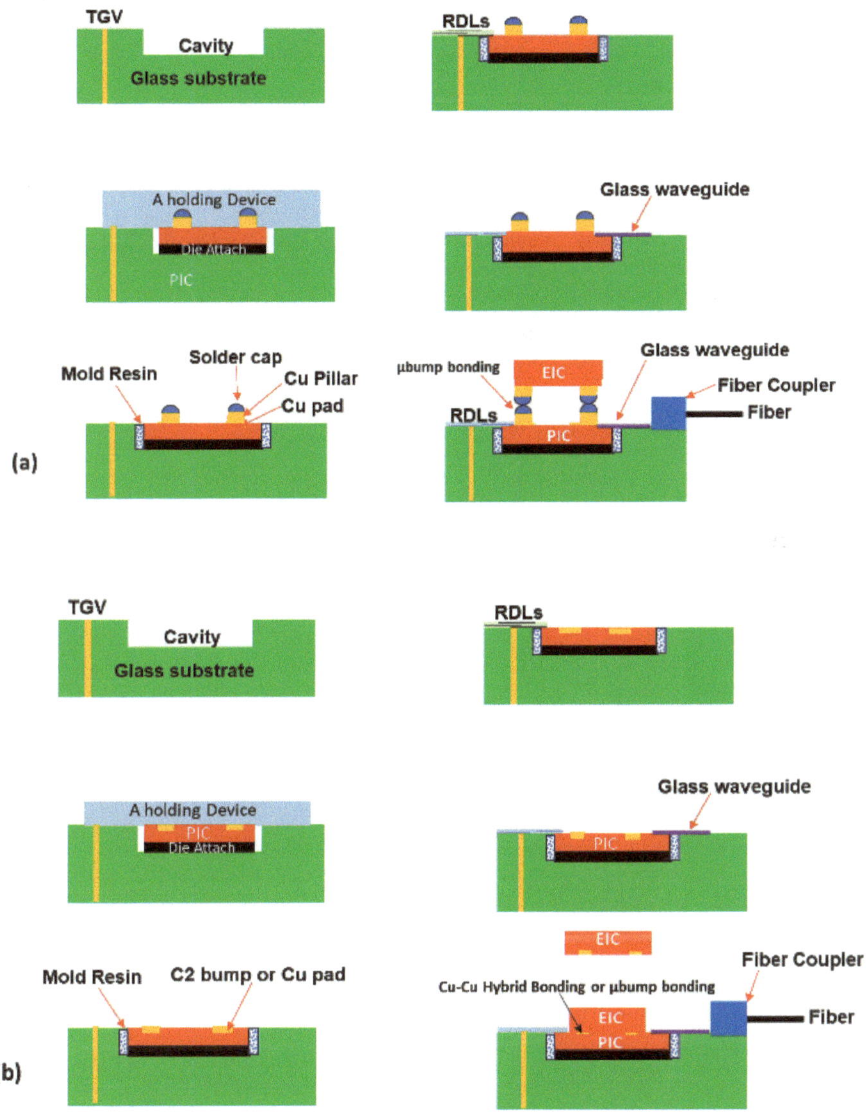

Fig. 6.27 Process in fabrication of the 3D heterogeneous integration of ASIC, EIC and PIC on a co-packaged substrate (glass interposer). **a** Microbump bonding. **b** Cu-Cu hybrid bonding

References

1. Chai, Y., J. H., Lau, et al. 2008. Development of a low cost 2.5 Gbps SFF optical transceiver using 0.18 *u*m CMOS ICs. *Proceedings* of *SPIE – Photonics Packaging, Integration, and Interconnects VIII*, San Jose, CA, January 19–24, 6899, pp. 0U.1–0U.9.

2. Lau, J. H., S. Erasmus, F. Sporon-Fiedler, S. Murphy, R. Herms, and S. Pan. 2001. Creep analysis of a Fiber-Optic transceiver housing. *ASME Paper No. 15868, ASME Inter-PACK'01*, Kauai, Hawaii, 8–13.
3. Lau, J. H, S. Erasmus, F. Sporon-Fiedler, and C. Chang. 2001. DSC, TMA, DMA, and TGA of epoxies for Fiber-Optic transceiver applications. *ASME Paper No. IMECE 2001/EPP-24720, ASME IMEC*.
4. Lau, J. H., C. L. Jiaa, and S. Erasmus. 2001. Corrosion analysis of a Fiber-Optic transceiver housing. *ASME Paper No. IMECE 2001/EPP-24721, ASME IMEC*.
5. Lau, J.H., et al. 2002. Corrosion analysis of a Fiber-Optic transceivers Housing. *ASME Transactions, Journal of Electronic Packaging* 124: 379–384.
6. Takemura, K., D. Ohshima, A. Noriki, D. Okamoto, A. Ukita, J. Ushida, M. Tokushima, T. Shimizu, I. Ogura, D. Shimura, and T. Aoki. 2022. Silicon-Photonics-Embedded interposers as Co-Packaged Optics Platform. The *Transactions of The Japan Institute of Electronics Packaging*. 15, pp. E21–012–1 – E21–012–13.
7. Ackermann, M., B. Shen, F. Merget, M. Wolz, and J. Witzens. 2022. Glass-molded optical interposers for wafer scale photonic integrated circuit packaging in 800G modules and co-packaged optics. *Proceedings of SPIE*, pp. M1–11.
8. Lee, B., N. Nedovic, T. Greer III, and C. Thomas Gray. 2023. Beyond CPO: A motivation and approach for bringing optics onto the silicon interposer. *Journal of Lightwave Technology*, 41, pp. 1152–1162.
9. Lars Brusberg, Aramais R. Zakharian, Jason R. Grenier, Barry J. Paddock, Daniel W. Levesque, Cliff G. Sutton, Robin M. Force, Lucas W. Yeary, Robert A. Bellman, Chad C. Terwilliger, Betsy J. Johnson. 2022. Glass substrate for Co-Packaged Optics. *Proceedings of IMAPS 55th International Symposium on Microelectronics*, pp. 236–241.
10. Noriki, A., A. Ukita, K. Takemura, S. Suda, T. Kurosu, Y. Ibusuki, I. Tamai, and D. Shimura. 2023. Demonstration of Silicon-Photonics hybrid Glass-Epoxy substrate for Co-Packaged optics. *Journal of Lightwave Technology* 41: 5078–5083.
11. Brusberg, L., J. Grenier, Ş. Kocabaş, A. Zakharian, L. Yeary, D. Levesque, B. Paddock, R. Bellman, R. Force, C. Terwilliger, C. Sutton, J. Clark, and K. Rousseva. 2022. Glass interposer for High-Density photonic packaging. *Proceedings of OFC (Optical Fiber Communication)*, pp. Tu3A.3 1–3.
12. Tan. M., J. Xu, S. Liu, J. Feng, H. Zhang, C. Yao, et al. "Co-packaged optics (CPO): status, challenges, and solutions", *Frontiers of Optoelectronics*, Vol. 16, March 2023, pp. 1–40.
13. Grenier, J., L. Brusberg, K. Wieland, J. Matthies, and C. Terwilliger. 2023. Ultrafast laser processing of glass waveguide substrates for multi-fiber connectivity in co-packaged optics. *Advanced Optical Technology* 12: 1–21.
14. Noriki, A., and T. Amano. 2023. Self-aligned optical connector assembly on polymer waveguide integrated package substrate for co-packaged optics. *Proceedings of IEEE/ECTC*, pp. 1892–1895.
15. Yeary, L., L. Brusberg, C. Kim S. Seok, J. Noh, and A. Rozenvax. 2023. Co-packaged optics on glass substrates for 102.4 Tb/s data center switches. *Proceedings of IEEE/ECTC*, pp. 224–227.
16. Gupta, P., R. Bernson, N. Nudds, S. Collins, K. Gradkowski, P. Morrissey, and P. O'Brien. 2023. Reference thermal chips for 2D and 3D Co-packaging process development. *Proceedings of IEEE/ECTC*. pp. 2165–2170.
17. Bhuvanendran, S., N. Gourikutty, L. Long, S. Wei, M. Jong, D. Ho, J. Wu, T. Lim, R. Mandal, C. Choong, C. Yee, X. Li, J. Liow, and S. Bhattacharya. 2023. A heterogeneously integrated Wafer-level processed Co-Packaged optical engine for Hyper-scale data centres. *Proceedings of IEEE/ECTC*. pp. 207–211.
18. Li, J., T. Hsu, M. Zhuan, S. Lin, T. Shih, N. Kao, and Y. Eang. 2023. Innovative Fan-Out embedded bridge structure for Co-Packaged optics. *Proceedings of IEEE/ECTC*, pp. 228–232
19. McDonough, C., S. Kruger, T. Ngai, S. Baranowski, H. Yang, S. Deckoff-Jones, C. Poulton, M. Watts, and D. Harame. 2023. AIM photonics demonstration of a 300 mm Si photonics interposer. *Proceedings of IEEE/ECTC*, pp. 233–238.
20. Integrated Optical Interconnect Systems (iOIS) for Silicon Photonics Applications in HPC

21. Hsia, H., S. P. Tai, C. S. Liu, C. W. Tseng, S. W. Lu, Y. Wu, C. C. Chang, Jason Wu, K. C. Yee, C. Y. Wu, C. H. Tung, D. Yu. 2023 Integrated Optical Interconnect Systems (iOIS) for silicon photonics applications in HPC. *Proceedings of IEEE/ECTC*, pp. 612–616.
22. Lu, M., S. Mu, C. Cheng, and J. Chen. 2022. Advanced packaging technologies for Co-packaged Optics. *Proceedings of IEEE/ECTC*. pp. 38–42.
23. Application Chou, B., B. M. Sawyer, G. Lyu, E. Timurdugan, C. Minkenberg, A. J. Zilkie, and D. McCann. 2022. Demonstration of Fan-out silicon photonics module for next generation Co-Packaged Optics (CPO). *Proceedings of IEEE/ECTC*, pp. 394–402.
24. Razdan, S., P. Dobbelaere, J. Xue, A. Prasad, and V. Patel. 2022. Advanced 2.5D packaging technologies for next generation Silicon Photonics in high performance networking applications. *Proceedings of IEEE/ECTC*, pp. 428–435.
25. Janta-Polczynski, A., and M. Robitaille. 2022. Optical fiber pigtails integration in co-package. *Proceedings of IEEE/ECTC*, pp. 410–421.
26. Maharry, A., L. A. Valenzuela, J. F. Buckwalter, and C. L. Schow. 2021. A PCB packaging platform enabling 100+ Gbaud optoelectronic device testing. *Proceedings of IEEE/ECTC*, pp. 1323–1328.
27. Nieweglowski, K., L. Lorenz, K. Bock, M. Catuneanu, K. Jamshidi. 2020. Electro-optical co-integration of chip-components in optical transceivers for optical inter-chip communication. *Proceedings of IEEE/ECTC*. pp. 139–147.
28. Lau, J. H., and T. J. Tseng. 2022. Package structure and optical signal transmitter. US Patent 11,860,428, Date of Patent: Jan. 2, 2024, Filing Date: Jun. 9, 2022.
29. Lau, J. H., and T. J. Tseng. 2023. 3D integration of electronic IC and Photonic IC. US Patent application filed on.
30. Lau, J. H., and T. J. Tseng. 2023. 3D heterogeneous integration of electronic IC and Photonic IC. US Patent application filed on.
31. Lau, J. H., and T. J. Tseng. 2023. Low-Cost heterogeneous integration packaging of electronic IC and Photonic IC. US Patent application filed on November 6, 2023.
32. Lau, J.H. 1995. *Ball Grid Array Technology*. New York: McGraw-Hill Book Company.
33. Lau, J.H. 2023. Recent advances and trends in multiple system and heterogeneous integration with TSV-Interposers. *IEEE Transactions on CPMT* 13 (1): 3–25.
34. Lau, J. H., C. Ko, P. Lin, T. Tseng, R. Tain, and H. Yang. 2022, Package structure and manufacturing method thereof. U.S. Patent No. 11,410,933, filed on
35. Lau, J.H. 2022. Recent advances and trends in advanced packaging. *IEEE Transactions on CPMT* 12 (2): 228–252.
36. Chiu, C., Z. Qian, and M. Manusharow. 2014. Bridge interconnect with air gap in package assembly. US Patent No. 8,872,349.
37. Mahajan, R., R. Sankman, N. Patel, D. Kim, K. Aygun, Z. Qian, et al. 2016. Embedded multi-die interconnect bridge (EMIB)—a high-density, high-bandwidth packaging interconnect. *IEEE/ECTC Proc.*, pp. 557–565.

Index

J. H. Lau, *Flip Chip, Hybrid Bonding, Fan-In, and Fan-Out Technology*,
https://doi.org/10.1007/978-981-97-2140-5

FSC
www.fsc.org
MIX
Papier | Fördert gute Waldnutzung
FSC® C083411

Kolibri 360 GmbH
Ferdinand-Jühlke-Straße 7
99095 Erfurt, Deutschland
produktsicherheit@kolibri360.de